Plants, Genes, and Crop Biotechnology

Plants, Genes, and Crop Biotechnology

SECOND EDITION

Maarten J. Chrispeels & David E. Sadava

Published in partnership with the
American Society of Plant Biologists and the
ASPB Education Foundation

ASPB *education foundation*

JONES AND BARTLETT PUBLISHERS

Sudbury, Massachusetts

BOSTON   TORONTO   LONDON   SINGAPORE

World Headquarters
Jones and Bartlett Publishers
40 Tall Pine Drive
Sudbury, MA 01776
978-443-5000
info@jbpub.com
www.jbpub.com

Jones and Bartlett Publishers Canada
2406 Nikanna Road
Mississauga, ON L5C 2W6
CANADA

Jones and Bartlett Publishers International
Barb House, Barb Mews
London W6 7PA
UK

PRODUCTION CREDITS
Chief Executive Officer: Clayton Jones
Chief Operating Officer: Don W. Jones, Jr.
Executive V.P. & Publisher: Robert W. Holland, Jr.
V.P., Design and Production: Anne Spencer
V.P., Sales and Marketing: William Kane
V.P., Manufacturing and Inventory Control: Therese Bräuer
Editor in Chief–College: J. Michael Stranz
Executive Editor: Stephen L. Weaver
Associate Managing Editor, College Editorial: Dean W. DeChambeau
Senior Production Editor: Louis C. Bruno, Jr.
Senior Marketing Manager: Nathan Shultz
Text Design: Anne Spencer
Copyediting: Linda Purrington
Composition: Graphic World, Inc.
Cover Design: Kristin E. Ohlin
Printing and Binding: Courier Company
Cover Printing: Lehigh Press

Library of Congress Cataloging-in-Publication Data
Chrispeels, Maarten J.
    Plants, genes, and crop biotechnology / Maarten J. Chrispeels, David E. Sadava ; with foreword by Gordon Conway.—2nd ed.
        p. cm.
Rev. ed of: Plants, genes, and agriculture. 1994.
Includes bibliographical references.
ISBN 0-7637-1586-7
    1. Crops—Genetic engineering. 2. Plant breeding. 3. Plant biotechnology.
4. Crop improvement. 5. Food supply. I. Sadava, David E. II. Chrispeels, Maarten J., Plants, genes, and agriculture. III. Title.
SB123.57 .C48 2002
363.8—dc21

2002067637

BRIEF CONTENTS

CONTENTS

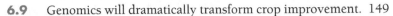

Chapter 7

Plants in Human Nutrition and Animal Feed 152

Chapter 8

The Genetic Basis of Growth and Development 182

FOREWORD

Just five decades have passed since Watson and Crick deciphered the double helix structure of DNA, the stuff of genes, and laid the groundwork for modern biotechnology. Today we know the full genetic sequence of more than sixty living organisms. The list includes human beings, over thirty important human pathogens, the photosynthetic plant *Arabidopsis thaliana,* plant pathogens, several *Archaea* from exotic environments such as ocean depths and hot springs, and nature's own genetic engineer, the bacterium *Agrobacterium tumefaciens,* which for millennia has been transferring bacterial genes into plants to create new plant traits *Agrobacterium* finds useful. Two draft sequences of the rice genome were recently published, and by the end of 2002 an international public-private partnership led by the Japanese, expects to put in the public domain the full genetic sequence of rice, the world's most important food crop and the genetic "Rosetta Stone" for all the cereals. Already the functions of many rice genes have been determined, and plant scientists throughout the world are using rice genome sequence data to improve productivity and disease resistance in rice and other crops.

In 2001, more than 5.5 million farmers worldwide planted about 52.6 million hectares of crops that were genetically manipulated (GM or transgenic crops, also called GMOs). This year the area has expanded further with India and Indonesia joining twelve other countries, including China, Mexico, South Africa, and Argentina, in approving the commercial planting of GM crops. But, agricultural biotechnology means a lot more than the creation of GM crops. It also involves the use of tissue culture to rapidly propagate disease-free seedling plants and to create new hybrids between plants that do not cross naturally, the use of sophisticated DNA-based genetic markers that allow breeders to follow and select for important traits more easily, and the use of DNA chips and other DNA-based diagnostic techniques to characterize pathogen populations for more effective deployment of resistant varieties. When we look globally we see that the real potential for public benefit from these technologies exists in developing countries where access to genetically improved crop varieties can mean the difference between hunger and a sustainable livelihood. These countries will need to draw on the best that science has to offer to help their farmers increase the productivity, nutritional value, stability of production, sustainability, and profitability of their crops. The task is to produce new crop varieties that are genetically altered to grow in poor soils and with less water, that use plant nutrients more efficiently, that can resist tropical pests and diseases—thus reducing yield losses—and that add micronutrients and essential amino acids to deficient diets. But, will this potential be realized?

Realistically, the question is no longer whether biotechnology will be used in food and agriculture, but rather how? We suspect the reason some people want to stop agricultural biotechnology, even when they understand its potential benefits, is that they do not believe those developing it will ever make the commitment to use it well and to use it equitably, and, therefore, they think biotechnology had better not be used at all. While we disagree with their conclusion, we believe they have legitimate concerns. We do need to establish more effective social and political processes to assure that biotechnology will be used wisely, that it will be made available to and used by public researchers to develop products that benefit people with limited purchasing power, and that these products will be made available for adaptation and adoption in developing countries.

While multinational corporations can help through public-private partnerships, for-profit companies will never make significant investments where there are no or limited markets. It is the international agricultural research centers of the Consultative Group on International Agricultural Research and the public agricultural research systems in larger developing countries that must assume responsibility for using these technologies to produce new crop varieties that will benefit poor farmers. These public sector agricultural research institutions need to be well supported both financially and technologically, including development of new strategies that will help them reestablish effective collaboration with advanced research institutes and universities in industrialized countries.

This second edition of *Plants, Genes, and Agriculture* with its new title *Plants, Genes, and Crop Biotechnology* provides the historical context, the global perspective, and the scientific information that students, scientists, development specialists, and knowledgeable people will find highly useful in better understanding plant genetics, the potential of agricultural biotechnology, and how biotechnology can help small-scale farmers everywhere sustainably meet their food and income requirements.

Gordon Conway, President
Gary Toenniessen, Director of Food Security
The Rockefeller Foundation
New York, New York

PREFACE

We have published the second edition of *Plants, Genes, and Agriculture* with a slightly changed title to provide teachers and students of introductory courses in plant biology and crop science with a resource that will allow them to see their respective disciplines from a new vantage point. Our objective is to present an integrated view of crop biology, leading to a broad appreciation of plant biology and biotechnology in agriculture (for plant biology courses), as well as the basic biological underpinnings of crop biology and biotechnology (for crop science courses). The population explosion of the 20th century had a dramatic effect on planet Earth: huge areas with natural ecosystems rich in plant and animal life were converted to agriculture with dramatic consequences for the environment. Humanity's challenge for the new century is to double food production and to do so in a sustainable way. To achieve these twin goals, people in all walks of life need to understand what it takes to grow crops, how progress in crop production was achieved in the past, and what the role of biotechnology will be in the future.

Many people in technologically advanced countries have become more interested in the ways that their food crops are grown and how they reach the markets where they shop. This interest results in part from a greater concern for nutrition and health and from a few scandals caused by breakdowns in the regulatory processes that assure the safety of our food supply. This concern about food presents us with an opportunity to educate people about the scientific principles that underlie crop production. We, therefore, view this text as being suitable for general education courses aimed at nonscientists.

Our topic is broad, and we discuss not only the natural sciences but the social sciences as well. Among the former are plant anatomy and development, plant physiology, molecular biology, genetics, plant breeding, evolution, ecology, soil science, pest and disease control, and biotechnology. Among the latter are such aspects as the economics of farming, trade policies, price supports, the funding of research, and the economic and social benefits of biotechnology.

The first chapter of this book deals with human population growth, and the news is fairly good. When we wrote the first edition in 1994, there was no end in sight to the human population explosion. Experts now agree that, barring unforeseen events, human population will level off at about 10 billion sometime during this century. This chapter, therefore, sets the stage for the challenge: doubling food production during the same time period.

The second chapter, written by two agricultural economists, describes the changes that have taken place in agricultural production as a result of developments that accompanied the industrial revolution, and emphasizes the role that science has played in this development. The tremendous economic return of investment in agricultural research justifies our subsequent focus on the scientific basis of increased crop productivity.

Chapter 3 looks at agricultural development from a biological vantage point and emphasizes the need to reach sustainability of this important human enterprise. The fact that directly or indirectly (as animal feed) plants are the basis of human nutrition leads to a discussion of the gradual intensification of plant agriculture throughout human history. The culmination of this intensification, known as the Green Revolution, raised food production in some developing countries to the point where they became food exporters; globally, the Green Revolution allowed food production to outstrip population growth in the last half of the 20th century.

The fourth chapter discusses a most difficult issue: Why are there 800 million hungry people in a world with adequate food for all? Clearly, our market driven economic system has not produced the result that many would like: equitable distribution of that food and food sufficiency for everyone. Purchasing power is the problem, but how we go about increasing it for the poorest of this world is a major challenge.

There are several reasons why we decided to devote an entire chapter (Chapter 5) to agriculture in sub-Saharan Africa. Sub-Saharan Africa is ecologically very diverse, and in Africa, the ecological dimension of crop production is most striking. In addition, the Green Revolution bypassed sub-Saharan Africa, in part because Africa has so many agroecological zones. Africa is the world's only continent where food availability per person has declined in the past 30 years, so it provides the greatest challenge to the agricultural system. Chapters 2, 4, and 5 are new in this edition, and together with Chapters 1 and 3, they constitute a discussion of economic, sociological, and ethical aspects of feeding the world that transcend the strictly biological problem.

The second part of the book comprises the basics of plant biology, beginning with a discussion of the molecular basis of crop improvement in Chapter 6. To understand both plant breeding and genetic engineering, one must understand how genetic information (DNA) is transmitted from one generation to the next and expressed in a plant as the outward appearance, or phenotype. The principles of genetics and molecular biology can now be used to transfer genes between unrelated organisms (genetic engineering). This powerful and unprecedented technology has resulted in genetically engineered crops, here referred to as genetically manipulated (GM) crops, which are widely used in several countries. Examples of GM approaches to crop improvement are discussed in many subsequent chapters.

Plants feed people, either directly or indirectly, but what are our nutritional requirements and how do our food plants satisfy them? Do different plants and plant parts have different nutritional values? Do vegetarians have to take special precautions to eat a healthy diet? How can plants be bred to be more nutritious? Is this an area where genetic engineering can play an important role? The role of nutrients such as vitamins, minerals, or proteins is well defined, but what about the role of non-nutrients such as antioxidants in our diets? These and other questions are dealt with in Chapter 7.

People eat a variety of plant parts: seeds (bread, rice, beans), roots (carrots, cassava), tubers (potatoes), leaves (cabbage), stems (sugar cane), fruits (bananas and plantains), and even flowers (artichokes). Chapter 8 describes how these plant organs and tissues are produced from a single cell, the fertilized egg. First a seed is produced, and after it germinates the vegetative body of the plant develops. Flowers and fruits come later. This orderly sequence of events is under exquisite genetic and environmental control.

Chapter 9, a new chapter for this second edition deals with the biology and technology of seeds. Seeds are not only our most important food source, but they are also agents of plant reproduction. Companies produce elite varieties of crops and sell these to farmers primarily as seeds, often as hybrid seeds or as GM seeds. Essentially the economic value of plant breeding and biotechnology is captured in seeds.

Plant growth depends on the assimilation of carbon dioxide from the atmosphere utilizing solar energy (photosynthesis) and on the uptake of minerals from the soil and their subsequent assimilation into molecules that underlie plant structure and function. These processes are the subjects of Chapters 10 and 11, respectively. In addition to dissolved minerals, soil is also the source of water needed to maintain the transpiration stream. The productivity of crops depends heavily on managing the physical environment by adding nutrients, changing the acidity of the soil or supplying irrigation water. Barring such management, the plants experience stresses, and stress management is important for optimizing plant productivity. Conventional plant breeding and GM approaches to alleviating these stresses are also discussed.

Plant biologists usually study processes such as plant development, photosynthesis, or nutrient uptake using plants grown in isolation and without considering the interactions of these plants with other living organisms. In nature, however, plants interact directly with numerous other organisms (pests, pathogens, and symbiotic microbes, for example), and they depend indirectly on the activities of other microbes that participate in the cycling of nutrients. Nowhere are such interactions more intense than in the soil, and "life together in the underground" is the subject of Chapter 12. The soil contains a complex ecosystem in which million of species interact with one another and with the roots of plants. This ecosystem derives its energy from root exudates and decaying roots and, therefore, from photosynthesis. Many of these soil processes are vital to the sustainability of agriculture, discussed later in the volume, and Chapter 12 marks the transition to the third part of the book, dealing with agriculture.

As a human activity, agriculture began millennia ago. Chapter 13 traces its origins and spread. People saved some of the seeds they harvested for the next year's crop planting. These saved seeds captured the gradual evolution of our crops from their wild ancestors to the landraces of yesteryear and the present elite lines.

This leads to a discussion in Chapter 14 of plant breeding, the principal mechanism of crop improvement in the past and in the future. "Genetics is King" when it comes to raising crop productivity, and plant breeders use a variety of tools to genetically improve crops. Pure-line selection, back crossing, hybridization, embryo rescue of wide crosses, mutation breeding, and genetic engineering are all tools of the plant breeder. In addition, the new science of genomics is certain to speed up plant breeding considerably in the 21st century.

In the field, crop plants face stiff competition from diseases such as molds, from insects, and from other plants (weeds). Minimizing this competition is the subject of Chapters 15, 16, and 17, respectively. Diseases take their toll on the amount of the crop that the farmer harvests and genetic improvement of crops remains the best solution (Chapter 15). Because disease organisms evolve rapidly, however, this solution is never permanent and the plant breeder's work is never finished. We are making rapid progress in understanding the molecular mechanisms involved in the plant's defense mechanisms against disease organisms, and this will lead to new GM-based approaches to combat diseases.

Insects and other pests such as mites and nematodes can damage the parts of the plant that people want to harvest (e.g., seed weevils), or they weaken the plants so that their growth is diminished (rootworms) or the plants fall over (stem borers). Other insects such as whiteflies suck the products of photosynthesis directly from the circulatory system of the plant. As discussed in Chapter 16, because of the enormous losses sustained by farmers, insect and nematode control are of paramount importance to them. Again there are different strategies: genetic improvement of the crop as part of an integrated pest management approach is the best approach. This may be combined with biological control measures that include the release of predators or with chemical sprays that simply kill the pests. The latter method invariably results in the appearance of resistant pest strains. Genetic engineering with toxin genes from bacteria now provides a targeted approach to the killing of specific pests.

There are many kinds of weeds, annuals and perennials, grasses and dicots; sometimes they closely resemble the crop, sometimes not. Weed control (Chapter 17) also takes several forms: manual removal (hoeing) is still very common on low-resource farms in developing countries, on organic farms in developed countries, and, when appropriate, on large farms. For many crops, chemical weed control has replaced mechanical hoeing in developed countries because it reduces labor costs. The newest refinement of weed control is to create GM crops that are resistant to certain herbicides. These crops are popular because they save the farmer money and reduce soil erosion.

Chapter 18, "Toward a Greener Agriculture," discusses the practices that make agriculture more sustainable. These practices view agriculture as part of a natural ecosystem, and their objective is to minimize the environmental impact of crop production. Adopting sustainable practices is usually tied to government policies that may reward farmers or discourage them from doing so. Sustainable agriculture is fully compatible with GM technology. Instead of adapting the environment to the crop plant, which has been a basis of agriculture throughout history, this technology has the potential to use new genes to adapt the plant to its environment, thus minimizing ecological impact.

At one time, agriculture supplied us not only with food but also with many other products that society needs. Many of these other products are now derived artificially, from petroleum or other industrial sources. For example, think of the replacement of cotton with synthetic fibers. This changeover was paralleled by the development of the chemical industry. Could (or should?) agriculture once again be used to produce the chemical feedstocks that industry needs, and what would be the impact of this on land utilization and on food production? Could plants be used to produce pharmaceuticals and vaccines? These fascinating questions are pursued in Chapter 19.

The final chapter in this book concerns the myths that have sprung up around GM crops. Groups opposed to this technology have introduced emotionally charged terms such as "genetic pollution" and "superweeds" into the discussion. Public polls conducted by reputable agencies have found that a substantial number of people believe that eating genetically engineered plants will cause their own DNA to be mutated. This chapter discusses some of these myths and forthrightly addresses some of the as yet unsolved problems associated with the adoption of GM crops.

We wish to thank the many people who made this second edition possible. First of all, the contributing authors of the chapters whose names are at the chapter headings responded enthusiastically to our call to update the first edition. This is as much their book as it is ours. Second, we are grateful to all the people who generously sent us photographs for inclusion in the book. Assembling the artwork from the far corners of the globe turned out to be a most arduous task, and we are thankful for all the help we received. We hope that they have all been acknowledged but fear that our record keeping is less than perfect. We received feedback from teachers, especially from Larry Grabau of the University of Kentucky, and from a number of colleagues who reviewed the chapters for accuracy. We are extremely grateful to our administrative assistant Milda Simonaitis who, for an entire year, worked tirelessly with the authors and with the publisher to help us go from concept to finished product. We are very thankful to Linda Purrington, the Golux, for making the writing styles of the various authors more uniform and polishing some of the rough edges when the writing was, well…, less than fluent.

Maarten Chrispeels would like to express his appreciation not only to his agricultural mentors, first at the School of Agriculture in Ghent, Belgium (1955–1960) and then at the University of Illinois (1960–1964) but also to his father Felix Chrispeels who imbued him with a love for cultivating the soil and caring for plants. David Sadava was inspired by his university teachers of plant biology, Margaret McCully and Frank Wightman.

We are especially thankful for the professionalism of the staff of Jones and Bartlett, including Louis Bruno, Anne Spencer, Kristin Ohlin, Stephen Weaver, and Dean DeChambeau. We hope that this book will help our fellow teachers design courses that will inspire a new generation of students to enter the basic and applied sciences that help feed the world.

Maarten Chrispeels
La Jolla, California

David Sadava
Claremont, California

CONTRIBUTORS

John H. Benedict, Ph.D.
Professor Emeritus
Department of Entomology
Texas A&M University System and Texas
Agricultural Experiment Station
Corpus Christi, Texas
Chapter 16

Andrew F. Bent, Ph.D.
Associate Professor of Plant Pathology
Department of Plant Pathology
University of Wisconsin–Madison
Madison, Wisconsin
Chapter 15

J. Derek Bewley, Ph.D.
Professor of Botany
Department of Botany
University of Guelph
Guelph, Ontario, Canada
Chapter 9

Kent J. Bradford, Ph.D.
Professor and Director of Seed
Biotechnology Center
Department of Vegetable Crops
University of California, Davis
Davis, California
Chapter 9

Maarten J. Chrispeels, Ph.D.
Professor of Biology
Division of Biology
University of California, San Diego
La Jolla, California
Chapters 3, 7, 8, 12, 19

Marc J. Cohen, Ph.D.
Assistant to Director General; Secretary,
Board of Trustees
International Food Policy Research Institute
Washington, DC
Chapter 4

Grant R. Cramer, Ph.D.
Associate Professor of Biochemistry
Department of Biochemistry
University of Nevada, Reno
Reno, Nevada
Chapter 11

Paul Gepts, Ph.D.
Professor of Agronomy
Department of Agronomy and Range Science
University of California, Davis
Davis, California
Chapter 13

T. J. Higgins, Ph.D.
Chief Research Scientist and Assistant Chief
Division of Plant Industry
Commonwealth Scientific and Industrial
Research Organization
Canberra, Australia
Chapter 7

Stephen P. Long, Ph.D.
Professor of Crop Sciences
Department of Plant Biology
University of Illinois
Urbana, Illinois
Chapter 10

Jesse Machuka, Ph.D.
Biotechnologist/Molecular Biologist
International Institute of Tropical Agriculture
Ibadan, Nigeria
Chapter 5

T. Erik Mirkov, Ph.D.
Professor of Plant Virology
Department of Plant Pathology
and Microbiology
The Texas A&M University Agricultural
Experiment Station
Weslaco, TX 78596
Chapter 6

John B. Ohlrogge, Ph.D.
Professor of Plant Biology
Department of Plant Biology
Michigan State University
East Lansing, Michigan
Chapter 19

Donald R. Ort, Ph.D.
USDA/ARS Professor of Plant Biology
Department of Plant Biology
University of Illinois
Urbana, Illinois
Chapter 10

Philip G. Pardey, Ph.D.
Professor of Science and Technology
Department of Applied Economics
University of Minnesota
St. Paul, Minnesota
Chapter 2

Todd W. Pfeiffer, Ph.D.
Professor of Plant Breeding/Genetics
Department of Agronomy
University of Kentucky
Lexington, Kentucky
Chapter 14

Idupulapati M. Rao, Ph.D.
Plant Nutritionist
International Center for Tropical Agriculture
Cali, Colombia, South America
Chapter 11

David E. Sadava, Ph.D.
Professor of Biology
Department of Biology
The Claremont Colleges
Claremont, California
Chapters 1, 3

Jonathan M. Shaver, Ph.D.
Assistant Professor of Plant and Soil Sciences
Department of Plant and Soil Sciences
Oklahoma State University
Stillwater, Oklahoma
Chapter 18

C. Neal Stewart, Jr., Ph.D.
Professor of Plant Sciences
Department of Plant Sciences
and Landscape Systems
University of Tennessee
Knoxville, Tennessee
Chapter 20

Patrick J. Tranel, Ph.D.
Assistant Professor of Molecular Weed
Science
Department of Crop Sciences
University of Illinois
Urbana, Illinois
Chapter 17

Sarah K. Wheaton
Graduate School of Oceanography
University of Rhode Island
Narragansett, Rhode Island
Chapter 20

Brian D. Wright, Ph.D.
Professor of Agricultural
and Resource Economics
Department of Agricultural
and Resource Economics
University of California, Berkeley
Berkeley, California
Chapter 2

Plants, Genes, and Crop Biotechnology

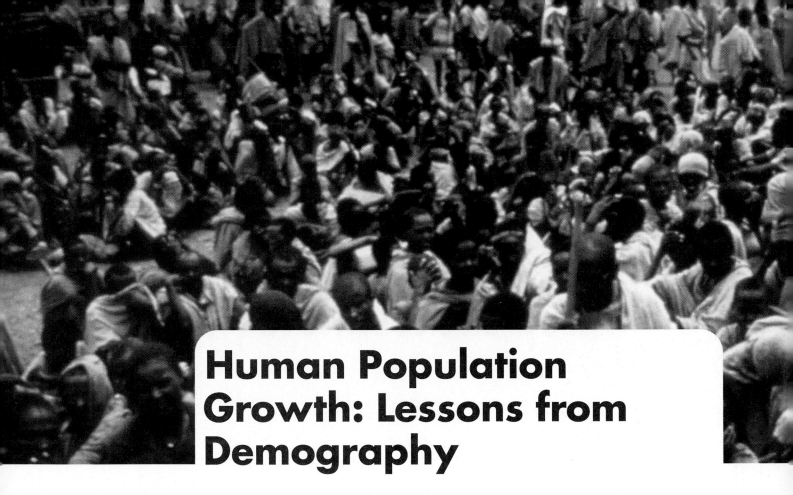

Human Population Growth: Lessons from Demography

David E. Sadava
The Claremont Colleges

On October 12, 1999, humanity reached a major milestone when the number of living people hit 6 billion. It took only 12 years to add a billion people, and by the time the 21st century dawned, people were adding 75 million every year. Feeding these millions has always been a major challenge to human ingenuity. Throughout history, times of abundant food have followed lean years according to changes in the environment. Agriculture, the application of science to food production, has tried to even out these uncertainties. Yet even now famines persist: The dramatic food shortages in North Korea during the late 1990s probably caused the deaths of 10% of its people (two million deaths), making it one of the worst such episodes in the 20th century.

Images of famine fill television screens and news magazines. Less obvious are the multitudes, estimated at about one person in seven, who are chronically underfed. But it could be much worse. The last half of the 20th century saw spectacular gains in food production capacity, which have allowed it to (just) keep up with the population growth. Now, with our increasing knowledge not only about the biology of food production, but also about the need to sustain our environment for the future, we have the potential for both feeding those who will join us over the next decades and of improving the nutritional situation for many of the world's poorest people.

1.1 The world population's rapid growth of the past 50 years is slowing.

Whereas various ancestors of humans existed over two million years ago, modern *Homo sapiens* probably emerged only in the past 50,000 years (see Chapter 13). At first the population increased slowly. Two thousand years ago the population reached 300 million, and now it stands at 6 billion, a 20-fold increase. Since then the human population has increased

20-fold (**Figure 1.1**). Why this sudden increase? Populations grow when their rate of increase (because of births and, for a particular region, immigration) is greater than their rate of decrease (caused by deaths and emigration). For humans as a whole, the ancient way of life of hunting game and gathering plant foods probably also entailed both high birth rates and death rates. This resulted in a very slow overall growth rate.

Then, about 10,000 years ago, agriculture gradually replaced hunting and gathering. With the more reliable food supply and more settled existence, birth rates rose and death

Figure 1.1 There are more than six billion people in the world today. Shown here is a large crowd at the fish market in Chikomey, Ghana. The population growth rate of Africa is still 2.5% per year and has remained unchanged for 30 years. The growth rate in Asia has declined to 1.5% per year. High growth rates are sustained by high birth rates, and educating these young people is a major challenge and key to slowing down population growth. Photo by P. Cenini, FAO.

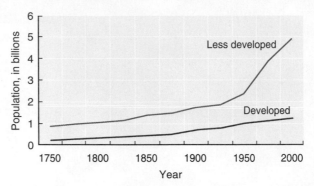

Figure 1.2 World population growth. The global population grew slowly until 1750, followed by an acceleration that continued through the 20th century. *Source:* Data from the United Nations Population Fund.

rates fell. This led to an increase in population, from 5 million to about 300 million by A.D. 1. For the next 1,800 years, high birth rates accompanied deaths from infectious diseases, famines, and wars. But the population did not "explode" until the past 200 years (**Figure 1.2**). This explosion in large part stemmed from the rise of modern science and technology, which reduced the death rate:

- Food production became more dependable. Improved transportation gave people access to food.

- Rising incomes allowed more people to afford the available food.

- Improved housing and public hygiene reduced the incidence of infectious diseases carried by rats, insects, water, and so on.

- Medical advances, such as the identification of disease agents (such as bacteria) and rational treatments (such as antibiotics), resulted in the control of previously lethal diseases.

In Europe and North America, these changes occurred over several centuries, so death rates there declined slowly. During the 19th century, Europe doubled its population; in North America, immigration from Europe and Africa fueled a 12-fold increase. When the 19th century began, about 25% of all people lived in the developed regions of the world (Europe, North America, and Japan) and 75% lived in what we now call the *developing countries*. By 1900, 33% lived in the developed countries and 66% in the less developed regions of the world (**Figure 1.3**).

In the first half of the 20th century, the events of the previous one accelerated. Improvements continued in agriculture, medicine (infant mortality—the death rate under 1 year of age—plummeted from 1 in 5 births to 1 in 20), and economics. In the United States, life expectancy at birth rose from 47 years in 1900 to 68 by 1950. With this dramatic reduction in the death rate, you might expect that the overall population growth rate would increase even faster than in the previous century. Instead, it gradually slowed down. The reason was a marked decrease in the number of children each woman had. Whereas in 1900 a typical U.S. woman had 5 children, by mid-century this number had fallen to 2.8.

Less developed regions of the world were not experiencing these types of changes. Indeed, change and lifestyle improvements have historically been slower in some regions of the world than in others. This has often led to explanations based on racial and cultural

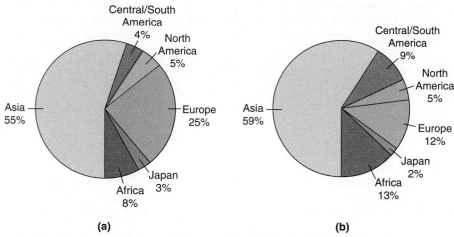

Figure 1.3 Distribution of the world's people, 1900 and 2000. During the past century, rapid population growth in the less developed regions resulted in their having the greatest share of the world population. This will continue in the current century. *Source:* Data from the United Nations.

superiority. But perhaps those in the currently developed regions were just lucky. The biologist Jared Diamond has shown that early hunter-gatherers in the Middle East had access to the right mix of plants and animals that they could tame from the wild. Few such organisms existed in any other region. Moreover, the invention of agriculture could spread quickly, because the climate is similar with respect to rainfall and temperature in, say, Spain and the Middle East, and there are no major geographic barriers to cultural exchange. In contrast, central and southern Africa have radically different climates, so a crop growing in one region will not grow in the other. Agriculture allowed the Eurasian population to grow and specialize, giving these people a "head start" in economic and cultural development, and they quickly widened their advantage.

By the dawn of the 20th century, most humans still lived with high death rates accompanied by high birth rates, just as they had for centuries and as people in the rich countries had previously. Large families (6.2 children per woman) were the rule, and infant mortality rates remained high, as did deaths of older people from infectious diseases. In 1950, a baby born in India had a life expectancy of 39 years.

In the developed world, the last half of the 20th century continued the trends of the previous 50 years. Fertility (the number of children per female of reproductive age) continued to decline, so that by the end of the millennium mothers in most of Europe, North America, and Japan were each bearing fewer than two children. Currently, in these areas two typical parents do not bear enough children to replace themselves, resulting in absolute population decline, excluding immigration. At the same time, death rates have continued to fall, although not as dramatically as in the first half of the 1900s. Improved hygiene, vaccines, and antibiotics were three positive measures that gave quick results; tackling the diseases of old age such as cancer and heart disease is more difficult.

Meanwhile, in the less developed regions of Asia, Africa, and Latin America (**Table 1.1**) many countries rapidly adopted these same three "simple" measures to increase life expectancy. Death rates plummeted. Coupled with high fertility, the overall population growth rates rose and the term population explosion became commonplace. By the 1980s,

Table 1.1	Comparison of developed and less developed regions		
Indicator	Developed	Less Developed	World
Population (millions), 2002	1,193	4,944	6,137
Annual percent growth	0.1	1.6	1.3
Life expectancy, years	75	64	67
People per room	0.7	2.4	1.9
Mortality under 5, per 100 births	0.8	6.1	5.6
GNP per person, US$	20,520	3,300	6,650
Grain production, millions of tons	810	1,259	2,069
Farmland/person, hectares	1.5	0.6	0.7

the proportion of humanity living in the more developed regions was 20%, down from 35% a century earlier.

But then, as the century ended, signs appeared that women in the developing regions were starting to have fewer children, as in the developed regions in the first half of the century. This drop was not uniform across all countries (no such trend is): In China fertility dropped, whereas in central Africa it did not. But as the new millennium began, the population growth rate in the developing regions was slowing significantly. The explosion was over.

Most population experts expect the trends of the 20th century to continue into the 21st and beyond. The UN Population Division makes projections of world population trends every two years. The latest version (2000) predicts that

- Fertility will continue to decline, especially in the less developed regions, to reach replacement level (a little over two children per woman) in 2050 (see **Figure 1.4**).

- Life expectancy will continue to improve (see **Figure 1.5**), most clearly in the less developed regions, as hygiene and living standards improve. However, HIV infection may affect these predictions (see later discussion).

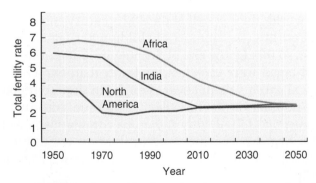

Figure 1.4 Declines in fertility. The total fertility rate is the average number of children a woman has as she passes through her childbearing years. Fertility fell in the developed regions after the "baby boom" following World War II. Fertility in the less developed regions has been falling since the 1980s and is predicted to keep falling during this century. *Source:* Data from the Population Division of the United Nations Department of Economic and Social Affairs.

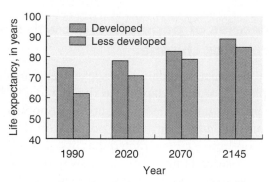

Figure 1.5 Life expectancies from birth. As sanitation and health care improved, life expectancy rose, first in the developed regions and then in the less developed. Over the next century this gap will nearly close. *Source:* Data from the United Nations.

- Overall world population will continue to grow because so many young people are entering reproductive age. But it will grow more slowly and will finally level off at 10.5 billion people by 2150 (**Figure 1.6**).

These projections differ dramatically from those made 20 or even 10 years ago. It is hard to predict scientific discoveries (or emerging infectious diseases); it is easier to foresee what these changes could do if applied to society as a whole. Scientists could predict the eradication of smallpox by the proper application of vaccine and the principles of public health, and this prediction was fulfilled. Put more broadly, changes in death rates can be explained and projected, barring the unforeseen. But how can you explain declining birth rates?

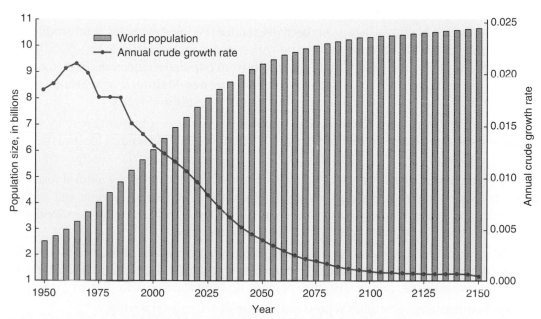

Figure 1.6 Projections of world population and annual growth rate. This is the median projection from the United Nations as of 2000. To calculate the doubling time for a population, divide the annual growth rate (such as 0.010) into 0.7 (in this case, the doubling time would be 70 years). Obviously, the smaller the growth rate, the longer the doubling time. *Source:* Data from the United Nations.

1.2   Different theories of population growth seek to explain why growth rates change.

Demography—the study of human populations—seeks to explain changes in population growth rates. These explanations often depend on the ideological leanings of the social scientists involved. The publication, in 1798, of political economist Thomas Malthus's *An Essay on the Principle of Population* started the population debate. After the French Revolution, some people optimistically predicted that living conditions for poor farmers would improve now that aristocrats no longer held sway over rural peasants. Malthus disagreed. He noted that the human population has the capacity to expand geometrically (1—2—4—8—16—), and (he believed) food production could increase only linearly (1—2—3—4—5—). As a result, he predicted, population growth would soon outpace food production; wars and famines would check population.

Malthus based his projections of food production on what he called the "qualities of the land" in England at the time. Even with what he called "great encouragements to agriculture," he felt there was little hope that farming could keep up with the growing population. He used his numeric scenario as a polemic device, perhaps overdramatizing to make his point.

Malthus reasoned as follows: Seven million people lived in England in 1800. If this population grew geometrically and needed X amount of food, a generation later (1825) 14 million people would need $2X$ food; because food production increased linearly, the $2X$ food would in fact be available. But look at the next generation: By 1850, 28 million people would be needing $4X$ food, but only $3X$ food would be available. So one quarter of the population would not have enough food. This food deficit would deepen as time went on. Malthus believed population increase promoted poverty, and he felt the solution was to postpone marriage (reduce the birth rate) and improve public health (reduce the death rate).

A basic tenet of Malthus's theory has been discredited: People now know food production can increase geometrically and can keep up with—and even rise faster than—population growth. Many of his dire predictions of the results from population increases have not been fulfilled. Yet his theory still finds strong support among neo-Malthusians, who agree with his general proposal that population increase tends to result in poverty.

The neo-Malthusians believe rapid population growth is a major threat to world economy, the environment, and political stability. They have extended Malthus's arguments in two ways:

- *Ecological Malthusians* stress that population growth undermines the natural resource base. Thus the growing population causes deforestation, soil erosion, water and air pollution, and other forms of environmental degradation, which in turn exacerbates poverty. And poverty may lead to further population growth.

- *Productionist Malthusians* stress that more people need more jobs and services from the government. This pressure severely strains government budgets to provide adequate education, housing, and health care, preventing economic growth. Poverty results.

Neo-Malthusians sometimes represent their views with the equation

$$I = P \times A \times T$$

where I = impact, P = population, A = affluence, and T = technology. Some argue that A or T is most important, but most contend that P is the most important component.

In contrast to this somewhat pessimistic view, non-Malthusians argue that population growth does not cause poverty and hunger. Instead, a common set of social causes links population growth to poverty and hunger: the lack of jobs, education, health care, and social stability. Thus population control cannot eradicate poverty unless people also deal with the fundamental causes of rising population. These include economic dependency, maldistribution of land, and unemployment. Some history supports the non-Malthusian viewpoint: In general, where incomes rise, population growth rates fall.

Some non-Malthusians, mostly economists such as the late Julian Simon, turn Malthus's argument on its head: They feel that, far from being the cause of human misery, population growth is a good thing, because humans are the "ultimate resource." As population grows, problems—and more importantly, solutions—are created. In terms of the equation, increasing P improves T, which improves I. In their argument, the need for more food stimulates better use of available resources through agricultural science, depletion of a resource through overuse stimulates the search for alternatives, and so on. In a famous exchange, Simon challenged a leading neo-Malthusian, Paul Ehrlich, to prove that five mineral resources in short supply in 1980 would become even more scarce during the late 20th century because of rising population, driving prices up. Simon proposed that human ingenuity would find either more of the resources or a substitute, and that the prices would fall. He won the bet. (The metals Ehrlich chose were copper, chrome, nickel, tin, and tungsten. All declined in price. For example, better methods of extracting nickel were found; more fiber-optic cables replaced copper wires; and so on.)

Given the slowdown in growth rate, dire neo-Malthusian predictions of population projections seem less urgent these days. In addition, data from demographic studies support the non-Malthusians. Observations by Kingsley Davis and other demographers show that industrialization and rising incomes play a major role in reducing population growth rates. This happened first in many countries in Europe, later in some regions of the developing world, and is projected for the rest of the developing regions over the next few decades. This change is called the *demographic transition*, and it has typically occurred in four stages (**Figure 1.7**):

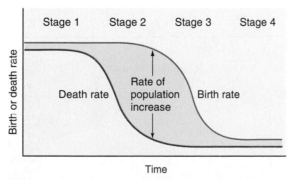

Figure 1.7 The demographic transition model. In stage 1, the birth rate and death rate are high, and the rate of population increase is small. In stage 2, better health care and sanitation reduce the death rate, but the birth rate stays the same, resulting in rapid population increase. In stage 3, the birth rate falls, and in stage 4 the rate of population increase is low. The population has stabilized once again, but at a much higher level than before. *Source:* Data from T. W. Merrick (1989), World population in transition, *Population Bulletin* 41.

Stage 1. Both birth rate and death rate are high, essentially canceling each other out so that the growth rate is low. Poor parents have many children because they provide cheap farm labor and care for the parents in their old age. Also, social factors such as religion and proof of male virility and female fertility can encourage couples to have many children.

Stage 2. Improvements in living conditions and health care reduce the incidence of disease and death. These advances are readily accepted by society. But social mores are slow to change, so the birth rate remains high, whereas the death rate drops and the overall growth rate increases.

Stage 3. As living conditions and education improve, the birth rate declines to near the level of the death rate. The population growth rate slows.

Stage 4. The birth rate and death rate are once again near each other, but now at a much lower level than in stage 1. The population growth rate returns to its previous low level.

Country after country has gone through these four phases. Most notably, fertility has declined as the well-being of the population improved. In Europe and North America, this happened in the 19th and early 20th centuries. In the developing world, it is happening now (**Figure 1.8**). How fast these stages occur depends on many factors, including culture, economic development, and political considerations.

Consider the example of Sweden, which has completed the transition. During the 18th and 19th centuries, the largely rural population had many children for the labor and social security reasons noted earlier. The death rate fluctuated because of periodic epidemics. As infectious diseases came under control in the late 19th and early 20th centuries, the death rate declined steadily. In the meantime, industrialization—coupled with more efficient farming—drew more and more people to the burgeoning cities. Children, once an economic asset, now contributed less to family income (extended education and child labor laws added to the time before children became productive). In addition, city hous-

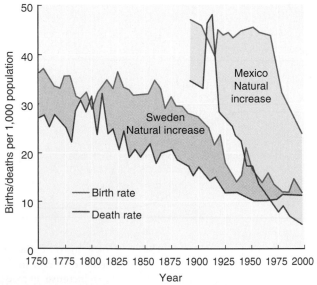

Figure 1.8 Demographic transitions in Sweden and Mexico. Although there were fluctuations, especially in death rates, as public health measures were gradually imposed, the general patterns in these two countries resemble each other and the theoretical pattern in Figure 1.7. The transition happened much faster in Mexico than in Sweden. *Source:* Population Reference Bureau.

ing was in short supply. Couples married later and relied on birth control to limit children. So fertility declined.

Consider the demographic history of Mexico, a country now in the middle of the transition. As in Sweden of 1750, in the early 20th century the birth and death rates in Mexico were both high. (Death rates spiked around 1925, during the Mexican Revolution.) But both parameters, and the growth rate of Mexico's population, began at a much higher level than Sweden's, so when the transition began in Mexico, it proceeded much faster. The rapid drop in death rates during the first half of the 20th century led to very rapid population growth; twice as many Mexicans were alive in 1950 as in 1930, and by the 1970s the population was growing by 3% a year. The population is still growing, but much more slowly now that economic development, land redistribution, and education have taken hold. Many families in Mexico now also have only two children.

All over the developing world, fertility rates are declining, which is a major cause for optimistic projections about the leveling off of the human population in the next 50 years. The varied reasons for these fertility declines include economic, social, political, and religious factors. In addition, fertility is inversely related to education. Societal changes in developing countries include the following improvements:

- Secondary school enrollments increased from 45% in 1990 to 58% in 2000.
- Adult literacy was 70% in 2000.
- Half of the women are using some form of contraception.
- Changes in economic systems and increased world trade have made more jobs available in cities.
- Infant mortality rates have declined significantly, ensuring the survival of more children.

A concerted effort by the government can alter fertility rates. A classic recent case is the Republic of Korea (South Korea; see **Table 1.2**). In 1960, Korea was a village society with high birth and death rates and a life expectancy of about 50 years (stage 1). By 1990, both birth and death rates had fallen precipitously and life expectancy had risen (stage 4). In the meantime, a rapid demographic transition took place. Two factors, both planned by the government, led to this transition. First, the population became urbanized, because jobs, many in high technology, were created to attract foreign investment. Second, the government gave high priority to birth control education and contraceptive use. The ethnic homogeneity of the Korean population and their high literacy rate were important factors aiding this effort.

Table 1.2	The demographic transition in Korea: 1965–1990	
	1965	**1990**
Total fertility rate	4.8	1.6
Infant mortality rate/1,000	75	24
Life expectancy (years)	55	73
Urban population (%)	28	75
GNP per person (US$)	130	4,500
Contraceptive use (%)	22	80

Source: Data from O. Kim and P. van den Oever (1992), Demographic transition and patterns of natural resource use in Korea, *Ambio* 21:56–62.

1.3   HIV infection is slowing population growth in Africa.

Population projections assume certain predictions about public health, development, and fertility. They cannot predict the unforeseen, such as wars, weather-induced famines, and new diseases. Although wars always seem to be in the world and people cannot do much about the weather except to have food "insurance," until recently most were confident humanity had the tools to defeat major diseases, especially those that afflict the young. AIDS has proven that confidence wrong.

In 2001 alone, 3 million people died of AIDS and 5 million more became infected with HIV. By 2002 the total number of infected people was 40 million. Although in the developed countries medicine has made some progress and has extended the lives of people with AIDS in the developed countries, these treatments are expensive and out of reach for 95% of the HIV-infected people who live in less developed regions of Africa and Asia. For these people, HIV infection is tantamount to a death sentence.

AIDS affects people of all ages, but in terms of population projections its effects are most dramatic for the young. Half of all the newly infected in 2001 took hold on people under 25. The vast majority of women with AIDS who get pregnant pass the virus to their offspring during pregnancy (rarely), delivery, or through breast milk. The mothers die, leaving their children orphaned. There are already 12 million such AIDS orphans, most of them infected with HIV.

The numbers are staggering for a disease that was unknown just 25 years ago. In sub-Saharan Africa, where nearly 70% of the global HIV/AIDS cases are found, 1 person in 30 is infected with HIV (**Figure 1.9**). In contrast to developed countries, where the infection

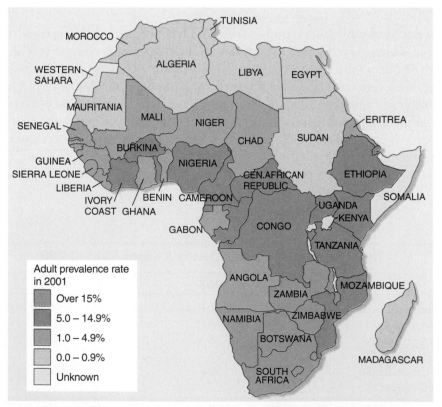

Figure 1.9 The distribution of HIV and AIDS cases in Africa. Obtaining affordable drugs for infected people and stopping transmission of the virus are major challenges facing poor African countries that lack health care delivery systems. *Source:* United Nations World Health Organization.

first spread among homosexual males and intravenous drug abusers, in less developed regions more women than men are infected with HIV. The reasons for this are not clear, but the female–male ratio is 13 to 10. There are many possible reasons for this ratio of infection; most are strain-specific differences. Also, exposure to HIV does not necessarily result in infection with HIV.

Because AIDS kills, its effects on population growth are profound. In 1990, life expectancy for a baby born in southern Africa had risen from 44 years to 60 years. AIDS threatens to wipe out this gain. By 2005, a newborn will have a life expectancy of only 45. One way to look at a population is to graph its age structure; that is, how many people are in each age group. Less developed countries with high birth rates and poor longevity have far more younger than old people, so the graph is shaped like a pyramid. This preponderance of young people of reproductive age signals a population momentum, so that even if fertility falls, merely replacing all these young people will keep the population rising. For some countries, AIDS is turning the pyramid graph into a chimney shape (**Figure 1.10**), with far fewer young people.

It appears now (2002) that African countries will receive cheaper drugs to fight AIDS, but poverty and lack of infrastructure in many regions will make it difficult to distribute and administer those drugs. In the absence of affordable treatment or a vaccine, the only way to slow this epidemic, which threatens populations all over the world—and which therefore becomes the responsibility of or concern for all populations— is to reduce the transmission of the virus from one person to another. The developing world has had several notable successes. In 1994, 13% of the Ugandan population was HIV positive. Political leadership by President Museveni (unlike Museveni, some leaders are reluctant to openly discuss AIDS), community programs, and the participation of religious leaders in disease control reduced the spread of IIIV by one third. At the same time, Thailand obtained similar results.

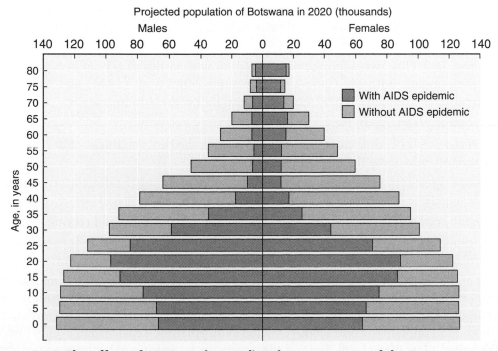

Figure 1.10 The effect of AIDS on the predicted age structure of the Botswana population, 2020. By killing so many young people, AIDS may turn the "population pyramid" into a "population chimney." *Source:* Data from UNAIDS.

The risk of exposure to HIV through sexual intercourse or birth is strongly related to poverty. Migrant workers, truck drivers, commercial sex workers, and children of infected mothers all have high rates of infection. Moreover, intravenous drug use is more common among poor people, and the virus can travel from one person to another by contaminated needles. Finally, AIDS, like many other issues in population growth, is an issue of gender.

1.4   Improvements in the status of women are essential for reducing HIV infection, population growth, and poverty.

Gender encompasses the economic, social, and cultural aspects of being male or female. In all societies, being a man or woman is not just a matter of sex, or the biological ability to produce eggs or sperm. There are particular social expectations about how the sexes should act, above and beyond biological differences. These roles range from clothing to work to relations with the opposite sex. They are not universal; for example, in much of the developing world, women are the farmers. But women's role in marketing the crops they harvest varies: In sub-Saharan Africa women play a major role, whereas in South Asia they do not. And where they do not market their crops, or where they do not control the finances, they benefit much less.

The roles of men and women must be taken into account in discussing population because different roles demand different approaches. Also, in all societies women have lower status than do men. Women usually have less education, less ownership of land, and less power to enter into or leave a marriage. Yet they often put in more working hours than men, most of them unpaid.

Take the case of HIV transmission, an increasingly important determinant of population growth in Africa. Within a stable relationship, the woman typically is not in control. She may obey the social dictum of monogamy, but her husband/lover often does not. The woman lacks the means or social authority to force her partner to engage in safe sexual practices.

During the 1990s, two major conferences dealt with issues of gender and population and development. Both the Cairo Conference on Population (1994) and the Beijing Conference on Women (1995) stressed the need for progress in four areas:

- Increase investments in the education of girls.

- Meet the needs of adolescent girls by encouraging the postponement of the first pregnancy and enhancing the status of women.

- Promote greater male responsibility in parenthood, reducing the disproportionate role of women in the home.

- Provide opportunities for women outside of childbearing and child rearing.

Besides satisfying the need for social justice, will these changes help reduce population growth? Many data show that the answer is yes.

The more educated a woman is, the more likely she is to have access to and use birth control methods. The last half of the 20th century saw a "reproductive revolution." The developments of the birth control pill, intrauterine device (IUD), simple sterilization methods, and implanted contraceptives made family planning easier than ever before. By 2000, half of all women of childbearing age were using some form of birth control, up from 10% a half-century earlier (**Figure 1.11**). There are differences among countries—in Mexico 65% use birth control, whereas in Mali only 7% do—but the trend is clearly up.

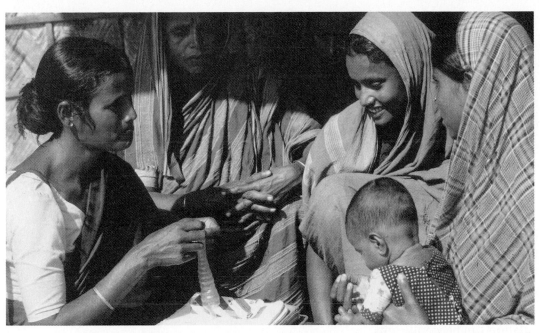

Figure 1.11 The reproductive revolution. A community health worker shows women in a Bangladeshi village how to use a condom. A major factor in the fertility decline has been the availability of and instruction in contraceptives use. *Source:* Ron Giling/Lineair/Peter Arnold, Inc.

Women's education is a powerful factor in reducing fertility. For example, in Togo, West Africa, a 1998 survey revealed that women who finished high school had an average of 2.7 children, whereas those with no education had 6.5. One reason is that educated women stay in school longer (and away from childbearing) and have more options. The net result is that they tend to postpone their first child until well into their 20s (**Figure 1.12**). Progress in educating girls has been steady for the past 20 years, although they still lag behind boys, typically being forced to drop out of school to work or get married. In 1980, 42% of boys and 28% of girls of school age in developing countries were enrolled in secondary school. By 2000, these numbers had increased to 55% and 45%, respectively.

Women usually marry at a younger age than do men. In the United States, average ages at marriage are 24 and 26 for women and men, respectively. However, in Niger the ages are 16 and 24, and in Bangladesh they are 18 and 26. This gap reinforces social customs

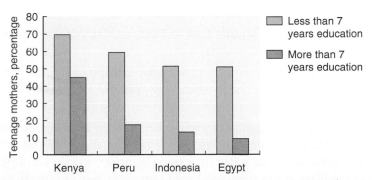

Figure 1.12 Mothers' education and teenage childbearing, 1995. The more education a woman has, the less likely she is to have a child while a teenager. Such births in less developed countries are often high risk for both the mother and child. In addition, women with more education have fewer children overall. *Source:* United Nations Population Fund.

of female subjugation in marriage. Reproductive choices are almost all made by the husband, and they usually tend toward having more children. As gender inequality lessens, it is possible that joint reproductive decisions will become the norm. Equality in making these decisions tends to result in smaller families.

1.5 Migration within and between countries is an important population issue.

Population growth is largely determined by birth and death rates, but for any given region or country migration can also play an important role. Two types of migration are

- From country to country (generally called *immigration* and *emigration*)
- From place to place within a country (often between urban and rural areas)

Moving is difficult for most people, and they do it only when benefits in income and social opportunities outweigh costs. In 2000, about 145 million people were living outside their native lands, and the number is increasing by 2 to 4 million every year. This is only 2% of the whole human population, yet the influence of migration on a region's population can be significant.

Historically, there have been times of great migrations among different areas of the world. Europeans colonized the Americas in the 17th and 18th centuries, and continued to come until the flow slowed in the second half of the 20th century. These were economic migrants. They brought millions of African slaves and other indentured servants who were essential in developing the New World. There are also sociopolitical migrants. When India became independent in 1947, millions of Moslems left the new country to form their own Islamic country, Pakistan, and millions of Hindus traveled from Pakistan in the opposite direction.

About half of all international migrations are between less developed countries. These people generally live on borders that are relatively poorly patrolled, or belong to ethnic groups present in both countries. Colonial powers originally drew many international borders without regard to ethnic group distribution.

Migrations from the less developed countries to the developed ones have more impact, because they usually involve great distances and cultural differences. Over the past decade, the major movements have been

- From South and Central America and Asia to North America. These movements account for 75% of U.S. immigration. The United States receives over a million legal and illegal immigrants per year, more than any other country.
- From North Africa and the Middle East to Europe. Many of these people have settled in England and France.
- From southern and eastern Europe to western Europe. This has occurred since the breakup of the Soviet empire. The most striking example is Germany, which has absorbed millions of immigrants from the former East Germany as well as from other republics.

The impact of these immigrants is most evident in their numbers compared to the low rate of natural increase (birth rate minus death rate) in the receiving countries. During the 1990s, immigration caused half of the population growth in the developed regions. The net emigration from the less developed countries hardly affects their population numbers because of the high growth rate. As more immigrants from different ethnic groups enter the developed countries, the ethnic compositions of the receiver countries change. Caucasians, who were by far the majority in California in 1960, will soon be a minority in the state.

All countries have laws regarding immigration, and many migrants enter their new home country under these laws. But because such laws usually limit the number of immigrants in some way, others try to enter illegally. These clandestine migrations, usually economic but sometimes sociopolitical, range from rickety African boats washing up on the shores of the Calabrian coast of Italy, to families crossing the desert from Mexico to the southwestern United States.

South Africa is a relatively wealthy country adjacent to much poorer countries. Every year, several hundred thousand economic migrants cross the poorly fenced border from Mozambique into South Africa. Many of these people are not Mozambican at all but come from more distant African countries. With a long border and no ocean or desert to cross, South Africa presents an attractive target for a destitute person seeking opportunities.

The reactions of the native South Africans to the newcomers in their midst vary. Politicians have blamed illegal immigrants for South Africa's soaring crime rate, for taking jobs away from natives, and for the high rate of HIV infection. The evidence for these claims is scant. Many employers rely on the illegal immigrants. Mining and agriculture employers, in particular, find them more willing to work harder than some natives, and highly skilled. Some citizens refer to the illegals derisively as *makwerewere,* meaning "those who jabber like grasshoppers." These three threads—blaming immigrants for social problems, accepting them because of their desperate willingness to work, and disliking foreigners—have intertwined in many countries.

Migrations forced by war or civil strife have been common throughout history, and the current era is no exception. Although their numbers are relatively small, refugees (people involuntarily living in another country who cannot return to their home) evoke strong sympathies (**Figure 1.13**). In 2000 there were about 12 million, with Palestinians

Figure 1.13 Refugees from the Rwandan civil war in Zaire. Wars always result in acute food shortages among civilian populations, in spite of international efforts to deliver food aid. *Source:* Courtesy of the United Nations High Command for Refugees. Photo by L. Taylor.

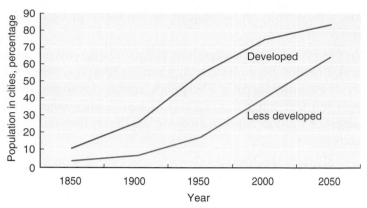

Figure 1.14 Trends for urbanization. The urban population in the world has grown three times faster than the rural population. Although the growth in developed regions is leveling off, more and more people in developing regions are migrating to cities, this trend is projected to continue in this century. *Source:* Data from the United Nations.

and Afghanis accounting for the largest fraction. The ultimate aim of any refugee is to return home. In recent years, mostly with UN assistance, about a million people have done so annually.

Most migrations take place within a country. In 1850, 11% of the people in the developed regions lived in cities. In the late 19th century, as agriculture became more efficient and industrialization offered the promise of urban jobs, people began to move en masse from rural areas to cities. Government policies hastened this transition by improving the infrastructure of roads, schools, and so on for the migrants. By 1900, cities held 26% of the population, and the percentage has since grown to its current 84% (**Figure 1.14**). As with the demographic transition, less developed regions have lagged behind in this rural-to-urban change, but they are catching up quickly.

In 1960, the three most densely populated urban areas in the world—London, Tokyo, and New York—were all in the developed regions. By 2000, Tokyo was joined by São Paulo, Brazil, and Mexico City. By 2015, demographers project, the three largest will be Tokyo, Lagos, Nigeria, and Mumbai (Bombay), India. In addition to these "megacities," the developing countries will have hundreds of new smaller cities each with fewer than a million residents.

What does urbanization mean for population growth? In general, people living in cities are better educated and have better health and life expectancy and higher incomes than their rural counterparts. As a result, urban fertility is lower. But as cities grow, so does the strain on the government to provide the basic services that are so necessary for demographic change.

1.6 Population policies are often inconsistent.

The reactions of the developed countries to the current wave of immigrants underline the changing nature of governmental population policies. These policies are actions designed to influence population growth, composition, or distribution. Basically, the immigration policy issues boil down to two questions: Should immigrants be welcomed in unlimited numbers, and should they have full rights as citizens? The United States has a long history of answering yes to both questions. In the 19th century, ships docking in New York harbor only needed to keep a log of who disembarked, so great was the need for settlers.

Subject to numerical limitation

- Close family relationship with a U.S. citizen or legal permanent resident: 226,000 per year
- Possessing needed job skills: 140,000 per year

Not subject to numerical limitation

- Spouses, children, and parents of citizens
- Refugees and people claiming asylum
- Spouses and children of people given permanent residence in the amnesty of 1986
- Amerasians born in Vietnam
- Aliens from countries adversely affected by the 1965 immigration law, which reduced their numbers (e.g., Ireland)

Figure 1.15 U.S. immigration policy. The only consistent internal population policy that the United States has concerns immigration. Rather than being strictly demographic, the policy has economic, political, and humanitarian goals.

Soon the numbers were so high that government designed tests to screen out the feeble in mind and body. Concerns about the influence of immigrants not from western Europe on U.S. culture, coupled with mistaken ideas of racial superiority, led in the 1920s to restrictive quotas, based on race, for immigrants from different world regions. Today, policies are more attuned with economic goals, and the laws have been broadened to accept families of those already in the United States (**Figure 1.15**).

In recent years, increased immigration from developing countries, coupled with the low natural growth rate of the Caucasian U.S. population, have renewed concern among some Americans. Germany, France, and Italy are grappling with the same issues. Immigrants in many European countries do not have full political rights; they are welcomed as labor but not as citizens.

Within a country, internal population policies also vary and evolve over time. Except in cases of deliberate genocide, most countries enact public health measures that reduce the death rate. Some countries have policies that regulate birth rate. For example, in the 1970s China had 500 million citizens under age 21. The government correctly feared that even if these people reproduced at the replacement level (and the rate at that time was above this level), by 2000 the huge population would be greater than the economy could support. So the government mandated a strict one-child policy. Between 1980 and 2000, fertility and population growth fell rapidly. In these terms, the policy was successful. But many critics felt it violated individual human rights. When Indira Gandhi's administration in India tried something similar with forced male sterilization in 1975–1977, it was thrown out of office and a backlash against any population policy resulted that took decades to overcome.

The evolution of 20th-century external population policies, in which a government acts to influence others, is a fascinating story of politics and religion. At first the main question was, Is there a global population problem? Shortly after it was founded in 1945, the United Nations established a commission to gather data on populations of member states. As the evidence mounted, two camps emerged: The United States, United Kingdom, Sweden, and India felt there was a serious problem of population growth. An odd coalition of the Catholic Church (which opposes any form of artificial contraception) and communist countries (who felt that the "problem" was the result of capitalism and would disappear under communism) joined to oppose the very idea of a population problem. This coalition effectively blocked any efforts at helping developing countries curb their birth rates.

By the 1970s, opposition by the communist states softened. A consensus emerged that there was indeed a population problem. This was a time of alarmist, apocalyptic books with titles such as *The Population Bomb* and *Famine: 1975!* Funders began to give modest assistance for population control, with the United States in the lead. But again two viewpoints emerged on how to proceed:

- *Incrementalists,* such as the United States and most Western developed countries, felt that individual countries must first curb population growth. Then they would be able to develop economically.

- *Redistributionists,* which included most developing countries, argued that the fastest way to improve their lot was to spread out the world's resources, which were (and are) concentrated in the rich countries. This economic development would reduce population growth.

The policies that resulted from this debate attempted to satisfy both sides but actually satisfied neither. Some redistribution of the world's wealth did occur beginning in the mid-1970s, but the recipients were a few oil-producing countries. Interestingly, as these countries got richer the results for population growth were mixed. Some countries reduced their growth rate; others, such as the Arab states, did not.

In 1984, the United Nations sponsored another population conference, at which delegates confronted an astonishing turnabout in U.S. policy. Under President Ronald Reagan, the United States, which had previously supported family planning programs, now opposed them. In fact, the U.S. delegation now took an almost redistributionist stance, favoring economic development before population control. The new approach was not quite redistributionist: Instead of favoring giving assets to the poor countries, the U.S. policy encouraged them to develop free-market economies on their own. Once the benefits of the free market stimulated economic well-being, the U.S. delegates predicted that fertility would decline. Three days after he took office in 1993, President Bill Clinton reversed this policy; the United States was once again a leader in global family planning efforts. However, when President George W. Bush took office in 2001, one of his first acts was to cut off U.S. aid to many family planning organizations.

Given these flip-flops on the international stage, it is not surprising that the United States does not have a consistent internal population policy. Over U.S. history, the government has tacitly encouraged fertility. But this approval is not as overt as in western Europe or in countries such as Canada, where the government gives parents an allowance for each child. Instead, in the United States the reward is indirect, as in the income tax deduction for children.

Some ecologists have pointed out that every child born in the United States uses far more global resources than a child anywhere else on Earth, and thus argue for limiting U.S. population growth. But there has been no policy to discourage fertility. However, the United States has always had an explicit immigration policy, as mentioned earlier.

What might a U.S. population policy entail? It could set goals for

- The ideal growth rate
- Birth rates and immigration policies to satisfy this growth rate
- Death rate targets, including childhood mortality, that vary in different parts of the country
- Distribution (and redistribution) of people around the country

Unfortunately, competing political, social, economic, and religious agendas make agreement on such a policy highly unlikely.

1.7 Increases in population have been matched by increases in food supply, but hunger persists.

Ever since agriculture began, an increased ability to produce food has accompanied the rise in human population. The recent history of food production shows a steady increase most markedly in the less developed countries, where the major increases in population have occurred (**Figure 1.16**). How has this food production increase been achieved? What foods have been important in the increases? Historically, there have been two ways to increase food production:

- Increase the amount of land used to produce food
- Increase the amount of food produced per season on the land already being used for agriculture

The amount of arable land presently being cultivated is about 1.5 billion hectares (ha). Researchers have estimated that cultivated arable land increased by 432 million ha between 1860 and 1920, and again by 419 million ha between 1920 and 1978. Thus a considerable proportion of the increased food production since 1860 has been achieved by plowing more land. Since 1978, however, the amount of cultivated land has remained more or less steady. Because the world population has added 1.7 billion people since then, the amount of cultivated land per person has dropped by about 25%. Clearly, the productivity of this land must have improved significantly to keep pace with the growing population, or we would be facing starvation on a massive scale.

Of the approximately 300 types of plants cultivated in the world, 24 supply nearly all our food. More than 85% of the human diet comes from eight species of plants, and over half comes from just three cereal grains: maize, wheat, and rice. Although the area devoted to these plants has not risen much in the past 50 years, the production index (total food produced per land area) has risen considerably. These impressive yield increases have been most pronounced in the developed countries, which can afford the technologies to coax the most out of crops. Farmers in poor countries have also seen significant improvements, especially since the Green Revolution.

How did these increases in cultivated area, yield per hectare, and population affect the amount of food available per person? The food index (food produced per person) has risen

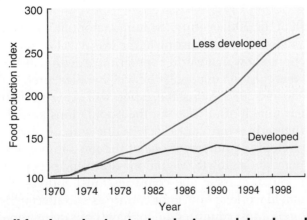

Figure 1.16 Overall food production in developing and developed countries. Notice that the developing countries, where most people live, have had more dramatic increases than the developed countries. However, the per person data look quite different. For purposes of illustration, production levels in 1970 are arbitrarily set at 100. *Source:* Data from United Nations Food and Agriculture Organization.

Figure 1.17 Food Index: Food production per person. The amount of food produced per person has remained the same in the developed regions for the past 25 years, whereas food production increases have outstripped the population rise in the developing regions. For purposes of illustration, production levels in 1970 are arbitrarily set at 100. *Source:* Data from U.S. Department of Agriculture.

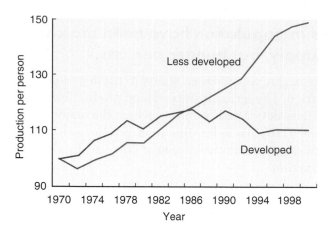

somewhat, although not nearly as dramatically as overall food production (**Figure 1.17**), once again with the gains most pronounced in the less developed countries. In Chapter 2 we discuss the relationships among these food increases, changing agricultural practices and technologies, and agricultural R&D (research and development).

Calculations invariably show that an adequate amount of food is produced to meet the needs of the global population. Yet hunger persists. The United Nations estimates that almost 800 million people are chronically undernourished, two thirds of them living in Asia and the Pacific. Children are especially sensitive to undernutrition or malnutrition. Surveys conducted in developing regions ranging from Asia to Africa to Latin America during the 1990s indicated that 4 in 10 showed stunting (low height for age), 2 in 10 were underweight (low weight for age), and 1 in 10 showed wasting (low weight for height). This malnutrition bodes ill for these children as they grow up. In Chapter 4 we discuss the reasons why food insecurity persists in a world of apparent plenty.

CHAPTER SUMMARY

During the second half of the 20th century, the human population increased rapidly to 6 billion, with most of the expansion occurring in the less developed countries, especially in Asia. Recent projections show that world population growth will slow dramatically in the next 25 years and that the human population will stabilize, probably at 8 to 9 billion, sometime between 2050 and 2075. This stabilization will occur as women in developing countries decrease their total fertility rate and the populations pass through the demographic transition, which brings birth rates back in line with death rates. Improving the status of women in developing countries is one of the most important factors in reducing the birth rate and population growth. Migration within and between countries has also emerged as a major population issue that countries must address. Worldwide, record numbers of undocumented workers from poor countries are entering richer neighboring countries in search of jobs. Most countries do not have consistent population policies to deal with these migrations. In the past 50 years, increases in food production have matched and outstripped increases in population. The result has been a steady rise in the food availability per person. Nevertheless, food insecurity and malnutrition have not been eliminated and one person in seven does not have food security.

Discussion Questions

1. What would be the best way for your country to reduce population growth in developing countries? Should your country be active in this arena, and why or why not?

2. Why does ethnic homogeneity in a developing country make it easier to slow the population growth rate, as happened in Korea?

3. How might regional traditions of a developing country hinder efforts at (a) reducing population growth and (b) improving the status of women?

4. What are possible reasons for higher rates of HIV infection in women than men in less developed regions of sub-Saharan Africa?

5. Compare and contrast neo-Malthusian and non-Malthusian schools of thought in the context of current undernutrition problems in Asia and Africa.

6. Immigration was very high in both the United States and western Europe during the 1990s. What factors caused these increases? How did these regions respond to immigration pressure?

7. Humanity currently produces enough food to give every person an adequate diet. What factors in your country cause an unequal distribution of food? How is the government of your country responding to this?

8. In the currently well-developed regions, the demographic transition occurred over many decades to centuries. Why is it occurring more rapidly in many less developed regions?

9. Thirty years ago, the human population was growing very rapidly with seemingly no end in sight. Now there is optimism that world population growth is slowing down and may level off during this century. List the major factors that led to this change.

10. Conduct a survey of your family and friends to determine what they know of human population and its growth. If they lack adequate knowledge (and most will), inform them of the facts. How can the public be better informed of these issues?

Further Reading

Birdsall, N., A. Kelley, and S. Sinding. 2001 *Population Matters: Demographic Change, Economic Growth and Poverty in the Developing World.* New York: Oxford University Press.

Bogin, B. *The Growth of Humanity.* 2002. New York: Wiley-Liss.

Brown, L., M. Renner, and B. Halwell. 2000. Vital Signs: *The Environmental Trends That Are Shaping Our Future.* New York: Norton.

Cuffaro, N. 2002. *Population, Economic Growth and Agriculture in Less Developed Countries.* London: Routledge.

DeRose, L., E. Messer, and S. Millman. 1998. *Who's Hungry? And How Do We Know? Food Shortage, Poverty and Deprivation.* Tokyo: United Nations University Press.

Diamond, J. 1997. *Guns, Germs, and Steel: The Fates of Human Societies.* New York: Norton.

Foster, P., and H. D. Leathers. 1999. *The World Food Problem.* Boulder, CO: Rienner.

Gould, W. T. S., and A. M. Findlay. 1994. *Population Migration and the Changing World Order.* New York: Wiley.

Livi-Bacci, M., and G. DeSantis. 1998. *Population and Poverty in the Developing World.* Oxford, UK: Clarendon Press.

Peterson, W. 1999. *Malthus: Founder of Modern Demography.* London: Transaction Publishers.

Population Reference Bureau. 1999. *World Population: More Than Just Numbers.* Washington, DC: Population Reference Bureau.

Simon, J. 1996. *The Ultimate Resource.* Princeton, NJ: Princeton University Press.

United Nations Department of Economic and Social Affairs. 1998. *Too Young to Die: Genes or Gender.* New York: United Nations.

United Nations Population Fund. 1999. *6 Billion: A Time for Choices.* New York: United Nations.

Agricultural R&D, Productivity, and Global Food Prospects

Philip G. Pardey
University of Minnesota
Brian D. Wright
University of California Berkeley

The works of great thinkers such as Malthus and Marx, as well as of Hardin and Ehrlich more recently, show how easy it is to be dead wrong about the productive potential of agriculture. When Malthus wrote his *Essay on Population* in 1798, in which he predicted that population growth would soon outpace food production (see Chapter 1), the practice of agriculture relied on local labor and natural resources, including land, seed, water, and organic fertilizer. Farm management was based on knowledge accumulated over centuries, and there were no obvious opportunities for rapid improvement if the land base could not expand in pace with population. Malthus had no way of anticipating how different the future path of food supply would be from its past. Karl Marx in *Das Kapital* predicted that agriculture would follow the experience of manufacturing, becoming an increasingly concentrated sector with many workers per farm, each worker specializing in a small fraction of the tasks involved in farm operation. The Soviet Union and China tried to implement this vision with collectivized agriculture, with calamitous results. More recently, eminent ecologist Paul Ehrlich, in *The Population Bomb* (1968) predicted that in the 1970s "the world will undergo famines—hundreds of millions of people are going to starve to death in spite of any crash programs embarked upon now. At this late date nothing can prevent a substantial increase in the world death rate" (p. xi). William and Paul Paddock's 1967 *Famine 1975! America's Decision: Who Will Survive?* had a similar message, advocating a triage approach to foreign aid. The "can't be saved" group, which should receive no aid, included India and the Philippines, both of which have since had years when the harvests was so large as to produce a glut. Biologist Garrett Hardin became famous for coining the term "the tragedy of the commons" to describe the very real problems that can arise from conflicts of interest when there is open access to exploitation of a natural resource. In 1977, he published *The Limits of Altruism* in support of a "tough-minded" approach recognizing that countries such as India had exceeded their "carrying" capacity.

Yet over the past century growth in productivity of both land and labor, domestically and internationally, has enabled world food supplies to outpace the unprecedented increase

in food demand caused by jumps in the growth rate of world income and by the doubling and redoubling of population. Waves of change in selection of plant varieties (**Figure 2.1**) and management of crops and pastures, improvement of animal breeds, mechanization of farm tasks, inorganic fertilizers, sophisticated genetics-based breeding techniques, and new methods of controlling pests and diseases have

Figure 2.1 Pulling up young rice plants for transplanting to a rice paddy in South Thailand. The dramatic increase in food production in Asia starting in the 1960s was made possible by new rice varieties. Researchers specifically bred the new rice plants to have short stems so they would not fall over ("lodge") when applying more nitrogen fertilizers caused the heads to grow larger. Cultivating paddy rice is often a two-stage process: Seedlings are started in a nursery (as shown here) and then transplanted out into a paddy.

actually resulted in such greatly increased supplies of food per person that, despite the increase in demand per person, food prices have fallen to their lowest levels ever. Innovation has also reduced the land required, and, by increasing feed efficiency, the waste products per unit of food. In contrast to the dire predictions just listed, food security has ceased to be a major popular concern as we enter the new millennium, although a substantial number of people are still food insecure (see Chapter 4).

The currently favorable dynamic balance between food supply and demand was not inevitable. It is the result of successful interactions among farmers, input suppliers, and an overwhelmingly publicly supported research and extension system that has furnished innovations and relevant knowledge for free. Continued strong performance in research and innovation is needed to maintain this balance over the next half century.

2.1   Dramatic yield increases during the past 50 years have made food cheaper and more widely available than ever before.

For thousands of years, farmers eked out yield gains collecting and selecting the best and most productive seeds and by improving techniques of cultivation and organic fertilization. However, expanding cultivated areas accounted for most of the total production increases. Over the past century, what many had feared became a reality. In the most populous countries, population growth outstripped the expansion in land for growing food. Yet food production continued to at least keep pace with the growth in demand. Indeed, global food production increased by 25% in the 1980s, and by about the same amount in the 1990s.

Starting in the late 19th century, yields of major crops in North America, Europe, and Japan began to increase at rates well beyond historical precedent. For example, beginning with an average wheat yield of 15 bushels per acre in 1866 (the earliest year for which data are available), it took 103 years, until 1969, for U.S. yields to double (**Figure 2.2**). Yield

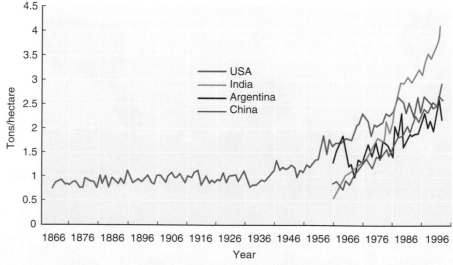

Figure 2.2 Long-run trends in wheat yields in the United States, Argentina, China, and India. *Sources:* Data from J. M. Alston and P. G. Pardey (2002), "Farm Productivity and Inputs," in S. Carter, S. Gartner, M. Haines, A. Olmstead, R. Sutch, and G. Wright, eds., *Historical Statistics of the United States—Millennial Edition* (Cambridge, UK: Cambridge University Press); and Food and Agriculture Organization of the United Nations (2000), "FAOSTAT Statistical Databases," available at <http://apps.fao.org> (accessed August 22, 2001).

Table 2.1	Percentage of area planted to modern varieties (semidwarf) of rice and wheat						
	Rice			Wheat			
Regions	1970	1983	1991	1970	1977	1990	1997
(percentage of area planted)							
Sub-Saharan Africa	4	15	na	5	22	52	66
West Asia/North Africa	0	11	na	5	18	42	66
Asia (excluding China)	12	48	67	42	69	88	93
China	77	95	100	na	na	70	79
Latin America	4	28	58	11	24	82	90
All developing countries	30	59	74	20	41	70	81

Sources: Byerlee and Moya (1993), Byerlee (1996), Heisey, Lantican, and Dubin (1999).

growth accelerated in the second half of the 20th century; it took only 43 years for U.S. wheat yields to double and reach the much higher 43 bushels per acre reaped in 1999.

Similar yield accelerations occurred in many other crops in the United States. For example, rice yields were just 1,114 pounds per acre in 1895 and 2,046 pounds per acre in 1945 (a compound rate of growth of 1.3% per year). By 2000, they had grown to 6,278 pounds per acre—a growth rate of 1.9% per year since 1945. Maize yields grew more slowly, by only 0.07% per year for the first half of the 20th century, but the rate jumped to 2.6% per year for the second half of the century.

Many crops in developed countries took a sharp upturn in yield performance in the middle of the century as an increasing number of genetically improved varieties, targeted to particular agroecological zones, became available. Beginning in the 1950s and continuing at an accelerated pace in the 1960s and 1970s, international and national agricultural research centers also made improved varieties available to many more farmers in the developing countries, and yields took off in those countries as well (for example, see Figure 2.2 for wheat). **Table 2.1** shows the rapid spread of modern (often semidwarf and higher yielding) rice and wheat varieties throughout the developing world, initially via adoption of breeding lines developed in international research centers over wide areas with favorable environments, and then by adapting these varieties to local ecologies and consumer preferences. Asia was quickest to embrace these new varieties; varietal change lagged in sub-Saharan Africa, partly because of the great diversity in agroecological zones (see Chapter 5).

Similar long-run patterns of yield growth have enhanced other food crops in many countries worldwide. Globally, yields of all major cereals have climbed steadily, at least since the 1960s. About 95% of the production gains since 1961 have come from increasing yields, except in Africa, where nearly 40% of the gains have come from expanding the cultivated area. Yields of major cereals have more than doubled in the past four decades. Indeed, area cultivated has actually begun to decline in some regions because of urbanization, road building, mining and industrialization, and agricultural mismanagement such as water erosion, wind erosion, and soil salinization. Even Africa, which has always relied heavily on cultivating new land for production increases, will increasingly need to count on yield gains to avoid the high financial and ecological costs for expanding into areas not yet cultivated. In South Asia, the per capita rice- and wheat-growing area shrank from 0.11 hectares in 1961 to about 0.07 hectares in 1998. Some researchers have recently claimed that the rate

Box 2.1

Labor Specialization Evolved Differently in Factories and Farming Operations

Starting about 150 years ago, agriculture moved beyond what scientists could achieve with farmers' cumulative knowledge and locally available inputs, and embarked on a new, science-based path using new products offered by other specialized input providers that organized during industrialization. In industrially developed countries such as the United States, as other industries coalesced into a small number of large firms much of agriculture remained highly competitive, with a multitude of independent producers. Despite major increases in land use and output, labor input per farm has remained almost constant, at the equivalent of about 1.5 full-time workers—about the same as in, for example, India. In addition, some farms employ temporary workers for specialized tasks such as harvesting.

Unlike factory workers, full-time farm workers cannot be highly specialized by task; they must take care of all the operations needed to run the farm. Nonetheless, there has been some significant specialization in farming. In advanced countries, and increasingly in others as well, mixed farming has given way to farms producing a more limited range of commodities, and the geographic specialization in commodities has also become more pronounced. For some commodities such as pig and poultry meat production and specialized horticultural crops such as canning tomatoes in the United States, contract forms of farming are now prevalent. Landholders cede many management and input decisions to integrators who provide feed and genetic inputs and process the output, in exchange for greater security about financial returns. But contract farmers typically also have noncontract production, and the range of tasks they perform on the farm tends to be larger than those handled by a specialized factory worker or a corporate manager. Thus, compared to manufacturing, farmers have experienced far less of the functional specialization identified by Adam Smith in *The Wealth of Nations* as a major source of efficiency than those employed in other sectors of the economy. It is also true (and likely to remain so for much of agriculture for some time) that few if any individual farms, or even multiholding farm operations, can reap much of the benefit of an innovation; however, their integrators are becoming increasingly concentrated and powerful in dealings with farmers, legislators, and consumers.

of growth of yields for some crops (such as rice and wheat) seems to have slowed in some regions. However, there is no uniform pattern among crops or regions. For example, in the United States wheat yields grew considerably faster during the 1990s than during the previous decade, whereas in the countries of the former Soviet Union and eastern Europe yields have declined because of a lack of inputs and other policy-related difficulties, rather than because of a ceiling on yield.

Indeed, despite the doubling and redoubling of crop yields seen in countries with favored environments, any absolute yield ceiling seems far off at present. Researchers have estimated yields that can be generated if a plant is given all the inputs it needs. For most cereals, potential yields are several multiples of the present average U.S. yields.

In addition to production gain from yield growth, another source of gain accrues from increased seed productivity. In medieval England, farmers had to save one quarter of their wheat harvest to use as seed for planting the next crop, leaving only three quarters of the harvest for food (and feed) consumption. This ratio has fallen sharply and is still decreasing; it was about 11% of the output in 1961, and fell to 6% in 1999 in the United States. For rice, the average planting rate was only about 5% in 1961, and fell as low as 3% in 1999. In addition, mechanization has released land formerly needed to feed draft animals (oxen, mules, horses) for producing food and fiber now. Thus yield growth actually underestimates the real net harvest gain from changes in technology. By reducing spoilage, improved storage and transport technologies have also increased food available to consumers from a given harvest. These dramatic increases in land productivity were not associated with any significant increase in number of full-time workers per farm, in sharp contrast to the experience in manufacturing (see **Box 2.1**).

Land Saved by Net Yield Increases. The world population today has increased by 80% since 1960. The environmental wonder of the past four decades is that today's farmers are feeding almost twice as many people far better from virtually the same cropland area. The world used about 1.4 billion hectares of land for crops in 1961, and only used 1.5 billion hectares in 1998 to get twice the amount of grain and oilseeds. Furthermore, the average citizen of the developing countries is getting 28% more calories, including 59% more vegetable oil (at twice the resource cost of cereal calories) and 50% more animal calories (which come, on average, at three times the resource cost of cereals). Except in countries devastated by AIDS or war, or disrupted by the collapse of the Soviet Union, people today also can expect to live much longer than those who lived 50 years ago in the wealthiest countries.

Producing today's world food supply with 1960 crop yields would probably require at least an additional 300 million hectares of land. In other words, through innovations in seeds, pesticides, fertilizers, crop management, confinement meat production and modern food processing, modern high-yield farming has reduced the cropland necessary to meet current food and feed needs by an area equal to the entire land mass of western Europe. Unprecedented and persistent advances in yields have confounded the predictions of experts, and provided a greater margin over subsistence needs for a greatly increased population. Food prices have declined to the lowest levels in history, to the benefit of consumers who are able to eat better while spending less and less of their budgets on food. Below, we examine in some detail how this happened, and what is needed now to satisfy food demands in the coming decades. But first let's look at the factors that will determine how those demands will evolve.

2.2 Income growth will replace population growth as the major challenge to world food production capacity in this century.

Demand for food is obviously influenced by the growth and movement of population. In addition, income growth, human resource development, lifestyles, and preferences are also very important in determining the effects "at the farm gate." In the next several decades, population growth will obviously contribute to an increased demand for food. Although population growth rates will continue to fall (see Chapter 1), about 73 million people, equivalent to the current population of the Philippines, will be added to the world's population on average every year between 2000 and 2020, increasing world population by 25% from 6 billion in 1995 to around 7.5 billion by 2020. About 97.5% of the increase in population is expected to occur in today's developing world.

Most of the population increase in developing countries is expected in the cities (see Figure 1.14 in Chapter 1). Urbanization will contribute to changes in the types of food demanded. The developing world's urban population is projected to double from 1.7 billion to 3.4 billion in 2020. Urbanization affects dietary and food demand patterns: Changes in food preferences caused by changing lifestyles, and changes in relative prices associated with rural–urban migration lead to more diversified diets. Food choices shift from

Figure 2.3 Vegetable vendor in a small town along the Mekong River in Vietnam. City dwellers in developing countries increasingly have more money to buy meat and a variety of fruits and vegetables. Local farmers respond to this demand.

coarse grains such as sorghum and millet to cereals such as rice and wheat that require less preparation and free women to exploit urban employment opportunities. Urban dwellers also tend to consume more livestock products, fruits, vegetables, and processed foods (**Figure 2.3**).

Although population growth has been the focus of world attention over the past 30 years, it will be supplanted, barring worldwide economic catastrophes, by income growth as the greatest challenge to world food production capacity. This will be true even though food expenditure will not keep pace with personal income growth. Since the 18th century economists have extensively studied the relationship between income and consumption of specific items. In perhaps the first empirical generalization about consumer behavior, based on consumption data of 153 Belgian families, Ernst Engel in 1857 proposed the famous hypothesis, now known as Engel's law, that the proportion of total expenditure devoted to food declines as income rises. This makes sense; the capacity of the stomach does not expand as income increases. Low-income people (those living on a dollar or less a day) can spend up to 70% of their income on food—rich people (say, those with average incomes in the United States) spend much more on food, but it amounts to less than 10% of their much higher incomes.

As the world's poor become more affluent, a shift will also occur in the composition of consumption, from subsistence diets comprised mainly of grains and roots/tubers and low in animal protein, to higher-quality diets comprised mainly of varied grains, meats, dairy products, eggs, and diverse fruits and vegetables. This pattern is evident in China, which more than doubled its meat consumption in the 1990s (**Figure 2.4**). Yet the average Chinese consumer still eats less than a third as much meat as the average Japanese consumer. As economic growth spreads further and deeper in these economies, the dietary shift will increase in both scope and pace, raising meat consumption further.

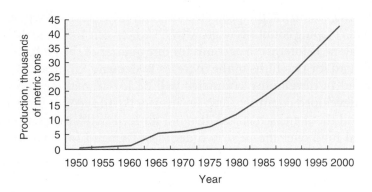

Figure 2.4 Pig meat production in China. The graph clearly shows the tremendous increase in pig meat production to meet the growth in consumption in China since 1970. Until 1980, the United States was the biggest consumer of meat in the world, but China now easily surpasses it. (China has, however, about four times as many people as the United States.)

We may look to Japan as a model for what to expect as Chinese income continues to rise. Economic growth in Japan has brought about a fundamental shift in Japan's dietary habits. Since 1965, Japanese consumers have reduced their consumption of rice by 37%, and they have increased dairy consumption by 123% and meat consumption by 220%. In all, the average Japanese consumer now eats about 55 grams of animal protein per day. And if the Japanese eliminated import restrictions, they would probably eat closer to 65 grams of animal protein per day. For comparison, Americans eat about 75 grams per day.

This pattern of increased meat demand is occurring throughout Asia, where nearly half the world's population lives. India's consumers have been adding one to two million tons of milk and dairy products to their national diet each year, despite feed shortages, high prices, and poor quality. However, on average, Asians still consume less than 20 grams of animal protein per day. By 2025, 4 billion Asians may each be consuming 55 grams of protein per day. That is nearly a 400% increase in the region's total meat consumption.

As most vegetarian cultures in the world, including China and India, move to a more affluent diet higher in animal protein, they will place increasing pressure on agricultural resources, which are strictly limited in Asia. It takes three to five times as many farm resources to produce a single calorie or a gram of protein of meat or dairy product, compared to cereal grains, legumes, or tuber crops.

As individuals consume more food, new food-related problems arise. Too little food is the scourge of poverty, but too much of the wrong sorts of food is also a major health problem (see Chapter 7). Obesity—excessive weight for a given height (often measured by a body mass index, weight in kilograms divided by the square of height in meters)—contributes to diabetes, hypertension, strokes, and cardiovascular diseases. The condition is increasingly prevalent in most developed countries, and is taking hold in parts of Latin America and Central Eastern Europe/Commonwealth of Independent States, although it is still largely absent from South Asia and Africa.

2.3 The growth in demand for grain for animal feed will outstrip the demand for grains used in human foods.

Researchers have projected that global demand for cereals will increase by about 33% between 2000 and 2020 to reach 2.5 billion metric tons, meat demand will increase by nearly 60% to 327 million tons, and demand for roots and tubers will increase by almost 40% to just over 900 million tons. Most increases in demand through 2020 are projected to occur in developing countries, which will account for about 85% of the increase in global cereal demand.

But in developing countries, the projected surge in meat demand will cause demand for feed grain in developing countries to grow by 85% between 1995 and 2020 to 432 million tons. China alone is forecast to account for one quarter of the global increase in demand for cereals and for two fifths of the increase in demand for meat. By 2020, 26% of the cereal demand in developing countries will be for animal feed, compared with 21% in 1995. In developed countries, feed for livestock will account for over 60% of the cereal demand, and the increased cereal demand for feed will far outstrip the increased demand for cereal food over the next two decades. Demand for feed will increase by 40% worldwide.

Thus, as income rises, demand for maize, mainly for animal feed, will increase much faster than for any other cereal, by a projected 2.39% per year between 1997 and 2020, compared with 1.61% for wheat and 1.25% for rice. An estimated 69% of the maize will go toward feeding livestock compared with 15% of wheat and 3% of rice in 2020. In China, where total demand for meat is projected to double between 1997 and 2020, demand for maize is forecast to increase by around 2.8% per year, whereas demand for rice, the most important staple for human consumption, is projected to increase by only 0.6% per year.

So how will the world meet the 21st-century food challenge? Already about 38% of the planet's total land area is devoted to agriculture (crops for food and feed, and pastureland), and it will not be easy to expand cultivated area significantly in the most populated regions. As people become more affluent, their food consumption becomes less and less responsive to the price rises that will occur if yield growth slows. Thus, if we are to save wildlife habitat, ecosystems, and biodiversity in the 21st century, we must meet the food challenge by raising yields on existing farmland even further, using means that do not degrade the environment.

2.4 A complete view of productivity changes includes the value of all inputs, not just land.

When economists measure productivity, they compare the quantity of one or more outputs to the quantity or value of one or more inputs used to produce the output(s). Thus far we have focused on crop yields in outlining the dramatic changes observed in modern agriculture in the last century. Yield is a partial productivity measure; it relates the quantity of output to just one input, land. But achievement of a yield increase usually comes at the cost of using more of other inputs such as labor, pest control programs, fertilizer, irrigation or improved plant varieties and animal breeds. A total productivity measure that expressed total output, relative to the total quantity of all the inputs used in production, would be very informative, but the data needed to measure the totality of inputs and outputs are rarely if ever available. Agriculture uses many unmeasured and often unpriced (or underpriced) inputs such as rainfall, natural soil nutrients and organic matter, and crop pollinators, and producers generate some nonmarketed outputs including "goods" such as pleasant rural landscapes and carbon sequestration to reduce global warming, and "bads" such as greenhouse gases, dust, and off-farm pollution of underground water and surface streams.

Multifactor productivity measures aggregate output, omitting outputs that are harder to measure or for which data have not been kept, relative to an aggregate of inputs. Even the best input aggregates generally omit, for example, the accumulation of highly localized (within a given farm) information on soil conditions, or improved planting, weeding, and harvesting operations that have important productivity consequences.

Management skill is another type of unmeasured input that accounts for some productivity growth.

Productivity Patterns in the United States. In 1990, in aggregate terms, U.S. agriculture produced more than three times the quantity of output in 1910, a compound growth rate of 1.61% per year. This rate of increase in output was achieved with an increase of only 0.06% per year in the total quantity of measured inputs. Thus multifactor agricultural productivity grew by the difference in these rates, 1.55% per year, a very rapid rate indeed, sustained over an 80-year period.

The remarkably small change in aggregate input use hides a good deal of variation across different categories of inputs, even when measured nationally, ignoring regional variations. In 1910, labor accounted for 29% of the total cost of inputs; but by 1997, the labor input accounted for only 11.9%. As a share of total input costs, energy has grown rapidly throughout. Fertilizer, lime, and pesticide expenses have generally accounted for between 4.6 and 13.8% of total input costs. Between 1950 and 1997, purchased intermediate inputs grew from 14.5% of input costs to 18.7% in 1997, with a general decline in inputs generated on the farm. Animals for traction, fodder and feed mixes for livestock, manure to fertilize the land, seeds for planting, or chicks for egg and broiler production have all been phased out of use, or are increasingly purchased from specialized input suppliers.

One weakness with these types of measures is inadequate adjustment for changes in quality. Simply counting machines used on farms does not capture the fact that machines are much better than they were 50 or even 5 years ago. Similarly, the composition of the labor force in agriculture has changed to include more experienced and better-educated farmers, and this means that "hours of work" in agriculture today means something quite different from what it meant in 1910. Nevertheless, labor-saving machinery represented an important element in the overall growth in farm productivity and also transformed the nature of much farm work. Important innovations in cropping were made when tractors replaced horses and self-propelled combines replaced tractor-drawn combines. In earlier periods, of course, the mechanical reaper and binder replaced the sickle and manual shocking.

For cereal crops, farm mechanization started in the 19th century in response to increasing rural wage rates, with important innovations and substantial continued progress in the first half of the 20th century. Cotton picking was not mechanized until after World War II, and innovations to mechanize the harvest for some other crops—such as tomatoes for canning, potatoes, and various tree crops and grapes—have been even more recent (**Figure 2.5**). Other important mechanical innovations include various technologies used to irrigate and (off the farm) to transport and process the harvest, including canning, refrigeration, and other ways to preserve food. Electrification was important in facilitating other innovations, particularly in dairying, where milking machines and refrigerated vats revolutionized the industry. Dryers fired by liquefied petroleum gas (LPG) are now used extensively to lower the moisture content of maize, reducing spoilage during storage.

A significant element in the aggregate productivity patterns has been genetic improvement, especially of crops. Genetic improvement is not the only factor in the dramatic increases in crop yields of the past century; people have made important improvements in chemical fertilizers, irrigation, and weed and pest control. But trials comparing modern varieties with previously popular alternatives indicate that genetics accounts for one third or more of the yield increase for many crops, especially for wheat, rice, and maize.

Figure 2.5 Potato harvester. For many crops, harvesters operated by a single person have replaced the seasonal workers who used to dig the harvest.

Genetic changes have led not only to improved yield potential but also to improved disease resistance, more uniform grain and fruit and other quality improvements; better tolerance of drought or waterlogging, or shorter growing seasons; better adaptation to particular climates or soil conditions; and greater suitability for mechanical harvesting (including more uniform ripening and the ability of the plant or its fruit to withstand mechanical processes). Not all these changes have improved the final product from the consumer's viewpoint. Recently demand has surged for old "heirloom" varieties, such as attractive and varied tomatoes that offer delicious flavors, and delightful colors and shapes, as a tradeoff for higher cost, shorter shelf-life, and restricted seasonal availability. Demand has also surged for higher-cost organic foods produced without artificial fertilizers, crop protection chemicals, or biotechnology. Such organic production is most feasible in highly favored ecologies.

Productivity growth has varied widely. Important improvements have been made in feed conversion genetics of poultry, swine, and dairy cows, and these tend to be less location specific as the improved animals tend to be free from environmental stresses. Significant innovations have also been made in stocking rates, disease control, and reproductive efficiency.

2.5 Agricultural land and labor productivity vary dramatically from country to country.

Yields of crops and animals vary by location and change over time. **Figure 2.6** gives an internationally comparable indication of the total value of agricultural output in 1996 per unit of agricultural land and per agricultural worker. These data are country averages, although both measures vary significantly within many countries. Globally, $266 (1989–1991 prices) worth of agricultural output was produced for every hectare of land in crops and permanent pasture, ranging from an average of $1,026 per hectare for

Europe to $69 in sub-Saharan Africa and just $51 in Australia and New Zealand combined. Areas with higher shares of irrigated land and hence higher cropping intensities (whereby multiple crops are grown on the same land over the course of a one-year season cycle), such as the intensively cultivated systems of East Asia (including China, Japan, and North and South Korea, among others), have the highest yields, producing $2,067

Value of Production per Hectare of Agricultural Land, 1995-97 Average

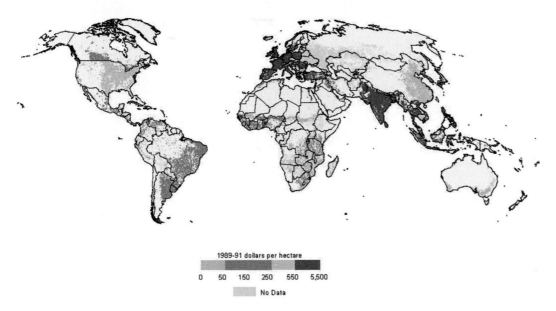

1989-91 dollars per hectare

0   50   150   250   550   5,500

No Data

Value of Production per Agricultural Worker, 1995-97 Average

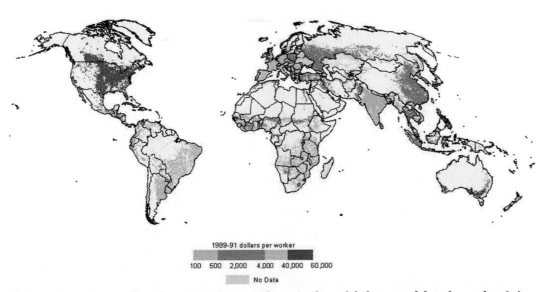

1989-91 dollars per worker

100   500   2,000   4,000   40,000   60,000

No Data

Figure 2.6 International comparisons of agricultural labor and land productivity.
Sources: Food and Agriculture Organization of the United Nations (1997 and 1999), "FAOSTAT Statistical Databases," available at <http://apps.fao.org> (accessed August 22, 2001); and World Bank (2000), *World Development Indicators* (Washington, DC: World Bank), cited in S. Wood, K. Sebastian, and S. J. Scherr (2000), *Pilot Analysis of Global Ecosystems: Agroecosystems* (Washington, DC: International Food Policy Research Institute and World Resources Institute).

| Table 2.2 | Input intensity indicators, 1995–1997 average |

Region	Agricultural Labor	Tractors[b]	Inorganic Fertilizer[a]				Irrigated Share of Cropland
			N	P$_2$O$_5$	K$_2$O	Total	
	(person per hectare)	(hectare per tractor)		(kilogram per hectare)			(percent)
North America	0.02	41	57.1	21.6	23.1	101.8	9.8
Latin America	0.28	102	26.7	18.3	17.1	62.1	11.3
Europe	0.15	14	89.7	32.2	36.5	158.4	12.5
Former Soviet Union	0.11	102	14.0	4.5	2.3	20.8	9.3
West Asia/North Africa	0.45	60	39.7	18.1	3.3	61.1	26.4
Sub-Saharan Africa	0.98	622	6.1	3.4	2.1	11.6	3.7
East Asia	3.58	47	130.7	51.1	83.2	265.0	38.7
South Asia	1.57	123	62.9	19.3	6.6	88.8	38.0
Southeast Asia	1.47	232	50.2	16.6	17.0	83.8	17.4
Oceania	0.05	138	17.7	25.5	6.8	50.0	5.2
World	0.85	57	53.2	21.0	15.5	89.7	17.5

Notes: Labor, fertilizer, pesticide, and tractor inputs are based on hectares of cropland (annual plus permanent crops).

[a] Includes only commercial inorganic fertilizers: nitrogen (N), phosphorus (P$_2$O$_5$), and potassium (K$_2$O).

[b] Tractors are defined here as all wheel and crawler tractors (excluding garden tractors) used in agriculture.

Source: Compiled from FAO (1999).

worth of agricultural output per unit area, compared with $375 for the countries of the former Soviet Union. The value of output per unit area also depends on how intensely other inputs such as water, labor, and fertilizer are used (**Table 2.2**). Land productivity is also sensitive to the mix of outputs and is lowest for the extensive livestock systems.

Comparing the upper and lower halves of Figure 2.6, you see that the geographic pattern of land productivity is virtually independent of the labor productivity pattern just discussed. In 1996, the United States ranked 90th out of 189 countries in terms of land productivity, but a clear first in terms of output per unit of agricultural labor—an estimated $51,850 of output for every person working in agriculture. Sub-Saharan Africa did poorly on both counts, with low land and labor productivity. Australia and New Zealand, which ranked low in terms of land productivity ($54 of output per hectare of land), ranked second in terms of labor productivity ($42,355 of output for each worker in agriculture). This part of the world is not well endowed with naturally productive soils and has low and erratic rainfall, but it has developed an extensive form of agricultural production (with exceptionally few agricultural workers per hectare) that generates a significant output per labor unit by international standards. In developed countries, labor productivity is not significantly determined by the agricultural environment; the major influence is the wage available in off-farm employment, adjusted for the costs of shifting to off-farm work.

Figure 2.7 tracks worldwide trends in labor (panel a) and land (panel b) productivity since 1961. Globally, land productivity doubled from 1961 to 1997, reflecting both the increased scarcity of land suitable for agricultural expansion, and successful research in increasing yields. Labor productivity grew more slowly, increasing by 50% over the same period. Land productivity grew fastest in land-scarce Asia, where agricultural output per hectare in 1997 was 164% greater than in 1961. In the former Soviet Union, which had

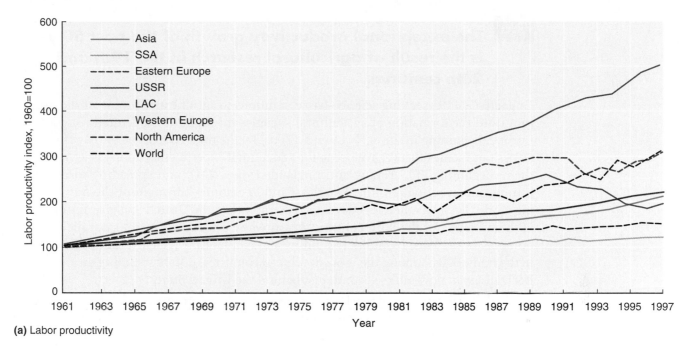

(a) Labor productivity

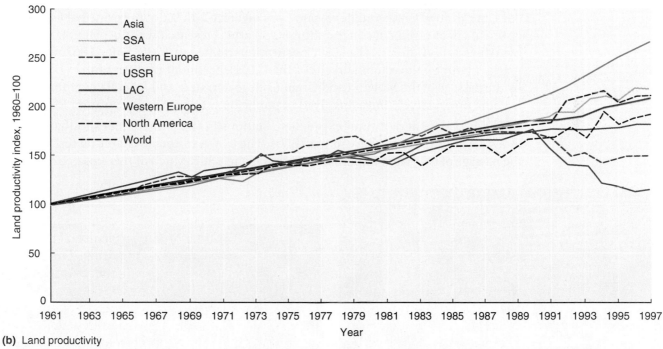

(b) Land productivity

Figure 2.7 **Trends in land and agricultural labor productivity in different regions of the world.** *Source:* Data from Food and Agriculture Organization of the United Nations (2000), "FAOSTAT Statistical Databases," available at <http://apps.fao.org> (accessed August 22, 2001).

the slowest gains, output per hectare was only 13% higher in 1997 compared with 1961; it declined markedly throughout the 1990s.

Labor productivity grew fastest in western Europe, the combined effect of a strong growth in agricultural output and an exodus of labor from European agriculture. Labor productivity also increased fairly rapidly in the United States (2.54% per year since 1961 compared with 4.45% for Europe) but barely budged in Africa, growing by only 0.24% per year since 1961.

2.6 **The exceptional productivity growth of the past 50 years is the result of agricultural research in the 19th and 20th centuries.**

Collective efforts seeking science-based solutions to agricultural problems did not take root until the formation of agricultural societies throughout the United Kingdom and Europe, beginning in the early to mid-1700s. By the mid-1800s, the efforts of these societies (and some others) gave rise to the agricultural experiment stations as we now know them; beginning in Germany and England (**Figure 2.8**), and spreading to the rest of Europe and eventually to the Americas, and to colonies throughout the now developing world. Japan, a much less developed country than the United States or Europe in the 19th and much of the 20th century, measured by per capita income, paralleled developments in the West by publicly funding and conducting agricultural R&D beginning in the mid-1800s. Among the more developed countries, public spending on agricultural R&D in Japan now ranks second, just behind the United States.

Until the second half of the 19th century, agricultural innovation in the United States was encouraged primarily by state and local governments and by farmer organizations, which provided prizes for and demonstrations of best practice at county fairs and such. The federal government subsidized collection and distribution of promising seed varieties, but relatively little public research was organized. Initially, the principal federal government encouragement of agricultural (and other) research was the patent law enabled by Article 1, Section 8, of the U.S. Constitution, ratified in 1788. Since 1862, which was marked both by the establishment of the U.S. Department of Agriculture (USDA) and by the passage of the Morrill Land Grant College Act, state and federal governments have become progressively more involved in public investments in agricultural R&D.

The first state agricultural experiment stations in the United States followed prototypes developed in Germany in the 1850s. The USDA funded extramural research after the passage of the Hatch Experiment Station Act in 1887. Much of this research took place

(a)

(b)

Figure 2.8 Rothamsted Agricultural Experiment Station in England. The Rothamsted Experiment Station, 50 miles north of London, officially dates back to 1843, when John Lawes founded it. **(a)** Two young men seated at the table, sorting grass samples. In the center of the room is the stove used for heating in winter. **(b)** Aerial view of the Broadbalk, experimental fields that have been in continuous use since 1843. *Source:* Photo courtesy of the Rothamsted Experiment Station.

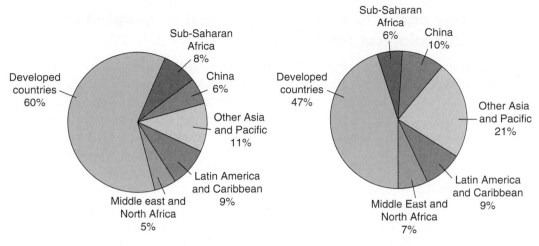

Figure 2.9 Global public agricultural research expenditures, 1976–1995. *Source:* Adapted from P. G. Pardey and N. M. Beintema (2001). *Slow Magic: Agricultural R&D a Century after Mendel.* IFPRI Food Policy Report. Washington, DC: International Food Policy Research Institute.

at state agricultural experiment stations, supported by a mixture of federal, state, and private funds, and generally located at the various land grant universities established across the nation. As an outgrowth of locally organized (and funded) efforts to provide information and technology transfer services to farmers in the United States, the Smith-Lever Act of 1914 created the Cooperative Extension Service and instituted a federal role in extension.

Agricultural science developed hand in hand with the institutions conducting the research. Darwin's theory of evolution, the pure line theory of Johannson, the mutation theory of de Vries, and the rediscovery of Mendel's laws of heredity all contributed to the rise of plant breeding in the beginning of the 20th century. Pasteur's germ theory of disease and the development of vaccines opened up lines of veterinary research. The effectiveness of these sciences in raising yields and solving farmers' production problems in developed countries became evident in the first half of the 20th century. This success encouraged similar developments in the newly independent, less developed countries, where agricultural research in the colonial era had been largely confined to export crops.

Over the past several decades, worldwide investments in publicly performed agricultural research have almost doubled in inflation-adjusted terms, from an estimated $11.8 billion (1993 international dollars) in 1976 to $21.7 billion in 1995 (**Figure 2.9**). In recent decades, the geographic balance of public research has shifted. R&D spending by developing countries, denominated in international dollars, grew from 40% of public-sector R&D spending worldwide in 1976 to 53% in 1995.

The $21.6 billion of public agricultural R&D spending is concentrated in just a handful of countries. In 1995, the United States, Japan, France, and Germany accounted for two thirds of public research done by developed countries, about the same share they had two decades earlier. Just three less-developed countries—China, India, and Brazil—spent 44% of the developing world's dollars committed to public agricultural research in 1995, up from 35% in the mid-1970s. The low-income developing countries invest the least intensively in research—about 0.6% of the value of farm production (designated by agricultural gross domestic product), compared with 2.6% in the high-income countries. However, these low-income countries often have unique agroecologies and cropping system challenges that only they can address (see following discussion and Chapter 5).

2.7 Many agricultural innovations must be adapted to local agroecological conditions.

Many agronomic technologies have a biological component that is sensitive to local climate, soil, and other biophysical attributes. For example, soybeans are day-length sensitive so different varieties must be developed for different latitudes. Likewise, many tropical soils are naturally acidic, a less prevalent problem in temperate areas. This local sensitivity distinguishes agricultural innovations from those in most other types of technologies, such as medicine, information, and mechanical technology, where applications seldom vary from Tijuana to Tokyo. Significant local adaptation is often required before agricultural technologies fit the local agroecology as well as the economic environment.

About 63% of the nontropical world's agricultural research occurs in developed countries, and developing countries such as Argentina, China, and South Korea that have broadly similar agroecological characteristics will tend to find this research relatively easy to adapt to local environments. Transferring technologies from nontropical regions to subhumid and moist semiarid tropical countries such as Brazil and India often require more local, adaptive research to modify varieties, as well as crop and livestock production practices, to fit local agroecological realities. For these reasons, grouping countries according to agroecological attributes offers a useful perspective on the potential pool of technologies that may be of common interest. One coarse, but nonetheless instructive, classification is to group countries in terms of tropical and nontropical, where tropical countries are those having a year-round, sea-level-adjusted, average temperature of greater than 18° Celsius. In 1997, an estimated 1.44 billion hectares (or 62%) of the world's agricultural land and 2.6 billion people (45% of the world's population) were in tropical countries. The share of the world's people and agricultural area in tropical countries greatly exceeds the tropical country share of public research spending, which is only about 28%, but 28% is almost exactly the share of agricultural output by value that comes from the tropics (**Figure 2.10**). If instead we classify countries as developed or developing, there is almost as close a match; in 1997, 59% ($935 billion) of the world's $1.3 trillion worth of agricultural output came from developing countries, quite close to the 54% of global R&D spending on agriculture that occurred in these countries. Yet as noted, in the poorest countries the relative investment in research lags significantly.

Figure 2.10 Tropical perspectives on agricultural R&D spending, 1995. *Source:* P. G. Pardey and N. M. Beintema (2001). *Slow Magic: Agricultural R&D a Century after Mendel.* IFPRI Food Policy Report. Washington, DC: International Food Policy Research Institute.

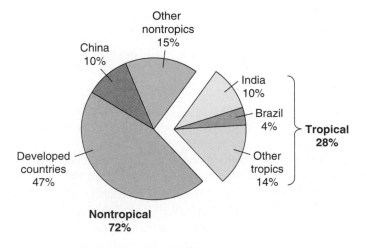

1995: $21.7 billion (1993 international dollars)

Figure 2.11 Aerial view of the International Rice Research Institute in the Philippines. This is one of 16 CGIAR institutes around the world that were initially funded by the Rockefeller Foundation and the Ford Foundation and that now receive public as well as private funding.

International Research. Internationally conceived and funded agricultural R&D—as distinct from colonial research largely funded by metropolitan governments in the United Kingdom, France, and Belgium—took hold in the mid-1940s and at an accelerated pace through the 1950s, as the Ford and Rockefeller Foundations placed agricultural researchers in some less developed countries to work alongside scientists in national research organizations on joint venture projects. These efforts became the model for many subsequent programs in international agricultural research, and later evolved into the International Rice Research Institute (IRRI) at Los Baños (**Figure 2.11**), the Philippines in 1960, and the International Maize and Wheat Improvement Center (CIMMYT) at El Batan, Mexico, in 1967. Soon after, other international centers were established at Ibadan, Nigeria, in 1967 (IITA), and at Cali, Colombia, in 1968 (CIAT) (**Table 2.3**).

These institutions joined the Consultative Group on International Agricultural Research (CGIAR, or CG for short) established in 1971. The CGIAR system grew rapidly until the 1990s, and included 16 institutions with a budget of $347 million in 1999 (**Figure 2.12**). The CG accelerated the spread of high-yielding varieties of wheat and rice and other technologies throughout the developing world (commonly called the Green

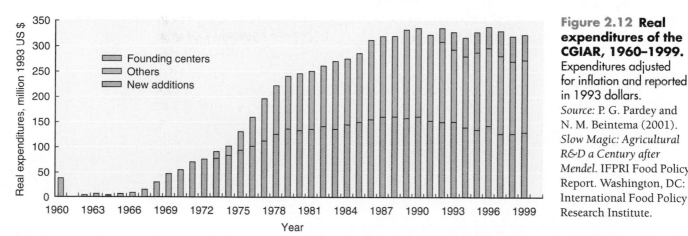

Figure 2.12 Real expenditures of the CGIAR, 1960–1999. Expenditures adjusted for inflation and reported in 1993 dollars. *Source:* P. G. Pardey and N. M. Beintema (2001). *Slow Magic: Agricultural R&D a Century after Mendel.* IFPRI Food Policy Report. Washington, DC: International Food Policy Research Institute.

Table 2.3 The CGIAR centers

Center	Date of Joining CG	Date of Foundation	Head-quarters Location	Main Areas of Focus Commodity/Activity	Main Areas of Focus Region/Agroecological Zone	1999 Budget
						(million US $)
IRRI, International Rice Research Institute	1971	1960	Los Baños, Philippines	Rice Rice-based ecosystems	World Asia	35.1
CIMMYT, Centro Internacional de Mejoramiento de Maiz y Trigo	1971	1966	El Batan, Mexico	Wheat, maize	World	37.4
CIAT, Centro Internacional de Agricultura Tropical	1971	1966	Cali, Colombia	Phaseolus bean, cassava Rice Tropical pastures	World Latin America Latin America/lowland tropics	30.7
IITA, International Institute of Tropical Agriculture	1971	1967	Ibadan, Nigeria	Farming systems Rice, maize, cassava, cocoyams, soybeans	Humid and subhumid tropics World	32.7
ICRISAT, International Crops Research Institute for the Semi-Arid Tropics	1972	1972	Patancheru, India	Farming systems Sorghum, millet, pigeonpeas, chick-peas, groundnuts	Semiarid tropics (Asia, Africa) World	23.2
CIP, Centro Internacional de la Papa	1972	1974	Lima, Peru	Potato, sweet po-tato, other root crops	World	21.6
IPGRI, International Plant Genetic Resources Institute[b]	1974	1974	Rome, Italy	Promote activities to further collection, conservation, evolu-tion, and use of germ plasm	World	20.4
WARDA, West Africa Rice Development Association[c]	1974	1970	Bouaké, Côte d'Ivoire	Rice	West Africa	10.9
ICARDA, International Center for Agricultural Research in the Dry Areas	1976	1976	Aleppo, Syria	Farming systems Barley, lentils, fava beans, wheat, kabali chickpeas	North Africa/Near East World North Africa/Near East	22.8

| Table 2.3 | continued |

Center	Date of Joining CG	Date of Foundation	Head-quarters Location	Main Areas of Focus Commodity/ Activity	Main Areas of Focus Region/ Agroecological Zone	1999 Budget
						(million US $)
ISNAR, International Service for National Agricultural Research	1979	1979	The Hague, Netherlands	Strengthen national agricultural re-search systems	World	9.7
IFPRI, International Food Policy Research Institute	1980	1975	Washington, DC, United States	Identify and analyze national and inter-national strategies and policies for re-ducing hunger and malnutrition	World, with primary emphasis on low-income countries and groups	20.1
ICRAF, International Centre for Research in Agroforestry	1991	1977	Nairobi, Kenya	Agroforestry, multipurpose trees	World	21.8
IWMI, International Water Management Institute[c]	1991	1984	Colombo, Sri Lanka	Water and irriga-tion management	World	8.8
ICLARM, International Centre for Livestock Aquatic Research Management	1992	1977	Metro Manila, Philippines	Sustainable aquatic resource manage-ment	World	12.4
CIFOR, Center for International Forestry Research	1993	1993	Bogor, Indonesia	Sustainable forestry management	World	12.7
ILRI, International Livestock Research Institute[a]	1995	1995	Nairobi, Kenya, and Addis Ababa, Ethiopia	Livestock production and animal health	World	26.5

Note: na indicates not applicable.

[a] ILRI became operational in January 1995 through a merger of the International Laboratory for Research and Animal Diseases (ILRAD, founded in 1973) and the International Livestock Center for Africa (ILCA, founded in 1974). ILRAD research focused on livestock diseases (world) and tickbone disease and trypsanomiasis (sub-Saharan Africa). ILCA did research on animal feed and production systems for cattle, sheep, and goats for sub-Saharan Africa.

[b] IPGRI was first established in 1974 as the International Board of Plant Genetic Resources (IBPGR). The board was funded as a CG center but oper-ated under the administration of FAO and was located at FAO headquarters in Rome, Italy. In 1993, IBPGR changed its name to IPGRI, and was es-tablished as a self-administering CG center in its own headquarters building in Rome. An International Network for the Improvement of Banana and Plantain (INIBAP) was established in Montpellier, France, in 1984. In 1992, INIBAP became a CG center; in 1994, INIBAP's functions were placed under the administration of IPGRI but it continues to maintain its own board.

[c] Until 1998, the International Irrigation Management Institute (IIMI).

Source: Updated table from Alston and Pardey (1999).

Revolution), but spends only a small fraction of the global agricultural R&D investment—in 1995, just 1.5% of the nearly $22 billion (1993 international dollars) in public sector agricultural R&D by national agencies, and 2.8% of research spending by the less developed countries. Two large French institutions, CIRAD and IRD, expend about half as much as the CGIAR (not all on agriculture), with a focus on tropical countries and a distinct set of commodities.

2.8 Private investment in agricultural research and development is substantial and concentrates on commercially attractive technologies.

Private investment in agricultural biotechnology, chemical, machinery and food-processing research is substantial and, at least until very recently, has been rising rapidly. Roughly one third of the $32 billion total public and private agricultural research investment worldwide is private (**Table 2.4**). Most private research was conducted in developed countries ($9.8 billion, or 94% of the global total), where privately performed R&D investment was about equal to the public research investment. In developing countries, the share is much smaller; only 5% of the agricultural R&D is private. Although the private presence has grown and is sizable, public funds are still the dominant source of support.

Private research is displacing public research in areas such as breeding of commercial crops with large and profitable seed markets, and various agricultural biotechnologies where expanded intellectual property rights have made private investments more attractive. Private firms concentrate on technologies that are easily transferable across agroecologies, such as food processing and other postharvest technologies, and chemical inputs, including pesticides, herbicides, and fertilizers. Thus private research is much more geographically concentrated and less agroecologically oriented than public research, but many of its fruits may be more easily transferred (given the right market incentives) across countries and even between developed and less-developed regions.

The type of R&D done by private firms has changed over time. For example, in the United States where time series data are available, agricultural machinery and postharvest food processing research accounted for over 80% of total private agricultural R&D in 1960. By 1996, these areas of research collectively accounted for only 42% of the total; the share of total private research directed toward agricultural machinery having declined from 36% in 1960 to about 13%. Two of the more significant growth areas in private R&D have been plant breeding and veterinary and pharmaceutical research. Spending on agricultural chemicals

Table 2.4	Estimated global public and private agricultural R&D investments, 1995					
	Expenditures			**Shares**		
	Public	**Private**	**Total**	**Public**	**Private**	**Total**
	(million 1993 international dollars per year)			*(percent per year)*		
Developing countries	11,770	609	12,379	95.1	4.9	100
Developed countries	9,797	10,353	20,150	48.6	51.4	100
Total	21,567	10,962	32,530	66.3	33.7	100

Source: Pardey and Beintema (2001).

research grew rapidly and now accounts for more than one third of total private agricultural R&D in the United States.

There is no prospect that private research will take over the task of developing, for the world's poor, new varieties of staple foods such as cassava, beans, yams, cooking bananas, millet, and others with little commercial value, nor will it contribute significant ecology-specific research related to their cultivation and management. Most of the world's population will continue to rely on public and nonprofit institutions, supplemented by some well-publicized private-sector donations, to develop the crops that constitute its staple foods.

2.9 The impact of agricultural research occurs after a considerable lag, but the returns are impressive.

Successful investment in agricultural research leads, as noted, to increases in agricultural productivity. These increases stem from developing new and better varieties of plants or annuals; new or improved outputs from new, better, or cheaper inputs; or new ways of doing things that let producers choose and combine inputs more effectively. Economic evaluations of research effects compare the size of investment in research to resulting productivity flows and economic benefits resulting from research. This requires procedures that account for differences in the timing of cost and benefit streams. The lag times between investing in R&D and reaping some return can be quite long, because some inventions bear fruit slowly. In addition, some innovations last a long time or are used in subsequent R&D, leading to further invention cycles and benefit streams. These lags are crucial in determining R&D benefits and may be an important reason for apparent underinvestment in research.

Figure 2.13 schematically represents the timing of benefit and cost flows from investment in a successful agricultural R&D project resulting in a particular innovation. The vertical axis represents the flow of benefits and costs in a particular year, and the horizontal axis shows years since the start of the R&D project. Initially the project involves expenditure without benefits so that during the "gestation" or research lag period (say, three to five years but often much longer, depending on the type of research), the flow of net benefits is negative.

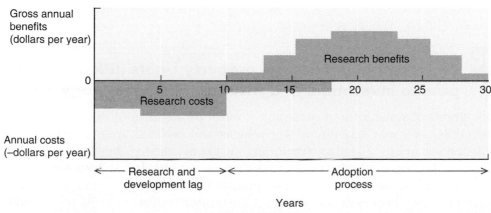

Figure 2.13 Flows of research costs and benefits. Notice that at the start (left) the costs exceed the benefits, and that after a lag of several years the benefits come on line. *Source:* J. M. Alston, M. C. Marra, P. G. Pardey, and T. J. Wyatt (2000), *A Meta Analysis of Rates of Return to Agricultural R&D: Ex Pede Herculem?* IFPRI Research Report No. 113 (Washington, DC: International Food Policy Research Institute).

Suppose the research is successful, leading to a commercially applicable result. After the research lag there may be further delays, including a development lag of several years and an adoption lag that can last many years. A conventional justification for agricultural extension services has been that they shorten the adoption lag so that benefits appear earlier. Eventually, as shown in the figure, the annual flow of net benefits from adopting the new technology becomes positive. In some cases, the benefits flow may continue indefinitely, but in many cases this flow eventually declines.

Figure 2.13 shows the flows over time of gross annual benefits attributable to the R&D project. They represent the sum of benefits across individuals in the society, accruing in each year, in contrast to what the situation would have been had the project not been undertaken. The right comparison is with and without the R&D, not simply before and after it. Why is this so? It may be that for a particular commodity, yields have not risen, yet yield-enhancing research has been successful and highly beneficial. In many cases, the relevant alternative is not constant yields but falling yields (or rising costs to maintain past yield performances as, say, increased pesticide, crop management, and labor costs are needed to counter the evolution of pests that are resistant to pesticides or attack earlier crop variety releases). Indeed, maintenance research—directed at maintaining yields and profitability in the face of countervailing pressures—is a major component of agricultural R&D.

Rates of Return to Research. To compare projects that have different time patterns of costs and benefits, economists use capital budgeting techniques to aggregate financial flows over time. Using such measurement methods, the calculated rates of return for agricultural research projects are impressive. An analysis of 1,772 such estimates, taken from 292 studies published worldwide since 1958, found the average annual rate of return among all studies to be an extraordinary 81%, but there is a large variation around this average. Not all research is scientifically or commercially successful—many projects have negative rates of return where research costs exceeded benefits. Other research is highly successful, leading to very large benefits compared with research costs.

There is now a popular sense that the easy gains in science have all been achieved—that it is now much more costly to innovate, and the returns are lower. The evidence does not support this view. Formal statistical analysis of estimated rates of return to research reveals that these rates have not declined over time—recent research seems as productive as research done four decades ago.

2.10 Protection of intellectual property rights (IPRs) promotes private investment in agricultural biotechnology.

Public grants of some form of monopoly control in the form of intellectual property rights (IPRs) over new agricultural technologies and products are nothing new. Utility patents on inventions related to farm inputs such as machinery, chemicals, and pharmaceuticals have been around for many years. The United Kingdom, which has the longest continuous patent tradition in the world, granted its first patent in 1449. The legislative basis of the U.S. patent system is the U.S. Constitution ratified in 1788. Since 1980, the Plant Patent Act has protected asexually reproduced plants, that is, plants such as grape vines, fruit trees, strawberries, and ornamentals that are clonally propagated through cuttings and graftings.

In 1980, a revolution began in patent protection, not in agricultural IPRs per se, but in extending the legal protection of plants or other genetic materials and methods increasingly used for plant breeding. In that year, the U.S. Supreme Court ruled in favor of utility patenting of life forms. In 1985, the U.S. Board of Patent Appeals ruled that utility patents could protect asexually and sexually propagated seeds, plants, and tissue culture (see Chapter 9).

Two other changes in the 1980s further fostered the proliferation of IPRs in the United States. First, federal patent law administration effectively made it easier to obtain and defend patents. Second, the passage of the Bayh-Dole Act in 1980 gave researchers the right to retain title to material and products they invented under federal funding in nondefense areas. These changes encouraged the profitable privatization of biotechnologies developed at universities and other public institutions, often via technology licensing, joint venture, or spinoff arrangements with private firms. Since then, the output of public researchers has been increasingly privatized, in the sense that others can use it only with the consent of the relevant property rights owner.

In the pre-1980 scientific environment, the post-1980 IPR revolution would have been almost irrelevant. Defense of patents owned by breeders or seed sellers requires proof of infringement. By serendipity, the revolution in analysis of genetic material ushered in by the Cohen-Boyer patent of 1980 produced a set of technologies well suited to detecting unauthorized reproduction or breeding via DNA analysis of seeds or leaves, or other genetic evidence. These methods have also been effective in enforcing state trade secret law as a protection of inbred parent lines used in hybrid maize breeding. A thicket of proprietary claims now controls the transfer and use of patented biotechnologies, limiting the freedom to operate public and private agencies alike. Proprietary claims now cover all sorts of biotechnologies, including

- Plant germ plasm.
- Trait-specific genes, which control specific "input" characteristics. These include the well-known "Roundup Ready®" herbicide tolerance trait, and genes from *Bacillus thuringiensis* (*Bt*) for insect resistance. Other genes confer traits such as tolerance of abiotic stress, fungal or viral resistance, cold tolerance, ripening, and output traits such as increased content of starch, oil, amino acids, proteins, vitamins, and minerals, or decreased content of traits that are harmful (for example, allergens) or contribute to environmental pollution (such as phytates that increase environmental damage from manure). Many of these genes have been patented, although few commercial cultivars have as yet been released.
- Enabling technologies, including
  - Transformation technologies, by which a gene that codes for a specific characteristic is inserted into plant cells
  - Promoters, used to control the expression of a gene in plants
  - Markers, genes used in conventional breeding or selectable markers used in production of transgenics to identify the presence of a desired trait
  - Gene silencing or regulating technologies, used to suppress or modify gene expression in plants
  - Genomics, the use of databases of information on plant genes and gene expression

This list will expand with time. Intellectual property protection is proliferating globally. As a condition for participating in the trade benefits of the World Trade Organization

(WTO), developing as well as developed countries must adopt intellectual property protections as delineated in the Trade Related Aspects of Intellectual Property (TRIPS) agreement. One exception is that plant varieties may be protected by an instrument such as a Plant Variety Protection Certificate (PVPC) instead of a utility patent. The latter does not constrain use for breeding new, distinct varieties, although it might prevent the sale of "essentially derived" varieties differing by, say, a single gene from a protected variety. (This issue is still an open legal question.) In general, plant breeders who wish to commercialize technologies in jurisdictions where they are protected by patents must sign patent licensing agreements, if they can get access at all. Farmers using proprietary technology in commercial seeds may be required to conform to the terms of licenses presented on seed bag labels (like software "shrink-wrap" licenses) or technology use agreements restricting seed application to one planting on a specific area of land, and facilitating inspections to enforce the restriction.

2.11 IPR protection can also hinder research and development.

Plant breeding is a cumulative science. As patents on research tools, processes and products proliferate, the restrictive monopolies these patents confer bear down on the next generation of research. The diversity of innovations used in modern cultivar development can balkanize competing claims, seriously hindering subsequent innovation. For example, rice rich in genetically engineered provitamin A, currently under development as Golden Rice, incorporates technology based on at least 70 patents with 32 owners. In cases where rights ownership is diffuse and uncertain, the multilateral bargaining needed to access all these rights can become difficult if not impossible. In the case of Golden Rice, major IPR holders have made their technologies freely accessible to poor farmers in developing countries. (Actually, most of the patents are not valid in those countries anyway.) But in the United States and some other developed countries, university research projects designed to produce new crops with modern biotechnology have been shut down because IPR-holders refuse to permit commercialization of varieties incorporating their intellectual property. Public and nonprofit institutions are at a particular disadvantage in bargaining over IPRs. They lack resources to license required technologies or to support the seven-figure expense of litigating a typical patent dispute.

In the private sector, the high costs of IPR transactions seem to encourage takeovers and mergers. Indeed, the agricultural input industries (seed, pesticides, and herbicides, and genetics) have undergone very rapid consolidation since 1995, raising concerns about the increased market power and even monopolization of these industries in the United States.

As IPR over plants have been extended in developed countries, nations that provided the domesticated seed varieties or landraces used in breeding are responding by attempting to assert their claims to the basic genetic material derived from their traditional varieties under the banner of farmers' rights. Exactly how such rights should be recognized is unclear. They do not seem amenable to protection by the usual IPRs that confer rights to individual inventors or institutions such as utility patents, PVPCs, or trade secrecy. Some form of collective rights seems more appropriate but is proving difficult to put into practice. Farmers are naturally unhappy about a system that gives the private sector free access to their landraces for breeding, but allow the private sector to charge farmers for the genetic modifications they add (see Chapter 13).

Future Prospects. The technology paths that private and public agricultural research will follow may break abruptly from recent trends. Modern methods of achieving recent productivity gains are increasingly controversial. People concerned with animal welfare denounce the confinement of animals in intensive livestock systems; people concerned about food safety challenge the use of growth hormones in dairy cows (injections of rBST to increase milk production) and of sex hormones and antibiotics in beef cattle, broilers, and hogs. Opposition to transgenic insect-resistant or herbicide-tolerant plants has been particularly strong in Europe (**Figure 2.14**) and is growing in the United States.

Under pressure from consumers, governments throughout Europe, and most recently in Australia, have adopted mandatory labeling of foods containing transgenic products. Marketers in the United States have resisted similar regulations for transgenic foods that are deemed essentially equivalent to other foods. The trend toward labeling is actually inevitable, even in the absence of legal mandates. It is part of the larger trend toward transforming homogeneous commodities into differentiated markets that satisfy changing demands as income increases and food security concerns fade in wealthy countries.

To implement product differentiation by source, innovations will be necessary to achieve "identity preservation" in the food marketing chain and to guarantee product specifications. The private sector will de-emphasize agronomic traits to serve niche markets

Figure 2.14 European opposition to GM crops. In the fall of 2000, Greenpeace put billboards in the railway stations in the Netherlands that read, "Your lettuce stays nice and fresh because we put genes from rats in them. Bon appetit!" The board carries the identifier "Genetic Research Centre, Texas, USA" and the logo of Texas A&M University. No GM lettuce is on the market anywhere, and there are no plans to create GM crops that express rat genes.

in organic foods, specialty foods, and livestock feeds; foods with particular real or alleged health benefits ("nutraceuticals" and "functional foods"), and production of drugs from plants ("pharming"), using a wide array of technological approaches. Overall assurance of food supply will be no more central to the plans of the private sector than is the health of the billions of poor people in less developed countries.

Although public investments in agricultural research have had extremely high rates of return relative to other government investments, and have helped feed the world's growing population better, public support for continued investments will probably continue to diminish. The reasons for this decline include

- Complacency because of recent sustained low food prices
- The belief that increased food supplies will just lead to increased population (Malthus revisited)
- The impression that the private sector, using modern biotechnology, can and will take over the responsibility for maintaining food supplies
- A decline in trust in public agricultural science, caused by obviously wasteful farm policies in the United States and Europe, and mismanagement of health emergencies ("mad cow" disease in England) and modern technologies

Complacency about food supply will, until the next world food crisis, continue to lead rich-country research administrators to divert agricultural research resources to other tasks such as improving rural income distribution in areas where most of the poorest people are not farmers. In less developed countries, international aid is likely to stay focused on the special problems of the poorest farmers in high-stress environments with low and erratic rainfall and/or poor soils (or steeply sloped areas with intensive rainfall that makes the land especially vulnerable to erosion), where currently science on its own has little to offer, and agricultural intensification often threatens the fragile ecology. (Paradoxically, many wealthy countries will continue to encourage farmers to withdraw marginally productive land from cultivation via conservation reserve schemes rather than encouraging production on such fragile ecologies.) There is little evidence that agricultural research is effective when diverted to these goals.

Complacency about food supply capacity can be dangerous as well as distracting. People have met the challenge of world food supply thus far, but have not permanently solved it. The role of the private sector in agriculture is increasing, but even more so than in biomedical research, private investment is (mostly) a complement to, not a substitute for, continued public and other nonprofit research. With current land and other inputs, people need the public and nonprofit research system to continue performing strongly if the world is to satisfy food requirements over the next several decades while sustaining and protecting the resource base. Foresight is needed, because the lag between investment and output is long.

C H A P T E R
S U M M A R Y

Great thinkers predicting the inevitability of world famine have been proven wrong by the dramatic increases in agricultural productivity in the 20th century: Food supply has outpaced population growth. In the 21st century, population will continue to rise, but the biggest challenge to the food supply will come from income growth, not population growth. Income growth allows people a richer, more varied diet, more dependent on animal products. Such diets consume more agricultural resources. As a result of increased demand for animal products, the demand for feed grain (e.g., maize) will rise much faster than the demand for food grains (such as rice). Without a continued improvement in yield, this increased demand will induce a demand for more land to be cultivated. Thus far, land productivity has increased quite rapidly for many years in most agricultural regions except Africa because of successful long-term public and nonprofit investments in basic and applied agricultural research, and because of private sector innovations in inputs such as fertilizers, farm equipment, and chemicals for crop protection. Labor productivity varies considerably from country to country, and in wealthy countries it depends mainly on the nonfarm wage rate.

Because agricultural technologies must be adapted to agroecological conditions, they are more location specific than are other technologies (such as many medical and information technologies). Since 1980, a revolution has taken place in protecting intellectual property rights for agricultural biotechnology, which has boosted incentives for private investment. However, the proliferation of patents is making it more difficult for public institutions and private start-ups to be active participants in biotechnology research. Moreover, the rights of indigenous peoples and low-resource farmers who have maintained many of the landraces still must be reconciled with the needs of industry and agricultural progress.

The private sector tends to be most interested in widely transferable or highly profitable technologies. Thus continued investment by the public sector in agricultural research is essential to support more basic and pretechnology innovations that are used most efficiently when made available to all without charge. Public sector support is also essential for research that aims to benefit consumers by lowering the price and/or increasing the quality of foods or fiber, or focuses on ecology-specific needs of poor farmers, and for research on the environmental and food safety consequences of agriculture, the social benefits of which cannot easily be appropriated and are therefore less attractive to private investors.

Discussion Questions

1. Discuss the sources of knowledge and information of modern farmers in industrialized countries, and compare them with those of low-resource farmers.
2. List on-farm inputs used 100 years ago, and show how purchased inputs have replaced them.
3. Why will urbanization and rising salaries, over and above the increase in the number of people, put pressure on the food production system? How does this explain that demand for maize will increase faster than demand for rice?
4. Rate of return on agricultural research has been estimated at 80% per year. Which group—farmers, agricultural input suppliers, marketers, or consumers—has benefited most from this research success? How?

5. Who is negatively affected by agricultural innovations, and how? Give some examples.

6. What is the purpose of patents? Under what conditions should it be possible to patent genes, which are not inventions but natural substances?

7. Should governments or companies compensate low-resource farmers for having grown landraces of crops for thousands of years when those landraces contain important genes? Why or why not?

8. Discuss the role of the private sector in agricultural research. Why can't we rely on the private sector alone to produce adequate research on food supply innovations?

9. How has agricultural innovation affected the number of workers per farm? Do farm wages in developed countries depend on the rate of agricultural innovation? Explain.

10. The share of agriculture in the world's gross value of output will likely decline over the next 20 years. Will more effective innovations in agriculture slow this process? Explain.

Further Reading

Alston, J. M., and P. G. Pardey. 1999. "International Approaches to Agricultural R&D: The CGIAR." Paper prepared for Office of Science and Technology Policy, Executive Office of the President of the United States, Washington, DC, February 9.

Alston, J. M., and P. G. Pardey. 2002. "Farm Productivity and Inputs." In S. Carter, S. Gartner, M. Haines, A. Olmstead, R. Sutch, and G. Wright, eds., *Historical Statistics of the United States—Millennial Edition*. Cambridge, UK: Cambridge University Press.

Alston, J. M., P. G. Pardey, and M. J. Taylor, eds. 2001. *Agricultural Science Policy: Changing Global Agendas*. Baltimore: Johns Hopkins University Press.

Alston, J. M., M. C. Marra, P. G. Pardey, and T. J. Wyatt. 2000. *A Meta Analysis of Rates of Return to Agricultural R&D: Ex Pede Herculem?* IFPRI Research Report No. 113. Washington, DC: International Food Policy Research Institute.

Byerlee, D. 1996. "Modern Varieties, Productivity, and Sustainability: Recent Experience and Emerging Challenges." *World Development* 24(4):697–718.

Byerlee, D., and P. Moya. 1993. *Impacts of International Wheat Breeding Research in the Developing World, 1966–90*. Mexico City: Centro Internacional de Mejoramiento de Maiz y Trigo (CIMMYT).

Ehrlich, Paul R. 1968. *The Population Bomb*. New York: Ballantine Books.

Engel, Ernst. 1895. *Die Lebenskosten belgischer Arbeiter-Familien fruher und jetzt*. Dresden: Heinrich.

FAO (Food and Agriculture Organization of the United Nations). 1997. "FAOSTAT Statistical Databases." Available at <http://apps.fao.org>. Accessed August 22, 2001.

———. 1999. "FAOSTAT Statistical Databases." Available at <http://apps.fao.org>. Accessed August 22, 2001.

———. 2000. "FAOSTAT Statistical Databases." Available at <http://apps.fao.org>. Accessed August 22, 2001.

Hardin, Garrett James. 1977. *The Limits of Altruism: An Ecologist's View of Survival*. Bloomington: Indiana University Press.

Heisey, P. W., M. Lantican, and H. J. Dubin. 1999. "Assessing the Benefits of International Wheat Breeding Research: An Overview of the Global Wheat Impacts Study." Part 2. In P. L. Pingali, ed., *CIMMYT 1998–99 World Wheat Facts and Trends. Global Wheat Research in a Changing World: Challenges and Achievements*, Mexico City: Centro Internacional de Mejoramiento de Maiz y Trigo.

Lucas, George R., and Thomas W. Ogletree, eds. 1976. *Lifeboat Ethics. The Moral Dilemmas of World Hunger*. New York: Harper & Row.

Malthus, Thomas Robert. 1817. *An Essay on the Principle of Population*, 5th edition. London: Murray.

Marx, Karl. 1977. *Capital: A Critique of Political Economy*. (*Das Kapital*, trans. Ben Fowkes.) New York: Vintage Books. (Originally published 1885)

Paddock, William, and Paul Paddock. 1967. *Famine 1975! America's Decision: Who Will Survive?* Boston: Little, Brown.

Pardey, P. G., and N. M. Beintema. 2001. *Slow Magic: Agricultural R&D a Century after Mendel*. IFPRI Food Policy Report. Washington, DC: International Food Policy Research Institute.

Pinstrup-Andersen, P., R. Pandya-Lorch, and M. Rosegrant. 1999. *World Food Prospects: Critical Issues for the Early Twenty-First Century*. IFPRI Food Policy Report. Washington, DC: International Food Policy Research Institute.

Smith, Adam. 1784. *Wealth of Nations*. London: Strahan and Cadell.

Wood, S., K. Sebastian, and S. J. Scherr. 2000. *Pilot Analysis of Global Ecosystems: Agroecosystems*. Washington, DC: International Food Policy Research Institute and World Resources Institute.

World Bank. 2000. *World Development Indicators*. Washington, DC: World Bank.

Wright, Brian D. 1998. "Public Germplasm Development at a Crossroads: Biotechnology and Intellectual Property." *California Agriculture* 52(6, November–December):8–13.

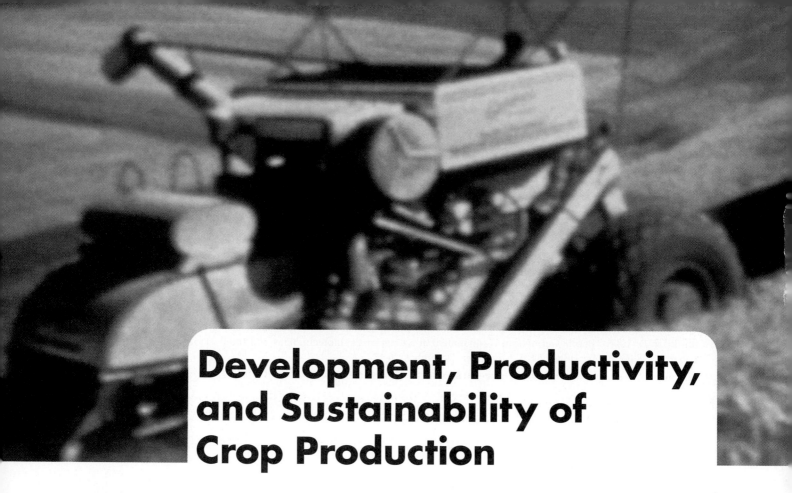

Development, Productivity, and Sustainability of Crop Production

Maarten J. Chrispeels
University of California, San Diego

David Sadava
The Claremont Colleges

In the previous chapter we discussed some of the broad outlines of agricultural progress in the 20th century: the remarkable increase in food production, the role of agricultural technologies and agricultural science, and the need for the public and private sectors to invest in such research. In this chapter, we look at some of these themes again in greater detail and consider not only the achievements, but also the problems that are looming on the horizon.

People are very much a part of Earth's ecosystems, and every ecosystem has both producers and consumers. The producers are the plants: They convert solar energy into chemical energy stored in food (see Chapter 10) that is available to the consumers. The consumers are the animals—from the smallest to the largest—and the microorganisms that break down organic matter and recycle nutrients. Humans are therefore consumers who directly or indirectly depend on plants for their food.

Before humans first practiced agriculture, ancient hunter-gatherers had evolved a complex relationship with their environment (see Chapter 13). They had an intimate knowledge of the plants and animals in their surroundings and used a wide variety of plant and animal foods. Aboriginal peoples relied largely on plants, but most had diets that included animal products. With the development of agriculture some 10,000 years ago, people narrowed their food selections. Instead of the many plant species once gathered in the wild, 24 cultivated plants now account for much of the food people eat. Three of them—the cereals: wheat, rice, and maize—make up about two thirds of the human diet **(Table 3.1)**.

Various cultures rely on different plants as their main food crop, or staple. The major food plants evolved at the same time as the societies that use them (see Chapter 13). Thus the Japanese have many different soybean-based foods and sauces, and Westerners eat wheat under many guises. Many Latin-Americans and Africans eat cassava, a tuber crop that is virtually unknown in North America and Europe.

3.1 Directly or indirectly, plants provide all of humanity's food.

A look at food sources in various regions of the world reveals another interesting feature: Some societies eat almost entirely plants, whereas others use a substantial amount of animal-derived foods. For example, in India plants supply 80% of dietary protein (cereals and legumes), but in the United States plants provide only 25% of dietary protein. The rest comes from animals and animal products (**Table 3.2**).

There are also differences among regions, social classes, and people of various religions. In developing countries, city dwellers typically eat more meat—historically regarded as a sign of affluence—than do farmers. For political reasons, governments often encourage meat consumption by providing agricultural price supports for feed grains, tax incentives for feedlot operators, guaranteed minimum prices, or government storage of surpluses.

Researchers estimate that 85% of the calories and 80% of the protein in the human diet now come directly from plants. However, this situation

Table 3.1	Humanity's most important food crops	
Category	Crop Name	Amount Produced, 2000 (millions of metric tons)
Cereals	Maize	596
	Rice	593
	Wheat	582
	Barley	136
	Sorghum	60
	Millet	27
Rootcrop	Potatoes	302
	Cassava	170
	Sweet potatoes	138
Legumes	Beans, dry	20
	All others combined	40
Vegetables	Tomatoes	100
	Cabbage	50
	Onions	50
Oil crops	Soybeans	162
	Oil palm fruit	98
	Coconuts	48
	Rapeseed (canola)	40
Starchy fruits	Bananas	58
	Plantains	30
Fruits	Oranges	66
	Apples	60
	Grapes	60

Notes: In addition to these, some 3,000 million tons of animal feed crops are harvested yearly.

Source: Data from Food and Agriculture Organization (FAO) of the United Nations.

Table 3.2	Comparison of diets in India and United States			
	Source of Calories		**Source of Protein**	
Food	**India**	**United States**	**India**	**United States**
Cereals, starchy foods	61%	23%	60%	21%
Sugars	6	12	—	—
Beans, lentils	11	4	18	4
Fruits, vegetables	2	6	1	4
Fats, oils	4	19	—	—
Milk, milk products	7	14	12	24
Meat, poultry, eggs, fish	9	22	9	47

Source: 1998 data from the Food and Agriculture Organization and U.S. Department of Agriculture.

is changing as people become more affluent. Rising affluence therefore puts additional pressures on the food system.

We have already noted, in Chapter 2, that urbanization and economic development, and the rise in income and expectations associated with them, make increasing demands on the food production system. People want to eat a more varied diet, and they want and can afford to eat more animal products. Global meat production has increased dramatically in the last 50 years, quadrupling since 1950 (**Figure 3.1**). Until about 1950 in the industrialized countries, and even today in many parts of the developing world, farmers who practiced a sustainable mode of farming integrated livestock rearing with food crop production. They rotated food crops (wheat, potatoes, and sugar beets) with feed crops (hay, clover, and alfalfa). People ate the former; farm animals ate the latter. Farmers spread animal waste (manure) over the soil as fertilizer, and leguminous feed crops (clover and alfalfa) added nitrogen to the soil (see Chapter 12).

Unfortunately, this integrated system is hard to sustain when demand for meat is high. Specialized animal production facilities, often located close to the population centers, feed animals soybeans and grains such as maize and sorghum. This system requires large areas of land just to raise food for animals. To understand why eating animal products puts additional demands on the food production system, also consider how matter and energy transfer from one organism to another in the ecosystem. In the food chain, plants are called *primary producers*. When an animal eats the plants, only about 10% of the plant matter and energy ends up as part of the growing animal's body. The other 90% is "lost": The animal uses a lot as energy to fuel its digestion and to refashion plant material into animal material. So if humans eat plants directly, the transfer efficiency is about 10%. Now consider what happens if an extra step occurs between plant and human: Let's say a cow eats grain, and then a human eats the cow. The efficiency from plant to human via cow is 10% ÷ 10%, or 1%. So people must grow 10 times the amount of grain to supply a meat-eating human with the same energy and matter as would be needed if the person ate the grain directly.

Actually, in some cases modern methods have improved the efficiency of transfer from grain to animal. Efficiency is higher in feedlots and chicken farms, where animals are

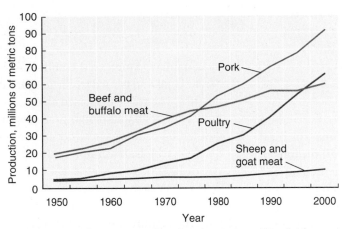

Figure 3.1 Increase in world meat production, 1950–2000. The fastest-growing meat source in the United States is poultry, whereas in eastern Europe it is pork. *Source:* Data from Food and Agriculture Organization.

confined and fed optimal diets. Still, it takes about 7 kg of grain to produce 1 kg of pork (15% efficiency), 5 kg of grain to produce 1 kg of beef (20%), and 2 to 3 kg of grain to produce 1 kg of egg, cheese, or poultry (33–50%).

Some people have argued that eating animal products wastes precious agricultural resources, but there is at present not enough economic demand for food grains (rice and wheat) to expand their production. Despite ecological inefficiencies, meat production is rising rapidly. The existence of several successful farming systems based on animal husbandry (for example, nomadic sheep, cattle, and goat herding as well as cattle and sheep ranching) confirms that under certain conditions the inefficient conversion of energy and protein is also a positive aspect of the food production system. For example, some 30 to 40 million pastoralists living in the world's driest lands produce animal products (milk, meat, and wool) in areas that are essentially unsuitable for crop agriculture. Nevertheless, meat consumption can sometimes have unintended negative consequences as well (see **Box 3.1**).

3.2 Land use patterns in agriculture show increased intensity of resource use on arable land.

With suitable temperature and rainfall, people can successfully use land to grow crops and produce foods. But where is the arable land (cropland) located, and how much exists? Certainly not all of Earth's land is used for crops. Of the total land area of Earth, people use about 11% for crops (arable land) and use another 24% for pastures; 31% is forested. The arable land is heavily concentrated in Canada, the United States, Europe (including Russia and the Ukraine), India, China, and Southeast Asia (**Figure 3.2a**). The rest of the land surface—about one third—is too cold, too dry, or too steep for plant growth. The unequal distribution of agricultural resources over the globe is shown clearly if we express the amount of cropland per inhabitant and the available freshwater per inhabitant for each continent (**Figure 3.2b**). Asia, which already has the most people, has the least agricultural land and freshwater per person.

Box 3.1

Meat, Manure, and Subsidies

Some choices are difficult. In the Netherlands, a small country in northwestern Europe, the meat and dairy industries produce 85 million tons of manure each year. Feeding animals is inefficient, and about 80% of the protein (and larger percentages of other nutrients) fed to animals shows up in manure in various forms (protein, urea). People could dispose of about 50 million tons of this manure in an environmentally acceptable way, but the other 35 million tons clearly are surplus. Export is not feasible because transport is expensive and the amount of manure is so great. A 25-mile-long manure train would have to leave the country every single day. So the nutrients (100,000 tons of phosphate each year, for example) continue to pollute the ecosystems.

Where does all the animal feed come from? In the past, cows ate grass and chickens ate grain. Until recently, the livestock, poultry, and dairy industries depended on locally produced animal feeds. The most recent rise in consumption of meat and animal products in Europe is driven by feed imports from abroad. Although Europe produces plenty of wheat and other grains (the countries of the European Union annually produce 165 million tons) and actually has a grain surplus of 35 million tons, the animal industries prefer to buy cheaper substitutes. The substitutes are cheaper, in part, because various governments heavily subsidize grain production (to the chagrin of other countries—such as Australia—that also have enormous grain surpluses). The grain substitutes that Europe buys—powdered cassava roots, maize gluten, soybean press cakes, and citrus pulp—are produced in the United States (40%) and in various developing countries (60%) and enter western Europe primarily through the giant harbor of Rotterdam. Much of the animal feed remains in the Netherlands and is used by the local meat and poultry industries. The rest is dispersed all over Europe by rail and boat. The livelihoods of many farmers and workers in the developing countries now depend on this export of animal feed, which arrives in Rotterdam in giant ships at the rate of 100,000 tons each day.

How can such a problem be solved? Consumers want cheap cheese and steak. Environmentalists want to reduce the manure output. European farmers want subsidies. The traders and those who promote "development" in the developing nations want to continue the trade, which earns hard currency for developing countries. Neither the United States nor the farmers of the developing nations want import taxes that would make European grain more competitive with their products.

Several principles should underlie the search for a solution. First, perhaps too much land is being cultivated in the developed countries. If subsidies are really necessary to give such farmers a decent standard of living, they should be maintained, but pinned to a lower total output. Land could then be taken out of production and be used for other purposes. Second, the price of steak or eggs should include the cost of disposing of the manure in an environmentally acceptable way. Third, policies in the developed nations should be examined for their effects on developing countries. This will involve value judgments. Who benefits if European pork is produced with cheap cassava from Thailand or with soybean press cake from the United States? Do policies in developing countries that promote the production and export of these new cash crops imperil the ability of these countries to produce food for their own people? Would it be better to take land out of production in the United States, rather than supply the Europeans with grain substitutes so they can have cheaper cheese, eggs, and meats?

The location of arable land is itself a function of climate, soil type, and type of vegetation that grew in the area before people cleared the land. Climate, soil type, and vegetation all interacted during the soil formation process and the most productive soils formed in areas of permanent grassland (such as the prairies of the midwestern United States) or hardwood forests (most of Europe and India). Grasslands produce rich soils with abundant organic matter that can remain productive for many decades after farmers convert them to cropland. These soils are quite suitable for annual row crops (maize, wheat, beans, and so on).

The process of converting natural ecosystems into pastures or cropland is now being repeated most dramatically in the tropical rainforests of South America. However, converting a tropical rainforest into an agricultural field can be very difficult. Most rainforests grow on relatively infertile and highly weathered soil (see Chapter 11). When people clear such land and burn the vegetation, they expose bare earth to the sun, soil temperatures rise, and the soil organic matter quickly decays, because the vegetation does not replenish it. Because of the high rainfall, the soil erodes easily.

Asia, Latin America, and Africa contain many examples of massive soil degradation caused by clearing away tropical forests and imposing agricultural practices on soils that are totally unsuitable for row cropping. People need new approaches for these soils (see Chapter 5).

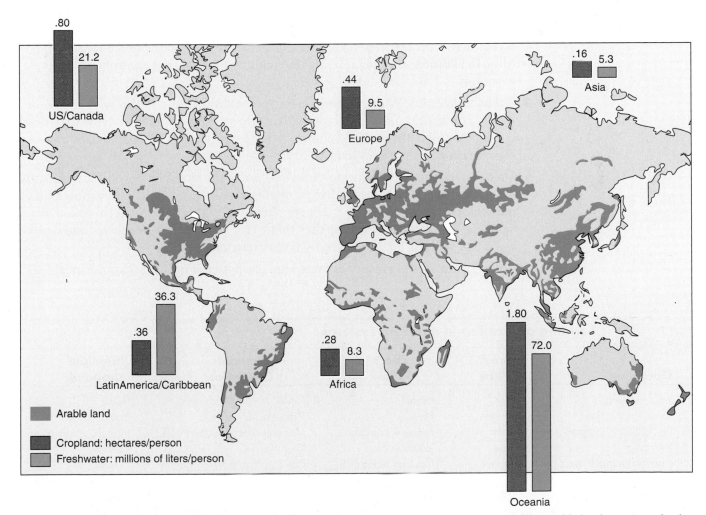

Figure 3.2 The world distribution of arable land and fresh water. Notice that most of the arable land is in Canada, the United States, Brazil, Argentina, Europe, Russia, western Asia, India, and China. Cropland and available freshwater are expressed on a per capita basis for each continent. Oceania includes sparsely populated Australia, New Zealand, and a number of small Pacific island states (total population about the same as one large Asian city).

Plowing more land may not achieve a greater total of cropland than the 1.5 billion hectares presently used. The FAO estimates that 5 to 7 million hectares are being lost each year to land degradation. One problem is that the areas with the greatest population pressures, such as the east coast of China and most of India, have the least land reserves that could be cleared for crops. Other areas—in the former Soviet Union, for example—have good arable land but not enough rainfall. Improved agricultural technologies could greatly expand food production in these regions (see Section 3.3).

The development of agriculture in different regions of the world has intensified land use. Each stage of intensity has its own characteristics and productivity (**Table 3.3**):

1. *Forest fallow.* Cutting and burning parts of a forest release nutrients contained in the plants; this fertilizes the land, allowing it to support crops. But this fertility lasts only until crops deplete the soil of its nutrients, and the farmers then move on to another part of the forest, repeating the cycle. In the meantime, they leave the original land fallow for up to 20 years, to restore its fertility.

2. *Short fallow.* As human population density grows, grassland replaces the forest and farmers shorten the fallow period to one or two years. This is not long enough to fully restore the soil, to let it support crops, so added fertilizer—often animal manure—is required. Farmers often plant special legumes to accelerate restoration of soil fertility. In addition, they achieve higher yields by manual pest control (for example, weeding).

3. *Annual cultivation.* With a still-increasing population, now coupled with market demand, farmers drop the fallow period and grow a crop every year. Now they need constant fertilization and pest control to keep up the yield. Crop rotations increase the sustainability of the system.

4. *Multiple cropping.* This is the most intensive use of the land. Not only do farmers grow several crops on the same land sequentially in one year, but also they grow two or more crops at the same time. Farmers often follow this latter pattern, called *intercropping,* for the same reason as they rotate crops: One species removes one type of nutrients, and the intercrop removes other nutrients.

Chapter 5 discusses the reality of these land use patterns in sub-Saharan Africa.

| Table 3.3 | Farming operations in different systems |

Operation	Forest Fallow	Short Fallow	Annual Cultivation	Multiple Cropping
1. Land clearing	Fire	None	None	None
2. Land preparation	None	Plow	Plow, tractor	Plow, tractor
3. Fertilization, compost	Ash	Manure, compost	Manure, compost, chemicals	Manure, chemicals
4. Weeding	Low	Intensive	Intensive	Intensive
5. Animals or machines, transport	None	Plowing, transport	Plowing, transport, irrigation	Plowing, irrigation
6. Percentage of world cropland	2%	28%	45%	25%
7. Grain yield (kg/Ha)	250	800	2,000	5,000

Sources: Adapted from H. Binswanger and P. Pingali (1988), Technological priorities for farming in sub-Saharan Africa, *World Bank Research Observer* 3:83. Data on yields and areas from P. Buringh (1989), Availability of agricultural land, in D. Pimentel, ed., *Food and Natural Resources* (New York: Academic Press).

3.3 | Modern agriculture depends on purchased inputs.

Historically, the transition from stages 1 to 3 in the preceding schema has been gradual, as populations grew slowly and the necessary technologies evolved over time. We discussed in Chapter 2 how purchased inputs gradually replaced farm-produced inputs such as hay or manure. Farmers now purchase the following inputs:

Farm Machinery. The process of mechanization started in the mid-19th century. New farming techniques depended on scientific advances and the industrial production of inputs. Manual labor and animal power were replaced by steam power and later by the internal combustion engine. Farmers used tractors to pull plows or the reaper-binder, which they later replaced with the combine harvester. Milking machines, cotton gins, cotton pickers, sugar beet and potato harvesters, and tomato-harvesting machines all have greatly reduced the need for manual labor. In developing countries small tractors have revolutionized agricultural production and greatly reduced the input of labor.

New and Improved Varieties. Initially, farmers selected suitable varieties of crops for planting the next year. These selected strains were the mainstay of agriculture until the science of genetics emerged in the 20th century. Then it became possible to deliberately interbreed different strains of a crop to consolidate desirable characteristics in a single strain. The first step is always to evaluate the field performance of known varieties. Scientists have systematically applied plant-breeding principles to improving rice and wheat, and have widely introduced new varieties in Latin America and Asia, displacing local varieties. This process, often called the Green Revolution, has significantly raised crop yields in an era when the amount of land cultivated has remained more or less constant. The introduction of hybrid rice in China and Southeast Asia is raising rice yields even further (**Figure 3.3**). About roughly 40% of all increases in crop productivity during the past 50 years stemmed from breeding new varieties. The other 60% improvement has come from managing the crop environment by inputs such as energy, fertilizer, and pesticides.

Figure 3.3 The availability of hybrid rice will increase rice yields beyond those experienced with the Green Revolution varieties. The bag on the right identifies the brand Proagro in different languages of India. *Source:* Courtesy of Eugene Hettel, IRRI.

Inorganic Fertilizers. Originally European farmers relied on manure and crop rotation to maintain fertility. Later they found that ground bones and rock phosphate enhanced crop production, as well as nitrates (long imported from Chile, in the form of fossil guano). The invention of a chemical process that combines nitrogen gas with hydrogen gas to form ammonia, allowed the widespread production of nitrogen fertilizers. This led to a tremendous increase in nitrogen fertilizer use worldwide. Fertilizer applications have been slowly decreasing in developed countries, but in developing countries they are still rising.

Herbicides and Pesticides. Herbicides replaced manual labor for weeding (**Figure 3.4**), and farmers used insecticides and fungicides to minimize crop losses. These changes in food production had a great impact on commercial grain farming (for example, in the United States, Canada, and Australia), on mixed crop and livestock farming (in Europe), and on intensive wetland rice farming in Asia and Africa. More complex approaches to weed control and pest control are now replacing heavy use of chemicals in agriculture.

Irrigation Technology. Irrigation has become a very important agricultural input in the past 40 years and accounts for much success in raising food production. Indeed, although only 18% of the world's arable land is irrigated, this land produces 40% of our food. Irrigated land is highly productive. However, this trend cannot be maintained: There simply is not enough water in many areas. Experts now see that water is the resource that will be most limiting for food production in the 21st century.

Information Technology. New innovations in agriculture that use information technology are often called *precision agriculture*. To obtain maximal yields, farmers of very large farms use remote sensing of their land and their crops by satellite or airplanes to adjust irrigation water, fertilizer application, and genetic varieties. Large tractors equipped with computers can now control row spacing and crop planting densities.

Inputs do not exist in isolation, but interact. Scientists often breed improved varieties to fit other production-enhancing inputs. For example, after scientists introduced hybrid

Figure 3.4 Woman farmers weed a communal field of groundnuts in Kashipa, Zambia. Hand weeding is a major agricultural activity in developing countries and is often done by women and children. Without weeding, the weeds will out compete the crop seedlings. Care must be taken not to disturb the root system of the crop seedlings. *Source:* Photo by P. Lowrey, Food and Agriculture Organization.

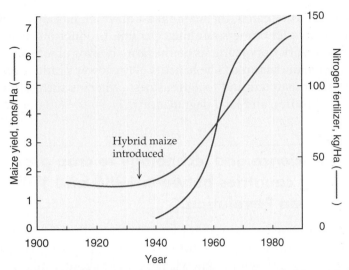

Figure 3.5 Interaction of inputs in crop production: Maize yields and the use of nitrogen fertilizers in the United States. The arrow points to the introduction of hybrid maize. By the time the planting of hybrid maize was complete (1950–1955), yields had increased significantly. The big yield increase parallels the use of nitrogen fertilizers and is caused by breeding strains that respond to nitrogen fertilizers. *Source:* Data from U.S. Department of Agriculture.

maize in the United States in the 1930s and nitrogen fertilizers became widely available, agronomists bred maize to respond to nitrogen fertilizers (**Figure 3.5**). After the invention of the automatic tomato harvester, tomatoes were bred to withstand the rougher handling to which such machines subject these fruit.

These technological advances have led to a style of agriculture that depends on purchased inputs. A survey of U.S. production methods for maize from 1910 to the present clearly shows this dependence (**Table 3.4**). The technological advances resulted from government-

Table 3.4	The technological basis of modern agriculture (inputs per hectare of corn in the United States)		
Input	**Hand Produced**	**1910**	**1980**
Labor (hr)	1,200	120	12
Machinery (kg)	1	15	55
Animal use (hr)	0	120	0
Fuel (L)	0	0	125
Manure (kg)	0	4,000	1,000
NPK fertilizer (kg)	0	0	316
Lime (kg)	0	10	426
Seeds (kg)	11	11	21
Insecticides (kg)	0	0	2
Herbicides (kg)	0	0	2
Irrigation (%)	0	0	17
Drying (kg)	0	0	3,200
Electricity (10^3 kcal)	0	0	100
Transport (kg)	0	25	326
Yield (kg)	1,880	1,880	6,500

Source: Data summarized by D. Pimentel, from USDA and estimates, in D. Pimentel, ed. (1989), *Food and Natural Resources* (New York: Academic Press).

sponsored and industrial research, the main thrust of which has always been to develop technological inputs that let farmers minimize per unit production costs. Many of these advances diminished work opportunities on the farm in favor of jobs in towns and cities where the inputs are manufactured. The benefits of the lower production costs (cheaper food) accrued to the consumers and to agribusiness, whereas society as a whole had to bear the penalties (pollution and land degradation).

3.4 Applying science and technology to crop production in developing countries between 1955 and 1985 resulted in the "Green Revolution."

Worldwide food production has been rising by 2.3% annually as a result of the development of high-input agriculture, which in turn results from applying agricultural research and technology. This use of scientific knowledge started about 150 years ago in the developed countries and has resulted in yields taking off spectacularly, as can be charted quite accurately for individual crops (see Figure 2.2 in Chapter 2).

Japan is a classic case of a country turning to high-yield agriculture. By 1900, this island nation was growing crops, mostly rice, on all its arable lands. But the population was still growing. The government was reluctant to become dependent on foreign food imports lest a hostile nation cut off food from Japan during an international crisis. Short of lowering the people's nutritional standard, the only thing to do was to increase the yield of the rice crop. The government mobilized the country's political, social, and scientific resources, and as a result the yield per hectare tripled between 1900 and 1965. One characteristic of the new crop varieties that raised Japanese rice yields and U.S. wheat yields was that they were short-stemmed (semidwarf) and responsive to nitrogen fertilizer. But as with most varieties, scientists had selected and then bred them specifically to grow in the soils and climate of developed regions. That is, a variety that thrived in the U.S. Northwest had the genetic makeup to do well there, but not necessarily in the drier, hotter regions of the less developed world. There, crop production still relied on local varieties (landraces).

The genetically improved varieties of wheat and rice that drove the Green Revolution resulted from a targeted crop improvement program at two CGIAR institutes: the International Rice Research Institute (IRRI) in the Philippines and the International Center for the Improvement of Maize and Wheat (CIMMYT) near Mexico City. Farmers adopted the new varieties over a period of about 10 years, repeating a phenomenon that had occurred in the United States some 30 years earlier, when farmers adopted hybrid maize.

Figure 3.6 Growth of wheat yields in Mexico after adoption of Green Revolution varieties. *Source:* Food and Agriculture Organization.

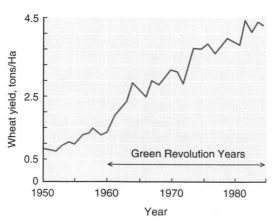

In the case of high-yielding varieties of wheat and rice, the farmers had to adopt not only the new seeds, but an entire technology package, including fertilizers, insecticides, weed killers, equipment for irrigation, and tractors to till the land. Indeed, the new varieties let farmers grow two or even three crops per year, potentially increasing the demand for labor. Thus labor-saving technologies (tractors and herbicides) had to accompany the new strains. Furthermore, farmers needed nitrogen fertilizer so the new varieties would yield up to their potential (without fertilizer, they produced the same yields or less, than the landraces they replaced). Adopting these new varieties of wheat doubled and tripled production in Mexico (**Figure 3.6**), India, and other countries in Asia. Rice production increased similarly.

But the yield increases on farms were never as great as those at research stations. Scientists at IRRI showed that the new varieties could outproduce the old ones by a factor of 3 or 4, yet most farmers in Asia realized increases of only 1.5. This discrepancy in production is not surprising and also occurs in developed countries, where farms seldom match yields obtained in experiment stations. Surveys of farmers who were using the new technology showed that the constraints on higher productivity were poor water control, inadequate nitrogen fertilizers, and diseases. Put another way, the major constraint was inadequate capital to purchase the input package needed to obtain the high productivity of which the seeds are capable. The fundamental fact of the Green Revolution is that the farmers must manipulate the environment to get the most out of the plants, and too often they have not been able to afford it.

Nevertheless, the Green Revolution allowed many developing countries to raise food production and even become food exporters. International agricultural experts saw breeding the high-yielding varieties and applying the industrial inputs as the quickest way to increase production. As a result, agriculture in developing countries came to resemble agriculture in developed countries: more heavily dependent on purchased outside inputs, more capital intensive, and less dependent on labor (**Figure 3.7**). Earlier, we noted that families needed the money generated from labor to buy food. Merely raising production is not enough to eliminate hunger. (But raising production is intended to create profits, not to eliminate hunger, as noted earlier.) Because the need for farm laborers decreases as agriculture becomes more technological, the displaced laborers must find other employment.

Figure 3.7 Agriculture in developing countries has come to resemble agriculture in developed countries. To get high productivity, farmers must use purchased inputs that include tractors, fertilizers, and pesticides. In this photo, a Filipino farmer is shown preparing the soil of his rice field. *Source:* Courtesy of Eugene Hettel, IRRI.

When farmers first apply a new technology—for example, fertilizer or pest control—yield immediately increases as the amount of input rises. A little bit is good, and more seems better! During this phase, the money spent on the technology results in a substantial return. Once the plant's inherent capacity to respond to the input is reached, however, the yield increase slows drastically. Adding fertilizer and other inputs follows the law of diminishing returns: less output for a given amount of input.

In the United States, farmers now apply 90–100 kg of fertilizer per ha of cropland, and it is not easy to raise the yield further by adding more. This figure is 65 kg/Ha for India, so there is still hope for improvement, but fertilizer use is only 4 kg/Ha in Ethiopia and 12 kg/Ha for Costa Rica. Gains in crop productivity should be possible in these countries if researchers can find the proper cropping systems and crop varieties that can take advantage of additional fertilizer. When all technologies are maximal, agriculture may reach a yield plateau. Some crop physiologists argue that people will always be able to find new crop varieties to escape the yield plateau and take advantage of new and existing technologies. Biotechnology has the potential to allow continual crop production increases. For example, scientists at IRRI and elsewhere are now producing hybrid rice varieties to raise rice yields even further (see Figure 3.3).

3.5 The CGIAR institutes are catalysts for agricultural research in developing countries.

The success of the Green Revolution, and the perceived need for more research on tropical and subtropical food crops, led to the establishment of other research institutes in different countries of the developing World (see also Chapter 2 for a listing). These institutes have a triple mandate: (1) to improve the crops assigned to them, (2) to study farming systems for these crops, and (3) to create gene banks in which the landraces of the crops under study can be preserved (in the form of seeds stored under conditions that ensure their longevity). Initially the institutes focused on the approach that was successful with high-yielding wheat and rice varieties in the 1960s and 1970s. That is, they concentrated on monoculture and on improving crop production by a package of improved varieties and associated technologies to modify the environment. More recently, this focus has begun to change, showing a greater concern for integrating agriculture more with the natural environment.

For example, scientists at ICRISAT in India have identified strains of groundnut (a drought-resistant legume) that produce more than 1,000 kg grain per hectare, even when stressed by drought, compared to the 500–800 kg per hectare the normal varieties produce. Similarly, they have found strains of sorghum that produce two to three times more grain than the present commercial types. One of their goals is to select drought-resistant crop plants to use not only in Asia, but also in the Sahel region of Africa.

Scientists at CIAT in Colombia have identified genes in wild varieties of beans that let them resist attack by bruchid beetles. These beetles do enormous damage to the seeds after harvest because they multiply in the dry, stored beans. By crossing a wild bean variety with a cultivated variety, CIAT scientists were able to introduce the bruchid-resistance genes into the cultivated variety.

When the CGIAR system of research institutes was first established, people saw biological and technological research problems as separable from political and social issues. The research institutes generated the necessary innovations and offered these as packages to various national and regional extension services or development authorities. Fine-tuning these packages to local conditions, and solving social and political problems that resulted from implementing new technologies, were the responsibilities of the implementing agencies or the governments of the countries in question. In this system, people saw agricultural development as an activity that came from the top (the CGIAR institutes) and trickled down to the bottom (the farmers).

This approach has worked reasonably well in some countries or areas, and poorly in others. Agricultural development has not often been a high priority for many governments of developing nations, except for developing cash crops that earn foreign exchange. In many African countries, national agricultural research efforts have been weak, in part because governments mistakenly assumed the CGIAR institutes would solve all their agricultural problems, and also because these high-profile institutes attracted all the international funds and the best scientists.

The Green Revolution has done more than raise overall food production. It has also caused the rapid disappearance of the landraces (see Chapters 13 and 14) and of indigenous farming systems that may have had much to offer to agricultural researchers. The top-down approach of development implies that agricultural researchers can learn little of value from subsistence farmers. For example, although the institutes have often stressed monoculture, many farmers have been growing more than one crop on the land (either by crop rotation or by intercropping) for centuries. These practices led to a more sustainable agricultural system than did monoculture.

Scientists at the CGIAR institutes know they must do more than create new strains to hand over to extension agents. They need to become involved in understanding the tropical agroecosystems at the farm level and to evaluate with the farmers the local varieties (genotypes) (**Figure 3.8**) and breed varieties for different environments. For example, IRRI is located at a place where rainfall is reliable and adequate—in the Philippines. But much rice in Asia is grown on land where rainfall is intermittent and unreliable. So how can a technology developed at IRRI be useful to a farmer who has very different needs? This type of problem led to the formulation of a second mandate for these institutes: to study local farming and make sure that proposed development uses the strengths of those systems (see Chapter 5). Biotechnology is an important tool, but only one of many, to raise crop productivity in the less developed countries.

Figure 3.8 Scientists from the Centro Internacional de Agricultura Tropical in Colombia, evaluating local varieties. These scientists and farmers are walking through an experimental bean field in which different varieties are grown. *Source:* Courtesy of CIAT.

3.6 Biotechnology will contribute to the continued rise of crop yields in the 21st century.

Between 1960 and 1985 the yield of the major cereals (wheat and rice) rose dramatically in many developing countries, resulting in a 25% rise in the per capita food availability. However, in more recent years (1985–2000) this trend has leveled off. As noted in Chapter 1, the rate of population growth is slowing, so it is hard to predict whether world crop production is on track to meet the needs of the future human population. But two relatively new factors have arisen to challenge crop producers. First, increased demand for feed grains has followed increased demand for meat. Second, people realize that high-input agriculture can degrade the environment, so that plants must be even better adapted to the natural system. Although people still need ever better genetically improved varieties in combination with new technologies, they also need to more thoroughly understand the sustainability of crop production. Researchers need to devise different strategies for the highly productive regions that are already intensely cropped and where yields have risen dramatically and for the more marginal soils where, as in sub-Saharan Africa, productivity has stagnated and per capita food availability has declined (see Chapter 5).

Classical plant breeding, more fully discussed in Chapter 14, has been a very important factor in the past successes. A new set of tools has become available in the past 20 years, that—combined with plant breeding—will allow people to produce the genetically improved varieties of the future. This set of tools, which comes under the general title of *biotechnology*, encompasses a variety of laboratory methods that include

- Cell, tissue, and embryo culture (**Figure 3.9**)
- Clonal propagation of disease-free plants
- Identification of chromosome regions (quantitative trait loci) that carry important multigenic traits
- Gene identification and isolation
- Genetic engineering for agronomic traits such as pest and disease resistance or better adaptation to environmental stresses (such as salinity, or water deficit)
- Genetic engineering for greater nutritive value (such as vitamin A, minerals, and better amino acid balance)
- Genetic engineering to reduce postharvest losses (**Figure 3.10**)
- Genetically engineered male sterility to facilitate hybrid seed production

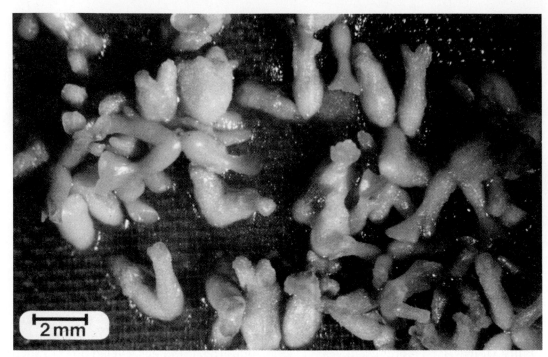

Figure 3.9 Cultured somatic embryos of alfalfa (*Medicago sativa*). Somatic embryos are embryos derived from ordinary vegetative cells of the plant body rather than from fertilized egg cells. The embryos shown are at the stage of development 18 days after scientists transferred vegetative cells to an embryo-inducing medium. The cells were originally derived from the stemlike portion of the leaf. Alfalfa is a cross-pollinating plant, and cross-pollination between plants from certified elite lines of alfalfa results in loss of yield and other important agronomic traits when the seeds of such crosses are planted. The quality of an elite line can be maintained by mass production of somatic embryos, which are genetically uniform because they derive from a uniform population of vegetative cells. *Source:* Courtesy of Derek Bewley, University of Guelph, Ontario, Canada.

Figure 3.10 Genetic engineering for reduced postharvest losses in tomatoes. The photograph shows the effect of down-regulating ethylene synthesis using gene technology. The GM tomatoes on the left produce about 5% of the normal level of ethylene. The tomatoes were harvested and kept for three weeks at room temperature. The control tomatoes on the right are spoiled, the GM tomatoes on the left are ripe and marketable.

Embryo culture is used to rescue the embryos produced when plants of different species are crossed. For example, a cross between the wild rice *Oryza nivara* and domesticated rice *Oryza sativa* is not normally viable. Researchers at IRRI used embryo rescue to introduce a trait for resistance to the grassy stunt virus present in this wild rice into domesticated rice. The horticulture industry routinely uses meristem culture and plant regeneration from small parts of the meristem to propagate virus-free planting materials.

The discoveries that the genetic material (DNA) of all organisms basically has the same structure and that genes from one organism can function normally after transfer to another organism have opened up the field of genetic engineering. In plants, gene transfer is made possible by a natural process discovered in a soil-dwelling plant pathogen that transfers a few of its genes to plant cells. Molecular plant biologists have manipulated this process so that the pathogen will transfer one or more genes of the scientist's choosing (see Chapter 6). This technique opens up unlimited possibilities for modifying crop traits. Plant breeders were previously limited to using the genes of the crop's closest relatives, but with genetic engineering they can now use any gene from any organism. It is important to emphasize repeatedly that the genetic engineering and biotechnology can make important

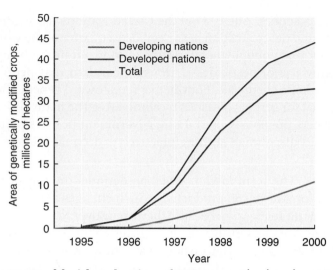

Figure 3.11 **Worldwide adoption of GM crops.** The data do not include China.

contributions but are not the silver bullet that will solve all food production problems. It is but one of the many technologies that people need.

Genetically engineered plants that resist certain insect pests or can tolerate herbicides so that farmers can destroy weeds more efficiently, are being grown in more than half a dozen countries already (Argentina, Australia, Canada, China, India, Mexico, and the United States). The rapid adoption of these new varieties (**Figure 3.11**) has pleased the biotechnology companies that produce them and the farmers who adopt them because it lowers their production costs, but has aroused some anxiety in the general public. Some countries in western Europe and a few localities in the United States have banned genetically engineered (or manipulated, GM) crops. The four major GM crops being grown are rapeseed (canola), cotton, maize, and soybean. Molecular techniques will also greatly facilitate chromosome mapping of important agronomic traits. This will make it easier to follow these traits in the progeny of crosses when the traits have no easily discernible phenotype (characteristic) in the field. It will also greatly speed up classical breeding, which is a rather slow and cumbersome process. Gene replacement (gene therapy) is not yet a reality, but is on the drawing board. It eventually will be possible to take a small chromosomal region that encompasses a whole set of genes, and simply replace it with another region from a plant that is more distantly related.

3.7 The effects of intensive agriculture on the ecosystems are causing concern about its sustainability.

Growing populations and increasing affluence demand ever-increased output from agricultural systems. This is true whether we are talking about the high-input agriculture of the United States and Western Europe, about nomadic herds in Africa, or about intensive rice cultivation in Asia. Concerns are now being raised about the sustainability of this ever intensifying productivity.

The World Commission on Environment and Development defined sustainable development as "development which meets the needs of the present without compromising the ability of future generations to meet their own needs." To understand sustainability, we must first look at the rate at which people are losing productive soils as a result of present agricultural practices. The FAO of the United Nations has estimated that salinization, soil erosion, and desertification have degraded a quarter of the world's arable land. Second, there may not be enough energy resources to maintain high-input agriculture. Third, the increasing use of pesticides is arousing concern that people are polluting ecosystems with synthetic chemicals. Fourth, the trend toward genetically uniform crops increases the potential for serious disasters by eliminating the many different strains of a given crop that farmers previously used. Fifth, government policies perpetuate conventional agriculture and discourage farming practices that could make agriculture more sustainable.

Land Degradation. The term *land degradation* denotes the physical and chemical changes that reduce long-term soil productivity. Such changes are sometimes very obvious but are often difficult to measure because they occur over long periods of time. Other improvements, such as more fertilizer or new genetic strains, often compensated for them, so the effects of land degradation do not always show up as decreased yield—at least, not in the short run.

Physical degradation takes the form of erosion, the carrying away of soil particles by wind and water. In an annual cropping system, the soil is alternately covered with plants, and then almost totally bare. When the soil is bare, it is exposed to higher wind velocities and to the force of raindrops, which destroy the structure of the uppermost layer of topsoil. The net result is muddy runoff and/or dust storms.

Chemical degradation can take several forms, including (1) acidification from acid rain, (2) alkalinization and salinization (buildup of salts) as a result of irrigation or the intrusion of saline groundwater into topsoil (**Figure 3.12**), (3) exhaustion of mineral nutrients when farmers don't use enough fertilizer to replace minerals removed with the crop, and (4) leaching of excess mineral fertilizers into streams, lakes, and groundwater.

Pesticides and Herbicides. Agriculture uses pesticides and herbicides extensively in the developed countries, and their use in developing countries has also increased rapidly. Many problems accompany their use, including the emergence of pesticide-resistant pests, adverse health effects on farm workers, pesticide residues on crops, and pollution of lakes, streams, and groundwater with pesticides and herbicides. Furthermore, the realization that pesticides have not diminished the proportion of the crop lost to pests is prompting many people to try alternate pest control methods. There is renewed emphasis on pest-resistant varieties and biological pest control (discussed in Chapters 16 and 17).

Loss of Crop Biodiversity Because of Genetically Uniform Crops. A main feature of the high-input system of agriculture is genetically uniform crops. These reliably produce high yields if used with appropriate inputs. But their use has several drawbacks. First, when farmers start planting seeds furnished by seed companies or state agencies, they stop using and usually discard their own varieties, which were well adapted to the microenvironment on their farm. Local varieties contain valuable genes that are thus lost to plant breeders. Second, genetic uniformity means crops are more susceptible to sudden outbreaks of disease. When a pathogen mutates and overcomes the plant's resistance, the entire crop over a wide geographic area can be ravaged. An excellent example is the potato blight that ravaged the Irish potato crop in the 1840s (see Chapter 13).

Government Policies. Sometimes government policies actively promote high-input agriculture and discourage sustainable practices and technologies. Such policies are difficult to change because large and powerful lobbies, including farm organizations and industry advocates, have vested interests in such policies.

Figure 3.12 Extensive salinity damage in abandoned cropland in California's Coachella valley. The land shown here was previously desert, converted to cropland through irrigation. Salts already in the soil dissolved and rose to the top via high evaporation/transpiration. The farmer has abandoned the large block in the foreground. Economics permitting, such land can be reclaimed by leaching off the salts with excess irrigation water.

To protect their resource base—the soil—and to decrease costs, farmers have begun to adopt alternative practices. Taken together, these practices constitute sustainable agriculture, which differs from conventional agriculture not so much by the practices it *rejects* (for example, heavy use of inorganic fertilizers), but by the practices it *incorporates* into the farming system. The more widely used term *organic farming* (or *biological farming*, in Europe) describes two major aspects of alternative agriculture: (1) substituting manures and other organic matter for inorganic fertilizers and (2) using biological pest control instead of chemical pest control. The objective of sustainable agriculture is to sustain and enhance, rather than to reduce and simplify the biological interactions on which production agriculture depends. Alternative agriculture is not a single system of farm practices, but encompasses many farming systems variously called *organic, biological, low-input, regenerative,* or *sustainable* systems. Such systems emphasize management practices as well as biological relationships between organisms; in addition, they take advantage of naturally occurring processes such as nitrogen fixation.

Alternative or sustainable agriculture is any food or fiber production system that systematically pursues the following goals:

■ More thorough use of natural processes such as nutrient cycles, nitrogen fixation, and pest-predator relationships in agricultural production

- Reduced use of off-farm inputs with the greatest potential to harm the environment or the health of farmers and consumers

- Greater productive use of the biological and genetic potentials of plant and animal species

- Improved match between cropping patterns and the productive potential and physical limitations of agricultural lands, to ensure long-term sustainability of current production levels

- Profitable and efficient production, with emphasis on improving farm management and conserving soil, water, energy, and biochemical resources

Studies of alternative agriculture show that the farms derive sustained economic benefits and are not necessarily at a disadvantage. Although yields are often lower, costs are also considerably lower.

3.8 Weather and climate profoundly affect crop production.

The weather is the one factor with which people are familiar that profoundly affects the growth of plants. Everyone knows that to grow plants need sunlight and rain and that both are quite variable around the world. Weather—defined as short-term (less than a few weeks) variations in the atmosphere—affects the year-to-year yield of crops. The longer-term variations that comprise climate determine the geographic conditions where crops can grow at all.

Short-term variations caused by the weather that affect crop yields occur even in the modern era of technology-based agriculture:

1. *Moisture stress* is caused by insufficient, too much, or ill-timed rainfall. Plants often have precise water requirements, in terms of both amount and time. If these requirements are not met, yields and/or quality suffer.

2. *Temperature stress* occurs commonly when the temperature is either too low or too high for optimal growth. A short period of stress can severely depress growth later. Sudden cold snaps injure plants because they do not have enough time to become acclimatized to the cold weather. Moisture stress often accompanies high-temperature stress.

3. *Natural disasters* occur, such as cyclones, hurricanes, and hailstorms. Although these are more common in such Asian countries as Sri Lanka and Bangladesh, they can also cause severe damage in developed countries by physically harming crops and the soil that supports them.

Yields of many crops vary considerably year to year in many locations. The probability of a staple crop yielding the same year after year can be quite low. Over much of North Africa and the Middle East, for example, yields fall significantly below expectations in half of the crop years. A similar situation exists in the countries of the former Soviet Union. Complex statistical models, proposed to explain crop yield variability in other situations, show that in many cases climate is the major factor. When things go well (the yield trend is a smooth upward or constant curve), it is because the weather is unusually benign.

Different regions of the world are affected differently by climatic variation. Some of these effects are as follows:

1. *Temperate regions.* The north and south temperate regions produce 75% of the world's wheat and maize. The most prominent climatic factors affecting yield are

moisture stress and temperature extremes. Wheat and soybeans are more drought resistant than maize.

2. *Humid tropics.* Ranging from fertile valleys to jungles to floodplains, these regions are heterogeneous and show great variability of year-to-year yields. The major determining factor here is rainfall. There are pronounced wet and dry seasons, with the wet season often being in the form of a monsoon, with very heavy, sustained rainfall. But the unpredictability of the intensity of the monsoon, combined with poor water-holding capacity of soils and high rate of evaporation, lead to unstable water supplies for agriculture. In addition, torrential rains on lowland areas, such as river deltas, inundate the crops.

3. *Semiarid tropics.* These regions (for example, the Sahel south of the Sahara, in Africa, and much of India) have a long history of periodic crop failures and resulting famines. The major problem here is rainfall or, more precisely, the lack of it. Most of the rain falls in a two- to five-month period (April–October in the Northern Hemisphere, October–April in the Southern Hemisphere), when temperatures are at their yearly peak. This leads to extensive evaporation and less water availability. The sequence of events in growing a crop here is exquisitely sensitive to the annual rainfall cycle. Farmers time the growth of the crop to coincide with the maximum available water. If this period is very short, so is the growing season. In these areas farmers often plant two crops: one that will yield some food if the rains fail to come, and one that will yield abundantly if they do come (**Figure 3.13**).

This relationship of agricultural practices to climate is an ancient one. Every culture has its deities and stories relating the cycle of the seasons to food production. Chapter 5 describes how Africans have adapted their agricultural practices to a variety of climatic zones.

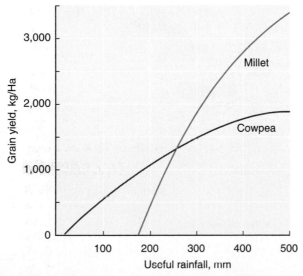

Figure 3.13 Grain yields of cowpea and millet respond differently to rainfall. In the Sahel region of Africa, farmers plant both cowpea and millet. Cowpea always produces some food, even when the rains fail completely; when the rains are abundant, millet easily outproduces cowpea. Planting both is a type of crop failure insurance. *Source:* Adapted from M. Ndoye, C. Dancette, M. Ndiaye, T. Diouf, and N. Cisse (1984), L' amelioration du niebe pour la zone sahelienne: Cas du programme national Senegalais. International Society for Research in Agriculture Publication. World Cowpea Research Conference, International Institute of Tropical Agriculture, Ibadan, Nigeria, November 1984.

3.9 Global climate change and global pollution will limit agricultural production in the future.

Human activities can substantially alter the global environment and thereby negatively impact our ability to grow crops. The public is now keenly aware of the effects of pollution on society, and generally considers industry to be the main source of pollutants. However, in the past 30 years agriculture has also become an important environmental polluter.

The book *Silent Spring*, written by biologist and science writer Rachel Carson in 1962, first awakened public opinion to unintended effects that pesticides can have on the environment. Although organochlorine insecticides then used have since been phased out, they did considerable damage both in developed and developing countries. These pesticides prevented crop damage by insects, but in Asian rice paddies pesticides kill fish, shrimp, and crabs, important sources of protein for poor people. (However, it is good to remember that these pesticides, used for mosquito control, also saved millions of people from contracting malaria.)

Burning vegetation to clear forestland or to encourage grass growth in savannas is an important source of global air pollution (**Figure 3.14**). When vegetation burns, some

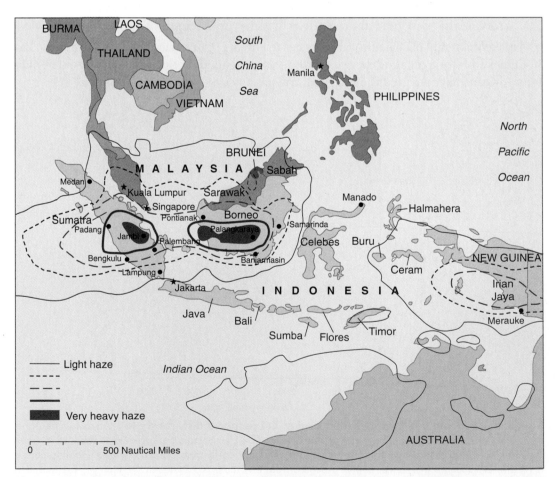

Figure 3.14 Burning of vegetation to clear land contributes to global air pollution and releases greenhouse gases. In 1997, excessive burning of forests in Indonesia led to an intense haze that spread over a huge area. Among other dramatic effects, agricultural production declined. *Source:* World Resources Institute; made by Y. M. Hardiono, Telepak, Indonesia, with data from NASA.

nutrients (potassium, phosphate, and calcium) return to the soil, but nitrogen, sulfur, and carbon disappear into the atmosphere as gases. In addition, burning releases particulates into the air. The sulfur dioxide and nitrogen oxides, in combination with those same two gases released from the burning of fossil fuels, help acidify soil via acid rain. In Europe and North America acid rain is killing some forests, and in China it has shrunk rice and wheat harvests, with most of the damage occurring at the beginning of the rainy season.

Burning biomass (vegetation) annually releases about 1 to 2 billion tons of carbon into the atmosphere as carbon dioxide. This amount adds to the 2 billion tons of carbon released by burning fossil fuels. Carbon dioxide is a greenhouse gas, which means it contributes to warming of Earth. Anaerobic decay of organic matter in marshes produces methane, another greenhouse gas. Rice paddies are an important source of methane worldwide, and increased rice cultivation in Southeast Asia has much increased methane production. These gases are responsible for the greenhouse effect, which warms the surface of Earth. If greenhouse gases were completely absent from the atmosphere, Earth's surface would be a chilly $-18°C$. Their presence maintains Earth at a higher temperature, making human life possible. The temperature of the planet has been rising slowly—about 0.3 to 0.6°C during the past 100 years, and is expected to rise another 0.4°C by 2020. A 2°-C rise by the end of the 21st century is possible. Although this slow global warming could be a natural phenomenon, most experts believe it is the direct consequence of the rise in emissions of greenhouse gases. A likely result of global warming is greater variability in the weather and more extreme weather conditions (droughts, floods, typhoons, frosts, heat waves, and so on). Rainfall may be more variable, making crop production even more variable than it is now.

Other gases, especially nitrous oxides and the now banned chlorofluorocarbons formerly used as refrigerator coolant, destroy the protective ozone layer in the upper atmosphere. Ozone shields Earth from destructive ultraviolet solar radiation. Closer to Earth, nitrous oxides emitted from cars or released by burning vegetation help create photochemical smog, of which ozone is a significant component. Even at quite low concentrations, ozone inhibits plant growth and recent experiments suggest that the ozone levels around cities and in some rural areas are now high enough to impair crop yields. Scientists may need to take this into account when breeding new crop varieties.

CHAPTER SUMMARY

In Earth's ecosystem, humans are consumers and consume primarily plants, although as their standard of living rises, they consume more animals and animal products. This lifestyle is "expensive" because it requires setting aside pastureland and arable land for producing feed crops. Some small human populations depend on animals that consume only plants grown in areas unsuitable for crop production.

During the past 10,000 years agricultural production has become gradually more intense. In the 20th century agriculturists have started applying scientific principles and technology to crop breeding and farming. This has resulted in yields taking off for the major staples (maize, rice, and wheat).

Since 1960, international research centers have been breeding crop varieties specifically for the climates and soils of developing countries. This activity generated the Green

Revolution, which dramatically increased staple production in Asia and Latin America. A network of international agricultural research institutions has been established to carry on with this task.

Modern agriculture depends heavily on purchased inputs. These include new genetic varieties, pest and disease control programs, irrigation technology, fertilizers, mechanical equipment, and even computers to access public databases containing relevant information.

During the past 40 years, scientists have used laboratory techniques (biotechnology) more and more as part of genetic crop improvement. This trend has developed into genetic engineering of crops. Genetic engineering involves transferring genes between unrelated organisms. So far, genetic engineering combines with traditional breeding practices to produce elite varieties adapted to local climate and soil conditions.

The practices of high-input agriculture are causing concerns about the sustainability of crop production. There are many negative effects on the environment, including pollution by pesticides, emission of greenhouse gases, soil degradation, air pollution by dust, and loss of landraces and other biodiversity. People need to develop new techniques that will keep agriculture both profitable for the farmer and make it sustainable for the future.

Discussion Questions

1. Discuss the economic demand for food versus the availability of food. How do they differ?

2. Discuss the use of resources to produce animal products. What would be gained by eliminating animal products from the human diet? How would this cause economic dislocation?

3. Discuss climate change (commonly called *global warming*), and its likely causes and effects on Earth and its food production capacity.

4. How should the political process treat farming? Should there be subsidies, and to whom should they go? How is farming subsidized in your country?

5. Discuss GM technology in relation to biotechnology in general and plant breeding. How has plant breeding been modified over the years?

6. How should the CGIAR system be funded? Explain your answer.

7. Discuss the relationship between agricultural and economic development in the developing countries and the security of your own country. How is your security tied to development elsewhere?

8. Discuss the differences between purchased inputs and inputs produced on the farm.

9. Plants can be "stressed out" just like people. Discuss plant stresses and what the response of the plants is.

Further Reading

Azam-Ali, S. N. and Squire, G. R. 2002. *Principles of Tropical Agronomy.* Oxon, UK: CABI Publishing.

Barrow, C. J. 1991. *Land Degradation.* Cambridge, UK: Cambridge University Press.

Brown, K., K. Hopkin, R. B. Horsch, and M. Mellon. 2001. "Genetically Engineered Foods: Are They Safe?" *Scientific American* 284:51–65.

Conacher, A. 2001. *Land Degradation.* Norwell, MA: Kluwer Academic Publishers.

Conway, G. 1997. *The Doubly Green Revolution.* Ithaca, NY: Cornell University Press.

Hillel, Daniel. 2000. *Salinity Management for Sustainable Irrigation.* Washington, DC: World Bank Press.

Kirpich, P. 1999. *Water Planning for Food Production in Developing Countries.* Lanham, MD: University Press of America.

Lomborg, B. *The Skeptical Environmentalist.* 2001. Cambridge, UK: Cambridge University Press.

Paarlberg, R. 2000. The global food fight. *Foreign Affairs* 79(3):24–38.

Pierce, J. T. 1990. *The Food Resource.* New York: Longman Scientific.

Rosensweig, C., Parry, M. C., Fischer, G., and Frohberg, K. 2000. *Climate Change and World Food Supply.* Oxford, UK: Oxford University Press.

U.S. Department of Agriculture. 1987. *Major World Crop Areas and Climatic Profiles.* Washington, DC: USDA.

Wittwer, S. 1995. *Food, Climate and Carbon Dioxide: The Global Environment and Food Production.* Boca Raton, FL: Lewis Publishers.

Food Security: Why Do Hunger and Malnutrition Persist in a World of Plenty?

Marc J. Cohen
*International Food
Policy Research
Institute*

In January 2001, the Indian government debated whether to dump surplus wheat and rice at sea, to make room for government-acquired reserves from the new harvest. This discussion occurred at a time when 208 million Indians were suffering from chronic undernutrition, including 150 million affected by severe drought, and with a majority (53%) of the country's preschool children afflicted by malnutrition. As the *Hindustan Times* newspaper put it, "Grain, grain everywhere, but not enough to eat." The problem of hunger amid plenty is worldwide, and in this chapter we explore its dimensions, causes, and possible solutions.

4.1 Food insecurity is widespread and pervasive in developing countries and is also found in developed countries.

According to the Food and Agriculture Organization (FAO) of the United Nations, global food production has been adequate every year since 1974 to meet minimum calorie requirements for every person on Earth. In 1995, about 2,600 calories were available per person per day in developing countries, compared to minimum requirements of 2,350 calories. The International Food Policy Research Institute (IFPRI) projects that availability will rise to 2,800 calories per person per day by 2020 (**Figure 4.1**). Increases in per capita food availability are expected in all regions. However, at less than 2,300 calories per person per day, average food availability in sub-Saharan Africa will fall below minimum requirements. Because available food will not be equally distributed, many Africans will actually have far less than the minimum required.

Despite adequate aggregate food availability, FAO reports that in 1997 (the last year for which data are available), 792 million people in developing countries (18% of the developing world's population) were food insecure. This represents a decline of 145 million people since 1980. In 1997, an additional 30 million people living in the newly independent states of the former Soviet Union and the countries in transition from centrally planned economies in central and eastern Europe experienced food insecurity.

Food security is a situation where every person has access to enough food to sustain a healthy and productive life, consistent with the principle that everyone has a basic human right to access to adequate food and nutrition and general human dignity. This status is generally conceived as having three dimensions: not just adequate food *availability* but also *access* to food and appropriate *utilization* of food. (The reverse situation is called *food insecurity*.)

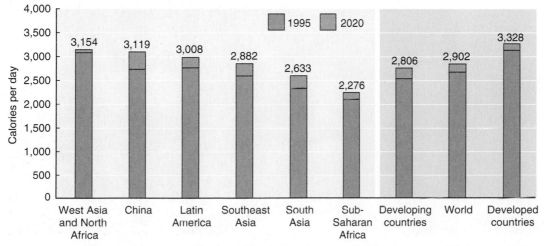

Figure 4.1 Daily per capita calorie availability, 1995 and 2020. The minimum needed for a healthy and active life is 2,350 calories per day. This will be barely reached in sub-Saharan Africa by 2020.
Source: IFPRI International Model for Policy Analysis of Agricultural Commodities and Trade (IMPACT) simulations, reported in P. Pinstrup-Andersen, R. Pandya-Lorch, and M. W. Rosegrant (1999), *World Food Prospects: Critical Issues for the Early Twenty-First Century, 2020 Vision Food Policy Report* (Washington, DC: IFPRI).

Two of every three undernourished people in the developing countries live in the Asia-Pacific region, and 44% live in China and India. In South Asia (India, Pakistan, and Bangladesh), 23% of the population is undernourished, compared to 13% in Southeast Asia and 12% in East Asia. The share of the population that was undernourished in the greater Asia-Pacific region fell from 32% in 1980 to 17% in 1997, but half a billion people remained food insecure in 1997.

In many parts of Africa, women farmers have less access to education and to labor, fertilizer, and other inputs than men, and often face limitations on their right to own or control land. In both Africa and Latin America, extension services and technical assistance focus primarily on male farmers. Total household agricultural output would increase if there were more equitable allocation of labor, services, and inputs. In South Asia, because families of daughters must pay bridegrooms a dowry, girls tend to receive less care and food than do boys. As a result, girls have higher mortality rates and are "missing"; there are only 950–970 females for every 1,000 males.

Sub-Saharan Africa is the developing region with the highest proportion of its population (34%) living in food insecurity. This is somewhat lower than the 1980 figure (38%). The number of food-insecure Africans jumped nearly 50% during this period, from 125 million to 186 million. In Africa, women grow most of the locally produced food, and they are essential to achieving food security on that continent. Women need access to research, extension services and credit from lending institutions to effectively contribute to the food security of their families. In developing countries generally, women play an important role in growing the food crops (**Figure 4.2**), marketing locally produced food, and cooking for their families. They are key to eliminating food insecurity.

(a)

(b)

Figure 4.2 Women have major responsibilities for growing, marketing and preparing food in developing countries. (a) These two women are transplanting rice in Hazaribagh, India. **(b)** Women cleaning cassava roots in Ghana. Cassava is an essential part of the diet for more than 500 million people. *Sources:* (a) Courtesy of Eugene Hettel, IRRI and (b) P. Cenini, FAO.

Even in affluent countries, segments of the population are vulnerable to hunger. The U.S. Department of Agriculture estimates that in 1999, 27 million people in the United States, including nearly 11 million children under age 18, lacked food security because of their households' inadequate resources. Being food insecure is not quite the same as being *clinically malnourished*. Among the factors behind these figures are high levels of income inequality; high costs for nonfood necessities such as shelter and health care; inadequate government outreach efforts to enroll low-income people in food, health, and other assistance programs for which they are eligible; and failure to maintain the purchasing power of the minimum wage. In the 1990s, about 9% of Canadians relied on private food assistance, because of cuts in federal social spending, high unemployment, and low wages for low-skilled workers. And high long-term rates of unemployment and pressure on governments to reduce social spending in order to remain competitive in the global economy and meet European Union fiscal policy requirements have forced expansion of the number of food banks and soup kitchens in Western Europe as well.

4.2 Food insecurity and malnutrition are unlikely to disappear in the next 30 years.

FAO projects that by the year 2015 the number of undernourished people in developing countries will fall to 576 million, or 10% of the people in the developing world (**Figure 4.3**). This is far short of the goal agreed on by the high-level representatives of 185 countries, including many heads of state and government, at the 1996 World Food Summit, who pledged "our political will and our common and national commitment to achieving food security for all and to an ongoing effort to eradicate hunger in all countries, with an immediate view to reducing the number of undernourished people to half their present

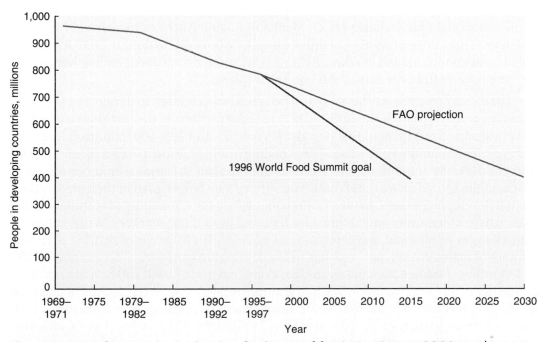

Figure 4.3 Food insecurity in the developing world, 1969–1971 to 2030. Food insecurity will decline slowly but will certainly not disappear by 2030. *Source:* Food and Agriculture Organization (FAO) of the United Nations, projected in 2000.

Table 4.1	Incidence of undernourishment in developing countries			
	1969–1971	**1979–1981**	**1995–1997**	**2030**
Sub-Saharan Africa	88 (34%)	125 (36%)	180 (33%)	165 (15%)
Near East North Africa	45 (25%)	22 (9%)	33 (9%)	35 (6%)
Latin America and Caribbean	54 (19%)	46 (13%)	53 (11%)	32 (5%)
South Asia	267 (37%)	337 (38%)	284 (23%)	82 (4%)
East Asia	504 (43%)	406 (29%)	240 (13%)	86 (4%)
Total	959 (37%)	937 (21%)	790 (18%)	401 (6%)

Note: Numbers are in millions of people. Percentages indicate percentage of the population in that region.

Source: Food and Agriculture Organization (FAO) of the United Nations.

level [to 400 million people] no later than 2015." The international community's failure to achieve adequate progress toward the summit goal is not just a political failure to keep this promise, but a moral failure as well. The summit's reaffirmation of the right of every human being to adequate food and freedom from hunger rings hollow for far too many of those human beings.

FAO forecasts that in 2015, 315 million South and East Asians will remain food insecure, although the incidence of undernutrition will be 10% or less in both subregions. In Africa, 184 million people (22% of all Africans) will be undernourished. The center of gravity for hunger will remain squarely in South Asia and sub-Saharan Africa, because those regions will be home to 61% of all food-insecure people, about the same share as in 1997 (**Table 4.1**).

People with real names and faces stand behind these rather dry statistics. They are such people as an elderly Egyptian man who says, "My children were hungry and I told them the rice is cooking, until they fell asleep from hunger." Or Kone Figue, who weeds and harvests her small rice farm in Côte d'Ivoire by hand, and seldom produces enough to feed her family of eight for a whole year. Or Medhanit Adamu Abebe, a 16-year-old Ethiopian girl, who writes in an essay about ending hunger, "Above all, to do this *PEACE* is necessary. In war conditions and in ethnic fighting you cannot dream about ending hunger and fighting poverty from the face of this continent (planet)."

Of particular concern are the malnourished preschool children in developing countries because good nutrition in the early years is necessary to assure that children reach their full physical and mental potential (see also Chapter 7). In 1995, 160 million children under age 5 in developing countries suffered malnutrition. Malnutrition is a factor in 5 million child deaths annually (a toll 10 times greater than the annual number of cancer deaths in the United States), and those who survive may be impaired in their physical and mental development (see Chapter 7). Food insecurity robs humanity of countless scientists, artists, community and national leaders, and productive workers. Without significant changes in national and international policies, IFPRI projects that by 2020, the number of malnourished preschoolers in the developing world will fall only about 16%, to 135 million (**Table 4.2**). This means that every fourth child will still be malnourished, compared to one in three in 1995. Child malnutrition is expected to decline in all major developing regions except sub-Saharan Africa, where the number of malnourished children is forecast to increase by 30% to reach 40 million by 2020. In South Asia, although the number of malnourished children will fall by 18 million, as many as two out of five children will still be malnourished in 2020. Child malnutrition is even more heavily concentrated in these two regions than is general food insecurity: These areas were home to

Table 4.2	Number (in millions) of malnourished children in developing countries: 1995 and 2020	
Region	**1995**	**2020**
South Asia	82	64
Sub-Saharan Africa	31	40
Southeast Asia	19	15
East Asia	18	10
West Asia and North Africa	6	4
Latin America	5	2
Total	161	135

Note: Projections based on the international model for policy analysis of agricultural commodities and trade (IMPACT) projections of July 1999.

Source: International Food Policy Research Institute (IFPRI) projections.

70% of all malnourished preschoolers in 1995, a level that is projected to rise to 77% by 2020. Many countries in these two regions are among the "least developed." They will require special assistance to avert widespread hunger and malnutrition in the years to come.

Child malnutrition for all developing countries is not expected to decline by as much as 25% over the period 1990–2020. Yet in 1990, the international community pledged at the World Summit for Children to work to cut severe and moderate malnutrition among preschool children by 50% by 2000.

New evidence shows that low birth weight, a sign of poor maternal nutrition, is a major predictor of child malnutrition. About one in four children in developing countries are born at low birth weight, usually as a result of poor maternal nutrition both before conception and during pregnancy. In effect, malnutrition is thus directly transmitted from one generation to the next. Child malnutrition is spread partly by poor access of women to food.

IFPRI research has examined the so-called Asian enigma, the paradox of significantly higher rates of child malnutrition in South Asia than in sub-Saharan Africa, even though on most indicators of human well-being (access to safe water, school enrollment, food availability per person, income per person, degree of democratic governance), South Asians fare better than Africans. But with respect to women's social status relative to men's, as measured by the female-to-male life expectancy ratio, Africa is doing better than South Asia. Some other possible factors in these malnutrition rates include climate, population density, and cultural norms that discourage sound child care and feeding practices.

Nutritionists generally agree that if a person takes in enough calories, he or she will also get the necessary protein. But this does not guarantee adequate intake of micronutrients— vitamins, minerals, and trace elements. According to the interagency Subcommittee on Nutrition, of the United Nations, as much as 80% of humanity experiences iron deficiency. About 2 billion people suffer from anemia, usually because of inadequate dietary iron, including 48% of all pregnant women and 56% of pregnant women in developing countries. In South and Southeast Asia, 76% of pregnant women and 63% of preschool children are anemic (**Figure 4.4**). Around 50% of the world's anemic women live in South Asia. Their risk of maternal mortality is 23% higher than that of nonanemic mothers. Their babies are more likely to be premature, have low birth weights, and die as newborns. Anemia can impair child health and development, limit learning capacity, impair immune systems, and reduce work performance. Even when iron deficiency does not progress to anemia, it can reduce work performance. These effects of iron deficiency result in annual economic

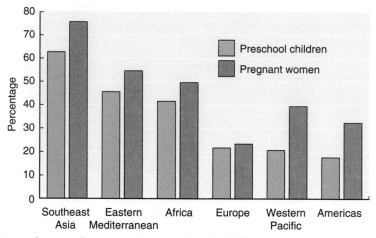

Figure 4.4 Prevalence of anemia in preschool children and pregnant women worldwide, by region, 1998. *Source:* UN Administrative Committee on Coordination/Subcommittee on Nutrition and IFPRI.

losses as high as 2% of gross domestic product in some developing countries. Nevertheless, high levels of iron deficiency anemia have persisted over the past two decades, and few high-priority public health programs tackle this problem.

Insufficient intake of vitamin A among children in developing countries is the leading cause of preventable severe visual impairment and blindness and contributes to infections and death. Pregnant women who are deficient in vitamin A face increased risk of mortality and mother-to-child HIV transmission. The World Health Organization estimates that 100 million to 140 million children are vitamin A deficient and that 250,000–500,000 of them go blind every year as a result. About half of the children who suffer from vitamin A blindness die within a year of losing their eyesight. National surveys indicate that trends in the incidence of clinical eye disorders are encouraging, but that vitamin A deficiency remains a serious public health problem in the developing world.

4.3 Causes of food insecurity include poverty, powerlessness, violent conflict, discrimination, and demographic factors.

A number of underlying factors result in hunger. These include poverty, powerlessness, violent conflict, discrimination, and demographic factors, such as population growth and urbanization.

Poverty. The most obvious factor behind food insecurity is poverty. Globally, 1.2 billion people live on the equivalent of less than US$1 a day and are too poor to meet their needs for food and other necessities on a sustainable basis. They spend as much as 50–70% of their incomes on food, and frequently lack access to land to produce food or other income generating resources. Two out of three of these absolutely poor people live in South Asia and sub-Saharan Africa, the regions that are most food insecure. Most (75%) people living in poverty are in rural areas, and the majority of poor people in developing countries will remain rural through at least 2035, although a majority of the overall population will be urban by 2020. Farmers who practice rain-fed agriculture, smallholder farmers, pastoralists, artisanal fisherfolk, landless laborers, indigenous people, people in

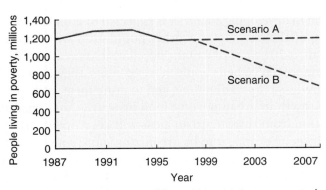

Figure 4.5 Poverty in the developing world, 1987–2008. Scenario A shows slow growth with continued or rising inequality; Scenario B shows "inclusive" growth. *Source:* World Bank.

female-headed households, and displaced people are the rural people most affected by poverty.

The World Bank projects that the number of poor people will remain unchanged over the next decade if growth remains slow and inequality increases from current levels. However, if countries adopt policies and interventions that foster inclusion and broad-based growth, the number of poor people could decline to 680 million by 2008 (**Figure 4.5**).

Powerlessness. Poor and food-insecure people frequently lack organizations that are accountable to them and capable of articulating their interests to policymakers and other power holders. As a result, policies tend to benefit people who are already well off, and policymakers tend to give low priority to the needs of poor and hungry people or to programs that would benefit them.

Violent Conflict. Since the end of the cold war, internal conflicts have proliferated in developing and transition countries, particularly in Africa. Fourteen million refugees have fled these struggles, which have displaced another 20–30 million people within their own countries. Uprooted people are vulnerable to malnutrition and disease, and depend on humanitarian assistance to survive. Postconflict reconstruction takes years. Not only does violent conflict cause hunger (**Figure 4.6**), but hunger can contribute to conflict, especially when resources are scarce and perceptions of economic injustice are widespread, as in Rwanda in 1994 or Central America in the 1970s and 1980s.

Figure 4.6 Violent conflicts in Africa and elsewhere have displaced millions of people who must live in refugee camps. Such conflicts disrupt food production and are a major contributor to poverty and malnutrition. Invariably women and children suffer most. Shown here are women and children in a camp in Guinea. *Source:* Photo by Trevor Rowe of the World Food Program.

Discrimination. Food-insecure people tend to be disproportionately female and young or elderly. Cultural practices and official policies that marginalize people on the basis of gender, age, race, and ethnicity frequently contribute to food insecurity. For example, in the Sudan, where Arab Muslim–dominated government forces have engaged in a civil war for the past 20 years with African Christians and animists in the country's south, rates of malnutrition in the conflict zones are among the world's highest, with up to 45% of the population affected.

Demographic Factors. World population will increase by more than 25% between 2000 and 2020, from 6 billion to more than 7.6 billion people. About 80 million people will be added each year, virtually all in developing countries. Six countries will account for fully half of this population increase—India, China, Pakistan, Nigeria, Bangladesh, and Indonesia, and India and China alone will account for one third. This population growth will stimulate substantial increases in demand for food and feed.

Urban population in developing countries is expected to double between 1995 and 2020, when the majority of the developing world's population will live in urban areas (see Chapter 1). In the next 20 years, 90% of the population increase will occur in the rapidly expanding cities and towns. When people move to cities, they tend to shift consumption to foods that require less preparation time, and to more meat, milk, fruit, and vegetables (see Chapter 2). Growth in urban poverty, food insecurity, and malnutrition and a shift in their concentration from rural to urban areas will accompany urbanization, although poverty and food insecurity appear to be urbanizing more slowly than the overall population of developing countries. Despite the growing consequences to urban well-being, policymakers do not yet have solutions that adequately reflect and respond to these challenges, which have long been conceptualized as rural problems. Urban dwellers depend more on money income, may have fewer opportunities to grow their own food, and require access to child care in order to pursue income-earning opportunities.

4.4 Environmental degradation is a major cause of food insecurity and of serious concern for the future.

In many developing countries, poverty, low agricultural productivity, and environmental degradation interact in a vicious downward spiral. This is especially true in resource-poor areas that are experiencing high rates of population growth and are home to hundreds of millions of food-insecure people, particularly in Africa. Agricultural growth, poverty alleviation, and environmental sustainability are not necessarily complementary, and achieving all three simultaneously cannot be taken for granted. Much depends on specific social, economic, and agroecological circumstances.

Some degradation in agricultural areas has been caused by the misuse of modern farming inputs (especially pesticides, fertilizers, and irrigation water). However, most environmental degradation, particularly soil degradation and deforestation, is concentrated in resource-poor areas that have not adopted modern technology. In these cases the problem typically is insufficient agricultural intensification, with yield growth failing to keep up with population growth. To raise more food the people expand their fields, putting more land to the plow. This usually means using poorer areas in ecologically fragile zones—higher up the slopes of cultivated areas or in hilly terrain, for example. Poor rural people in developing countries tend to depend on annual crops (which generally degrade soils more than perennial crops) and common property lands (which generally suffer greater degradation than privately managed land). They often cannot afford to invest in land improvements. Degradation and lack of access to high-quality land frequently push poor people into clearing forests (**Figure 4.7**) and pastures for cultivation at the expense of wildlife habitat, contributing to further degradation.

IFPRI and the World Resources Institute have found that depletion of soil organic matter in developing countries is widespread, leading to significant economic losses and reduced fertility, moisture retention, and soil workability. Degradation also leads to increased carbon dioxide emissions from agricultural land, contributing to global warming. Cropland and pasture management strategies that result in improved soil organic matter content also increase carbon sequestration capacity, and thus help reduce agriculture-induced greenhouse gas emissions.

Preharvest losses to pests (insects, animals, weeds, and plant diseases) reduce the potential value of farm output by 40%, and postharvest losses cost another 10%. In developing countries, losses greatly exceed agricultural aid received. Developing countries' share of the global pesticide market is expected to increase significantly during the early 21st

Figure 4.7 Erosion on hillsides. Putting more land to the plow does not always have a good outcome. If the land is on a hillside, great care must be taken to prevent erosion or severe land degradation may result as in this view of Laos. The white patches in the green forest are completely denuded hillside where all the topsoil has been lost. *Source:* Courtesy of Eugene Hettel, IRRI.

century. Insecticides now used in developing countries are often older and acutely toxic, and are generally banned in developed countries except for export.

Unless properly managed, freshwater may well emerge as the key constraint to global food production. Although aggregate supplies of water are adequate to meet demand for the foreseeable future, water is poorly distributed across and within countries, and between seasons. Demand for water will continue to grow rapidly. Since 1970, global demand has increased by 2.4% annually. IFPRI projects that global water withdrawals will increase by 35% between 1995 and 2020. In developed countries, most increased demand will be for industrial use. Developing countries are projected to increase their withdrawals by 43% during this period, with the share of domestic and industrial uses in total water demand doubling at the expense of agriculture. In 1995, agriculture accounted for 72% of global water withdrawals and 87% of developing country withdrawals. Growth in irrigated area is expected to slow significantly. Although Africa's irrigated area is projected to increase by 50%, it will remain very low.

4.5 Strong social and economic forces will affect future food security.

A number of global factors will affect food security over the next 20 years, for good and ill. These include globalization, falling aid, technology, global climate change, and health issues.

Globalization. Globalization offers significant new opportunities for economic growth in most developing countries, but it may also carry significant risks. These include the inability of many developing country domestic industries to compete in the short term, the potential destabilizing effects of uncontrolled short-term capital flows, increased exposure to price risk, and worsening inequality, because many poor people and backward regions may get left behind. Managing these risks while exploiting growth opportunities will be a key challenge for developing countries in the years ahead. Most developing countries have liberalized food and agricultural trade since the 1980s. Developed countries have not reciprocated, instead maintaining barriers to high-value imports from developing countries such as beef, sugar, peanuts, dairy products, and processed goods. Most developed countries also continue to subsidize their own agricultural exports. Many developing countries lack administrative, technical, and infrastructural capacity to comply with the global trade rules defined primarily by developed countries.

Declining Aid. As the cold war was the principal force driving foreign aid, since 1991 official development assistance has become a much lower priority for governments of industrialized countries. Aid from the principal donors fell about 17% between 1992 and 1997 (**Figure 4.8**). Subsequently, it appears that, in the aggregate, the volume of aid has stabilized, but at a level below that of the mid-1990s.

Between 1986 and 1997, aid to agriculture and rural development shrank by almost half, and the share of total aid going to agriculture dropped from 25 to 14%. Although aid to agriculture increased somewhat in the late 1990s, in real terms it remains well below what was provided 15 years ago. IFPRI research shows that donors are being short-sighted. Aid to developing country agriculture not only is effective in promoting sustainable development and poverty alleviation, but also leads to increased export opportunities for developed countries, including increased agricultural exports, as agricultural growth spurs more general economic growth and demand for food products.

Donors have also scrimped on aid to education. Only 5 (Denmark, Finland, Germany, the Netherlands, and Sweden) of the 21 main donors who provide aid directly, provide more than 2% of their aid to basic education. The international community has failed to deliver on the commitment made at the 1990 World Conference on Education for All to universal primary education, for girls and boys alike, by no later than 2000. Nor will the revised goal of universal primary education by 2015 come close to being met, at the current rate of progress. As noted, education, especially for girls, is crucial in reducing malnutrition and fostering broad-based agricultural and rural development.

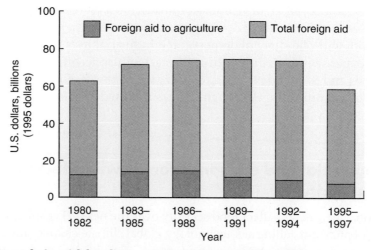

Figure 4.8 Trends in aid funding, 1980–1982 to 1995–1997. *Source:* Food and Agriculture Organization (FAO) of the United Nations.

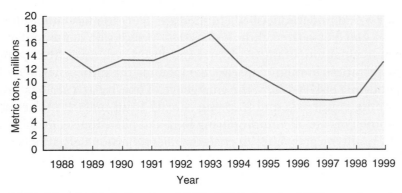

Figure 4.9 Global food aid deliveries, 1988–1999. Data for 1999 are estimated. *Source:* World Food Program.

Food aid levels have fluctuated considerably since the mid-1990s (**Figure 4.9**). Late in the decade, following major reductions in 1994–1996, the United States (which remains the largest donor) expanded food aid substantially. The sharp peaks and valleys stemmed far less from developing country needs for either humanitarian or trade liberalization adjustment assistance than from U.S. domestic market conditions, as the United States inflexibly continues to tie its food aid to U.S. farm products. In contrast, the European Union and its member countries provide food purchased in the recipient country or neighboring countries for a large share of their food aid. In this way, their food aid does not depend on the availability of donor country surpluses, which may not always be available at times of greatest need.

Technology. Technological advances achieved through agricultural research and development substantially contributed to the spectacular increases in food production witnessed during the 20th century. But rapid changes are taking place in the financing, management, and organization of agricultural research, the proprietary nature of the agricultural sciences, and the nature of the biological sciences themselves. These changes are placing an increasing share of agricultural research and the ownership of new technologies in the private domain, raising concerns about the extent to which agricultural R&D will help eliminate hunger for the world's poor people in the decades to come. At the same time, biotechnology, modern information and communication technologies, and energy technologies offer new opportunities that could benefit poor people, their food security, and natural resource management, if policies are in place to assure that poor people can reap these benefits.

Global Climate Change. Global climate change is raising average temperatures and sea levels and destabilizing weather patterns, bringing more frequent and severe droughts and flooding. These changes will profoundly and often negatively impact developing countries, with many of the poorest countries being most vulnerable. Great controversy rages within the scientific and policy communities over the extent of, and future trends with regard to, this phenomenon. Nevertheless, the United Nations-sponsored Intergovernmental Panel on Climate Change recently reported that anticipated temperature increases mean that in the tropics, where some crops are already near their maximum temperature tolerance and where dryland agriculture predominates, even minimal increases in temperature would generally shrink yields. Where rainfall decreased greatly, crop yields would shrink even more. Drier soils and heat could also reduce crop production in some parts of the North American "breadbasket."

Health. Hungry and undernourished children are likely to miss more school days because of illness, and diet-related chronic diseases—perhaps linked to undernutrition *in utero*—take individuals out of the workforce and absorb resources from primary health services. The relentless burden of disease compounds the difficulties of malnutrition. The tragic pandemic of HIV/AIDS, the emergence of obesity as a serious health risk in many developing countries, and ongoing threats from malaria, tuberculosis, and other health problems are all compromising food and nutrition security in developing countries. The result is a global health crisis that disproportionately afflicts the poor and undernourished.

4.6 Broad-based agricultural and rural development is needed to alleviate food insecurity.

Appropriate national and international policies are essential for achieving food security. By mustering the political will to make food security a higher priority, national governments and the international community can do much to hasten progress toward food security for all. Recent protests and activism may help push food security higher on the agenda of the international community.

Because the center of gravity for poverty—and therefore food insecurity—will remain rural during the first 35 years of the 21st century, broad-based agricultural and rural development must be at the center of any strategy to achieve food security in the developing world. Even when rural poor people are neither farmers nor farmworkers, their livelihoods depend on activities closely related to agriculture. Low agricultural productivity in developing countries results in high per-unit costs of food, poverty, food insecurity, poor nutrition, low farmer and farmworker incomes, little demand for goods and services produced by poor nonagricultural rural households, and urban unemployment and underemployment. IFPRI research has shown that for every new dollar of farm income earned in developing countries, income in the economy as a whole rises by up to US$2.60, as growing farm demand generates employment, income, and growth economy-wide. Agricultural growth also helps meet rising food demand stemming from population and income growth and urbanization (**Figure 4.10**).

Small farmers in developing countries face many problems. Low soil fertility and lack of access to fertilizers, along with acid, salinized, and waterlogged soils, contribute to low yields, production risks, and natural resource degradation. Inadequate infrastructure, land tenure biased against poor people, poorly functioning markets, and lack of access to credit and technical assistance add impediments.

Sound public policies are needed to guarantee that agricultural and rural development is indeed broad-based, creating opportunities for small farmers and other poor people. The development of well-functioning and well-integrated markets for agricultural inputs, commodities, and processed goods, along with the supporting institutions and infrastructure (such as roads and storage, especially in rural areas) will contribute enormously to poverty alleviation, food security, and the overall quality of life in developing countries. An economic system in which decisions about the allocation of resources and production are made on the basis of prices set through voluntary exchanges among producers, consumers, workers, and owners of the factors of production will lead to more efficient creation and distribution of goods and services. This in turn will create new economic opportunities. However, it is the role of governments to assure that poor people have access to those opportunities and that the benefits are equitably distributed. The private econ-

(a)

(b)

Figure 4.10 Transportation systems and well-functioning local markets are essential for achieving food security. (a) River transport of rice in southern Vietnam. *Source:* Courtesy of Eugene Hettel, IRRI. **(b)** Fruit and vegetables on sale at a market in Asmara, Eritrea. Well-functioning markets create demand for products and encourage food production. *Source:* Photo by R. Faidutti, FAO.

omy by itself cannot, and should not, be expected to assure equity. Key public policies and investments include the following:

- Assuring poor farmers access to yield-increasing crop varieties (including drought- and salt-tolerant and pest-resistant varieties), improved livestock, and other yield-increasing and environment-friendly technology

- Assuring poor farmers access to productive resources, including land, water, tools, fertilizer, and pest management

- Removing institutional barriers to the creation and expansion of small-scale rural credit and savings institutions and making them available to small farmers, traders, transporters, and processing enterprises

- Assuring access to agricultural extension services and technical assistance (**Figure 4.11**)

- Particular attention to the needs of women farmers, who grow much of the locally produced food in developing countries

- Primary education, health care, clean water, safe sanitation, and good nutrition for all

These investments must be supported by good governance—the rule of law, transparency, sound public administration, and respect for human rights—as well as trade, macroeconomic, and sectoral policies that do not discriminate against agriculture and that favor poverty reduction and food security. Policies must also provide incentives for sustainable natural resource management, such as secure property rights for small farmers.

Development efforts must engage low-income people as active participants, not passive recipients. To assure responsive policies, poor people need accountable organizations that articulate their interests. In addition, poor people can often gain empowerment effectively by making alliances with groups of nonpoor people. For example, in Brazil, Citizen Action is an antihunger organization composed of urban unemployed people and landless rural people,

Figure 4.11 Farmers need access to schooling and extension services, as in this classroom for the continuing education of farmers in Bhutan. *Source:* Courtesy of Eugene Hettel, IRRI.

as well as writers, artists, professionals, and businesspeople. It lobbies the government to enact policies that support food security and administers community-level antipoverty programs. In Zimbabwe, the Organization of Rural Associations for Progress (ORAP) is a nongovernmental organization that provides technical assistance and advocacy support to development initiatives devised by poor rural communities themselves.

At present, however, developing countries are underinvesting in agriculture. On average, they devote 7.5% of government expenditures to agriculture, and the figure is even lower in sub-Saharan Africa, where agriculture's contribution to gross domestic product (GDP) is 30–80%. Although there are no reliable regional data on military spending in sub-Saharan Africa, researchers estimate that some of the region's poorest countries are devoting between 4 and 15% of GDP to military spending. In South Asia, the regional average is 2.4% of GDP, and if India is excluded, the figure jumps to more than 3%.

For their part, aid donor countries must reverse the overall aid decline and the precipitous fall in aid to agriculture. They must also rethink their 20-year emphasis on reducing government's economic role, which has contributed to developing countries' public disinvestment in agriculture.

4.7 Greater public investment in agricultural research and better management of natural resources will be required to eliminate food insecurity in developing countries.

Public investment in agricultural research that can improve small farmers' productivity in developing countries is especially important for food security. It must join all appropriate scientific tools and methods—including agroecology, conventional research methods, and genetic engineering—with better use of indigenous knowledge. For example, in India efforts to intensify hill tribe agriculture and boost incomes combine traditional soil and water conservation techniques with cultivation of new, high-value cash crops. It is important that poor farmers have access to insights into agricultural development from the full range of approaches to tackling their problems. For instance, in developed countries farmers do not plant potatoes from material they saved from the previous year, because viruses readily infect potatoes. Instead, farmers obtain virus-free planting materials from producers of seed potatoes. The same approach would raise production of sweet potatoes in developing countries three- to fourfold. Alternatively, GM technology could be used to make sweet potatoes virus resistant so that farmers could continue to plant the potatoes they harvest (**Figure 4.12**).

Research should focus on productivity gains on small farms, emphasizing staple food crops and livestock. More research must be directed as well to appropriate technology for sustainable intensification of agriculture in resource-poor areas.

The private sector is unlikely to undertake much research needed by small farmers in developing countries because expected gains will not cover costs (see Chapter 2). So public sector agricultural research is needed to focus on the problems of small farmers and consumers. However, some multinational agricultural biotech companies are developing noncommercial projects for crops of developing countries. The idea is that the technology will be free as long as it is applied to the crops or crop varieties of low-resource farmers. For example, Monsanto has given three genes for virus resistance in potato to Mexico for incorporation into varieties favored by poor farmers (Norteña and Rosita). However, should Mexican entrepreneurs wish to incorporate these genes into commercial cultivars grown by large producers for exports, they will need to negotiate a license. This is one way the private sector can help developing countries. The public sector can also encourage private sector involvement by agreeing to purchase the rights to the results of private research for distribution to small farmers.

Agricultural research produces high gains for society and poor people. Social rates of return to agricultural research investment exceed 20% per year, compared to long run real interest rates of 3–5% for government borrowing. Yet the average share of farm production invested in agricultural research in low-income developing countries is much smaller than in higher-income countries. Average annual growth rates of public agricultural research expenditures in developing countries in the 1980s and 1990s fell below those of the 1970s (see Chapter 2).

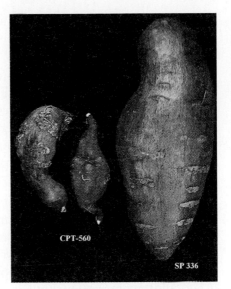

Figure 4.12
Comparison of virus-infected and virus-free roots of sweet potatoes. The large root (right) is from a healthy plant, the much smaller one (left) is from a virus-infected plant.
Source: Courtesy of W. K. Kaniewski, Monsanto Company.

The indirect impacts of agricultural research on poverty and food insecurity may be even more important than the direct benefits (or losses) from on-farm technological adoption and changes in agricultural employment. As a result of widespread adoption in Asia of high-yielding cereal varieties, developed at international agricultural research centers and national agricultural research institutions (the Green Revolution), output increased dramatically. This in turn lowered food prices and, over time, increased nonfarm employment opportunities, to the great benefit of poor people. Lower food prices not only benefited nonfarm poor consumers but also helped many poor farmers who were net purchasers of food. For farmers who produced more than they consumed, lower prices did not necessarily mean losses where technical advances allowed farmers to reduce production costs.

Sustainable Natural Resource Management. A high degree of complementarity among agricultural development, poverty reduction, and environmental sustainability is more likely to be achieved when agricultural development has a broad base and includes small- and medium-sized farms; is market-driven, participatory, and decentralized; and is driven by technological change that enhances productivity but does not degrade the natural resource base. To achieve this complementarity, agricultural research must attend more to sustainability features of technology, to broader aspects of natural resource management at the watershed and landscape levels, and to problems of resource-poor areas. Polluters and degraders should be taxed and regulated. Public institutions that manage and regulate natural resources in developing countries need increased budgets, larger staffs with enhanced technical skills, and improved authority to enforce regulations.

Low soil fertility and lack of access to reasonably priced fertilizers, along with past and current failures to replenish soil nutrients in many countries, must be rectified through efficient and timely use of organic and inorganic fertilizers and improved soil management. Policies should support an integrated nutrient management approach that seeks to both increase agricultural production and safeguard the environment for future generations. Chemical fertilizer use should be reduced where heavy application is harming the environment. Governments should remove fertilizer subsidies that encourage excessive use. However, governments may still need to subsidize fertilizer in regions where current use is low and soil fertility is deteriorating. In such areas, increased fertilizer use can help boost production and reduce land degradation. Policies should raise the value of forests and pastures, offer incentives for sound management, and help create nonfarm employment opportunities.

Until recently, developing country governments and aid donors encouraged use of chemical pesticides. Now, consensus is emerging on the need for integrated pest management, emphasizing alternatives to synthetic chemicals, except as a last resort. Alternatives include use of natural predators and biological pesticides, as well as breeding pest-resistant crops (see Chapter 16).

Comprehensive water policy reform can help save water, improve use efficiency, and boost crop output per unit of water, while reducing the risk of armed conflict between countries sharing scarce surface water or groundwater sources. Such reforms will be difficult to carry out, because of long-standing practices and cultural norms in many places that treat water as a free good. Also, vested interests benefit from current arrangements. The nature of the needed reforms will vary from country to country, depending on the level of economic development, institutional capability, degree of water scarcity, and the intensity of agricultural production. Key elements of needed reforms include providing secure water rights for individual users or groups of users. In some countries and regions, these rights should be tradable, thereby increasing incentives for efficient water use. Devolving irrigation infrastructure and management to water user associations, combined with secure access to water, will provide incentives for efficient use, for bargaining with government water agencies, and for improved operations and management. Reducing or eliminating explicit or implicit water subsidies can also improve efficiency of use. Privatization and regulation of urban water services can, for example, slow the rate at which cities take more water from agriculture. In the industrial sector, increased water prices can encourage investment in water recycling and conservation technology. Funds freed by eliminating general subsidies will make it possible to provide targeted subsidies to underserved poor urban dwellers and farmers. As incentives are introduced for water conservation, appropriate technology, such as drip irrigation, needs to be available.

4.8 Eliminating food insecurity will require other initiatives, such as tackling child malnutrition, resolving conflicts, and implementing equitable globalization.

Appropriate policies are needed to address the underlying causes of hunger and the social and economic factors that are likely to contribute to hunger in the coming years. If the needed steps are taken, then we can greatly accelerate progress toward eliminating hunger.

Tackling Child Malnutrition. IFPRI has found four critical reasons why child nutrition improved in the developing world between 1970 and 1995. Improvements in women's education accounted for 43% of the total reduction in child malnutrition during this period, followed by improvements in per capita food availability, in the health environment, and in women's status relative to men. Together, improvements in women's education and in per capita food availability accounted for nearly 70% of the reduction (**Figure 4.13**). Therefore, aid donors need to boost the share of the resources they provide that go to the agriculture and education sectors, and developing country governments need to make these sectors a higher public investment priority. Enrollment of girls and boys alike must be a key educational policy priority.

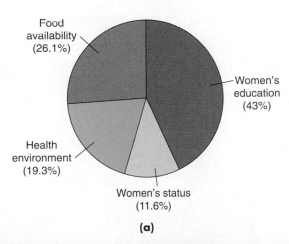

(a)

(b)

Figure 4.13 Women's education and child malnutrition. (a) Pie chart showing that changes in women's education are the single most important factor that accounted for the reduction in child malnutrition between 1970–1995. **(b)** A girls' classroom in Bangladesh. The education of women is also the most important factor in bringing down the birth rate. *Sources:* (a) L. C. Smith and L. Haddad (2000), "Overcoming Child Malnutrition in Developing Countries," 2020 Vision for Food, Agriculture, and the Environment Discussion Paper No. 30 (Washington, DC: IFPRI); (b) Mark Edwards/Still Pictures/Peter Arnold, Inc.

Fighting Micronutrient Deficiencies. We should take a variety of approaches to reduce the huge numbers of people who suffer from what is often called "hidden hunger": These approaches include fortifying, supplementing, and diversifying diets; changing eating habits; and breeding nutrients into staple crops through conventional plant breeding or biotechnology. The various approaches must be posed as complementary, not either/or choices.

Preventing and Resolving Conflicts. It is essential to include conflict prevention in food security and development efforts, and to link food security and economic development to relief. Savings from conflict avoidance should be included and calculated as monetary returns to aid and development spending. Humanitarian assistance provided in emergency situations must be linked to longer-term efforts to foster agricultural and rural development, secure livelihoods for the affected people, and sustainable social and agricultural systems. For example, assistance should not only help people affected by conflict to meet their immediate food and medical needs but should also help reconstruct

Figure 4.14 Street scene in Phnom Penh, Cambodia. Population growth throughout the world will occur primarily in mid-sized and large cities, putting even greater demands on the food production system because city dwellers demand more and better food. *Source:* G. Bizarri, FAO.

schools, clinics, roads, and water works. When peace has been achieved, refugees returning to their rural homes can be engaged in reconstruction activities and paid wholly or partly in food until the first postconflict harvest comes in.

Responding to Demographic Forces. Policies that allow poor people to achieve economic security are the best way to assure that birth rates will decline. It is also crucial to facilitate women's access to reproductive health services, consistent with individual consciences and beliefs. As food insecurity urbanizes along with the general population, it will be increasingly important for governments to improve livelihoods and employment among urban poor people, support environmentally sound urban agriculture, promote healthy physical environments and adequate caring and feeding practices, and design more participatory urban programs and strategies. Urban people rely on food purchases for 90% or more of their food needs (**Figure 4.14**).

Making Globalization Work for Poor People. Developing countries must be encouraged to participate effectively in global agricultural trade negotiations, pursuing better access to industrialized countries' markets. Coalitions with certain groups of higher-income countries may help improve the bargaining position of developing countries. However, without appropriate domestic economic and agricultural policies, developing countries in general and poor people in particular will not fully capture potential benefits from trade liberalization. Domestic distribution of productive assets will largely determine the distribution of benefits from trade within countries. Low-income countries must seek to strengthen their bargaining position and pursue changes in both domestic policies and international trade arrangements if they are to capture fully potential benefits from trade liberalization. For example, the following are needed:

- Domestic policy reforms in developing countries that remove biases against small farmers and poor people while facilitating access to benefits from more open trade

- Elimination of industrialized countries' export subsidies, taxes, and controls (**Figure 4.15**)

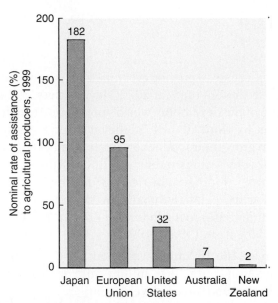

Figure 4.15 Industrialized countries' agricultural subsidies vary widely and put developing countries at a disadvantage in trade. The level of internal farm subsidies varies greatly among the wealthy countries. The numbers shown here are the "nominal rates of assistance" in 1999 in five different countries and country groups. The nominal rate of assistance is the level of support to internal producers through prices paid to them above world market prices, plus agricultural support in the national budget, divided by the value of production at world prices, multiplied by 100. It includes direct subsidies to farmers, services such as research, extension, marketing, and inspection, and the impact of import quotas and duties that keep domestic prices high and hurt developing country exports. Experts estimate that direct subsidies in the industrialized countries account for about one third of the amounts shown. U.S. support payments in 1998, 1999, and 2000 included $5.5 billion each year in "market loss assistance" to certain U.S. farmers to compensate them for low world prices of maize, wheat, oats, barley, grain sorghum, rice, and cotton. Developed country export subsidies also hurt developing countries; in 2000, European Union members together spent $4.2 billion on such subsidies. Some developed countries, such as Australia and New Zealand, have virtually phased out government subsidies for agriculture. Developing countries generally have very low nominal rates of assistance, but hard data are not available. *Source:* Organisation for Economic Co-operation and Development.

- Technical assistance and financial support for poor countries' agriculture

- Strong animal and plant health standards and technical support to help developing countries produce for developed country markets

- Adequate levels of food aid targeted to poor groups in ways that do not displace domestic production

There is nothing inevitable about the rather pessimistic outlook for food security. It is possible to meet and even exceed the World Food Summit's goal. Doing this will require concerted and committed action by governments, citizen groups, and the international community to empower poor people; mobilize new technological developments to benefit poor and hungry people in developing countries; invest in the factors essential for agricultural growth, including agricultural research and human resource development; and harness the political will to adopt sound antipoverty, food security, and natural resource management policies. Failing to take these steps will mean continued low economic growth and rapidly increasing food insecurity and malnutrition in many low-income developing countries, environmental deterioration, forgone trading opportunities, widespread conflict, and an unstable world for all.

CHAPTER SUMMARY

Global food supplies are adequate to provide everyone with minimum calorie requirements, and will remain so at least through the year 2020. Nevertheless, 792 million people in developing countries (18% of the developing world's population) are food insecure, mainly because they are too poor to afford all the food they need on a sustainable basis, and lack access to the resources to grow their food. Even larger numbers of people suffer from micronutrient deficiencies, with serious public health consequences. The number of undernourished people in developing countries has declined since 1980. Most hungry people live in the Asia-Pacific region. Sub-Saharan Africa is the region with the highest proportion of its population (34%) living in food insecurity. The number of food insecure Africans jumped nearly 50% from 1980 to 1997.

Of particular concern are the 160 million malnourished preschool children in developing countries. Malnutrition factors in 5 million child deaths annually, and those who survive face impaired physical and mental development. By 2020, if business-as-usual continues, child malnutrition will increase substantially in sub-Saharan Africa.

The causes of hunger include lack of jobs, lack of political power on the part of hungry people, violent conflict, discrimination (based on gender, age, race, and ethnicity), and environmental degradation. A number of global forces will affect food security over the next 20 years: globalization, falling aid, technology, global climate change, population growth and urbanization, and health issues.

Appropriate national and international policies are essential for achieving food security:

- Because 75% of food-insecure people live in rural areas, broad-based agricultural and rural development is central to food security. Small farmers in developing countries face many problems: low yields, production risks, and natural resource degradation resulting from low soil fertility, lack of access to plant nutrients, and acid, salinized, and waterlogged soils; inadequate infrastructure; land tenure biased against poor people; poorly functioning markets; and lack of access to credit and technical assistance.

- Policies must assure that agricultural and rural development is indeed broad-based, creating opportunities for small farmers and other poor people. At present, developing countries are underinvesting in agriculture. Aid donor countries must reverse the overall aid decline and the precipitous fall in aid to agriculture.

- Public investment in agricultural research that can improve small farmers' productivity in developing countries is especially important for food security. It must join all appropriate scientific tools and methods—including agroecology, conventional research methods, and genetic engineering—with better use of indigenous knowledge. The private sector is unlikely to undertake much research needed by small farmers in developing countries, because expected gains will not cover costs. Public agricultural R&D can lead to productivity gains, which in turn increase output, lower food prices, and, over time, increase nonfarm employment opportunities.

- Between 1970 and 1995, improvements in women's education and in per capita food availability accounted for nearly 70% of the reduction in child malnutrition. Hence, developing country governments and aid donors need to pay much greater attention to agriculture and education as public investment priorities.

- It is essential to include conflict prevention in food security and development efforts, and to link food security and economic development to relief programs.

- Policies that allow poor people to achieve economic security and give women access to reproductive health services are the best way to assure that birth rates will decline.

- High complementarity among agricultural development, poverty reduction, and environmental sustainability is more likely when agricultural development has a broad base and includes small- and medium-sized farms; is market-driven, participatory, and decentralized; and is driven by technological change that enhances productivity without degrading natural resources. Agricultural research must focus on sustainability, natural resource management, and resource-poor areas.

- To participate effectively in global agricultural trade negotiations, developing countries must have better access to industrialized countries' markets. Coalitions with higher-income countries may help improve their bargaining position. However, without appropriate domestic economic and agricultural policies, developing countries and poor people will not fully benefit from trade liberalization.

- Governments and international agencies must address child nutrition and micronutrient deficiencies as a key part of any comprehensive effort to achieve food security and reduce poverty.

Discussion Questions

1. If lack of access is the key food insecurity problem, rather than lack of food availability, then why is increased agricultural productivity such an important part of the solution?

2. How does agricultural research contribute to reducing hunger? Is it mainly by making more food available? Explain.

3. Compare and contrast food insecurity among urban dwellers and rural people.

4. Some people say that hunger persists in a world of plenty because of failures of "political will." What are some concrete examples of such failures? How can these best be addressed? Explain.

5. What mix of policies would be most effective in reducing child malnutrition? Why is good nutrition so crucial for preschoolers?

6. What sorts of policies can promote complementarities among agricultural development, poverty reduction, and environmental protection? How does each sort contribute?

7. Discuss the relationship between population growth and hunger. Is family planning the key to ending hunger, or will sustainable development lead to lower birth rates? Explain.

8. Given the broad-based growth in the U.S. economy in the 1990s, why has hunger persisted in the world's most affluent country?

9. In what ways are gender issues involved in food security?

10. Proponents of globalization say, "Trade, not aid" will end hunger and poverty, whereas critics call for "deglobalizing and re-localizing the food system." How can globalization contribute to food security, and what risks does it pose?

11. Discuss "foreign aid," also called "development assistance." Include discussion of the following: Who benefits from it? Is cash transferred to other countries or goods as well? Who produces those goods? What different routes might there be to channel such assistance?

Further Reading

Alston, J. M., P. G. Pardey, and V. H. Smith, eds. 1999. *Paying for Agricultural Productivity*. Baltimore and London: Johns Hopkins University Press for IFPRI.

Cohen, M. J., ed. 1994. *Causes of Hunger: Hunger 1995*. Silver Spring, MD: Bread for the World Institute.

Delgado, C. L., J. Hopkins, and V. Kelly. 1998, "Agricultural Growth Linkages in Sub-Saharan Africa." Research Report No. 107. Washington, DC: IFPRI.

Delgado, C. L., M. W. Rosegrant, H. Steinfeld, S. Ehui, and C. Courbois. 1999. "Livestock to 2020: The Next Food Revolution." 2020 Vision for Food, Agriculture, and the Environment Discussion Paper No. 28. Washington, DC: IFPRI.

Diaz-Bonilla, E., and S. Robinson, eds. 1999. "Getting Ready for the Millennium Round Trade Negotiations." 2020 Vision Focus No. 1. Washington, DC: IFPRI.

Flores, R., and S. Gillespie, eds. 2001. "Health and Nutrition: Emerging and Re-emerging Issues in Developing Countries." 2020 Vision Focus No. 5. Washington, DC: IFPRI.

Food and Agriculture Organization of the United Nations (FAO). 2000. *Agriculture Toward 2015/2030, Technical Interim Report*. Rome: FAO. Available at http://www.fao.org/es/ESD/at2015/toc-e.htm. Accessed 8/22/01.

Garrett, J. L., and M. T. Ruel. 2000. "Achieving Urban Food and Nutrition Security in the Developing World." 2020 Vision Focus No. 3. Washington, DC: IFPRI.

International Fund for Agricultural Development. 2001. *Rural Poverty Report 2001*. Oxford, UK: Oxford University Press.

Messer, E., M. J. Cohen, and J. D'Costa. 1998. "Food from Peace: Breaking the Links Between Conflict and Hunger." 2020 Vision for Food, Agriculture, and the Environment Discussion Paper No. 24. Washington, DC: IFPRI.

Pender, J., and P. B. R. Hazell. 2000. "Promoting Sustainable Development in Less Favored Areas." IFPRI 2020 Vision Focus No. 4. Washington, DC: IFPRI.

Pinstrup-Andersen, P., ed. 1993. *The Political Economy of Food and Nutrition Policies*. Baltimore and London: Johns Hopkins University Press for IFPRI.

Pinstrup-Andersen, P., R. Pandya-Lorch, and M. W. Rosegrant. 1999. *World Food Prospects: Critical Issues for the Early Twenty-first Century*. 2020 Vision Food Policy Report. Washington, DC: [IFPRI].

Riches, G., ed. 1997. *First World Hunger: Food Security and Welfare Politics*. New York: St. Martin's Press.

Sen, A. 1999. *Development as Freedom*. New York: Knopf.

Smith, L. C., and L. Haddad. 2000. "Overcoming Child Malnutrition in Developing Countries." 2020 Vision for Food, Agriculture, and the Environment Discussion Paper No. 30. Washington, DC: IFPRI.

U.N. Administrative Committee on Coordination/Subcommittee on Nutrition (ACC/SCN) and IFPRI. 2000. *4th Report on the World Nutrition Situation*. Geneva: ACC/SCN, and Washington, DC: IFPRI.

U.N. Development Programme. 2000. *Human Development Report 2000*. Oxford, UK: Oxford University Press.

Wood, S., K. Sebastian, and S. J. Scherr. 2001. *Pilot Analysis of Global Ecosystems: Agroecosystems*. Washington, DC: World Resources Institute and IFPRI.

World Bank. 2000. *World Development Report 2000/01*. Oxford, UK: Oxford University Press.

Yudelman, M., A. Ratta, and D. Nygaard. 1998. "Pest Management and Food Production: Looking to the Future." 2020 Vision for Food, Agriculture, and the Environment Discussion Paper No. 25. Washington, DC: IFPRI.

Note: Most IFPRI in-house publications may be downloaded from http://www.ifpri.org/.

Developing Food Production Systems in Sub-Saharan Africa

Jesse Machuka
International Institute of Tropical Agriculture

We noted in earlier chapters that global food production has outpaced population growth and that there are good reasons to be optimistic about the future (although solving food insecurity will be very difficult). Nevertheless, this agricultural revolution has bypassed Africa south of the Sahara desert (sub-Saharan Africa), which has the highest human population growth rate (about 3% per year), inherently poor soils, and the lowest per capita food production in the world. The new wheat and rice varieties of the Green Revolution were targeted to the agroecological zones and cropping systems of Asia and Latin America. About three quarters of Africans depend solely on income from the agricultural sector, which in turn constitutes 40% of the gross domestic product (GDP) of African nations. Although per capita food production in the world has increased by about 18% over the last 30 years, Africans are nutritionally worse off today than they were 30 years ago, because per capita food availability there has steadily declined. Food production shortfalls have led to widespread hunger and malnutrition aggravated by local conflicts. To arrest and reverse this trend, we must develop sustainable food production systems by developing agricultural systems that produce enough food for the present and are sustainable in the future. We devote this entire chapter to African agriculture so that you may better understand the situation in this part of the world.

5.1 Knowledge of agroecology is a prerequisite for developing appropriate technologies and farming systems that foster expansion of food production.

The huge landmass of sub-Saharan Africa (SSA) covers an area of approximately 2.27 billion ha, or 82% of Africa's total land area. To meet food production needs in this region through sustainable agriculture, we need to consider first its ecology, because the ecology defines the capacity and limitations of the natural environment in which crops must be

produced. Understanding the ecology is a prerequisite for developing appropriate technologies for small-scale rural farmers while at the same time preserving their environment.

The climate of any region is determined by amount and distribution of rainfall, temperature, latitude, and altitude. These factors and soil-forming processes (see Chapter 11) influence natural vegetation and crops that can be cultivated. Moreover, interactions among climate, soil-forming processes, natural vegetation, and agricultural activities affect the natural fertility of soils. In Africa, temperature and day length generally favor plant growth because of the continent's latitude (centered on the equator) and the limited extent of mountain ranges. Rainfall, however, is highly variable, with maximum rainfall occurring in the equatorial zone and minimum rainfall in the regions on either side of the equator. The Saharan-Sahelian zone, the Horn of Africa and the southwestern edge of the continent are very dry, whereas the Congo basin is extremely wet. Eastern Africa and parts of West Africa have a unique bimodal rainfall pattern that imposes special requirements for crop growth.

The relative lengths of the wet and dry seasons are a function of either one of two primary air masses, namely a dry continental air mass known as the *harmattan*, or the northeast trade winds, and the moist, maritime air mass—the southwest monsoon. Where precipitation is very low (<50 cm) and average temperatures high (>30°C), vegetation is sparse. This is typical of the arid and semiarid areas of Africa. The reliability of the "short" and "long rains" varies at relatively short distances, causing irregular droughts such as those experienced in East Africa and the Horn of Africa throughout 2000. At slightly higher rainfall (50–150 cm), the typical vegetation is savanna if the temperature is high (20–30°C) or grassland if the temperature is lower (5–18°C). The higher the rainfall, the greater the likelihood that the savanna or grassland will have woodland. Forests occur in places where annual rainfall is above 150 cm, provided the mean annual temperature is above 4°C. There are rainforests in areas with the highest rainfall, and seasonal forests if the rainfall does not exceed 250 cm.

**Figure 5.1
Agroecological zones in sub-Saharan Africa.** This map includes the agroecological zones in the central portion of Africa. Further mapping in southern Africa and the Horn of Africa are needed to complete the picture. *Source:* IITA GIS Database, Jagtap & Ibiyemi (1998).

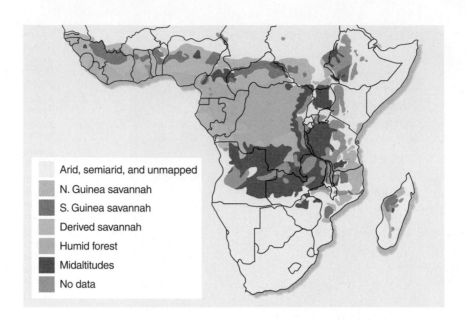

Arid, semiarid, and unmapped
N. Guinea savannah
S. Guinea savannah
Derived savannah
Humid forest
Midaltitudes
No data

As noted, one of the major hallmarks of SSA is its diversity in agroclimatic zones (farming and climatic zones). The agroecological conditions largely determine crop production and self-sufficient population capacity of a region. For this reason, researchers use agroecological characteristics to compare farming systems as well as crop production and to investigate alternatives. Perhaps the best agroecological zone system for SSA is that developed by the International Institute of Tropical Agriculture (IITA), based in Ibadan, Nigeria (**Figure 5.1**). The region has four major zones, namely arid and semiarid zone (ASZ), the moist savanna zone (MSZ), the humid forest zone (HFZ), and the mid-altitude zone (MAS). A fifth "zone" consists of inland valley systems/swamps (IVSs), which are distributed throughout the four major zones. The IVSs are relatively shallow and narrow valleys along upper reaches of streams and rivers. **Table 5.1** summarizes the characteristics of the five agroecological zones and their subzones, corresponding altitudinal zones, and their associated agricultural potential under rain-fed agriculture.

Table 5.1 Classification of land with rain-fed crop production potential according to agroecological zones in SSA

Altitude[a]	Agroecological Zone	Moisture Regime[b] (months)	Area (million ha)
Lowland	Dry semiarid (ASZ)	2.5–4	89
Lowland	Moist semiarid	4–6	179
Lowland	Moist savanna (MSZ)	4–6	116
Mid-altitude	Semiarid	5–6	85
Lowland	Moist savanna	6–7	100
	Moist savanna	7–9	163
Mid-altitude	Savanna and woodlands (MAS)	6–9	124
Mid-altitude	Savanna and woodlands	9–12	23
Lowland	Humid forest (HFZ)	9–12	171
Lowland	Inland valleys (IVS)	9–12	130

[a]*Lowland:* <800 m; *Mid-altitude:* 800–1200 m; *Highland:* >1200 m.
[b]Moisture regime is the number of months that moisture permits crop growth. This period varies according to rainfall.

5.2 A clear understanding of traditional farming systems is essential for developing sustainable crop production systems.

During the 1970s and 1980s, farmers greatly increased food production in Asia and Latin America by applying Green Revolution technologies (see Chapters 2 and 14). This crop-centered approach used genetically improved strains of specific crops, especially wheat and rice. In contrast, the SSA agricultural development movement that began during the late 1970s has taken place in the broad context of "farming systems" research. This approach takes into account the diversity of existing farming systems and the agroecological and socioeconomic circumstances that characterize them. Only when researchers clearly understand the biological and chemical processes of traditional farming systems can they develop alternative farming systems based on traditional practices. What are the general principles and key features of traditional SSA modes of cultivation? What are the strengths and weaknesses of these modes? What pressures have been brought to bear down on them, and with what consequences? Which alternative farming systems and technologies hold promise for a successful agricultural transformation to ensure increased sustainable food production in SSA?

Traditional Land Use Practices and Farming Systems. Until recently, the rural SSA population adapted to the varying agroecological conditions just outlined (see Section 5.1) by evolving several different land use and farming systems. The relatively low population densities and abundant land kept land use at sustainable production levels. Among many tribes, land tenure relied on inheritance as a means of securing agricultural land. To date, most SSA farmers still hold their land under customary tenure systems. Such systems often split up farms into increasingly smaller and scattered units. The system also makes it difficult for investors (tribal or nontribal) to procure enough land for large-scale farming.

Although variations exist, the general pattern of traditional African farming systems can probably be best depicted as a series of concentric rings of cultivation, as shown in **Figure 5.2**. In the center of this ring is the homestead, consisting of houses and gardens. Farmers in villages use fields close to their compounds to grow crops year-round, hence these farms are often called *compound farms*. In this zone, a wide range of crops are grown for different uses by the family such as production of cooking oil, fiber, drugs, seasoning, firewood, and fresh vegetables. Women control the activities in this compound zone. They keep the soil very fertile by frequently adding organic matter, mainly household refuse and de-

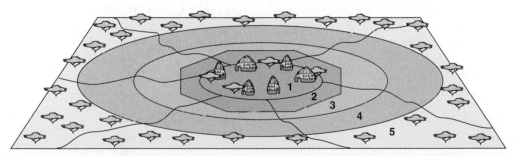

Figure 5.2 Typical spatial organization of land use in G'ayene, Senegal: (1) houses and gardens; (2) permanent cultivation; (3) semipermanent piece of land left to rest after a cultivation; (4) intensive cultivation; (5) bush and extensive shifting cultivation. *Source:* IITA Publications, 1993.

composed animal waste. Farming intensity decreases as you move away from the home-stead to more distant fields. This area is continuously farmed. Semipermanent cultivation, shifting cultivation, and other cropping systems are carried out mainly by men, who have the responsibility of clearing bushes and forests for cultivating starchy staple crops. In these remote fields or in sparsely populated areas, people still use the more traditional farming practices, namely shifting cultivation and bush fallows, to which we now turn for more detailed descriptions.

Shifting Cultivation. SSA agriculture has evolved to its present stage from shifting cultivation, which in its true form involves the gradual shifting of entire communities from one piece of land to another. The communities grow a variety of crops for a few seasons, after which they let the land revert to its natural vegetation until it is ready for cropping again. Farmers have mostly carried on this mode of cultivation in the humid forest and moist savanna zones, especially in parts of West and Central Africa. In the tropical forest regions, soils are strongly weathered (see Chapter 11) and acidic, and have rather low levels of plant nutrients, so shifting cultivation allows forest regrowth during the fallow period, regenerating soil fertility. In these regions, people grow cassava as the main staple, but plantain-/banana-based systems, as well as yam-, rice- and coffee-/cocoa-based cropping systems, are also prominent. In the mid-altitude and highland savanna woodlands, farmers practice shifting cultivation using an array of crops such as cassava, maize, sorghum, millet, sweet potatoes, and bambara groundnut. When there is limited natural

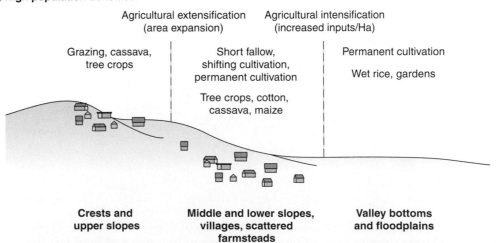

(a) Low population densities:

Unused or grazing, hunting, timber land

Long fallow, shifting cultivation

Upland rice, cassava, maize, yam

Unused or dry season gardening

Fishing, hunting

(b) High population densities:

Agricultural extensification (area expansion)

Agricultural intensification (increased inputs/Ha)

Grazing, cassava, tree crops

Short fallow, shifting cultivation, permanent cultivation

Tree crops, cotton, cassava, maize

Permanent cultivation

Wet rice, gardens

Crests and upper slopes

Middle and lower slopes, villages, scattered farmsteads

Valley bottoms and floodplains

Figure 5.3 Stylized representation of the relationship between land use intensity and landscape in African farming systems. *Source:* Modified from G. Sambrook (1992), IITA Publications.

vegetation—as in Zambia, for example—farmers must clear 5 to 20 times the land area that will be farmed. Wood and plants from the large area are piled up for burning, once they have dried, on the smaller area to be farmed. When the trees are burned, some of the nutrients contained in the vegetation from a large area are thus deposited into a small area. Unfortunately, nitrogen and sulfur are lost and escape as gases during the burning. In shifting cultivation, farmers spread the ash on the land, where it supplies nutrients to the crops. Research has shown that shifting cultivation is a sustainable use of land at low population densities *if* the value of the virgin forest cover is disregarded. Shifting cultivation is fast disappearing and is now rarely used in Africa, because increasing population pressure, urbanization, and commercialization of farming have forced most farmers to settle down. **Figure 5.3** depicts the evolution of traditional African farming systems under low through high human population pressures.

Bush Fallow Cultivation. As noted, true shifting cultivation is now rare in Africa and most of what remains can best be described as bush fallow cultivation. Fallow land is a piece of land allowed to lie idle after a period of cropping. Farmers cultivate land using very simple tools. During the fallow period, a variety of vegetation grows on the land, such as herbaceous plants, shrubs, and creepers. Soil fertility is restored during the fallow period through the gradual decay of organic matter provided by secondary growth of the natural vegetation and the nitrogen-fixing activities of symbiotic bacteria in legume roots. The fallow period also suppresses weeds and reduces pest incidence (**Figure 5.4**).

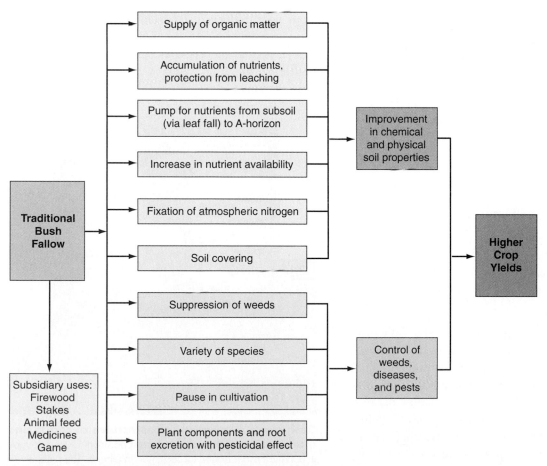

Figure 5.4 Functions of traditional bush and fallow systems. *Source:* Modified from D. Prinz (1986), Increasing the productivity of small holder farming systems by introduction of planted fallows, *Plant Research & Development* 24:31–56.

Box 5.1

Both Population and Market Forces Drive Land Use Patterns in Africa

Population increases in the developing countries of Africa require increases in crop production, which can come about in two ways: by plowing more land and by intensifying crop production on land already cultivated. As more people take jobs in the cities, and as trade develops with other countries, market demand for food expands. So what drives crop production and land use, the market or the population? In West Africa, two thirds of agricultural land use is driven by population, and one third is driven by the market. Research by African scientists shows that there are four scenarios for land use. When the scientists disaggregated the data and looked at the expansion of arable land and the intensification of existing production, they found the results shown in the pie chart: Over half the land use is population-driven intensification (54%), and about one third is market-driven intensification (31%); only 12% is population-driven expansion, and even less (3%) is market-driven expansion. What is remarkable is that these different types of land use have very different agricultural characteristics as well.

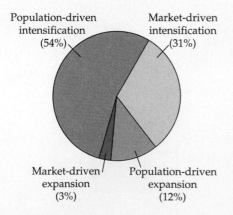

Population-driven intensification (54%)

Market-driven intensification (31%)

Market-driven expansion (3%)

Population-driven expansion (12%)

Population-driven intensification entails typically subsistence farming by low-resource farmers who have no money to purchase inputs (fertilizer, for example) and whose farming depletes and often erodes the soil. Continuous cropping and short, or no, fallow periods cause a buildup of weeds and pests. This type of farming depletes the resource base and is not sustainable. A similar scenario prevails with population-driven expansion. All these farmers depend on natural resources to sustain agriculture and prefer technologies that economize on purchased inputs and labor. The new technologies developed for these farmers must obey these rules and may even have to be tailored specifically to men or to women, because of distinct gender roles in crop production.

Market-driven intensification and expansion entail the production of a profitable cash crop, using purchased inputs. "Cash crop" in this context does not mean a crop to be sold overseas, but can also be food crops (such as cowpea, maize, or cassava) that are sold in the cities or exported to other African countries (see figure). This type of land use depends on good transport, access to wholesale markets, and government policies and technologies that ensure sustained production and a reasonable return to the grower. The farmers readily adopt new technologies that demand purchased inputs and labor, because of the gains that will accrue from selling their farm produce.

This research makes it clear that new technologies must fit the problems to be solved and the constituency to be served. This tailoring is also relevant when new crops are introduced. The overall challenge is to ensure that improved technologies offer some visible gain to the farmer, such as increased income, more food, or better health and nutrition, while at the same time ensuring conservation of land resources.

Cash crops require an infrastructure of roads and markets. This truck loaded with cowpeas from Ouagadougou in Burkina Faso is bound for Techiman in Ghana. The picture was taken at the Ghana–Burkina Faso Border Post at Paga. *Source:* Courtesy Augustine Langyintuo.

The number of years the land lies fallow is variable, depending on the agroecosystem, but typically, the duration of the fallow depends on the following factors:

1. The quality, variety, and abundance of plants found during the fallow period; for example, legume plants restore soil fertility more efficiently than other species of crops, hence require shorter fallows

2. The quantity of seeds, tree stumps, and roots of the fallow plants that remained alive during cultivation; a long period of cultivation is usually followed by a long fallow period

3. The amount of farmland available

4. The soil type, whether sand, loam, or clay

5. The cultural methods used in cultivation; degraded, nutrient-poor, and eroded soils are more difficult to regenerate than is well-nourished soil

Few communities practice bush fallow exclusively and usually use the system in association with both sole and mixed cropping.

SSA farmers have developed the current, dynamic farming methods to adapt to changes in farming objectives brought about by several factors. These include population and market pressures, introduction of new crops and associated methods of production, readiness to adopt scientifically proven alternatives, and technologies of the Green Revolution (**Box 5.1**). Despite the scarcity of arable land, mixed cropping under the bush fallow system and its variant improved versions (described in Section 5.5) is still the predominant method of cultivation. What crops are produced by these cropping systems, and how are the systems adapted to the different agroecological zones described earlier?

5.3 Intercropping is the major mode of crop production in sub-Saharan Africa.

Except for very few regions (particularly in eastern and southern Africa) where the colonial powers established plantations of single crops, intercropping is by far the most dominant SSA crop production pattern. Intercropping means that more than one crop is grown on the same piece of land. The crops can be either in rows (as in alley farming, see below) or less regularly placed on the field (**Figure 5.5**). Row intercropping is uncommon in SSA, except where farmers use tractors or animal traction. Research suggests that intercropping has several advantages over sole cropping. Some of these are

1. More efficient use of land resources, such as intercropping low-vigor cassava varieties with early-maturing maize or grain legume varieties

2. Higher labor productivity

3. Lower risk of crop failures resulting from stress, such as drought

4. Better management of weeds and insect pests

New or alternative technologies and farming systems for sub-Saharan agriculture should aim to improve indigenous intercropping practices, as opposed to replacing them with modes of cultivation from temperate zones. Over the past three decades agricultural scientists and farmers have tried to develop improved, sustainable cropping systems under prevailing population and market pressures to suit the different agroecological zones. Before we examine cropping patterns according to agroecology, it is useful to summarize the major SSA crops. As shown in **Figure 5.6**, cereals and coarse grains cover the

Figure 5.5 Inter-cropping is the rule in Africa. In this plot peanuts, which are nitrogen fixing legumes, are planted between mounds of yam. *Source:* Photo by P. Cenini, FAO.

largest arable land area, followed in descending order by oil crops, root and tuber crops, pulses, fruits, bananas and plantains, and fibers and vegetables. Maize, sorghum, millet, and rice are the major cereals. Cassava, yams, sweet potatoes, and taro (cocoyam) are the major root and tuber crops. (Note that sweet potato, taro, and pumpkin leaves are also a major source of soup greens and hence vitamins.) Cowpeas and dry beans constitute the bulk of grain legumes. Nigeria, the most populous African country, is the leading producer of staple cereals and roots and tubers, and Uganda alone produces nearly half of all the plantains and bananas in SSA.

In the *humid forest zone*, the predominant cropping systems are cassava-based (42% of agricultural land), rice-based (16%), coffee-/cocoa-based (21%), and banana-/plantain-based (9%). In this zone, cassava is the major food staple, and in many cases it is also the main cash crop. Yam and maize are also important; farmers increasingly intercrop maize with cassava to meet specific food security needs early in the growing season. In the *moist savanna zone*, the cropping system is dominated by short-to-medium annual crops such as maize, cowpea, sorghum, millet, and cotton in the drier areas. Some of these crops are also important in wetter areas, in addition to long-cycle annuals such as yam and semiperennials such as cassava. In West and Central Africa, the predominant cropping systems are based on cassava (28%), yam (14%), maize (14%), sorghum (12%), and cotton (9%). The *mid- and high-altitude savannas and woodlands* include the largest maize-growing areas in SSA, particularly in highland areas of eastern and southern African countries such as

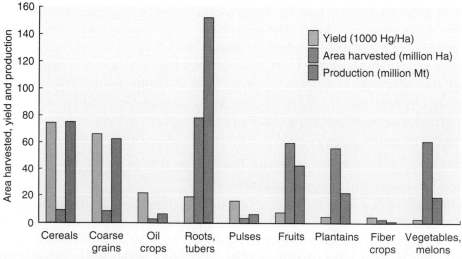

Figure 5.6 Yield and production estimates per agricultural land area harvested for major crop categories in sub-Saharan Africa. Note that cereals and coarse grains (sorghum) occupy the largest harvested areas, but roots and tubers have the highest yield per hectare. Fruits, plantains, and vegetables also have high yields. Protein-rich pulses (legumes) have low yield in a small area. *Source:* Food and Agriculture Organization of the United Nations, 1998.

Kenya and Zimbabwe. Diversified cropping by smallholder farmers is gaining ground in the savanna zones, involving intercropping of maize with other stabilizing crops such as cassava, sweet potato, cowpea, beans, bananas, as well as soybeans for oil production. In regions of Uganda, Rwanda, Burundi, and the Bukoba/Kilimanjaro highland regions of Tanzania, bananas are the staple food for about 50% of the population (Figure 5. 6 and **Table 5.2**). To supplement daily staple food needs, farmers also grow maize, sweet potatoes, cowpeas, potatoes, dry beans, squash, and other vegetables. Trees and perennials feature more extensively in East African than in West African mixed cropping systems. For example, on the East African coast, particularly the coastal districts of Kenya, farmers may intercrop cashew with coconut, banana, citrus, mango, pawpaw (papaya), and cassava.

Although farmers intercrop a large number of crops in *inland valleys*, paddy rice is the most common (sole) crop, particularly during the rainy season when the valleys are flooded. In most areas, cropping is only possible in the valleys during the dry season. At such times, farmers grow several annual and perennial crops, including cassava, sweet potato, maize, grain legumes, sugar cane, and vegetables. They grow perennials such as bananas and plantains mainly on the valley fringes. Despite being the most fertile and productive areas, only 10–25% of SSA valleys are cropped, presumably because of problems related to land settlement, disease prevalence, flooding, and clearing of the vegetation.

The *arid and semiarid regions* of sub-Saharan Africa have three dominant traditional farming systems: agro-silvi-cultural, agro-silvi-pastoral, and silvi-pastoral. As the name suggests, agro-silvi-cultural systems involve intercropping of the natural vegetation of trees and shrubs with food crops (such as maize and millet) and cash crops (such as groundnuts and cotton). This practice is common in the southern parts of the Sahel. Agro-silvi-pastoral systems place greater emphasis on crop–livestock integration and use of trees that produce browse and fodder for cattle and small ruminants. In the silvi-pastoral system, migratory pastoralists are the key players. The practice is common in the northern fringes of the arid and semiarid zone, where intrusive pastoralists move into the zone in search of water, crop residues, and forage (trees, shrubs, and grass). In this process, they also practice a little cropping.

The foregoing survey of food production systems has highlighted some general characteristics, rather than constraints that hinder countries in SSA from realizing high potentials. Besides sociopolitical issues, the other constraints are abiotic (climate and soil-related), biological, technological, and socioeconomic. These are discussed in the following section, with emphasis on abiotic and biotic constraints.

Table 5.2	Annual production of the major SSA food crops in millions of metric tons
Cereals	
Maize	27
Sorghum	18
Millet	13.7
Paddy rice	11.4
Wheat	2.2
Other	3.15
Roots and Tubers	
Cassava	92
Yams	37
Sweet potatoes	9.2
Taro (cocoyam)	7
Potatoes (white)	4.2
Legumes	
Cowpeas	3.1
Beans (*Phaseolus*)	1.8
Broad beans	0.4
Other	0.8
Plantains	22.6
Bananas	6.3

Source: Data from the Food and Agriculture Organization of the United Nations, 2000.

5.4 Traditional and modern farming and cropping practices face serious constraints that hinder increased food production by small-scale farmers.

As noted, traditional agriculture has failed to adequately feed the African population. Nor have Green Revolution technologies based on high-yielding varieties and purchased inputs such as fertilizers, tractors, pesticides, and irrigation been successful. We begin to

Table 5.3	SSA land use
Land Area	**Area (million ha)**
Total land area	2,270
Agricultural land	890
Permanent pastures	730
Arable land	160
Nonpermanent crops	140
Permanent crops	20

Source: Data from the Food and Agriculture Organization, 1998.

explain these failures by first broadly examining the figures pertaining to SSA land use (**Table 5.3**). Typically, farms are very small, as shown in **Table 5.4** for Ghana. In many countries, for more than 80% of farmers the average farm size is less than 1 hectare. With mounting population pressures and given existing land tenure laws, more and more landholdings become fragmented and used so intensively for agriculture that soil degrades.

Climatic Constraints. Rainfall in Africa tends to be unreliable in terms of onset, duration, and intensity. Unpredictable periods of drought, floods, and storms are also frequent, with disastrous consequences similar to the 1998–2000 eastern and southern African El Niño floods. These floods destroyed livestock and subsistence crops such as rice, cassava, and maize, leaving thousands of rural farmers in countries such as Mozambique homeless and dependent on food and emergency aid, mostly from the World Food Program. Prolonged droughts frequently exacerbate soil degradation, particularly within the arid and semiarid zone. Many regions in Africa, especially in this zone and parts of the moist savanna zone, experience uniformly high temperatures (sometimes over 40°C) with relatively little seasonal variation. Such temperatures are detrimental to crops and livestock introduced into these regions. High temperatures also cause high rates of decomposition of organic matter and evapotranspiration. These in turn deplete soil moisture and lower the groundwater table. Nearly 50% of the water available per capita is concentrated in humid central Africa, compared with 4% in the Sahel. Because of poor irrigation, water harvesting, and conservation resources, several SSA countries are classified as water scarce, including Burundi, Kenya, Malawi, Rwanda, the Sudan, and Somalia. Such countries do not have enough water for crops, livestock, and humans. In SSA, agriculture alone accounts for approximately 85% of overall water use.

Soil Degradation. Soil degradation, discussed in greater detail in Chapter 11, reduces the capacity of soils to produce crops. Researchers classify approximately 26% of African soils as strongly degraded, meaning the land is unlikely ever to be reclaimed. The main causes of soil degradation are overgrazing (49%), cropping (24%), deforestation (14%),

Table 5.4	Size and distribution of farm holdings in Ghana				
Size of Holdings (acres)*	**Number of Holdings**		**Percentages of Total**		
0.0–1.9	246,100	**745,000**	30.6		**92.5**
2.0–3.9	194,200		24.1		
4.0–5.9	105,200		13.1		
6.0–7.9	71,800		8.9		
8.0–9.9	42,000		5.2		
10.0–14.9	55,000		6.8		
15.0–19.9	31,600		3.9		
20.0–29.9	27,200	**45,100**	3.4		**5.6**
30.0–49.9	17,900		3.2		
>50.0	14,000		1.8		
TOTAL	805,000		100		

*1 acre = 0.4 hectare.

Source: A. Karem (1981), Small scale versus large scale farming: Which way? *Agricultural Bulletin* (Ghana–German Agricultural Development Programme, Tamale, Ghana) 30:8–16.

Figure 5.7 Devastating erosion caused by poor land management practices in the lowlands of Lesotho in southern Africa. Erosion is often caused by overgrazing, which removes the ground cover, exposing the soil to the forces of wind and rain. *Source:* Photo by P. Lowrey, FAO.

and overexploitation of vegetative cover (13%). In highland areas, such as parts of Ethiopia, Kenya, Uganda, and Madagascar, soil erosion is particularly acute because of the steep slopes. In fragile soils, water erosion leads to loss of topsoil, to run-off, and to gully formation (**Figure 5.7**). In the arid and semiarid zone, wind action—the major agent of erosion—is usually aggravated by human activities that reduce vegetation cover, such as overgrazing and plowing to prepare soils for dryland crops. Water (46%) and wind (36%) erosion together account for 82% of soil degradation in Africa.

Chemical processes, which contribute about 9% to soil degradation, involve salinization, acidification, and gradual loss of nutrients (nutrient mining) or of organic matter. Nutrient mining mostly takes place on hilltops, plateaus, and upper slopes. The drier forest and sub-humid moist savanna zone, especially in Central Africa, have highly erodable soils with little available water, low to medium fertility, and poor phosphate fixation (ability to retain phosphate fertilizer farmers have applied to the soil). Soils of the humid forest zone are strongly acidic, with high aluminum toxicity, low fertility, and medium phosphate fixation. Characteristic soils within the arid and semiarid zones are very fragile and infertile because they are low in nitrogen, phosphates, and organic matter, and cannot bind nutrients well. These soils also retain water poorly and form hard clay pans. In the Sahel, rampant chemical deterioration arising from nutrient depletion has occurred on 25 million ha (compared to 2.5 and 1.5 million ha degraded by salinization and acidification, respectively).

Physical processes that cause soil degradation include compaction, crusting, sealing, and waterlogging. Soil compaction can be particularly acute in SSA during the dry season when soil is heavily trampled by livestock being moved from place to place in search of water and food. Under these circumstances, the little water that seeps into the soil is lost again through evaporation. Moreover, rainwater fails to enter the narrow pores of strongly compacted soils, leading to flooding and topsoil erosion during the wet season. In many parts of sub-Saharan Africa, land once sustainable under traditional farming practices has become unsustainable because nutrients are not replenished by organic or mineral fertilizers. In intensely cultivated regions, poor nutrient husbandry, diminished activities of nitrogen-fixing bacteria, mycorrhizal phosphorus uptake, and reduced soil structure have all exhausted the soil, so that yields have stagnated or dropped.

Figure 5.8 When Striga strikes, farms become battlefields. Here, the struggle between the weed *Striga* and a maize crop shows the victor: *Striga* in full bloom. The farmer will have to burn the field to try to get rid of *Striga*. *Source:* Courtesy of Alan Watson, McGill University.

Biological Constraints. Biological constraints on increased food production in SSA include weeds (mainly parasitic plants such as *Striga, Alectra,* and *Imperata*), diseases, and field and storage pests. These biotic stresses seriously threaten cereal and grain legume production. For example, according to researchers at the International Centre of Insect Physiology and Ecology (ICIPE) based in Nairobi, stemborers and *Striga* (**Figure 5.8**) account for losses of 15–40% and 10–20% in the eastern and southern African regions, respectively. When the two pests attack simultaneously, farmers can lose their entire crop. More intense land use and monocropping has increased the population of weeds, diseases, and pests. These forces, along with the reduction of prolonged rotational fallow systems, have led to the loss of biodiversity in indigenous food crops and landraces.

Socioeconomic and Technological Constraints. Many socioeconomic problems hinder food production in SSA. These include rapid population growth (3% annually), unfavorable land tenure systems, lack of credit (for example, to purchase fertilizer, tractors, or other inputs needed to raise production), decline in commodity prices, poor transport infrastructure, crude farming tools, inadequate postharvest and processing methods, and inadequate funding and policy frameworks for agricultural research. Desertification (caused by increasing human and animal pressures on land) and the emergence of "cash" cropping now very acutely threaten the natural resource balance in the arid and semiarid ecosystems. Along forest margins, smallholder "slash and burn" shifting agriculture is the main cause of forest degradation. The transition from completely forested through intermediate stages of forest depletion, to shrub and short fallow almost always entails soil degradation and loss of biodiversity.

New technologies and cropping systems have been developed and could be tested, if capacity to do so existed. A major problem is inadequate human resources and institutional capacity for formulating and implementing effective agricultural policies and strategies. National agricultural research centers are weak throughout SSA and ineffective in developing appropriate technologies for sustainable agriculture based on traditional knowledge of crop genetic resources and farming systems. Traditional and emerging technologies have also been very poorly integrated.

Although this chapter will not address sociopolitical constraints, note that environmental degradation often reflects desperate competition for access to resources under unstable political and social conditions. It is almost impossible to progress toward sustainable food production unless these constraints are removed, as shown by ongoing civil strife in countries of the Great Lakes region such as the Democratic Republic of Congo, Burundi, and Rwanda. A good but sad example that illustrates Africa's food production constraints in a nutshell is based on the success story pertaining to sweet potato cultivation in Rwanda

Box 5.2

Sweet Potato Cultivation in Rwanda: Constraints and Opportunities for Africa's Agriculture in Microcosm

Root and tuber crops (R&T) such as cassava, sweet potato, yams, and cocoyams are the essential subsistence staples for much of sub-Saharan Africa. These crops fit well into traditional agricultural systems and are generally more tolerant of harsh environments characterized by drought, poor soils, pests, diseases, and weeds. R&T, long considered "the poor person's crops," were not the focus of any research projects. In Rwanda, research efforts to build better-yielding R&T and indigenous capacity have faced different cultural, technical, and political challenges. These challenges illustrate the importance of having the correct "environment" for available technologies to become useful in enhancing sustainable food production in sub-Saharan Africa.

Rwanda, the "country of a thousand hills," is a small, very densely populated nation in Central Africa. Most of the 5.6 million Rwandans are subsistence farmers, and they are already farming all the suitable land. Population pressures are pushing agriculture into less suitable land such as the wet bottomlands and the semiarid eastern lowlands. The farms are very small—1 ha on average—yet a single farm may grow 20 crops. Food crops are grown on 95% of the arable land, and root crops are especially important because cassava and sweet potatoes are the major staples for 70% of Rwandans.

Although sweet potatoes, introduced in the 18th century, are a relatively new staple, Rwandans now eat more sweet potatoes per person than anyone else. They grow sweet potatoes high up on the slopes during the wet season and low down in valley bottoms during the dry season. Fields are prepared by hoe. Most farmers plant sweet potatoes on ridges, beds, and mounds for formation of tubers. They don't use fertilizers, and rarely weed (only once if at all). A Rwandan proverb says, "Only those with nothing to do weed sweet potatoes"! About 80% of the crop is eaten directly, the remainder is marketed. Major constraints include *Alternaria* leaf blight, a virus complex, and the sweet potato weevil (*Cylas* spp.), which tunnels into tubers, paving the way for fungal and bacterial rot underground and in storage. To address these constraints, a sweet potato improvement program involving the Institut des Sciences Agronomiques du Rwanda (ISAR), working in collaboration with Canada's International Development Research Center (IDRC) and IITA, was started in 1983. The breeding program focused, through farmer participation approaches, on developing early-maturing and disease- and pest-resistant varieties, as well as on processing, marketing, and taste characteristics. This research produced superior varieties of sweet potato. At least 10 new varieties matured within 4–5 months, instead of 5–8 months in older types, thus allowing two crops to be grown per year. Sadly, the political events in 1994 that led to the genocide of more than a million Rwandans, with at least another million displaced from their homes, and the resulting moral and social trauma, have left this small country shattered. The clock for Rwandan agriculture, as for almost everything else, was reversed, and now desperately needs a fresh restart.

(**Box 5.2**). Sadly for Africa, the constraints just discussed are preventing people from moving forward even though there are more options and opportunities for development of improved farming and cropping practices than ever before.

5.5 The potential for sustainable food production in sub-Saharan Africa can be realized if people implement existing opportunities for improving current production systems.

Estimates by the FAO of the United Nations indicate that agricultural production in SSA must grow at an average rate of 4–5% per annum if Africans are to close the food gap in the next 10–15 years. Clearly, this very optimistic projection can serve only as a target for African governments, scientists, and farmers. But how will this growth be achieved?

According to some experts, to close the gap African countries need only make heavy financial investments in existing Green Revolution technologies, such as irrigation, fertilizer, and development of rural infrastructures, especially roads. Others, perhaps more realistically, argue that African governments must focus on developing a broad-based strategy involving investments in agricultural research and extension systems to generate and transfer sustainable technologies that may be lower yielding than Green Revolution technologies but that nevertheless will close the food gap. Experience and research in Africa over the last four decades tend to favor the latter view. We therefore describe next some of the technologies that hold promise for increasing food production in SSA.

Alley Farming as an Alternative to Slash and Burn. Alley cropping involves cultivating food crops between hedgerows of multipurpose trees and shrubs. This dual-purpose agroforestry technology sustains soil fertility and offers an alternative to the slash-and-burn method. Farmers plant trees or shrubs (usually legumes) in rows and plant food crops in the "alleys" between trees. They periodically prune the hedgerows and use the cut materials to mulch the strips between rows. Leguminous trees are often used in alley cropping (**Figure 5.9**). Legumes have deep roots that help draw up soil nutrients from deeper soil layers, and they harbor symbiotic nitrogen-fixing bacteria in their root nodules. Farmers can use the nitrogen-rich leaves of the trees as fertilizer or as livestock feed. This technology, initially conceived by agronomists in colonial times and further developed at IITA, has been adopted in African countries where conditions are appropriate: villages characterized by high land-use pressure, decline in soil fertility, soil erosion problems and firewood, and animal fodder scarcity. Alley farming also reduces the amount of land that must be cleared for farming, thus helping preserve natural vegetation. Recent surveys conducted in Nigeria, Benin, and Cameroon show that farmers who adopted the system are now beginning to modify it by planting fruit and commercial trees such as ba-

Figure 5.9 Alley cropping sustains soil fertility by imitating the self-renewal process of the natural forest. In this plot, double rows of pigeon pea bushes alternate with six rows of maize. The farmer is holding the pigeon pea pods. Pigeon pea is a drought-resistant legume that fixes nitrogen and enriches the soil. *Source:* Photo by A. Conti, FAO.

nana, coffee, and cocoa, and by including fallow periods. Factors that have hindered more widespread adoption so far include high labor demands, lack of knowledge about management, and scarcity of stocks of hedgerow legume trees and shrubs.

Alley farming is also practiced in parts of the humid and subhumid tropics of Southeast Asia, the South Pacific, and Latin America where agroecological problems are similar. For the future, it is important for researchers to integrate modifications by farmers, to make alley cropping flexible and adaptable for the farmers. Researchers should target modified systems to areas where farmers need incentives for changes in land use, such as areas where land or fuelwood are scarce. The already functional Alley Farming Network for Tropical Africa can continue to promote and disseminate the technology.

Improved Fallow Systems. In improved fallow systems, farmers grow specific plants during the cropless fallow, for a period of one to several years, rather than let the system lapse into as unmanaged bush fallow. This new approach is a superior way of restoring soil fertility. Traditional bush fallows take at least 10 years to build up organic matter and fertility sufficient for 2 years of cropping. Hence a key objective of improved or "planted" fallows is to achieve the aims of the traditional bush fallow in a shorter time. A good fallow plant species must

- Grow and close canopy quickly so as to suppress weeds and minimize erosion
- Produce high biomass
- Pump nutrients from deep soil layers to the topsoil
- Be removed fairly easily
- Occupy a different niche from the cultivated crop
- Resist accidental dispersion
- Lend itself to multiple uses such as firewood, fruit, feeds, drugs, stakes to support field crop plants, tool handles, pestles and mortars, building materials, hedging plants, and dyes

In collaboration with farmers, researchers have studied improved fallow systems at the International Centre for Research in Agroforestry (ICRAF) in Nairobi, at the IITA, and at national agricultural research systems (NARSs). Various kinds of improved fallow systems are recognized, depending on what kind of plants farmers grow or how they plant the fallows. These systems include tree/shrub, herbaceous, successive, and simultaneous fallows.

When alley cropping incorporates a fallow period, it is considered an *improved tree fallow system*. Usually farmers allow two years of leguminous fallow. Then they clear land, harvest firewood, burn twigs, and use the leaves for mulch. In the humid forest zone, farmers prefer legumes that flower all year, because they permit bee-keeping. Acceptance of the tree-/shrub-improved fallow in SSA will depend on its profitability, land availability, and establishment costs. Where land is scarce, as in western Kenya, farmers will find it difficult to adopt these systems unless the benefits are overwhelming.

Improved herbaceous fallow systems are similar to improved tree/shrub fallows, except that they use herbaceous plants, again mostly legumes. In West and Central Africa, fallows based on velvet beans (*Mucuna pruriens*) (**Figure 5.10**) have become very popular because this legume can restore soil fertility, reduce insect pests, and smother speargrass, the noxious weed that has seriously impoverished soils in West and Central Africa. Basically, farmers cultivate the beans as ground cover or live mulch.

Figure 5.10 *Mucuna*, **a tropical legume used as a cover crop for soil improvement.** *Mucuna* is also called velvet bean because it is so hairy (leaf, stem, and pod hairs). The reddish colored one on the left causes itching because of toxic substances produced by the hairs. It is not used by the farmer. The green one on the right does not cause itching and is used by the farmers. Velvet bean (*Mucuna*) is also used in Latin America and other areas as a cover crop. *Source:* Photo courtesy of M. Flores Barahoma and the Centro Internacional de Información Sobre Cultivos de Cobertura.

Normally, farmers plant velvet beans 30 days after the food crop (which is mainly maize). After the maize harvest, the beans remain in the field until the end of the cropping season. In the following dry season the plants die, allowing the farmer to plant maize once again. The key to the success in adopting this technology in the early 1990s was the on-farm (farmers participating) experimentation in countries such as Benin and Rwanda. Farmers greatly helped refine this technology, showing that their participation was invaluable in bringing forth the benefits they were seeking.

In *successive fallow systems,* farmers plant fallow plants and food crops consecutively on the same piece of land, and the fallow period usually is shorter than the cropping period. For example, a three-year successive fallow system may have two years of cropping and one fallow year. Rwandan farmers have obtained significant yield increases by growing legumes in the fallow year.

In *simultaneous fallows,* farmers plant fallow plants around crops in the same field either before, during, or after planting of crops. Alley cropping, described earlier, is an example of a simultaneous fallow. Another version involves growing nonfood legumes between rows of food crops shortly after planting the food crop. Farmers do not let the tree or shrub legumes grow higher than the crop until after the crop fruiting phase. They cut the legume leaves during the dry season, work them into the soil, and use stems as firewood. In one example from Nigeria, *Sesbania rostrata,* an annual legume with nitrogen-fixing nodules on its stems, increased rice yields by up to 50% when used as mulch. Simultaneous cropping with another legume (*Pueraria phaseolides*) as a cover crop was also very successful in preventing or retarding soil degradation without fertilizing. Nitrogen fixation, nutrient cycling from the subsoil, and stimulation of the activity of soil biota all improve soil fertility. *Pueraria* is drought tolerant and very effectively controls soil erosion and weeds by overgrowing weeds at the onset of the wet season. Farmers plant *Pueraria* seeds at the same time as the cassava/maize intercrop. Later, they cut *Pueraria* down to 30 cm (1 foot) in height. Because of its slow growth, it does not compete with the food crops. The second year involves an entire fallow of *Pueraria* regrowth. Simultaneous cropping with *Pueraria* leads to higher cassava and maize production than do sole cassava/maize intercrops. The phosphate that maize and cassava take up is compensated for by phosphate that *Pueraria* takes up from the subsoil and brings to the surface layers. Such rotations are now ready for trials in farmers' fields throughout SSA.

5.6 Multistrata systems and integrated crop-livestock-fish systems are two other sustainable food production systems.

Multistrata systems increase productivity of smallholder farmers by mimicking forests. They contain a complex but integrated mix of different annual and perennial crops and natural forest vegetation where the vegetation forms several layers or strata. Farmers are usually reluctant to invest their land and labor in tree-based systems, because the harvest lies too far in the future. Therefore, researchers are developing multistrata systems to include annual food crops at the lower strata, while perennial crops such as cocoa, oil palm, and fruit trees occupy the upper canopy. A simple system is to plant beans underneath banana or plantain trees (**Figure 5.11**). Since 1995, scientists and farmers working in various sites within the humid forest zone in Cameroon have investigated the effects of introducing nitrogen-fixing cover crops, legume hedgerows, and timber trees with the use of chemical fertilizers, on productivity, and on farmer preference of multistrata systems. The results will be compared to traditional systems where farmers plant crops such as cocoyam (taro) and plantain in forest gaps created by fallen trees. Results so far have been encouraging. For example, in the case of a timber-banana combination in Cameroon, heavier banana bunches were harvested under tree stands with 65% canopy cover, compared with solely timber stands with 15% canopy. In addition, the incidence of black sigatoka

Figure 5.11 A multistrata system of beans and banana trees. In this picture from Rwanda, the soil underneath the banana trees has been readied to plant beans. For this project seeds were donated by the United Nations Food and Agriculture Organization. Farmers normally save seeds from one year to the next but after the Rwandan civil war nearly a million families needed seeds to get farming started again. *Source:* Photo by G. Diana, FAO.

fungal disease of banana declined, and the activity of soil fauna was enhanced. The challenge for future researchers is to find suitable combinations of food crops adapted to multistrata environments and early-yielding fruit trees (such as mango, citrus, guava, and avocado), export plantation crops (such as coffee, cocoa, and oil palms), and timber trees to ensure continuous revenue and reduce the need to exploit natural forests.

Integrated crop-livestock systems have been shown to sustainably increase food, fodder, and livestock production throughout SSA. In the dry savannas of West and Central Africa, subsistence farms cultivated with traditional varieties of sorghum, millet, maize, cowpea, and groundnut were very unproductive, partly because of lack of inputs and minimum crop-livestock integration. Joint research by the International Crops Research Institute for the Semi-Arid Tropics (ICRISAT), the International Livestock Research Institute (ILRI), the International Fertilizer Development Centre (IFDC), IITA, and national agricultural research centers in SSA has led to the development of improved dual-purpose crop varieties and farming systems, which show 100–300% productivity increases over traditional systems. In areas where animals are not grazed, such as central and coastal Kenya and the Usambara mountains in Tanzania, farmers maintain intensive food crop/animal feed gardens. Trees and shrubs provide feed and bedding for the livestock. Farmers apply manure from the livestock to crops and fodder shrubs. However, to be sustainable the system also requires chemical fertilizers.

Integrated crop-livestock/fish systems have been adopted in some African countries. For example, farmers in parts of Nigeria are successfully using a poultry-fish-crop system where poultry manure provides food for fish in ponds. After the fish harvest, farmers use the nutrient-rich pond mud or manured pond water to fertilize crops (especially vegetables). They may also use crop residue, weeds, and rotten fruit from vegetable farms to fertilize ponds. This system and similar ones such as rice-fish and pig-fish systems need to be promoted for wider adoption by farmers in Africa.

5.7 Integrated nutrient management and pest management are important tools of sustainable agriculture.

Perhaps the central challenge for improving agricultural productivity in sub-Saharan Africa is how to maintain soil fertility under the intensive annual cropping systems of low-income, smallholder farmers. Integrated nutrient management incorporates several methods for maintaining soil fertility. We have already discussed some methods for maintaining soil fertility, such as alley cropping, planted fallows, use of mulches or crop residues, and cover crops. These are cost-effective ways of enhancing soil fertility. Several studies have also shown that good land clearing and plowing (for example, with shear blades and tractors), with minimal disturbance to the soil, help reduce soil erosion. Minimum-tillage or no-tillage (zero) farming with ground cover crops, mulches, and herbicides also helps maintain soil fertility. Inorganic fertilizer regimes that enhance crop growth without causing soil acidification or toxicity are also an option. The type of fertilizer should be tailored to the conditions of smallholder farmers—for example, who may apply fertilizer during or just after planting—or to fit rainfall patterns. Local production of fertilizers can help reduce costs. A key component of integrated nutrient management (INM; see Chapter 11 for full discussion) is the integration of both chemical fertilizers and organic manure with suitable cropping and farming patterns. Organic manure helps reduce the negative effects—especially acidification—of inorganic fertilizers. Acidification has afflicted

Box 5.3

Two Examples of Successes in Integrated Pest Management in Sub-Saharan Africa

■ The cassava green mite can cause up to 80% yield losses in infested fields of cassava. This noxious insect, first sighted in Uganda in 1971, is thought to have hitchhiked from Colombia on an infested shipment of cassava planting material. It is now found in the entire cassava belt of Africa. In 1983, the IITA began a research project with the goal of using biological control to keep this pest at bay. By 1991, three predatory mites imported from northern Brazil were established in a range of African cassava habitats. By 1994, biocontrol based on these predators was combined with improved, high-yielding and locally adapted cassava lines that had increased resistance to cassava green mite and improved agronomic qualities and food quality traits. In 1995, Dr. Hans Herren, an entomologist working at the IITA, was awarded the prestigious World Food Prize for his contributions to the biocontrol of cassava green mite. The success of this single control program spawned new programs to attack other pests that devastate African crops.

■ Swarms of migratory locusts have devastated crops since biblical times, and from the mid-1950s to the mid-1980s the principal means of control was to spray persistent synthetic pesticides such as DDT. Such pesticides harm both people and the environment. In 1989, four international agencies joined the IITA to start LUBILOSA, an acronym for the French for "biological control of locusts and grasshoppers." The researchers formulated an oil-based biopesticide from an African fungal pathogen of locusts. The biopesticide, nicknamed *Green Muscle* because of its powerful but environmentally friendly effects on a wide range of grasshoppers and locusts, has no adverse effects on mammals. It controls variegated grasshoppers, rice grasshoppers, Sahelian grasshoppers, and the menacing migratory desert locusts, all pests that seriously damage major cereals, grain legumes, and tuber crops. Because of the need for specialist production facilities, Green Muscle will be produced and marketed by Biological Control Products in South Africa and National Plant Protection in France.

---

post–Green Revolution Asian farms that relied heavily on inorganic fertilizer applications. INM should also use appropriate soil and water conservation practices, such as crop rotation, contour stone and earth ridges, stone terraces and dikes, and water storage pits, pools, or cisterns to collect rain runoff. The sloping agricultural land technology originally developed for upland regions in the Philippines to control erosion and maintain soil structure and fertility can also be used. This technology involves planting dense, double hedgerows of legume trees on very steep slopes, which characterize much terrain in Rwanda.

Integrated pest management (IPM, see Chapter 16 for full discussion) relies on a combination of techniques to minimize the impact of crop pests (insects, mites, nematodes, fungi, viruses, weeds, and so on). IPM incorporates biological control, host plant resistance (for example, through crop genetic improvement), farming and cultural practices, and soil and natural habitat management. Farmers use synthetic pesticides sparingly if at all. The impact of IPM is measured in terms of yield gains as well as environmental and social effects. An example of successful IPM involves controlling Striga in maize and sorghum using resistant crop varieties and cultural practices such as rotation with nitrogen-fixing legumes (cowpea and soybean) for suicidal germination of Striga seeds. IPM for cowpea and soybean involves insects, diseases, and use of varieties resistant to Striga, in combination with insecticidal plant extracts as biocontrol agents. (See **Box 5.3** for two other successes in IPM.)

5.8 Varieties improved through conventional breeding and biotechnology hold promise for the future of SSA agriculture.

Improved crop varieties adapted to the different agroecological conditions, farming systems, and consumption patterns of SSA farmers are becoming increasingly available. High-yielding hybrids with enhanced resistance or tolerance to abiotic stresses (such as drought, soil toxicity, and low nitrogen and phosphorus) and biotic stresses (such as diseases, insects, and weeds), as well as increased nutritional quality and processing characteristics, are very desirable. Farmers are now cultivating many improved varieties of major staple cereals, grain legumes, root and tuber crops, and banana and plantain (**Figure 5.12**). Where droughts are frequent (in the arid and semiarid, and moist savanna zones), a priority is the development of sorghum, pearl millet, maize, cassava, and dual-purpose grain legume (such as cowpea) varieties that do well in these marginal areas. The West African Rice Development Association (WARDA) is leading efforts to develop improved rice varieties for the lowland rain-fed environments of the inland valleys and swamp zone. Target traits include resistance against rice yellow mottle virus, rice blast, and African rice gall midge, adaptation to fluctuating water tables, competition with weeds, and resistance to drought. Breeders have generated improved upland rice varieties from an interspecies cross between Asian rice (*Oryza sativa*) and African rice (*O. glaberimma*). Such genetic improvements must be accompanied by improved agronomic practices and soil and water management.

The role of international and national agricultural research centers is crucial in helping achieve these goals. In particular, international centers should spearhead the application of biotechnology to crop improvement, because capacity is very weak in national SSA research programs. The International Maize and Wheat Improvement Center (CIMMYT) in Mexico is currently developing insect-resistant transgenic maize expressing *Bt* toxin genes (see Chapter 16). Scientists are genetically engineering maize lines adapted for cultivation in eastern and southern Africa. Other interesting ongoing projects

Figure 5.12 Bicycles loaded with plantain (*Musa paradisiaca*) or cooking bananas. Plantains differ from the much sweeter "desert" bananas consumed in the developed countries. Plantains are a major food source in the SSA. *Source:* Photo by R. Faidutti, FAO.

involve transforming sweet potatoes to enhance protein levels and resistance to weevils and viral diseases. Scientists at CIMMYT, CIAT (Centro Internacional de Agricultura Tropical, in Colombia), and IITA are also using molecular markers to speed up breeding programs in major African crops such as banana, cassava, cowpea, maize, and yam. Applying biotechnology tools will overcome some major food production constraints in SSA, but it will also be necessary to develop an infrastructure for biotechnology. Regulatory frameworks pertaining to biosafety and intellectual property rights issues still need to be developed in most countries. Applying biotechnology to the needs of SSA, it will be crucial to incorporate from the outset traditional knowledge, current farming and cropping systems, minor crops, and the needs of smallholder SSA farmers. A top-down approach is unlikely to succeed in SSA.

5.9 Nations must encourage sustainable food production by establishing necessary social policies.

To achieve sustainable food production, SSA governments need to establish effective agricultural policies with food security as a priority. Favorable environments must be created for developing, implementing, and disseminating improved technologies. Political and social stability will be a prerequisite for success. SSA also needs to

- Increase agricultural commodity prices
- Provide credit
- Establish efficient methods for production and distribution of planting materials
- Increase investments in irrigation, livestock, forestry, fisheries, postharvest, and fertilizer resources
- Improve traditional land tenure systems and rural infrastructures
- Increase investments in R&D in agricultural biotechnology, land use systems, and extension services
- Improve the status and education of women farmers and marketers

On a regional level, organizations such as the Southern Africa Root and Tuber Crops Research Network (SARRNET), East Africa Root Crops Research Network (EARRNET), West and Central Africa Maize Research Network (WECAMAN), Association for Strengthening Agricultural Research in East and Central Africa (ASARECA), and Ecoregional Program for the Humid and Subhumid Tropics of sub-Saharan Africa (EPHTA) will continue to play key roles in networking, collaborating, and disseminating research results.

CHAPTER SUMMARY

In many sub-Saharan African countries, agriculture is in a crisis and food security remains one of the biggest problems facing governments. Unfavorable sociopolitical conditions have aggravated the situation. Traditional agricultural systems have failed to satisfy the food requirements of the increasing human population. Although Green Revolution agriculture has increased some yields, its technologies are largely beyond the reach of smallholder farmers, especially in marginal, dryland areas. Many abiotic and biotic constraints restrict food production. Improving this critical situation depends on researchers thoroughly understanding traditional agricultural production systems in SSA and on their integration into the agroecology of the different regions. A number of sustainable improved systems have been devised, such as alley farming and planted fallows, that permit substantial yield increases. Researchers and farmers developed them from the ground up. Biotechnology can improve SSA crop production if it is integrated into successful local production systems and if the goal is not to replace present systems. The way forward for African governments is to focus on developing a broad-based strategy involving investments in agricultural research and extension systems that generate and transfer sustainable technologies that may result in lower yields than do Green Revolution technologies, but that nevertheless are more likely to lead to food self-sufficiency in SSA by the year 2020.

Discussion Questions

1. Discuss the agroecological complexity of sub-Saharan Africa in relation to the difficulties of raising food production. How do these zones compare with those of your own country?

2. Discuss the advantages and disadvantages of intercropping, the major mode of food production in SSA. Why don't farmers practice intercropping more in the developed countries?

3. Despite a poor food production record, many aspects of SSA agriculture can be described as sustainable. Discuss these, and also discuss those aspects that make SSA agriculture unsustainable.

4. Discuss the differences between top–down and bottom–up innovations in agriculture. Who knows most about crops among subsistence farmers, and how do we involve them?

5. In the United States and the European Union, cropping systems have evolved out of traditional systems. How should SSA cropping systems be developed, from traditional systems or from top–down innovation? Explain.

6. Does African agriculture need biotechnology? In discussing this question, distinguish between genetic engineering and biotechnology.

7. Developed countries moved away from integrated crop-livestock systems in the second half of the 20th century. What are the advantages and disadvantages of integrated systems?

8. If sustainable agriculture is geared toward smallholder farms, can SSA expect to reach food security by 2020, given population and demographic predictions?

9. If SSA is not helped to develop food security, what do you anticipate will be the social costs to its population? To the rest of the world? To the developed (aid donor) world?

10. What social policies need to be put in place to make SSA capable of developing sustainable agriculture?

Further Reading

Badejo, M. A., and A. O. Togun. 1998. *Strategies and tactics of sustainable agriculture in the tropics,* Vol. 1. Ibadan, Nigeria: College Press.

Board on Science and Technology for International Development. 1996. *Lost Crops of Africa.* Vol. 1: *Grains Board on Science and Technology for International Development National Research Council.* Washington, DC: National Academy Press.

Bunders, J. F. G., and J. E. W. Broerse. 1991. *Appropriate Biotechnology in Small-scale Agriculture: How to Reorient Research and Development.* Wallingford, U.K.: CAB International.

Devries, J., and Toenniessen G. 2001. *Securing the Harvest. Biotechnology, Breeding and Seed Systems for African Crops.* Wallingford, U.K.: CAB International.

Dupriez, H., and P. De Leener. 1988. *Agriculture in African Rural Communities: Crops and Soils.* Nivelles, Belgium: Terres et Vie.

FAO (Food and Agriculture Organization of the United Nations) and IITA (International Institute of Tropical Agriculture). 1999. *Agricultural Policies for Sustainable Management and Use of Natural Resources in Africa.* Ibadan, Nigeria: IITA.

FAO (Food and Agriculture Organization of the United Nations), United States Agency for International Development (USAID), and World Bank. 1999. In C. Haan, H. Steinfeld, and H. Blackburn, eds., *Livestock & the Environment: Finding a Balance.* Fressingfield, Suffolk, U.K.: WRENmedia.

FAO (Food and Agriculture Organization of the United Nations). 1995, 1998, 2000. FAOSTAT Database. Rome, Italy: FAO.

IITA (International Institute of Tropical Agriculture), Ibadan, Nigeria, and FAO (Food and Agriculture Organization of the United Nations). 1995. In B. T. Kang, I. O. Akobundu, V. M. Manyong, R. J. Carsky, N. Sanginga, and E. A. Kueneman, eds., *Moist Savannas of Africa Potentials and Constraints for Crop Production.* Proceedings of an International Workshop. Ibadan, Nigeria: IITA.

IITA (International Institute of Tropical Agriculture). 1993. *Sustainable Food Production in Sub-Saharan Africa.* Vol. 2: *Constraints and Opportunities.* Ibadan, Nigeria: IITA.

The Molecular Basis of Genetic Modification and Improvement of Crops

T. Erik Mirkov

The Texas A & M University Agricultural Experiment Station

For thousands of years agricultural people slowly improved their crops, either unwittingly or sometimes quite consciously, by setting aside part of the harvest as the planting material for the next crop (see Chapter 13). Crop improvement was greatly accelerated in the 20th century by the rediscovery of the laws of inheritance, first formulated by Gregor Mendel in 1865, and after scientists established that the chromosomes contain the information for specific traits encoded in DNA. The rise of molecular biology, starting in the 1950s with the discovery of DNA's structure, led to a series of further discoveries showing how traits are encoded as proteins in the cell and the plant. This understanding, and new techniques that allow scientists to manipulate DNA in the laboratory, are revolutionizing plant breeding.

In the future, increasing yield for traditional crops, improving nutritional or postharvest quality of these crops, adapting them to more stressful environments, domesticating new crops, and converting existing ones to plant factories that produce chemicals for industry will all flow from applying the principles of plant breeding, genetics, molecular biology, and genomics. "Classic" plant breeders—those who perform mostly field work—will only succeed in making major improvements by exploiting more and more of the tools used in molecular biology. Collaboration between plant breeders, plant pathologists, molecular biologists and geneticists will be essential to expedite the long process of producing new and improved varieties.

6.1 Genes are made of DNA.

The 30-year period from the 1930s to the 1960s was a very exciting time for scientific discovery, and must go down in history as the period when one of the most fundamental questions science ever asked was answered: What exactly are genes made of?

As you will see in later sections, careful observation of how chromosomes separate during mitosis, and the discovery that sex cells contain only half the number of chromosomes (one set instead of two) that other cells have, led scientists to postulate that Mendel's inherited units, later called *genes*, are organized on the chromosomes. Chromosomes are the filamentous structures in nuclei (**Figure 6.1**). Geneticists knew that chromosomes were composed of both protein and DNA (<u>d</u>eoxyri<u>bonucleic</u> <u>a</u>cid). For many years, biochemists incorrectly believed that genes consisted of protein because they knew proteins were large, complex molecules (they consist of countless sequences of 20 different amino acids) that show great specificity in their activity (mostly enzymatic activity). Biochemists thought that only protein molecules could contain enough information to specify inherited traits.

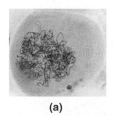 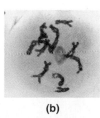  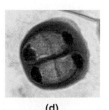

(a)   (b)   (c)   (d)

Figure 6.1 Structure of chromosomes in microsporocytes of lily. Microsporocytes are the precursors of the sex cells or gametes that are formed in flowers anther. The cells shown here have been treated with a dye that binds to DNA and allows all DNA-containing structures to be seen clearly. In **(a)** the chromosomes have just started to contract and are visible as long thin threads. In **(b)** contraction is at its maximum and each chromosome is clearly distinguishable. In **(c)** the chromosomes are separating from each other during the second meiotic division, and in **(d)** the formation of the four gametes is nearly complete. *Source:* D. Sadava (1993), *Cell Biology* (Boston: Jones and Bartlett), p. 487.

In contrast, they thought DNA was a small and relatively simple molecule because it contains only four nucleotides—surely DNA could not contain enough information to account for the great diversity of life forms on Earth. Several elegant and decisive experiments conducted with bacteria and viruses by Frederick Griffin, Oswald Avery, Alfred Hershey, and Martha Chase showed that genes consist of DNA and have the necessary information to direct the synthesis of proteins. The researchers showed that mutation of bacterial genes results in the loss of specific enzymatic activities, indicating that genes are responsible for the presence of proteins in the cell. This discovery later led to the "one gene, one enzyme" hypothesis that a distinct gene controls the presence of each enzyme in the cell.

Now that biologists were convinced DNA, not protein, was the genetic material, biochemists wanted to answer another fundamental question: What precisely is the structure of DNA? Because they knew this discovery would go down in history, researchers were racing each other to unravel the structure of DNA.

With new chemistries developed in the 1940s, Erwin Chargaff determined in detail the base composition of DNA. Scientists already knew that DNA consists of four different types of nucleotides: two with the purine bases adenine and guanine, and two with the pyrimidine bases thymine and cytosine. As shown in **Figure 6.2**, the purine bases have a double ring structure, whereas the pyrimidine bases have a single ring. These four nucleotides—the building blocks of DNA—consist of a phosphate group, the sugar

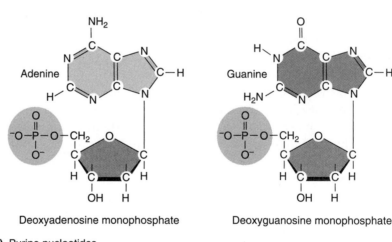

(a) Purine nucleotides

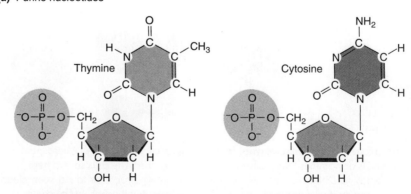

(b) Pyrimidine nucleotides

Figure 6.2 The building blocks of DNA. DNA molecules are polymers, built out of monomers called *nucleotides*. **(a)** The double-ring purine nucleotides and **(b)** the single-ring pyrimidine nucleotides each have a phosphate, the five-carbon sugar, deoxyribose, and one of four nitrogen-containing bases.

deoxyribose, and one of the four nitrogen-containing bases (A, C, G, or T). Chargaff showed that in some species, such as *E. coli* and maize, the DNA consists of about 25% of each type of nucleotide, but that in most species, such as humans, this is not the case. Thus DNA shows the variability among species that you might expect of the genetic material of life. Within a species, however, Chargaff's experiments showed that the amount of A always equals the amount of T, the amount of G always equals the amount of C, and that the percentage of A plus G equals 50% and the percentage of C plus T equals 50%. These data suggested that A is always paired with T and that G is always paired with C.

Rosalind Franklin studied the structure of DNA using X-rays and found that DNA has a helical structure. Using Chargaff's and Franklin's data, and possibly some of Linus Pauling's data, James Watson and Francis Crick constructed a model for the structure of DNA (**Figure 6.3**), for which they won the Nobel Prize in medicine in 1962. They concluded that DNA is a double-stranded helix with the adenine in one strand always

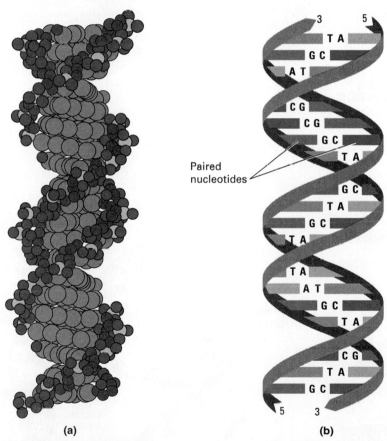

Paired
nucleotides

(a)   (b)

Figure 6.3 The structure of DNA. (a) A space-filling model in which each atom appears as a sphere. DNA molecules are polymers made by linking the phosphate on one nucleotide to the oxygen on a specific carbon of the deoxyribose of the next nucleotide in the chain. This produces a molecule that consists of a backbone of alternating phosphate and deoxyribose groups, with the bases protruding. Molecules of RNA are built in a similar fashion, but in a complete DNA molecule (as shown here), two nucleotide chains pair to form a double helix. **(b)** The diagram shows a segment of double-stranded DNA that contains only 18 paired nucleotides; however, DNA molecules normally can contain thousands. A small gene may contain 2,500 nucleotides; a large one, 25,000 or more. The deoxyribose–phosphate backbone of each chain is on the outside, and the bases are on the inside of the double helix. The bases of one chain are paired with those of the other chain according to the base-pairing rules: thymines always pair with adenines and cytosines with guanines. *Source:* D. Hartl and E. W. Jones (1998), *Genetics: Principles and Analysis*, 5th ed. (Boston: Jones and Bartlett), p. 9.

hydrogen-bonded to the thymine in another strand; similarly, cytosine in one strand is always hydrogen-bonded to guanine in the other strand. Together the two strands form the double helix with the sugar–phosphate backbones on the outside and the paired bases on the inside. The two strands of the DNA molecule are said to be *complementary strands* because if you know the base sequence of one strand, you can deduce the base sequence in the other complementary strand. If one strand contains the sequence ATTGCC, then the other strand must in the same region have the sequence TAACGG, because of the base-pairing rules (A opposite T, and G opposite C).

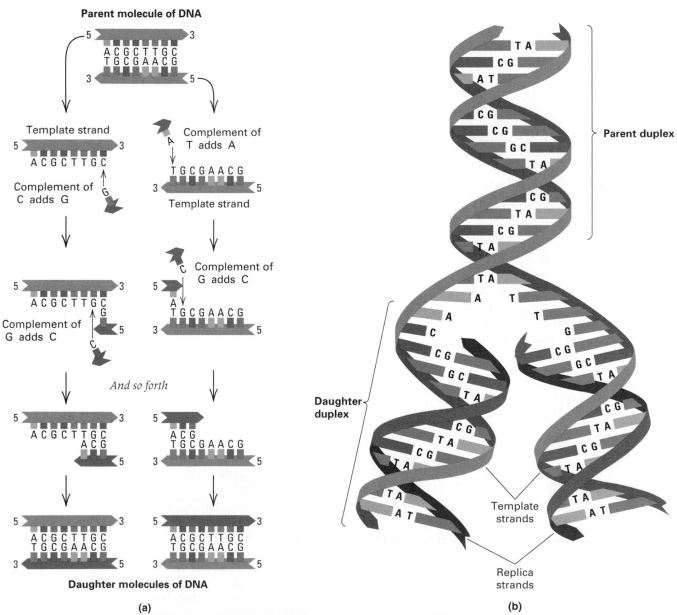

Figure 6.4 Replication of DNA. (a) When DNA replicates, the two strands of the molecule separate and each serves as a template for the synthesis of a new complementary strand, which grows in length by the successive addition of a single nucleotide. **(b)** The obligatory pairing of adenines with thymines and guanines with cytosines ensures that the two daughter molecules will be duplicates of the original. *Source:* D. Hartl and E. W. Jones (1998), *Genetics: Principles and Analysis,* 5th ed. (Boston: Jones and Bartlett), p. 11.

Box 6.1

Plant Genomes Contain Mobile Elements That Contribute to Genome Evolution

At the time Barbara McClintock was studying inheritance in maize, geneticists thought that each gene had a fixed location on a chromosome. Her experiments showed that what she called "controlling elements" could move from one place to another on the chromosome. When such an element ended up in the middle of a gene, it blocked the expression of that gene. Her work showed that movement of these elements could account for the various pigment patterns found in Indian maize. Scientists now call her controlling elements *mobile genetic elements* or *transposons*.

When McClintock published her results in the 1950s, the scientific community dismissed them as anomalies unique to maize. Since then, molecular geneticists have discovered transposons in bacteria, yeast, other plants, flies, and humans. Scientists now know that transposon movement can cause localized mutations, carry a copy of host genes when they jump, and leave copies of themselves and host genes before jumping. These movements have

been linked to a variety of mutations, including some that cause diseases and others that add desirable diversity to genomes. Transposons also contribute to genome expansion resulting in repetitive DNA called "junk DNA." In fact, more than 50% of the maize genome consists of transposons. Recent work indicates that in certain grasses transposons can also contribute to substantial DNA losses. Because transposition has had such a powerful effect on genotypes and phenotypes, it has certainly played a significant role in the evolution of plant genomes. Transposons will undoubtedly continue to contribute to keeping plant genomes dynamic rather than static. For her discovery of transposons, Barbara McClintock, at the age of 81, was awarded a Nobel Prize in medicine in 1983.

Another requirement of the genetic material is that it must replicate and then transfer to daughter cells. Watson and Crick proposed a mechanism for the copying of DNA based on the specific base-pairing rules, and scientists have now confirmed that DNA replicates by complementary base pairing. DNA copying is called *semiconservative replication* because one of the old strands is present in each new daughter molecule. When DNA replicates, the two old strands separate and the complementary strands then synthesize using the existing strands as templates. When the two new strands are synthesized, A is always opposite T and C opposite G, so that two identical molecules form (see **Figure 6.4**). Because the information is contained in the exact sequence of the bases, DNA replication makes a new and exact copy of the information.

Most DNA in a plant cell is found in the nucleus and the totality of this DNA is called *nuclear genome*. Two cellular organelles, the chloroplasts and the mitochondria, have their own DNA—their own genomes—which encodes a number of genes. Mitochondrial DNA encodes about 15 genes, and chloroplast DNA, about 100 genes. Chloroplasts and mitochondria are the evolutionary descendants of endosymbiotic bacteria, and they still share many properties with present-day bacteria. Mitochondrial DNA and chloroplast DNA are inherited in a uniparental manner, usually with the female sex cell and not in Mendelian fashion (see following discussion). Transforming plants and creating genetically modified (GM) crops usually involves introducing DNA into the nuclear DNA of plants (see Section 6.8). Recently it has also become possible to introduce new genes into chloroplast DNA.

At one point it was thought that the nuclear genome of an organism is completely stable except for the occasional mutations caused by chemicals or irradiation. Now we know that the chromosomes contain mobile elements (see **Box 6.1**) that move from place to place and may cause duplications or deletions of genes. DNA is constantly being reorganized by the shuffling of these mobile elements.

6.2 Genes code for proteins via the molecular trilogy: DNA, RNA, and amino acids.

Proteins are linear molecules of 20 different amino acids, usually between 100 and 10,000 of them, folded in a three-dimensional configuration that is often globular in over-all shape. The individual amino acids join together by linkages called *peptide bonds* (**Figure 6.5**), so a protein molecule is often called a *polypeptide*. This string of amino acids then winds into a helix or can take the shape of a pleated sheet. When the entire chain is folded, some parts will be helical, other parts will be pleated, and yet other portions of the chain will form a random coil (**Figure 6.6**).

Proteins may interact with other molecules in the cell in very specific ways: as enzymes, as regulators of gene activity, or as transporters of ions and water. This specificity resides in their exact three-dimensional shape, which is in turn determined by the sequence of amino acids. Thus each kind of protein (and some cells probably have more than 5,000 different proteins) has a characteristic amino acid sequence. The information that specifies the order in which the amino acids must assemble when a protein is synthesized, is contained in the DNA, specifically in the sequence of bases attached to the sugar–phosphate backbone.

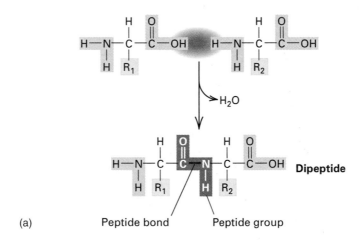

(a)

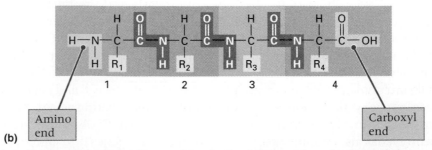

(b)

Figure 6.5 Properties of a polypeptide chain. (a) Formation of a dipeptide by reaction of the carboxyl group of one amino acid (top) with the amino group of a second amino acid (bottom). A water molecule is removed to form the peptide bond (red line). **(b)** A tetrapeptide showing the alternation of α-carbon atoms (black) and peptide groups (blue). The four amino acids are specified by the R groups. *Source:* D. Hartl and E. W. Jones (1998), *Genetics: Principles and Analysis* (Boston: Jones and Bartlett), p. 414.

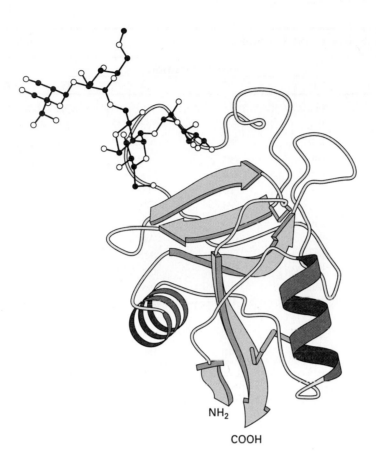

NH₂

COOH

Figure 6.6 A ribbon diagram showing the ways in which a polypeptide can be folded. Arrows represent parallel β sheets, and helical regions are shown as coiled ribbons. The protein in this example is a protein that binds to carbohydrates. The stick figure in the upper left is a small oligosaccharide consisting of several sugars bound to the protein. *Source:* D. Hartl and E. W. Jones (1998), *Genetics: Principles and Analysis* (Boston: Jones and Bartlett), p. 415.

At first sight this coding might seem to be a problem, because DNA has only four different bases and proteins have 20 different amino acids. This coding problem is solved by having a sequence of three bases specify one amino acid. When four different nucleotides are arranged in groups of three, 64 different combinations ($4 \times 4 \times 4$) are possible, more than enough to specify 20 amino acids. In the 1960s, Marshall Nirenberg, Heinrich Matthei, and Philip Leader figured out which three-base combination or codon specifies a particular amino acid. The correspondence between a three-base combination and an amino acid is called the *genetic code*. The genetic code is universal: It is the same in all organisms. Most amino acids are represented by more than one codon. Of the 64 different possibilities, 61 codons specify amino acids, and 3 specify "stop" signals marking the end of a protein-coding segment of the DNA (**Table 6.1**).

Cells have an elaborate machinery for translating the nucleotide sequences in the DNA into amino acid sequences in proteins. It would be simple if amino acids could recognize their own codon and just line up in the right order on the surface of the DNA to be linked together by an enzyme, but nothing so simple evolved. Rather, cells use a different class of nucleic acids, RNAs (ribonucleic acids), to help translate the information contained in the DNA. RNAs are also strings of nucleotides, and consist of a sugar–phosphate backbone with a base attached to each sugar group, except that the sugar is ribose. Three of the bases in RNA are the same as in DNA (C, A, and G) but instead of thymine (T), RNA uses uracil (U).

Gene transcription proceeds via the formation of a messenger RNA (mRNA) molecule. The mRNA carries the information that will specify the amino acid sequence from the nu-

| Table 6.1 | The genetic code | | | | |

First Position	Second Position				Third Position
	U	C	A	G	
	Phe	Ser	Tyr	Cys	U
	Phe	Ser	Tyr	Cys	C
U	Leu	Ser	Stop	Stop	A
	Leu	Ser	Stop	Trp	G
	Leu	Pro	His	Arg	U
	Leu	Pro	His	Arg	C
C	Leu	Pro	Gln	Arg	A
	Leu	Pro	Gln	Arg	G
	Ile	Thr	Asn	Ser	U
	Ile	Thr	Asn	Ser	C
A	Ile	Thr	Lys	Arg	A
	Met	Thr	Lys	Arg	G
	Val	Ala	Asp	Gly	U
	Val	Ala	Asp	Gly	C
G	Val	Ala	Glu	Gly	A
	Val	Ala	Glu	Gly	G

Note: A sequence of three nucleotides forms the nucleic acid codon for a single amino acid. The four nucleotides U, C, A, G can produce 64 different three-nucleotide combinations. All the amino acids except methionine (Met) and tryptophan (Trp) have more than one codon. The "stop" codons UAA, UAG, and UGA do not code for amino acids but signal the end of a protein. All proteins start with methionine.

The codons are given as they appear in messenger RNA. The four bases in the nucleotides of ribonucleic acids are uracil (U), cytosine (C), adenosine (A), and guanine (G). The amino acids specified by the genetic code are alanine (Ala), arginine (Arg), asparagine (Asn), aspartic acid (Asp), cysteine (Cys), glycine (Gly), glutamine (Gln), glutamic acid (Glu), histidine (His), isoleucine (Ile), leucine (Leu), lysine (Lys), methionine (Met), phenylalanine (Phe), proline (Pro), serine (Ser), threonine (Thr), tryptophan (Trp), tyrosine (Tyr), and valine (Val).

cleus, where the nucleotide sequence of the DNA contains the message, to the cytoplasm of the cell, where this message is translated and the protein synthesized. Translation of the message involves transfer RNA (tRNA) molecules as well as ribosomes, specialized structures on whose surface new proteins are assembled. As the amino acid chain lengthens, it begins to fold up, first in a helical configuration, and then in a three-dimensional globular shape.

During transcription, the two strands of the DNA separate and ribonucleotides assemble on one of the DNA strands in such a way that the bases will pair in the same way as for DNA replication, except that A (adenine) in the DNA is paired with U (uracil) in the RNA being synthesized (**Figure 6.7a**). Thus, a sequence that is CGATC in DNA becomes GCUAG in RNA. When the mRNA reaches the cytoplasm, it binds to the surface of a ribosome where tRNAs with amino acids attached line up the amino acids in the correct order along the length of the mRNA, starting always with the amino acid methionine. A tRNA recognizes the codon in the mRNA by means of an anticodon, which aligns according to the base-pairing rules. As the tRNAs with their respective amino acids line up on the mRNA one by one, the amino acids link together by an enzymatic reaction and the tRNAs are released to react with new amino acids.

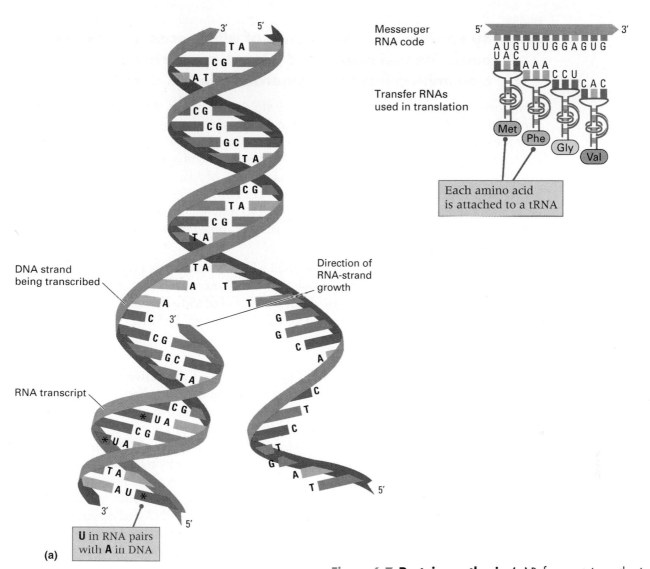

Messenger RNA code

Transfer RNAs used in translation

Each amino acid is attached to a tRNA

DNA strand being transcribed

Direction of RNA-strand growth

RNA transcript

U in RNA pairs with **A** in DNA

(a)

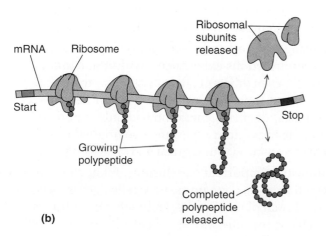

mRNA   Ribosome

Ribosomal subunits released

Start

Growing polypeptide

Stop

Completed polypeptide released

(b)

Figure 6.7 Protein synthesis. (a) Before protein synthesis can begin, the genetic information (sequence of bases) in the DNA must be copied into an RNA molecule. The two strands of the DNA molecule separate, allowing one strand to be copied into a messenger RNA (mRNA). This process, called *transcription,* takes place in the nucleus. The mRNA moves from the nucleus to the cytoplasm, where protein synthesis occurs. This coincides with a maturation process in which parts of the mRNA that do not code for protein are removed. In the cytoplasm, specific enzymes join each of the amino acids to the appropriate transfer RNA. **(b)** The messenger RNA and the transfer RNA for the first amino acid of the protein form a complex with the ribosome. Then the ribosome effectively moves along the messenger RNA molecule. As it moves, the "anticodon" on the transfer RNA for each successive amino acid recognizes the corresponding codon on the messenger RNA and brings the amino acid into position for joining to the growing protein chain. In this way, the information encoded in the linear sequence of nucleotides in the DNA of the genes translates into the linear sequence of amino acids in a protein. *Source:* D. Hartl and E. W. Jones (1998), *Genetics: Principles and Analysis* (Boston: Jones and Bartlett), pp. 14–15.

6.3 By studying the inheritance of all-or-none variation in peas, Mendel discovered how characteristics are transmitted from one generation to the next.

More than 130 years ago, Gregor Mendel, a Moldavian monk, formulated the two fundamental laws that govern heredity. People had suggested various ideas about heredity before Mendel began his experiments with peas, but none of these was supported by experiments. When Mendel began his work, most breeders agreed that both sexes contributed to a new individual. They thought that parents of different appearance always produced offspring with an intermediate appearance. Thus, a cross between a plant with red flowers and one with white flowers would lead to only plants with pink flowers. When future generations produced red and white flowers, breeders mistook this to mean that the genetic material was unstable.

Mendel conducted a series of experiments in which he crossed two varieties of pea plants with contrasting characteristics such as white and red flowers or round and wrinkled seeds. Remarkably, the characteristics he used can still be found today, precisely because they all have all-or-none variation. He published his observations in 1865 in a local scientific journal, but his findings went largely unnoticed until other scientists made the same observations 35 years later.

Mendel, who was very familiar with peas, knew that they always breed true (the offspring are exactly like the parent plant and like each other), because peas are self-fertilizers and the pollen fertilizes the pistil (female reproductive organ) even before the flower opens. However, by cutting away the anthers (male reproductive organ) and brushing on pollen from another plant, Mendel figured out that peas could be cross-pollinated (**Figure 6.8**). Mendel had many types of true-breeding peas growing in his garden, and he wanted to know what would happen if he cross-fertilized lines of his peas with contrasting characteristics. One line of pea always produced smooth, round peas, but another one produced wrinkled peas. When the latter peas dried out at the end of seed maturation, some of the inner tissues collapsed, providing the seed with a wrinkled appearance (phenotype).

When Mendel crossed round and wrinkled peas, he observed that the first generation (the F_1 generation) consisted entirely of round peas. When these round peas were allowed to sprout, grow, and flower, and when the plants were allowed to set seed, most seeds of the F_2 generation (about 75%) were round and a minority (about 25%) were wrinkled (**Figure 6.9**). None were in between; none were just a little wrinkled or nearly round. Thus a characteristic that disappeared in the first generation reappeared in the second.

Mendel repeated these experiments using seven other discontinuous characteristics (for example, green and yellow seeds) and confirmed his experiments with round and wrinkled peas. Interestingly, he found that these seven different characteristics were not linked, but were transmitted to the next generation independently. Thus, when he crossed a round yellow pea with a wrinkled green pea, the first generation had only round yellow peas. However, in the second generation he found four types: the two original types and two additional combinations of the characteristics (round green and wrinkled yellow).

From such experiments, Mendel drew two important conclusions. First, characteristics or traits transmit to the next generation as discrete units, now called *genes*. Second, an individual must contain two copies of each of these units, and each parent transmits only one copy to the next generation. That is the only way Mendel could account for the disappearance and subsequent reappearance of a characteristic. The implication is that the unit (gene) is always present, but may not be *expressed*, as is the case with the wrinkled

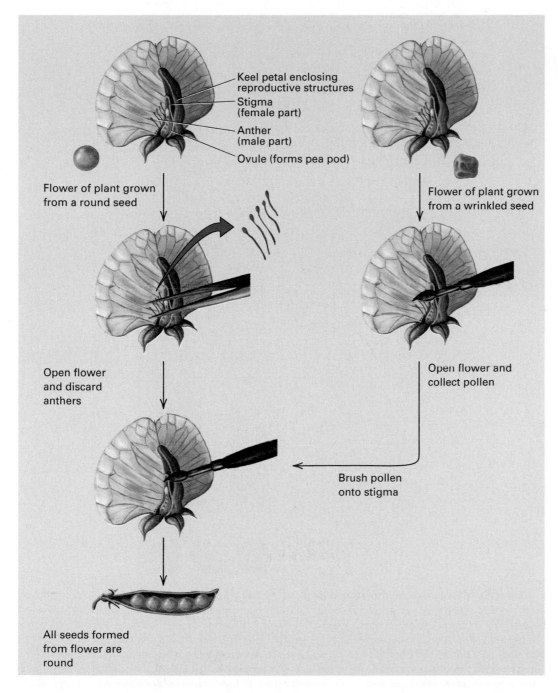

Figure 6.8 Making crosses (hybrids) with the garden pea. Anthers from one flower are removed before they produce pollen. The female part is not removed and is fertilized by brushing pollen grains obtained from another flower. *Source:* D. Hartl and E. W. Jones (1998), *Genetics: Principles and Analysis* (Boston: Jones and Bartlett), p. 33.

or green characteristics in the first generation of crosses of round and wrinkled or green and yellow peas.

Although Charles Darwin and Mendel were contemporaries (Mendel published his work in 1865, fifteen years before Darwin's death), Darwin did not know of Mendel's work. Darwin knew that characteristics could disappear and reappear, but he did not understand their mode of inheritance. In his book, *On the Origin of Species by Means of Natural Selection,* Darwin discussed the notion that a specific trait can be inherited by one child but not by another and that traits of grandparents sometimes appear in the grandchildren although they did not appear in the parents. He did not understand this skipping of a generation, which Mendel's experiments so beautifully explained.

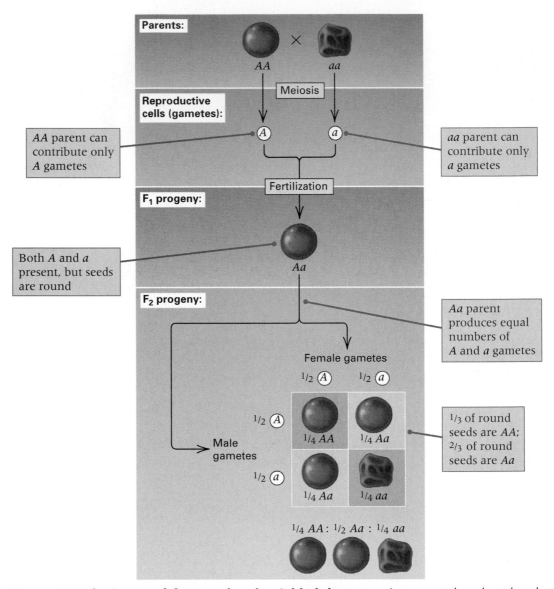

Figure 6.9 Inheritance of the round and wrinkled characters in peas. When plants that always produce round seeds are mated with plants that always produce wrinkled seeds, the plants in the first generation (F$_1$) all produce round peas. However, the F$_1$ plants still have the wrinkled gene. When the F$_1$ plants are crossed with one another, one quarter of the seeds in the F$_2$ generation are wrinkled and three quarters are round. This pattern of inheritance can be explained by assuming that each plant carries two copies of the genes for each character, and that being round is dominant over being wrinkled. The original parents, which breed true, are either AA or aa, and when these plants are crossed, gametes are either A or a, and all offspring are Aa. When these F$_1$ plants are crossed, both can produce A and a gametes and there are more possibilities for the offspring: there will be two Aa plants for every one that is either aa or AA. However, because A is dominant, three quarters will be round (AA and Aa) and only one quarter will be wrinkled (aa). *Source:* D. Hartl and E. W. Jones (1998), *Genetics: Principles and Analysis* (Boston: Jones and Bartlett), p. 37.

Why are there smooth and wrinkled peas? Biochemists have recently discovered that wrinkled peas lack one of the important enzymes for starch synthesis. During their development, peas import sucrose from the rest of the plant. This sucrose is quickly converted to starch, which makes up 60% of the weight of a mature pea seed. Wrinkled peas lack one of the enzymes for starch synthesis, so when the seeds mature they contain a lot of sucrose and water and much less starch than normal. Then, when the seeds dry out, they wrinkle. The absence of the enzyme is caused by an alteration in the gene that encodes the information for synthesizing this enzyme. A single gene usually controls all-or-none variation, as in the case of the smooth and wrinkled peas.

Many agronomically important characteristics, such as yield or protein content of seeds, show continuous variation, and the inheritance of such traits is controlled by many genes. Such traits are referred to as *multigenic* or *quantitative* traits. Thus, the ability of a plant to take up soil nutrients, photosynthesize, transport photosynthate to the seeds, and withstand drought all affect yield. And each of these characteristics is controlled by many genes, making yield truly a multigene trait. The inheritance of multigenic traits is discussed in Chapter 14.

Most organisms have two copies of every gene in every cell, except in the sex cells (sperm and egg cells), which each only have one copy of each gene. In humans, each cell has 22 pairs of chromosomes and 2 sex chromosomes, for a total of 46. Maize has 10 pairs of chromosomes, and *Haplopappus,* a plant that thrives in dry areas, has only 2 pairs. Because cells have two copies of every chromosome, they also have two copies of every gene, one copy on each of the chromosomes that make up a pair of chromosomes. When these two copies are identical, the organism is said to be *homozygous* for that gene. If one of the two gene copies has mutated, they will be different, or *heterozygous.* These different forms of the gene are called *alleles.* Plants that normally self-fertilize, such as peas and beans, are homozygous for a very large number of genes, whereas plants that normally outcross, such as maize, may be heterozygous for most of their genes.

6.4   Mitosis and meiosis are important cell processes during which genetic information is passed on from cell to cell or generation to generation.

As noted earlier, genes are located on filamentous structures, called *chromosomes,* in the cell nucleus. Most complex organisms such as flowering plants, insects or mammals have in their cells some 20,000–60,000 genes but only a small number of chromosomes. Thus thousands of genes are linearly arranged on each chromosome. The chromosomes play a vital role in passing on genes, and therefore phenotypic characteristics, from a mother cell to the daughter cells during cell division, and from the parents to the offspring during reproduction. Cell division, which in plants occurs in meristems, is preceded by mitosis, a process of chromosome duplication, and the subsequent separation of the two sets of chromosomes. Thus for a short while a cell contains four copies of every gene, but as the new chromosomes separate, each daughter cell again contains two copies of every gene. During the formation of egg cells and sperm cells, a different type of cell division occurs. A single cell undergoes one round of chromosome replication followed by two rounds of chromosome separation and cell division. Four cells are formed in this process, which is called *meiosis,* and each cell ends up with a single copy of each gene. If the organism is heterozygous for that gene, then half the sex cells will have one allele of the gene and the other half will have the other allele.

When two sex cells fuse during fertilization to start a new organism, the cells of this new organism will again contain two copies of every gene. A plant homozygous for a given gene will produce sex cells, all of which have identical forms of that gene. That is, in terms of that gene, it produces only one type of reproductive cell. However, a plant that is heterozygous for a certain gene, with (for example) one normal and one mutant copy, will produce two types of reproductive cells. Half the sex cells will carry the normal gene, half will carry the mutant gene. The fact that many plants are normally heterozygous for many genes creates many possibilities for variation in the offspring, because random assortments of genes are brought together when the sperm cell fertilizes the egg cell.

Let us consider a plant with a haploid chromosome number of 2. Its normal diploid cells will have two pairs of chromosomes in each cell. Such a cell is shown (Figure 6.10a and 10b) with two heterozygous genes. T and t are alleles of the same gene and are located on the long chromosome (note that T came from the male parent and t from the female). A and a are alleles of a different gene and are located on the short chromosome. Before mitosis begins, the cell has four chromosomes in its nucleus: a pair of each of two types. Mitosis involves three steps (**Figure 6.10**):

- *Duplication* of the chromosomes' genetic material (DNA); this actually occurs during the time that the cell nucleus is visible and chromosomes cannot be seen as distinct structures.

- *Lining up* of the duplicated chromosomes, usually at the center of the cell.

- *Separation* of each member of the duplicated chromosomes from its partner, such that each of the two new cells gets one copy of the pair.

Because the two members of the duplicated chromosome are genetically identical, the two new cells that are formed contain the same genes (T and t, A and a) as the cell that produced them.

The earliest part of meiosis (Figure 6.10) is identical to mitosis: The chromosomes are *duplicated* and each one becomes a double strand. The chromosomes *line up* in the center of the cell; however, in this case, they line up as homologous pairs. Now, *separation* occurs such that the duplicated chromosomes separate from each other, one duplicated member going into each of two new cells. The resulting cells clearly have only half the number of chromosomes (in this case, 2) as the originating cell (4). These two chromosomes are double, so that in meiosis II, the two members of the doublet separate, just like in mitosis. Now there are four cells formed, each with a haploid chromosome number (2).

Figure 6.10 shows an additional feature of meiosis that is essential for genetic variability. At the lineup and separation stages of meiosis I, the only requirement is that each new cell gets one of the two members of the homologous pair. So it is equally probable that one of the new cells will get the "A" and "T" chromosomes and the other "a" and "t" as it is for them to get the "A" along with "t" and "a" with "T." So in our simple case, there are four possible chromosome combinations in the gametes: AT, at, At, and aT.

Few plants have just two pairs of chromosomes (diploid number of 2). For example, corn has 10. As we have just seen, the number of possible gametes formed with two chromosome pairs, just with regard to paternally and maternally derived chromosomes, is 4 (2^2). For 10 chromosomes, the number is 1,024 (2^{10}). If we consider that there are often allelic differences between the two chromosomes (see our example, T and t, and A and a), there are 1,024 genetically different gametes possible. These combinations mean that in sexual reproduction, the offspring will not be genetically identical to the parents. This provides much genetic variation, the raw material of evolution by natural selection. It also provides a challenge to plant breeders, who seek to minimize variation in offspring. Additional genetic variation is created when chromosomes exchange portions during crossing over in meiosis, a phenomenon that is explained in Chapter 14.

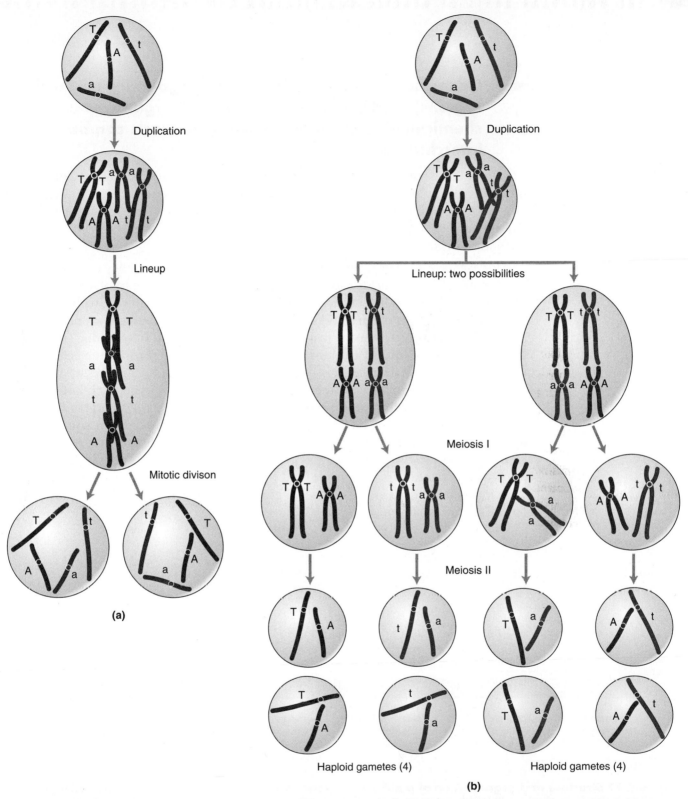

Figure 6.10 Mitotic and meiotic cell divisions. (a) The mitotic method of cell reproduction occurs in plant meristems. One member of each gene pair (e.g., T and t) is located on one member of a homologous chromosome pair (e.g., the large chromosome). One of the pair came from the male gamete (e.g., the blue chromosome) and the other from the female gamete (e.g., the red chromosome). Note that in mitosis two cells are formed, each one genetically identical to the original cell. **(b)** The meiotic method of cell reproduction results in the formation of sex cells (pollen or egg cells). Note that four haploid gametes are formed from a diploid cell. The key is the separation of the duplicated homologous chromosome pairs during meiosis I and the separation of each member of the duplicated chromosomes during meiosis II. Depending on how the homologous pairs line up during meiosis I, different combinations of genes end up in the haploid gametes that are formed.

6.5 Genes have a multipartite nature (regulatory and coding regions), and because the DNA of all organisms is chemically identical, genes and gene parts can be interchanged.

In Chapter 8, we will discuss the role of genes in development. Orderly development of an organism requires that the correct proteins be synthesized at the correct time and place during development, and in response to the proper stimuli. For example, the enzymes that cause softening of cell walls in the ripening fruit are only synthesized in the fruit and as a result of an increase of the hormone ethylene.

So far, we have described genes only in terms of the information they contain to specify a sequence of amino acids and the structure of a protein. This part of a gene is called the *protein-coding region*. However, a gene is more complex (much longer, really) than just a protein-coding string of nucleotides. First of all, the protein-coding portions, called *exons,* may be interrupted from one to a dozen times by very short to very long stretches of DNA, called *introns*. Introns are transcribed when RNA is first made in the cell nucleus, but they are later removed when the initial product of transcription is processed into a mature mRNA that is transported to the cytoplasm. These interruptions in the coding sequence are characteristic and different for every gene, and can also be involved in regulating the gene's expression. Second, a DNA segment on each side of the protein-coding region specifies when and where the gene is to be activated during development, to which stimuli (hormonal or environmental) the gene should respond, in which cells it should be active, how much mRNA is to be made, and when the gene is to be turned "on" and "off." These *regulatory regions* are, of course, of great importance for the orderly development of a cell and an organism (**Figure 6.11**).

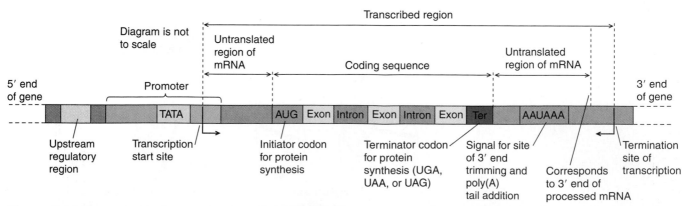

Figure 6.11 Structure and organization of a eukaryotic gene. A gene has several sections. The transcribed region functions as a template for synthesizing RNA, which then translates into the protein product of the gene. The transcribed region is interspersed with noncoding sequences that partition the region into coding sections (exons) and noncoding sections (introns). The transcribed region is flanked on either side by noncoding sequences that play a role in regulating the gene. Most of the regulatory sequence elements are in the 5′ flanking region. The first 1,000 bp or so of the 5′ flanking region is called the *gene promoter,* because it contains sequence motifs important for the "promotion" of transcription. The most highly conserved part of the promoter is the TATA box, which is usually located within the first 50 bp of the transcription start site. The TATA box coordinates the recruitment of RNA polymerase to the gene. *Source:* R. Buchanan, W. Griussem, and R. Jones (2000), *Biochemistry & Molecular Biology of Plants* (Rockville, MD: American Society of Plant Physiologists), p. 340.

Thus, genes that encode enzymes for the synthesis of chlorophyll are "on" in the leaves, but "off" in the roots and the flowers. Furthermore, they are "off" in the dark and "on" in the light. When a root finds itself in soil that is rich in nitrate, genes are turned "on" that encode proteins needed to take up nitrate, its transformation into ammonia, and the use of this ammonia in amino acid biosynthesis. When the spore of a plant pathogen germinates on the surface of a leaf and tries to penetrate into the leaf cells, defense genes are rapidly turned on in the cells of resistant plant varieties. These defense genes encode enzymes that synthesize toxic compounds that will kill the invader. When light strikes a seedling that has been growing in the dark underneath the soil surface, hundred of genes that were completely off or barely on are turned on and the cells start making hundreds of new proteins that let the chloroplasts develop, and photosynthesis and autotrophic growth begin. These are only a few examples of the gene activity regulation in response to specific stimuli.

Regulating gene activity is the responsibility of proteins called *transcription factors* that bind to the regulatory regions of genes, and these transcription factors are themselves the products of genes. Thus, turning genes "on" and "off" is not a simple matter and may involve a regulatory cascade in which the product of gene A activates gene B, whose product activates gene C, whose product activates gene D by binding to its regulatory region.

Plant breeding and genetic engineering both involve transfer of genes from one organism to another. In plant breeding, the breeder selects for an ultimate outcome or a phenotype, thus ensuring that the whole gene regulation pathway will operate correctly. Let us take the example of transferring resistance to a specific pathogen from a wild relative of wheat to a cultivated wheat variety. A resistance gene may encode a protein that lets the plant detect the invader quickly so the plant can turn on its defenses. The gene that encodes this detector protein responds to a stimulus from the pathogen—perhaps a chemical or metabolite made by the pathogen—and this response involves several other genes. Thus when the breeder transfers resistance to that specific pathogen from a wild variety of wheat to a domesticated wheat by selecting for pathogen resistance in the field, the entire regulatory cascade must work correctly. If the new combination carries the resistance gene but is not correctly expressed, then the new variety would not resist the pathogen.

Genetic engineers transfer one gene at a time and need to understand how the gene they transfer is regulated. First, they need to transfer not only the protein-coding part of a gene but also its regulatory region. Or they can equip the gene with a new regulatory region that ensures correct expression. For example, regulatory regions of bacterial genes generally do not work in plants because the gene transcription machinery of bacteria differs from that of plants. Similarly, a regulatory region of the gene from a monocotyledon (maize) may not work in a dicotyledon (bean). Thus, a full understanding of the regulation of gene expression is of great interest not only to plant biologists who want to know how plant development is regulated, but also to genetic engineers who transfer genes between organisms.

6.6 Restriction enzymes and bacterial plasmids permit the manipulation of genes in the laboratory.

Since the 1960s, molecular geneticists have made tremendous advances that have led to the development of exciting new areas of science in DNA technology and plant biotechnology. Biotechnology is the use and manipulation of living organisms, or substances obtained from these organisms, to make products of value to humanity. Although *biotechnology* is a relatively new term, this idea is not new—people have bred plants and animals to express particular traits since the beginning of civilization (see Chapter 13). However, biotechnology based on manipulating DNA outside of the living cell is a newer and much more powerful application of the technology. These new techniques let scientists isolate genes from one organism and then insert them into another, to produce an organism with a new trait. Before these developments, plant breeders were largely limited to genetic exchanges within a new species and between closely related species. Now scientists can transfer genes (genetic engineering) and therefore inherited traits between very different organisms. This ability to transfer genes among humans, plants, and bacteria has revolutionized biotechnology.

Progress in genetic engineering depends on a new technology that lets scientists isolate, identify, and clone (produce multiple copies of) genes. This recombinant DNA technology, called *gene cloning,* was developed in the 1980s, but originated with the discovery of plasmids and restriction enzymes in bacteria in the 1970s. Scientists make recombinant DNA (rDNA) by joining, or "recombining," DNA segments from different sources: plant and bacterial DNA, or plant and animal DNA. Because DNA strands from all organisms have the same chemical structure, they can be cut into segments and the segments linked together again in new and different ways.

Cutting the DNA is done with special enzymes called *restriction enzymes,* which occur in bacteria as part of a natural defense system against invading viruses. Restriction enzymes are highly specific to the nucleotide sequence, and each enzyme only cuts at one specific short base sequence in the DNA. For example, the enzyme *Eco*R1, found in the intestinal bacterium *Escherichia coli,* cuts only at GAATTC (the complementary strand, CTTAAG in this case, is assumed), and such a sequence occurs on the average only once every 4,000 nucleotides of DNA. An interesting feature of most restriction enzymes is that they make a staggered cut across the two strands of DNA, producing what molecular biologists call "sticky ends." Another enzyme, DNA ligase, can rejoin (or ligate) those sticky ends together again (**Figure 6.12**), and this can happen whether the DNA is from the same organism or from different organisms. Such manipulations of DNA in the laboratory are often called *gene splicing.*

Scientists carry out the job of cloning—of producing multiple copies of a gene or DNA segment—with the help of what is called a *vector.* The most common vectors are plasmids, and these are the delivery vehicles scientists use to introduce recombinant DNA into a host cell. Plasmids are (small) double-stranded circular DNA molecules that replicate in bacteria. An important property of plasmids is that they can sometimes be transferred from one organism to another. When a foreign gene is introduced into a plasmid, the plasmid can still replicate. By using a restriction enzyme, a scientist can cut the circular plasmid open. If DNA from another organism has been cut with the same restriction enzyme so that the same sticky ends were created, DNA ligase can be used to reform circular molecules that contain a segment of foreign DNA spliced into the site where the plasmid was cut. The new plasmid is transferred to bacteria that have been specially treated for this purpose; when the bacteria are allowed to multiply, the plasmids are copied within the bacteria. Millions of copies of the plasmid and the inserted gene can be reisolated from these bacterial cells (**Figure 6.13**).

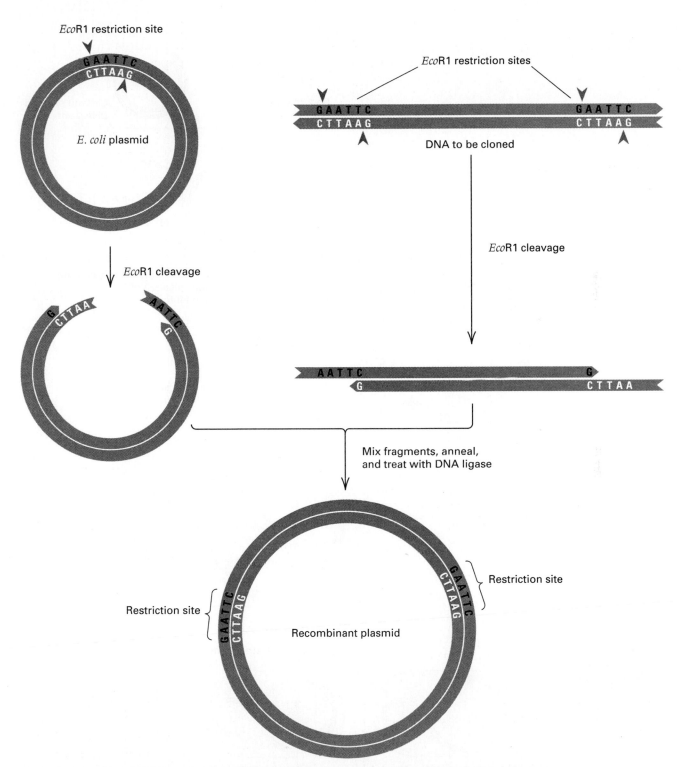

Figure 6.12 Construction of recombinant DNA molecules. A specific restriction enzyme— *Eco*R1, for example, as shown here—makes a staggered cut producing "sticky" ends. The single-stranded end of one fragment can therefore recognize and bind to the end of any other fragment produced by the same enzyme, even if the fragments originally came from the DNAs of different species. The two fragments can then be joined and circularized by the action of an enzyme called *DNA ligase*. Restricting and ligating DNA in this way forms the basis of recombinant DNA technology. The red arrowheads indicate the *Eco*R1 cleavage sites. *Source:* D. Hartl and E. W. Jones (1998), *Genetics: Principles and Analysis*, 4th ed. (Boston: Jones and Bartlett), p. 367.

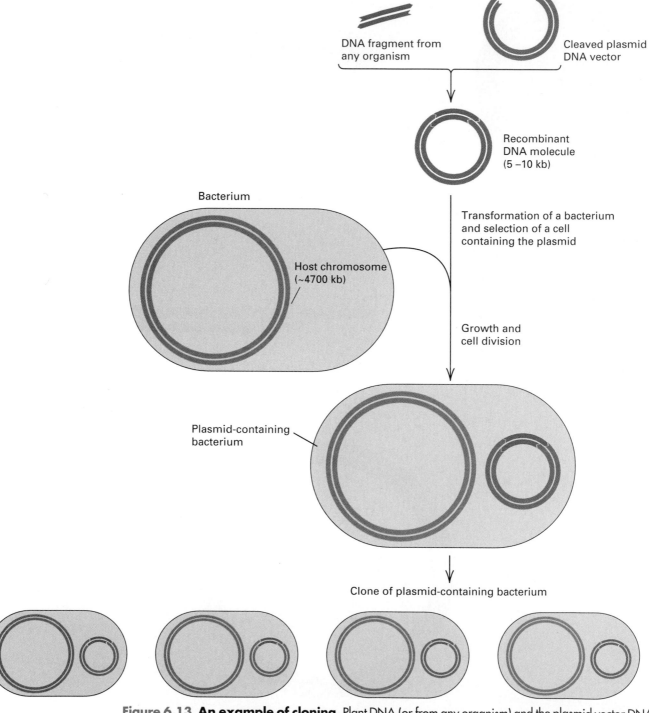

Figure 6.13 An example of cloning. Plant DNA (or from any organism) and the plasmid vector DNA are cleaved by the same type of restriction enzyme and spliced together by DNA ligase. Gene cloning occurs when the host bacterium takes up the recombined plasmid and the plasmid is reproduced in all the progeny bacteria. *Source:* D. Hartl and E. W. Jones (1998), *Genetics: Principles and Analysis,* 4th ed. (Boston: Jones and Bartlett), p. 363.

6.7 Plant regeneration technology underlies the generation of genetically engineered plants.

When small pieces of plant tissue are put in sterile culture on a solid nutrient medium, the cells proliferate and a callus forms, and if auxin and cytokinin are present in the correct amounts, shoots will form (**Figure 6.14**). Shoots normally arise from small groups of rapidly dividing cells within the callus. This discovery led scientists to ask a fundamental question: Can a whole plant be grown from a single cell, given the correct nutritional and hormonal environment? Many years ago, German, Japanese, and U.S. scientists discovered independently that this is indeed the case, and this discovery let them draw important conclusions about genes and how they function.

(a)

(b)

(c)

(d)

Figure 6.14 Embryogenic callus and plant regeneration. For most plants, the levels of two hormones, auxin and cytokinin, in the medium and the ratio of one to the other determine whether a small piece of tissue only forms callus (when both hormones are at high levels), sends out shoots (high cytokinin, low auxin), or sends out roots (high auxin, low cytokinin). **(a)** Proliferation of embryogenic callus derived from young leaf tissue. **(b)** Shoots began to regenerate from pieces of callus after the biologist reduced the concentration of cytokinin in the media. **(c)** When shoots have elongated, the biologist cut them off the callus and placed them on a media with a high auxin concentration. **(d)** Roots develop on media with a high auxin concentration. In this example, the plant is sugarcane. *Source:* Courtesy of James E. Irvine and T. Erik Mirkov.

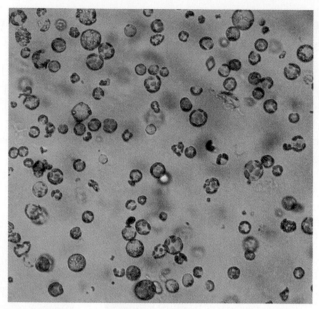

Figure 6.15 Freshly isolated protoplasts. Plant cells from which the cell wall has been removed are called *protoplasts*. The most common procedure to make protoplasts involves enzymatic cell wall degradation using a mixture of enzymes extracted from fungi, snail gut, and termite gut (all organisms that can digest cell walls). The protoplasts shown here were isolated from citrus leaves. *Source:* Courtesy of Eliazar Louzada.

Making an entire plant depends on the correct expression of at least 10,000 genes, perhaps more. When a living plant cell is isolated from a mature tissue, it can be induced to start dividing again and all the genes necessary to make an entire organism can be induced to function again in the correct sequence. This ability of a single mature plant cell to give rise to an entire organism is called *totipotency*. Because plant cells are interconnected by their cell walls, in a kind of honeycomb, it is not so easy to isolate single cells. It is much easier to digest the cell walls of a small piece of plant tissue with enzymes and then isolate the protoplasts (naked cells without walls). The protoplasts (**Figure 6.15**) are quite fragile, but they too can be cultured and regenerated into entire plants.

Regenerating whole plants from single cells (whether a protoplast or a cell that is still part of a tissue) is an important aspect of genetic engineering. Genetic engineering involves introducing a gene, usually from a different plant or an unrelated organism such as a bacterium, into a plant. If you can get the gene incorporated into the genome of just one cell and then encourage that cell to divide and form a whole organism, then you can get that gene into all the cells of that new plant. Two methods of introducing a new gene are commonly used, *Agrobacterium*-mediated gene transfer or a gene gun (see next section). With either method only a few cells are transformed; most cells remain untransformed. The latter must be prevented from growing into whole plants, otherwise you won't know which new plants are transformed and which are not.

To kill all the untransformed cells, scientists use a herbicide or an antibiotic in the culture medium, and let the transformed cells survive by linking them to a second gene that inactivates the herbicide or antibiotic as soon as it enters the cell. This gene is called the *selectable marker*. Genetic engineers usually introduce two genes at the same time: the gene that encodes the novel trait they wish to introduce—also called the *gene of interest* and the selectable marker gene. Molecular manipulations later eliminate the molecular marker gene. However, during the tissue culture phase the selectable marker gene lets the cells synthesize the enzyme that breaks down the antibiotic or herbicide, so only transformed cells survive and generate transformed plants. Plants produced with these techniques are usually called *genetically modified (GM)* or *genetically enhanced (GE) plants*. We use this terminology here for consistency, recognizing that plants improved by classical breeding are also genetically modified or enhanced.

6.8 Plant transformation depends on a "promiscuous" bacterium that transfers its DNA to the plant genome, or on the direct introduction of DNA using a "gene gun."

The goal of GM technology is to isolate one or more specific genes and introduce these into plants. GM crops (or foods) are crops produced by using molecular techniques to introduce new genes, followed by several years of plant breeding (see Chapter 14). For many plant species, one or a few genes can be introduced via the natural gene transfer system of the pathogenic soil bacterium *Agrobacterium tumefaciens,* which causes tumors (called *crown galls*) in many plants. When these bacteria infect a wound site, usually on the stem close to the ground, the infection disturbs the normal healing process. Instead of making a protective tissue that covers the wound, cells proliferate into cancerous growth. Cells from this tumor can be grown in tissue culture, and unlike normal plant cells, they continuously proliferate even when hormones are absent from the culture medium.

Molecular biologists in Belgium, the Netherlands, and the United States found that after the bacteria attach themselves to the wound site, they transfer some of the genes on the Ti plasmid to the plant cells. The bacteria have a large plasmid of 200,000 bases (200 kilobases or 200 kb) (also called the *Ti plasmid,* for "tumor inducing"), and when the bacteria infect a wound site a portion of the plasmid, called the *T-DNA,* is cut out by enzymes, coated with proteins and then becomes integrated in a plant chromosome (**Figure 6.16**). This integration is random, meaning that scientists cannot yet direct the T-DNA to a specific plant chromosomal location.

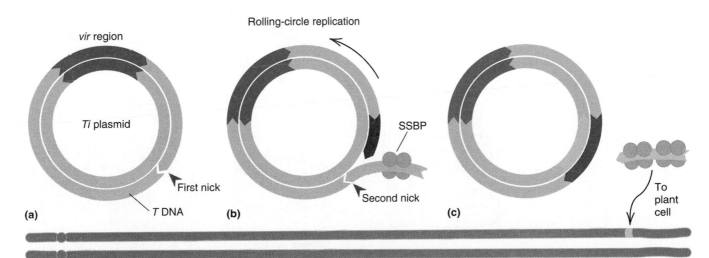

Figure 6.16 Genetic transformation of a plant genome by T-DNA from the Ti plasmid. **(a)** A nick forms at one end of the T-DNA after the bacteria have attached to a wound site on the plant. **(b)** Replication elongates one end and displaces the other end where the first nick was formed. The single-stranded binding protein (SSBP) that is coded by one of the *vir* genes then stabilizes this displaced piece of single-stranded DNA. A second nick terminates replication. **(c)** The single-stranded DNA that is coated with the SSBP is transferred into a plant cell and integrates into the plant genome. *Source:* D. Hartl and E. W. Jones (1998), *Genetics: Principles and Analysis,* 4th ed. (Boston: Jones and Bartlett), p. 382.

Genes thus transferred carry the information necessary to make auxin and cytokinin. The abnormally high levels of these hormones produced in the plant cells prompt cancerous cell growth. This resembles the situation in culture where the continuous availability of the same two hormones in the culture medium causes cell proliferation and callus formation. After the genes have been transferred from a single bacterium to a single plant cell and integrated into a chromosome of the plant cell, the hormones in the culture medium induce the cell—which is now transformed with the bacterial genes—to proliferate. So the bacterial genes will be present in every cell derived from the transformed cell. This transfer of DNA between the different species is sometimes called "promiscuous," because DNA is usually transferred only between individuals of the same species, as part of the process of sexual reproduction. Nevertheless, this transfer of DNA between totally unrelated organisms is a natural process. Soon molecular biologists realized that if they substituted other genes for the genes that the bacterium transferred to the plant cell,

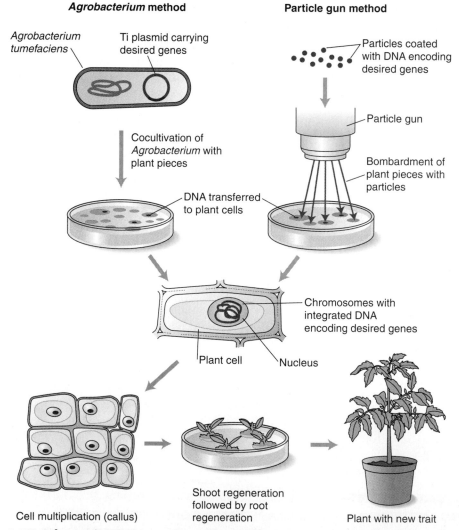

Figure 6.17 Schematic representation of two different ways to create transgenic plants. In the *Agrobacterium* method (left), the biologist inserts DNA carrying the desired genes into the tumor-inducing (Ti) plasmid of the bacterium, and when the bacterium infects a wounded tissue, it transfers this DNA to a cell nucleus and integrates it into the chromosome. In the particle gun method, metal particles coated with DNA become integrated into the plant chromosome. When a new plant regenerates from a single transformed cell, all the cells in the plant carry the new genes. *Source:* Adapted from "Transgenic Crops" by C. S. Gasser and R. T. Fraley, *Scientific American,* June 1992. Copyright 1992 by *Scientific American, Inc.* All rights reserved.

they could obtain transformed plant cells that carried any gene they wished to introduce into the plant.

Not all types of plants can be easily transformed with *Agrobacterium* (especially the monocots), so researchers have developed other methods to introduce foreign DNA into the genomes of important crops such as maize, wheat, sorghum, sugarcane, beans, and soybeans. John C. Sanford and his colleagues at Cornell University developed a micro-projectile gun (or "gene gun") to shoot DNA into plant cells directly, through the cell wall and the cell membrane into the nucleus. They coated small (about 1 micron in diameter) gold or tungsten particles with DNA and constructed a gun that accelerated the particles with enough speed to enter the first cell layer of a tissue. For reasons not yet completely clear, DNA that enters the cells in this way is integrated into the chromosomes and passed on to the cells' progeny. By using tissue culture techniques, as described earlier, researchers can generate a whole new plant from this transformed cell (**Figure 6.17**).

6.9 Genomics will dramatically transform crop improvement.

Gene technologies and tissue culture techniques have dramatically increased the possibilities for crop improvement. It is no longer necessary to rely only on the genes of close relatives of a particular crop because biologists can use genes from any source and can introduce a particularly useful gene into many crops. Using the same bacterial gene, for example, they can make 10 different crops tolerate the same herbicide (see Chapter 17). In the 1990s a new science emerged: genomics, or the study of all the genes of an organism. Genomics is concerned with the nucleotide sequence of all the genes, the functions of all the genes, and the expression (when, where, in response to which stimuli) of all the genes. Scientists use high-throughput methods to analyze 20,000 or more genes at once!

Advances in DNA-sequencing technology and computer science have now made it possible to sequence entire genomes. Advances in plant transformation technology make it possible to produce tens of thousands of random insertion mutants (see **Box 6.2**) that allow scientists to study the characteristics of plants in which genes are inactivated by the insertion of a large piece of DNA. Such a collection of "knockout" mutants with a different gene inactivated in every plant constitutes a valuable resource for molecular geneticists. A collection of such mutants is already available for *Arabidopsis* and may be available for rice as soon as 2005. Agricultural biotechnology companies are already racing to produce such collections of rice plants.

Another new technology, called *expression profiling,* allows scientists to determine the level of gene expression (amount of mRNA) of all the genes at once and to study the effect of environmental perturbations (lack of nutrients, water deficit, shade, pathogen attack) on gene expression. This ability will quickly lead molecular biologists to manipulate the pathways of gene expression, either with new biodegradable chemicals or by subtly altering a few genes of the plant. For example, a plant may resist a pathogenic fungus because it can quickly mount a defense response by inducing expression of several defense genes. A normal plant susceptible to fungus can't defend itself fast enough. Once scientists understand which gene is the master switch for this response, they may be able to speed up the response, allowing the plant to mount an effective defense. This type of genetic engineering will require subtle changes in genes and will resemble the normal process of evolution more closely than do current genetic technologies that introduce completely new genes.

Box 6.2

Plant Insertion Libraries Permit Scientists to Discover the Function of Unknown Genes

Twenty years ago, few researchers had ever heard of a small weed called *Arabidopsis thaliana*, but now hundreds of scientists all over the world are analyzing its genes. An international program to determine the base sequence of each and every one of its about 26,000 genes (about 116 million nucleotides total) has recently been completed. Why such a sudden interest in a seemingly useless weed? In the 1980s, molecular biologists discovered that this plant has the smallest genome known. Geneticists use the term *genome* to describe the sum total of all the genetic information of a species. Other plants, such as tobacco and wheat, probably have only twice as many genes, but these genes are embedded in 20 and 60 times more DNA respectively. This large amount of "excess" DNA makes it much harder to find the genes and figure out what they do. The genome of *Arabidopsis* is only about 25 times larger than that of the intestinal bacterium *Escherichia coli*, yet the plant is incredibly more complex. With its small genome, *Arabidopsis* does all the same things that plants with much larger genomes do. Indeed, *Arabidopsis* and humans have about the same number of genes.

Molecular geneticists were the first to realize that many things people want to know about important crops can be figured out much more easily by working with *Arabidopsis*. Now that the complete nucleotide sequence of the *Arabidopsis* genome is complete, scientists need to determine what all the genes do. Unfortunately, the nucleotide sequence data can't give all the answers—so far geneticists can only assign probable functions to about half of the genes—they have no idea what the functions of the other 15,000 or so genes are. How can scientists determine the function of all the unknowns? Kenneth A. Feldmann developed one very elegant method to help determine gene function. This method relies on the natural gene transfer system of *Agrobacterium tumefaciens*. As discussed in Sections 6.7 and 6.8, this bacterium transfers some of its own DNA, called *T-DNA*, into the plant genome. Sometimes the T-DNA lands in the middle of a gene, creating a mutant that can no longer express the protein. This is called a *knockout mutant,* and the gene is now "tagged" with the T-DNA. Researchers have generated many *Arabidopsis* mutants using a seed transformation process. If the mutant plant has a novel phenotype as compared to the nonmutant plant, the phenotype offers a clue as to the function of the knocked-out gene. Because the T-DNA has tagged the gene, it is relatively simple to clone the corresponding gene. By 2010, variations of this method and methods that use plant viruses to knock out gene expression should provide clues about the function of most of the genes in *Arabidopsis* and by extrapolation in other plants as well.

CHAPTER SUMMARY

Crop improvement through plant breeding can be readily explained at the molecular level. Plants have thousands of individual traits that are transmitted from one generation to the next because the information for each trait is contained in one or more genes, short segments of DNA. Each cell contains two copies of each gene, and when a cell divides, its DNA faithfully replicates, and two copies of each gene go to the daughter cells.

Genes contain the information to string amino acids together in the correct sequence to make specific proteins. This information must first be transcribed in an RNA molecule, which is then translated to become a protein or polypeptide.

Crop improvement invariably involves changing a plant's genetic makeup. Biologists can do this with conventional breeding techniques or by combining such techniques with molecular gene manipulation. Introducing a new gene into a plant gives the plant new information to synthesize a new protein that establishes a new trait. Understanding the molecular basis of heredity, and progress in molecular biology, together have allowed crop improvement through gene manipulation. Plant genetic engineering achieves in a very quick and precise way what breeders try to accomplish by years of backcrossing: the introduction of only (1) that segment of the DNA that has the necessary gene(s) to confer the new trait and (2) the necessary regulatory region or genes to confer the correct expression of that gene.

Genetic engineering depends on our ability to isolate genes, clone them (produce thousands of copies), and then introduce them into the plant genome using *Agrobacterium*-mediated gene transfer or a gene gun.

Genetic engineering will not replace plant breeding. After gene manipulation and tissue culture have introduced a single gene into a crop, several years of plant breeding are always needed to make sure that the new plant has the right agronomic characteristics. In the future, the molecular science of genomics will contribute substantially to crop improvement because genomics will help identify hundreds of plant genes that regulate all the processes of growth and development, adaptation to stress, and defense against insects and pathogens. Molecular techniques will improve crops by altering the regulation of those genes.

Discussion Questions

1. What are the requirements for the genetic material—what must it be able to do?

2. What are the differences between DNA and RNA? Which one accounts for the inheritance of traits? Explain.

3. What did Mendel think he discovered? Did his experiments help explain the blending concept of inheritance? If so, how?

4. Why can genes and parts of genes be transferred from one organism to another? What is required to have a foreign gene expressed in a different organism?

5. What methodology is used to produce recombinant DNA for gene cloning?

6. How do traditional plant breeding and genetic engineering differ?

7. Discuss the steps required to (a) introduce a new gene into a crop plant by traditional methods, and (b) introduce a new gene into a crop plant by genetic engineering.

8. When genetic engineering introduces a new gene into a plant genome, will it be passed on to progeny following Mendel's rules? Explain.

9. Are foods produced by traditional plant breeding methods safer to eat than those produced by genetic engineering? Why or why not?

10. Do you think foods produced by genetic engineering should be labeled as such? Why or why not?

Further Reading

Bruening, G. 2000. Transgenes are revolutionizing crop production. *California Agriculture* 54:38–46.

Conway, G., and G. Toenniessen. 1999. Feeding the world in the twenty-first century. *Nature* 402:C55–C58.

Dandekar, A. M., and N. Gutterson. 2000. Genetic engineering to improve quality, productivity and value of crops. *California Agriculture* 51:49–56.

Hartl, D. L., and E. W. Jones. 1998. *Genetics: Principles and Analysis*. Boston: Jones and Bartlett.

Keller, E. F. 1983. *A Feeling for the Organism: The Life and Work of Barbara McClintock*. San Francisco: W. H. Freeman.

Lander, E. S., and R. A. Weinberg. 2000. Genomics: Journey to the center of biology. *Science* 287:1777–1782.

Moffat, A. S. 2000. Transposons help sculpt a dynamic genome. *Science* 289:1455–1456.

Tanksley, S. D., and S. R. McCouch. 1997. Seed banks and molecular maps: Unlocking genetic potential from the wild. *Science* 277:1063–1066.

Plants in Human Nutrition and Animal Feed

T. J. Higgins
CSIRO Plant Industry

Maarten J. Chrispeels
University of California San Diego

The purpose of raising most crops is to produce food—but what is food? Food is any substance that provides an organism with energy and nutrients. Plants that have become food sources contain energy-rich compounds and nutrients that are digestible and are not loaded with toxic compounds. Humans, like other monogastric animals (such as pigs, chickens, fish), readily digest starch and protein, but cannot digest the large carbohydrate molecules such as cellulose that make up plant cell walls. However, ruminant animals such as cows and sheep can digest these molecules. Mammals also differ in how they handle toxic compounds. Acorns are rich in starch and protein, but contain such high levels of toxic tannins that humans can't use acorns unless they remove the tannins. Pigs, in contrast, eat unleached acorns quite happily. Historically, experiments with small mammals (especially rats) identified the essential nutrients needed for normal human development and this research formed the basis of dietary recommendations. Nutritionists aimed these guidelines at preventing clinical nutrient deficiencies. When epidemiological studies made clear that the prevalence of certain chronic diseases such as atherosclerosis and cancer was related to diet, scientists refined these recommendations. For example, in addition to recommending a certain number of calories per day, U.S. government guidelines now say that less than 30% of those calories should come from fat, and less than 10% from saturated fatty acids. Finally, human foods also contain many chemicals such as phytoestrogens and anti-oxidants that are not nutrients, but nevertheless can have positive effects on health.

7.1 Plants and animals differ fundamentally in their nutritional requirements.

Plants are autotrophic ("self-feeding") organisms with simple nutritional requirements. They need water, carbon dioxide (CO_2), and a variety of inorganic chemicals derived from soil minerals (see Chapter 11) as well as oxygen (O_2). The most important feature of plants,

the biosphere's major autotrophs, is that they use light energy from the sun to provide fuel for their own energy needs through the process of photosynthesis (see Chapter 10). By using sugar (the product of photosynthesis), water, and inorganic nutrients such as nitrate and phosphate from the soil, plants can synthesize all the complex organic molecules, such as amino acids and vitamins, that they need to grow and reproduce. Although plants and animals use many of the same organic molecules for their basic metabolic processes, only plants can synthesize these molecules using sugar and inorganic nutrients alone.

Heterotrophic ("other-feeding") organisms have more complex nutritional requirements. Rather than being able to use the sun's energy directly, they must ingest energy-rich organic molecules (mostly fats and carbohydrates) to sustain life. All heterotrophs, whether they eat plants (herbivores), animals (carnivores) or decompose dead organic matter (decomposers), ultimately depend on plants to provide them with energy-rich molecules. The exact nutritional requirements of most animals are not known because they have not been studied. Humans require the following organic nutrients (in addition to certain minerals, water, and oxygen):

1. Ten of the 20 amino acids that make up proteins—namely leucine, isoleucine, lysine, methionine, phenylalanine, threonine, tryptophan, valine, and (for infants) arginine and histidine—must be provided by proteins in our food.

2. The fatty acid linoleic acid must be provided by fats in the diet.

3. The vitamins A, B_1, B_2, B_3, B_6, B_{12}, C, D, E, K, pantothenic acid, biotin, and folic acid must be provided by various foods.

When nutritionists estimate the daily needs for food for people, they consider energy needs and nutrient needs differently. Energy requirements vary, depending on such factors as body weight and physical activity, so nutritionists recommend an average consumption (see

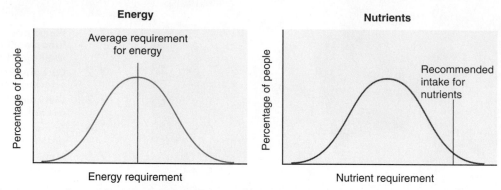

Figure 7.1 The rationale for requirements for energy and nutrients. Depending on body weight, activity, and genetic background, energy requirements in a population vary. An average requirement can be set, but it clearly does not apply to all individuals. Overconsumption should be avoided. Nutrient requirements (such as vitamins and proteins) are set at a level high enough to meet the needs of all people.

Figure 7.1). But body tissues cannot grow and renew if the supply of even one nutrient is inadequate. Nutritionists therefore set the RDA (recommended daily allowance) of nutrients well above the average requirement.

7.2 Carbohydrates and fats are our principal sources of energy.

The energy and nutritive values of plants as food depend on their chemical compositions, and foods vary in their ability to supply these dietary components. Carbohydrates are the source of most food energy for people. Not surprisingly, carbohydrates are abundant in staple food plants such as wheat, rice, maize, cassava, millets, and potatoes. Simple carbohydrates consist of one or several sugar molecules, whereas complex carbohydrates (also called *polysaccharides*) are made up of hundreds or even thousands of units of sugar molecules.

Simple sugars, such as glucose or arabinose (**Figure 7.2a**), or small oligosaccharides, such as sucrose (cane or beet sugar) or lactose (milk sugar) (**Figure 7.2b**), are readily soluble in water and are easily digested and taken up from the intestinal tract into the bloodstream. As a result, they provide a source of quick energy. Although most adult Caucasians can digest lactose, many Asians and Africans cannot because they lack the enzyme lactase, which splits lactose into its two component sugars. Such people are lactose intolerant. When they eat milk or milk products, the lactose is fermented by their intestinal bacteria, causing severe cramps and discomfort. Babies generally have high levels of lactase, so they have no problem digesting milk sugar.

Polysaccharides, or complex carbohydrates, such as starch and cellulose, are macromolecules that consist of linear or branched chains of sugar units (**Figure 7.2c**). They can consist of a single type of sugar, as in cellulose and starch, or different types of sugars, as with hemicelluloses, the major polymers found in soluble and insoluble fiber. Starch and cellulose, although both polymers of glucose, have different chemical and physical properties and play different roles in the plant. Cellulose forms fibers and is found in the cell walls of plants where it provides strength. Wood is primarily cellulose. Starch, consisting of amylase and amylopectin (Figure 7.2c), in contrast, is deposited by plants as food stored in specialized organs such as roots, tubers and seeds, for the plants to use whenever they need a source of energy.

Monosaccharides or simple sugars

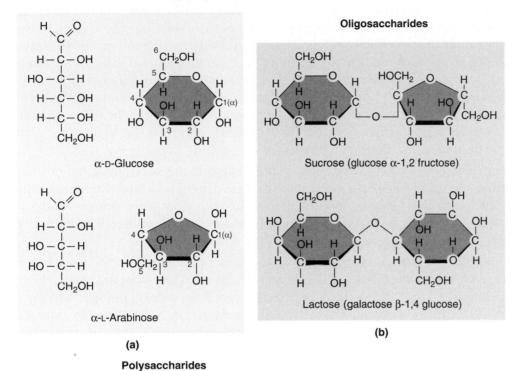

α-D-Glucose

α-L-Arabinose

(a)

Oligosaccharides

Sucrose (glucose α-1,2 fructose)

Lactose (galactose β-1,4 glucose)

(b)

Polysaccharides

Cellulose

Amylose

Amylopectin

Glucose

Xylose
Galactose
Fucose

Xyloglucan, a hemicellulose polysaccharide found in plant cell walls

(c)

Figure 7.2 Structures of some important sugars (a, b) and polysaccharides (c).

Complex carbohydrates are not soluble in water and must be broken down by digestive enzymes into simple sugars before the human body can use them. Breaking down starch requires four different enzymes, which are present in the human digestive system. However, people cannot digest cellulose or hemicellulose, the other major polysaccharides present in plants. So the sugar units in these macromolecules pass unused through human bodies and provide no food energy. Nevertheless, cellulose and hemicellulose do benefit human health. They constitute the insoluble and soluble fiber that promotes peristalsis of the bowels and prevents constipation. A high-fiber diet may lower the risk of colon cancer, but this needs further research. Fiber certainly increases the well-being of many people despite its contribution to flatulence.

Some microorganisms have the enzymes necessary to digest cellulose and hemicellulose and can therefore use the sugar units in these materials as food. Ruminant animals have such microbes living in a portion of their digestive tracts—the rumen—and get energy from the release of sugars from these polysaccharides, either directly or indirectly after the microbes convert sugars to volatile fatty acids and methane.

Lipids, or fats, oils and certain other molecules are present in all organisms, being essential structural components of all cellular membranes. Their most important chemical property is their insolubility in water, which allows them to form physical barriers that control the exit and entry of many of the chemicals in cells and tissues. To enter a cell, glucose requires the presence of a specific glucose carrier protein in the plasma membrane. Lipids occur in large amounts in certain animal tissues such as brain, eggs, and some oil-rich seeds such as canola, soybeans, and peanuts.

Chemically, there are many different types of lipids. Three important classes of lipids are triacylglycerols (fats and oils), phospholipids, and cholesterol.

1. Triacylglycerols have three fatty acids (**Figure 7.3a**) connected to glycerol (**Figure 7.3b**), a simple three-carbon molecule. Triacylglycerols are a major storage form of energy in animals and plants.

2. Phospholipids (**Figure 7.3c**) are present in cellular membranes. A membrane consists of a double layer of phospholipid molecules with embedded proteins that permit the passage of small molecules through the membrane.

3. Cholesterol is structurally very different from triacylglycerols and phospholipids. Cholesterol and other molecules like it, such as vitamin D and the steroid hormones, are made out of four linked rings of carbon atoms (**Figure 7.3d**). Humans synthesize cholesterol to make bile acids and hormones such as testosterone and estrogen. Plants have similar molecules (such as phytoestrogens).

Nutritionally, the most important lipids are the triacylglycerols (fats and oils). The size of the fatty acids (number of carbon atoms in the chain) and the number of double or unsaturated bonds (Figure 7.3a) determine the chemical and physical properties of the fat molecules as well as their metabolic fate in the human body. The more saturated the fatty acids and the longer the chains, the more solid the fat will be at body temperature. Fats with long, saturated fatty acids are commonly found in animal products (meat, milk, and cheese) and in some plant products (cocoa butter). Fats that are liquid at room temperature are called *oils* and are found in seeds and fish. Oils can be converted into fats by a process called hydrogenation, in which hydrogen atoms attach to the double bonds that link adjacent carbons. Hydrogenation produces *trans* fatty acids, so called because the new hydrogen atoms attach on different sides of the carbon chain. Fats in the diet also help the body absorb certain vitamins that are not readily soluble in water, such as vitamin A. Triacylglycerols from which one or two fatty acids have been removed are called

(a)
Common fatty acids

These are carboxylic acids with long hydrocarbon tails. Hundreds of different kinds of fatty acids exist. Some have one or more double bonds and are said to be unsaturated.

This double bond is rigid and creates a kink in the chain. The rest of the chain is free to rotate about the other C—C bonds.

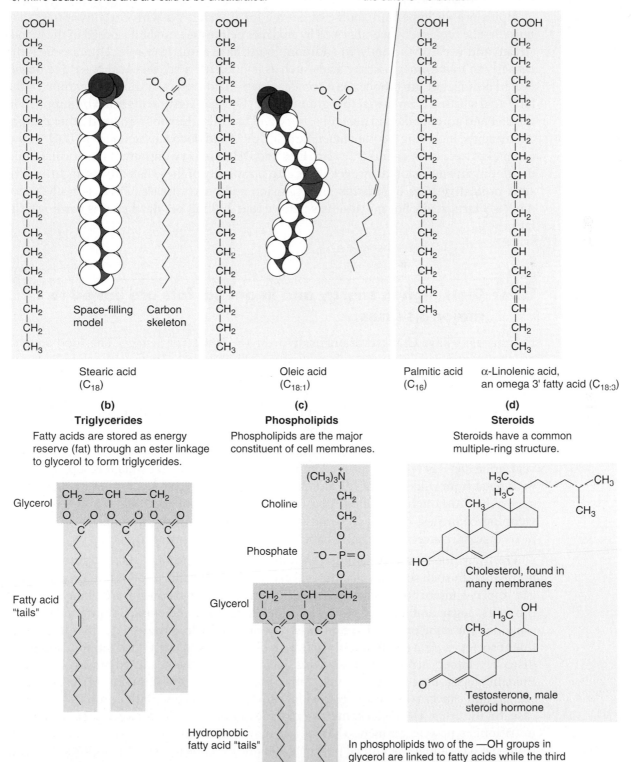

Stearic acid
(C_{18})

Space-filling model   Carbon skeleton

Oleic acid
($C_{18:1}$)

Palmitic acid
(C_{16})

α-Linolenic acid, an omega 3' fatty acid ($C_{18:3}$)

(b)
Triglycerides

Fatty acids are stored as energy reserve (fat) through an ester linkage to glycerol to form triglycerides.

Glycerol

Fatty acid "tails"

(c)
Phospholipids

Phospholipids are the major constituent of cell membranes.

Choline

Phosphate

Glycerol

Hydrophobic fatty acid "tails"

In phospholipids two of the —OH groups in glycerol are linked to fatty acids while the third —OH group is linked to phosphoric acid. The phosphate is further linked to one of a variety of small polar head groups.

Phosphatidylcholine

(d)
Steroids

Steroids have a common multiple-ring structure.

Cholesterol, found in many membranes

Testosterone, male steroid hormone

Figure 7.3 Structures of some important cellular lipids.

diglycerides and *monoglycerides,* respectively. Strictly speaking, they are not fats, and they are the ingredients of "fat-free" margarine.

Lipids are an important source of energy for humans. As with carbohydrates, they must first be broken down or digested by enzymes before they can be absorbed in the bloodstream and used by the body. In addition, lipids have a nutritive role. Humans lack the capacity to make omega-3 fatty acids such as α-linolenic acid; this acid has three unsaturated bonds, but the number 3 in "omega-3 fatty acid" refers to the third carbon of the fatty acid starting from the methyl group (CH_3). Omega-3 fatty acids must be supplied in the diet and are abundant in plant oils, nuts, and fish oils. Humans also need other omega-3 fatty acids, including eicosapentaenoic acid (EPA) and docoshexaenoic acid (DHA), but can make these from α-linolenic acid. EPA and DHA are very important for normal brain development and vision. However, the recent popularity of these fish oils stems from possible protective roles in arthritis, hypertension and heart disease. There is no RDA for omega-3 fatty acids, but nutritionists believe that 200 mg per day benefits cardiovascular health.

7.3 Diets high in energy and in animal fats are linked to major diseases.

Human diets have changed dramatically over the course of history. The food sources used by hunter-gatherers were extremely varied. Most collected many different plants and supplemented a largely vegetarian diet with meat or fish when available. These mobile societies often followed their food sources as seeds, fruits, or roots of different plants became available throughout the seasons. The major cause of illness or death in these societies was infection caused by a lack of sanitation.

The human diet changed with the transition to agriculture (Chapter 13). A still largely vegetarian diet that reflected the predominance of a few cultivated plants was supplemented with meat from wild or domestic animals. Many people in the developing world still eat such a diet. Although infectious diseases and undernutrition are still too prevalent in developing countries, the diseases that cause most deaths in the developed countries—heart disease, cancer, stroke, and diabetes—are only now emerging there.

The transition to modern affluent societies a hundred years ago led to a radical change in the diet. In such societies, plants, especially vegetables and whole-grain products, began to play a minor role and people greatly increased their consumption of meat, animal products, sugar, and alcohol. However, people's nutrient and energy requirements have remained the same or even have decreased because they lead more sedentary lives in affluent societies. As a result, obesity and high blood pressure are very common, and heart disease, cancer, stroke, and diabetes have become the leading causes of death. Epidemiologists have found a positive correlation between high fat intake, on the one hand, and certain cancers (especially colon and prostate cancer) and with coronary heart disease, on the other. Having become aware of the important links between diet and health, nutritionists now recommend that we eat more plants (five portions of vegetables or fruits per day) and fewer animal products and meat. In this respect it is important to start with a healthy diet early in life, because changing one's diet is extraordinarily difficult.

The relationship between fat intake and coronary heart disease is incredibly complex (see **Box 7.1**), and many studies yield contradictory results. The popular press has given

Box 7.1

Good Fats and Bad Fats

To understand the notion of good fats and bad fats, we must first look at the role of fat in a disease called *atherosclerosis* or hardening of the arteries. This disease is life threatening when it affects the coronary arteries that supply the heart with blood. The hardening is caused by the deposition of plaque, a fatty material that consists of cholesterol; small blood cells, called *platelets*; and collagen. The level of cholesterol that circulates in the blood seems to determine the extent of the hardening, and people with high levels of blood cholesterol are at greatest risk of heart disease. Because cholesterol is a fat, it is not soluble in water; it circulates in the bloodstream as a cholesterol–protein complex, a small protein-coated fat globule.

Two types of these complexes exist: low-density complexes (a little bit of protein surrounding the fat) and high-density complexes (a lot of protein coating the fat globule). The low-density lipoprotein (LDL) complexes seem to be the major culprits in hardening of the arteries and in heart disease. The nutritional advice given by the American Heart Association and other agencies is predicated on the assumption that lowering LDL (and cholesterol in general) will lower the risk of heart disease. It is also good to remember that obesity, smoking, and hypertension contribute to heart disease. Obesity and hypertension are also linked to poor diet.

Most people can lower blood cholesterol simply by eating less cholesterol, a fat only found in animal products. However, the human body synthesizes its own cholesterol and specific drugs can inhibit this synthesis. Foods heavy in certain saturated fats raise cholesterol levels. Especially bad are the saturated fats of milk products,

whereas other saturated fats (in meat or cocoa butter) have less or no effect on LDL levels. Foods rich in monounsaturated or polyunsaturated fats (many vegetable oils and fish oils) may actually lower blood cholesterol levels. People on high olive oil diets (rich in monounsaturates) generally have lower serum LDL levels, but does that mean olive oil lowers blood cholesterol, or is there another reason? Research does not yet have the answer. Peoples whose diets are rich in fish oil (with omega-3 unsaturated fatty acids) also have a low incidence of heart disease. However, such peoples (for example, the Inuit of Alaska) usually also have a completely different lifestyle from that, say, of affluent Westerners.

In the United States, the historical increase in consumption of *trans* fatty acids in hydrogenated margarine closely parallels the dramatic increase in coronary heart disease. During the same period (1900–1960), the intake of animal fat declined. Several studies have shown that *trans* fatty acids decrease serum levels of HDL and increase LDL levels.

To complicate matters even more, atherosclerosis is greatly influenced by the antioxidant vitamin E, a fat-soluble vitamin found abundantly in minimally processed liquid vegetable oil. To cause atherosclerosis, LDL needs to undergo an oxidative modification, a step that might be blocked by vitamin E. Men and women with a high intake of vitamin E (mostly from supplements) have a lower incidence of heart attacks than those with a low intake.

much attention to the Mediterranean diet (high fat intake, mostly as olive oil) and the Japanese diet (low fat intake, mostly from fish) in contrast to the U.S. diet (high fat intake, mostly from dairy products, hydrogenated oils, and meat). The first two diets are associated with a much lower incidence of cardiovascular disease. After reviewing all the studies, W. C. Willett from the Harvard School of Public Health concluded, "Intake of partially hydrogenated vegetable fats and saturated fats, particularly those from dairy sources, should be minimized . . . and that consumption of a substantial proportion of energy as monounsaturated fat would not be harmful and might even be beneficial" (Willett, 1994, p. 534).

7.4 To make proteins, animals must eat proteins.

Protein is a major constituent of all organisms, and a given cell may contain more than 5,000 different proteins. As noted in the previous chapter, proteins are long strings of 20 different amino acids (**Figure 7.4**), and a given protein has 100 to 1,000 of them, with their specific order and string length determining the protein's properties. The strings are folded in specific three-dimensional structures, unique for each protein (see Figure 6.6).

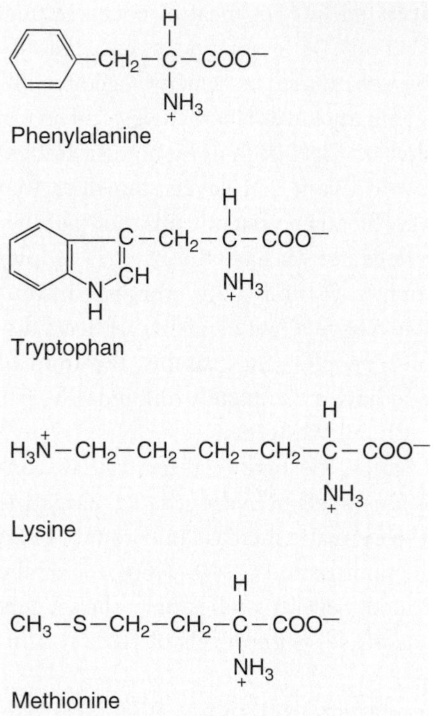

Proteins play important roles in cells: (1) as enzymes that catalyze biochemical reactions, (2) as regulators of gene activity, (3) as transporters of small molecules through membranes, (4) as transducers of signals from the environment and other parts of the plant, (5) as structural components of the cell, (6) as toxic agents to defend the plant against pests and pathogens, and (7) as reserves of nitrogen, sulfur, and carbon for later use. Most food proteins are structural proteins such as those found in muscle cells of meat and fish, or play a nutritive role, as in milk, egg white, or seeds.

As autotrophs, plants can synthesize all 20 amino acids from the sugars made in photosynthesis and the nitrate and sulfate taken up from the soil. The amino acids are then linked together in a specific order using the information in the genes (see Chapter 6). Mammals, which are heterotrophs, cannot make all the amino acids and need to get some of them in their food. The amino acids that they cannot make are called the *essential amino acids*. For adults, there are eight essential amino acids, and Figure 7.4 shows the structures of four of these. The nutritional value of protein-rich foods depends not only on their digestibility (not all proteins can be digested equally well), but also, and especially, on the relative abundance of the eight essential amino acids. Nutritionists use these criteria to evaluate proteins for their dietary value to humans or other mammals.

Human milk protein and egg protein are defined to have protein scores of 100, which means that they contain the essential amino acids in the exact proportions required by the human body. **Table 7.1** gives the protein scores for various foods. Animal proteins generally have a higher protein score than the major dietary plant proteins, because the latter tend to be low in the essential amino acids tryptophan, methionine, and lysine.

A marked deficiency in even one essential amino acid drastically lowers the protein score of a protein. Maize flour, which is low in tryptophan and lysine, has a protein score of 49, and navy beans, which are low in methionine, have a protein score of 44. Foods with protein scores lower than 70 are generally considered unsatisfactory for human growth and maintenance. This does not mean, however, that maize protein and bean protein have no nutritional value for humans. Foods that are deficient in different amino acids can complement each other when eaten together. A meal of maize tortillas with beans has a much higher protein score than either food alone.

As with calories, the amount of protein you require each day depends on many variables (see **Table 7.2**). Rapidly growing young people, physically active people, and pregnant and lactating women need more dietary

Figure 7.4 Structures of four essential amino acids.

| Table 7.1 | Protein scores of selected foods |

Food	Protein Score
Hen's egg	100
Beef	80
Cow's milk	79
Chicken	72
Fish	70
Rice	69
Soybeans	67
Wheat	62
Maize	49
Beans	44
Potatoes	34

Note: Egg is considered to have a perfect protein, and other sources are rated in comparison with it to give a protein score.

Source: Food and Agriculture Organization (1970), *Nutritional Study No. 24* (Rome: FAO).

Table 7.2	**Recommended daily intake of energy and protein for people in East Africa**		
	Age, Stage, Lifestyle	**Kilocalories per Day**	**Protein (grams per day)**
Children	1–2 years	1,000	40
	5–6	1,400	50
Girls	11–12	2,200	65
	13–17	2,500	70
Boys	11–12	2,000	60
	15–18	3,000	80
Women	Sedentary	1,800	55
	Very active	2,500	65
	Pregnant	Add 400	85
	Lactating	Add 900	95
Men	Sedentary	2,200	60
	Very active	3,000	70

Source: M. C. Latham (1965), *Human Nutrition in Tropical Africa* (Rome: Food and Agriculture Organization), p. 243.

protein. If your daily protein intake is very high (over 100 g per day), you metabolize excess amino acids and excrete the nitrogen (as urea), because your body has no way to store amino acids. In contrast, carbohydrates have limited storage as the polysaccharide glycogen in liver and muscle, and lipids are easily stored in fatty deposits. In low-caloric diets, body proteins are broken down for energy, because the body always satisfies its energy needs first. Physical activity, as in sports, greatly increases the body's need for energy, but when muscular activity doubles your body's caloric requirements, it increases your need for protein by only 5%.

If a woman does not eat enough protein during pregnancy, the intellectual development of the child may be slowed. This finding has focused attention on the importance of proper nutrition during pregnancy and lactation. Fetal growth (especially during the final six months of development) and milk production require that the woman's normal diet be supplemented with additional calories, proteins and other nutrients (essential fatty acids, vitamins and minerals). Women who are pregnant or lactating need to increase their daily protein intake by 20 or 30 g, respectively. Women who are undernourished therefore pass this condition on to their babies (see Chapter 4).

7.5 Plant breeding and genetic modification can improve the protein value of animal feed.

An adequate supply of essential amino acids is important both for human nutrition and for the efficient production of livestock that supply meat, milk, and wool. For intensively fed livestock such as pigs and poultry in developed countries, the compounding (mixing) of feeds to correctly balance essential amino acids is a highly developed scientific operation. It usually involves mixing components from a number of plant sources (cereal and legume grain) with some animal sources such as fishmeal or the unusable waste from slaughterhouses. In developing countries, in contrast, people often have to rely on a single staple such as maize that does not provide an adequately balanced diet. Research has focused on improving the nutrient quality of such diets.

Maize and other cereals are deficient in lysine and tryptophan, and one of the first efforts to create seeds with a better amino acid balance (higher protein score) resulted in the identification of a mutant maize line, called Opaque 2, that has a higher lysine content in its seeds (32% more lysine than the control lines). Research in the 1960s clearly demonstrated the nutritional advantage of this type of maize, but farmers did not widely adopt Opaque 2 maize, because it had poor agronomic properties (yield per hectare was 15% less). This discovery led to a 30-year major breeding effort to introduce this trait into elite maize lines, culminating in the production of quality protein maize lines. The lysine content of these lines is not quite as high as in the original Opaque 2 lines (20% more instead of 32% more), but the agronomic properties of the new lines are excellent and the nutritional benefit is still substantial. Farmers in Africa and South America, where maize is an important staple for humans, have widely adopted quality protein maize lines.

In the case of ruminant livestock such as cattle and sheep, the nutritional situation is more complicated because of the nature of ruminant digestive systems. The rumen of these animals functions as a large microbial fermentation vat in which ingested materials (cellulose, lignin, protein, fat) are digested, hydrolyzed, and converted to microbial mass (protein, cell walls) and fermentation products (methane, fatty acids). Thus ruminant animals are ideally suited to convert materials that humans cannot use (grass and other forage) into useful products (meat, milk, wool). Some human populations (such as nomadic herdspeople) depend on this conversion (**Figure 7.5**). Increasing the efficiency of this conversion process is a major goal of animal husbandry research. After the microbial mass is formed in the rumen, it continues its journey in the animal's gastrointestinal tract and is in turn hydrolyzed to its constituent sugars and amino acids, which are taken up into the animal's bloodstream through the wall of the intestine. The amino acids are used to synthesize new proteins for the animal's tissues. However, microbial protein has a significantly lower content of the sulfur-containing essential amino acid methionine, and there is a net loss of this amino acid during the conversion of plant protein to microbial protein in the rumen. As a result, when feed supplies are scanty, the lack of methionine limits production of milk, muscle (meat), and wool.

In times of feed shortage, the diet of ruminants in pastures may be improved with high-protein legume grains, but these are also a poor source of methionine. Efforts are now underway to create methionine-rich legumes using GM technology. Researchers have identified a protein in sunflower seeds that has an unusually high content of sulfur-containing amino acids. A further property of this protein is that it resists breakdown by microbial flora in the animal's rumen. It therefore passes undegraded into the animal's intestine, where the digestive enzymes can hydrolyze it into its constituent amino acids, including methionine, that the animal then uses to synthesize protein. Australian researchers introduced the gene coding for this protein into lupin (a legume) and targeted it for expression in the seeds. The result was a 100% increase in the methionine content of the seed protein, and when this grain was fed to sheep their weight gain increased by 7% and wool production by 8% compared to sheep fed unmodified seeds. The success of this approach has prompted researchers to introduce a leaf-specific form of the gene for this high-methionine sunflower seed protein into the leaves of pasture plants, with the ultimate goal of improving the balance of essential amino acids available to grazing ruminants.

Figure 7.5 Pastoralists with ruminant animals produce valuable food resources in arid regions unsuitable for agriculture. This camel herder from Mauretania collects milk for a small dairy that produces pasteurized milk and camel cheese. Most pastoralists feed their families and sell some products (meat or cheese) produced by their herds. Pastoralism is present in regions with little rainfall and poor quality land not suitable for crop production. *Source:* Photo by I. Balderi, FAO.

7.6 Vitamins are small organic molecules that plants can synthesize, but mammals generally cannot.

Vitamins are important organic molecules that plants and bacteria can synthesize, but, as with the essential amino acids, the human body cannot (except for vitamin D). Therefore, people must obtain vitamins in their diet. Vitamins are relatively small molecules, comparable in size to amino acids or sugars. Whereas daily protein requirements are measured in grams, vitamin requirements are generally measured in milligrams (mg). But even in trace amounts they play important roles in your body.

Several vitamins are coenzymes, molecules essential for making an enzyme function. For example, you need vitamin C for the enzymatic formation of collagen, a substance important in wound healing and in the stability of the joints. You need vitamin A for synthesizing an eye pigment. Vitamin E is an important antioxidant, preventing tissue damage by chemicals. Vitamins are present in foods and are taken up in the body dissolved either in water (vitamins B and C) or in fats (vitamins A, D, E, K). Vitamin D can also be synthesized by the human body when the skin is exposed to sunlight.

Although all living organisms contain nearly all the vitamins, some are particularly rich in certain vitamins and so are especially valuable food sources. Vitamin C is abundant in citrus fruits and is found in fresh vegetables. The B vitamins are most abundant in meats, wheat germ, and yeast. Cod liver oil is a good source of the lipid-soluble vitamins A and D. Vitamin A is also found in yellow vegetables (squash, carrots, sweet potatoes), but vitamin D is not abundant in plants. Vitamin E is especially abundant in green, leafy vegetables and unprocessed plant oils.

Box 7.2

Using GM Technology–Enriched Food Crops for Vitamins

Vitamin A

In a brilliant application of genetic engineering to plants, scientists developed a genetically modified line of rice that has increased provitamin A (β-carotene) content. Polished rice is extremely low in vitamin A and in countries in Southeast Asia, Africa, and Latin America where rice is a major component of the diet, vitamin A deficiency is widespread, particularly in children. This deficiency can cause blindness and exacerbate conditions such as diarrhea, respiratory diseases, and measles. Researchers estimate that 124 million children worldwide are deficient in vitamin A, and a quarter of a million go blind each year as a result. Scientists increased the vitamin A level by introducing into rice the genes for the last three enzymes in the biosynthetic pathway leading to the synthesis of provitamin A. They then added control sequences to ensure that the genes would be synthesized in the rice endosperm. Therefore the vitamin is not lost when milling removes the bran. The remaining seed (largely endosperm) is now a golden color due to the presence of the β-carotene (see figure). The announcement of this breakthrough was accompanied by so much hype that its limitations were overlooked: At the present level of provitamin A expression, a normal rice diet can only satisfy about a quarter of a person's vitamin A requirement.

Tomatoes are also an excellent source of provitamin A. A single tomato provides 23% of the recommended daily allowance. Gene technology has recently almost doubled the provitamin A content of tomatoes. Moreover, a mixed diet containing carrots or green, leafy vegetables such as chard, curly kale, or spinach supplies provitamin A in abundance. In Asia and Africa, many other often local vegetables also supply vitamin A, but poor city dwellers cannot always afford to buy them.

"Golden Rice." This new variety of rice produced by GM technology is yellow in color because provitamin A is produced in the entire grain (instead of just in the outer layers, as in traditional rice). Typical servings of rice will not satisfy daily vitamin A requirements, but the rice could be a significant factor in reducing vitamin A deficiency in the poor rice-eating countries. *Source:* Courtesy of Ingo Potrykus, Swiss Federal Technical University.

Vitamin E

The group of lipid-soluble *tocopherols,* known collectively as vitamin E, are essential antioxidants in human and animal diets. They are made only by green plants. Vegetable oils from soybean, maize, canola, and palm oils are a major source of this vitamin in human diets. Although most diets, at least in developed countries, supply the RDA for this vitamin group, an intake well in excess of the RDA is correlated with decreased risk of cardiovascular disease, improved immune function, and slowed progression of several degenerative conditions. A vitamin E deficiency in sheep results in white muscle disease, reduced fertility, and sudden death from heart failure. These conditions often occur in grazing animals, such as sheep, when green pasture is unavailable and supplementary cereal grain is the main energy source.

In most oil seeds the major tocopherol is γ-tocopherol, a relatively inactive precursor of α-tocopherol, which is the active form of this vitamin. Converting γ- to α-tocopherol involves adding a methyl ($-CH_3$) group. Recently, scientists have isolated the gene that codes for the enzyme catalyzing this methylation reaction in green plant tissues. Genetic engineers equipped the gene with a seed-specific promoter and introduced it into a test plant (*Arabidopsis thaliana*). The result was a plant in which 95% of the tocopherol was in the active, methylated form: an 80-fold increase in active vitamin E level. Applying this concept to oil seed crops promises to greatly improve vitamin E supply for both humans and their livestock.

Vitamin K

Another potential application of genetic engineering to human diet is raising the level of vitamin K. Vitamin K is important in bone formation; supplementary doses of vitamin K decrease postmenopausal bone loss in women, and may play an important role in maintaining bone mass in astronauts. Some green plants such as broccoli are rich in vitamin K, and it is possible that gene modification could raise levels of this vitamin in other plant sources, as with vitamins A and E.

The minimum levels of vitamins (recommended daily allowance, see Figure 7.1) are set high enough to prevent specific deficiency diseases. For example,

- Vitamin A deficiency causes blindness and increases childhood mortality.
- Vitamin B_1 deficiency (thiamine) causes beriberi, characterized by weak muscles and paralysis.
- Vitamin B_3 deficiency (niacin) causes pellagra, characterized by skin lesions, diarrhea, and mental apathy.
- Vitamin D deficiency causes rickets, characterized by weak and misshapen bones.

Unfortunately, the cereal grain staples that make up two thirds of the human diet do not offer abundant vitamins. Moreover, grain milling removes the vitamin-rich outer layers. For example, beriberi became much more common in Asia when mills started polishing rice, a process that removes the thiamine-rich outer layers of the grain. In North America, beriberi spread when bread made from white wheat flour became popular during the late 1800s. Milled grain is also deficient in folic acid, a vitamin the fetus requires for developing the neural tube and the spinal cord. Maize contains little tryptophan, an amino acid that can act as a precursor for synthesizing vitamin B_3. Moreover, the high levels of the amino acid leucine in maize proteins seem to block conversion of tryptophan into vitamin B_3. These two characteristics combine, so people who eat maize as a staple often suffer from pellagra.

There are four ways to solve vitamin deficiencies. First, if the right foods are available, people can eat a more varied diet, and thus eat foods rich in the vitamins they lack. Second, people who can afford to can buy vitamin supplements, or governments can distribute free supplements. Third, manufacturers can add vitamins to foods when they are processed. This process, termed *fortification,* has often been very successful. For example, in many developed countries both white and brown wheat flours are fortified with vitamins such as thiamine, riboflavin, niacin, and pyridoxine when the grain is milled. This process actually makes these flours, and products made from them, a better source of vitamins than the original whole grain. Another familiar example is the addition of vitamin D to milk. This fortification has greatly reduced the incidence of rickets. Fortification provides a large population with essential nutrients such as vitamins in a convenient way and is a cheap and effective way of improving a population's nutritional status. Fourth, agronomists can genetically modify food plants to contain higher levels of certain vitamins. A recent success story in this regard is the production of Golden Rice, so called because it is yellow, reflecting its high level of provitamin A (see **Box 7.2**).

Many people believe organically grown fruits and vegetables are richer in vitamins and minerals than conventionally grown produce. However, no data substantiate those beliefs. Nevertheless, organically grown (Europeans use the term "biological") products are a fast-growing sector of the food market. Because these products are more expensive than conventionally grown crops, the organic food markets cater to wealthier consumers (**Figure 7.6**).

Figure 7.6 The popularity of organic produce is rising in many countries. Many people believe organic produce to be nutritionally superior, which accounts for its rise in popularity. In this market in Santa Barbara, California, produce was sold under five different labels, "Certified Organic," "Registered Organic," "Organic," "Chemical free," and "Biological," or without a specific label. It is doubtful that the buying public fully understands the agronomy behind such designations.

7.7 Minerals and water are essential for life.

At least 18 different minerals are essential for human life and must be taken in the diet (**Table 7.3**). You need some of them, such as calcium and phosphorus, in large amounts. You need others, such as iron or magnesium, in smaller amounts. Still others—such as copper, cobalt, and molybdenum—you need in trace amounts. Because many of these minerals are quite common in the liquids people drink and the foods they eat, nutritionists may not pay enough attention to them. The minerals of special concern are those known to be deficient in certain diets: calcium, phosphate, iron, and iodine.

Calcium, phosphate, and magnesium are called the "bone builders" because people need them in large amounts for bone formation. To incorporate these minerals into bones, the body requires vitamin D. Milk and milk products contain abundant calcium and phosphate, which are also in grain products, meat, and a variety of vegetables. A 1985 nutritional survey of the United States revealed that over 40% of the people surveyed took in less calcium than the recommended allowance. Urban households on low incomes tended to have the most calcium-deficient diets. A lack of calcium causes osteoporosis or loss of bone density, a serious disease of the elderly.

Sodium, potassium, and chloride are present in all your cells and fluids as electrically charged ions or electrolytes. These electrolytes maintain blood pressure, play an important role in the acid–base balance of the fluids (also called the pH), and are vital for transmission of nerve impulses and muscle contraction. An important source of sodium and chloride is table salt (NaCl); potassium is found in all plant foods. Deficiency of these minerals occurs in cases of chronic diarrhea, where fluid losses are excessive. Such loss can lead to congestive heart failure and death.

Iron is the most important micromineral, and iron deficiency leads to anemia, a disease characterized by a general weakening of the body because it cannot synthesize hemoglobin needed to carry oxygen in the bloodstream. Anemia is more prevalent among women than men, because women must replace the blood lost during menstruation, pregnancy, and

Table 7.3	Some mineral requirements in the human diet	
Mineral	**RDA (mg)**	**Functions**
Calcium	1,000–1,300	Bone and tooth formation
Phosphorus	700–1,250	Bones and teeth formation; some metabolic functions
Magnesium	320–400	Bone formation, enzyme activation
Sodium	500–3,000	Bone, electrolyte formation
Chloride	750–3,000	Electrolyte formation
Potassium	2,000	Electrolyte formation
Iron	10–15	Hemoglobin, enzymes formation
Zinc	12–15	Enzymes, insulin formation
Copper	1.5–3	Enzymes formation
Iodine	0.15	Thyroxine formation
Manganese	2.5–5	Enzymes formation
Selenium	0.04–0.07	Fat metabolism
Chromium	0.05	Glucose metabolism

Sources: RDAs from Food and Nutrition Board, U.S. National Research Council (1989), *Recommended Daily Allowances,* 10th ed. (Washington, DC: NRC) and (1998) *Dietary Reference Intakes* (Washington, DC: NRC).

childbirth. Therefore, the RDA for iron is higher for women (18 mg) than men (10 mg). Normal diets provide between 10 and 15 mg of iron per day, which is not enough for women who lose a substantial amount of blood during menstruation.

The iron requirements of pregnant women are very high, and anemia is especially prevalent among them. Iron deficiency anemia affects 400 million women of childbearing age (15–49 years old) mostly in developing countries. One third of all mothers' deaths at childbirth result from iron deficiency anemia; the babies are often stillborn or underweight. A newborn child contains about 400 mg of iron; placenta growth and blood loss during delivery use up about another 300 mg. The total iron requirement for a pregnancy is therefore about 700 mg, or 2.5 mg per day, in addition to the basic metabolic requirement for iron. Because the body normally absorbs only 10% of the dietary iron, the diet should contain an extra 25 mg per day, for a total of 40 to 45 mg per day. Iron supplements can usually satisfy this high need for iron. But, as with vitamins, scientists can genetically engineer food crops to contain more iron. They have made GM rice that can supply 80% of the iron RDA, compared to 10% for the same amount of non-GM rice consumed (150 g). This rice is not yet available, and further testing is needed to confirm that it is both nutritionally effective and safe.

Although the human body needs only small amounts of iodine per day, the World Health Organization (WHO) estimates that many millions of people suffer from goiter, a disease caused by iodine deficiency. Goiter is characterized by an enlarged thyroid gland, a sluggish metabolism, a tendency to obesity, and enlarged face and neck. The iodine content of many foods is variable, and in many areas of the world it is so low that a normal diet does not provide the body with enough iodine. Iodine deficiency can be most easily prevented by fortifying table salt with iodide.

People take water so much for granted that they often do not realize how crucial it is for life. The loss of 10% of body water is very serious, and a loss of 20% usually results in death. You must take in water continuously to prevent your body from dehydrating and to maintain the proper balance of salts in body fluids. Water is the medium in which all biochemical transformation of the other nutrients takes place. Water helps carry nutrients from your gastrointestinal tract into your bloodstream, because the nutrients are dissolved in water. Water is also important in excretion; waste products are dissolved in it and excreted as urine. Water helps regulate your body temperature by absorbing the heat released by the respiratory activity of all your tissues; much of the heat is used to transform liquid water into water vapor during the process of perspiration. An average person (60 to 70 kg) needs to take in 1,800 to 2,500 g of water each day. About half of this is excreted as urine, and the other half leaves the body as perspiration or in expired air.

7.8 Food plants contain many biologically active chemicals whose effects on the human body remain to be discovered.

To defend themselves against pathogens and pests, plants synthesize many unusual chemicals—more than 50,000 different ones. The effects of only a few of them—usually potent poisons—are known, but researchers have never studied most of these chemicals. Plant biologists refer to them as *secondary metabolites* or *phytochemicals*. Classified by their biological activities, they could be considered pesticides, fungicides, bacteriocides, and so forth. The overwhelming majority occur in nonfood plants, but hundreds also occur in food plants. Quite a large number are carcinogenic, and nearly the entire carcinogenic "load" in our food comes from these types of chemicals, not from synthetic pesticides or other human-made chemicals.

These plant defense chemicals can affect our cells and metabolic processes, but often have no effect on plant cells. In some cases they need to be partially metabolized by the microorganisms in the intestinal tract to become bioactive; in other cases they are stored in separate cellular compartments of plant cells (for example, in vacuoles), where they are harmless. Only when chewing disrupts the plant tissues are the chemicals released. Some cause serious health problems. For example, the fava bean (*Vicia faba*) contains chemicals (vicin and convicin) that cause favism, an acute anemia that affects individuals in Mediterranean countries with a genetic deficiency in one particular enzyme (glucose-6-phosphate dehydrogenase).

Cyanogenic glucosides are abundant in some varieties of cassava and need to be removed by extensive washing of cassava pulp because otherwise they would cause cyanide poisoning. Residues of the toxic compounds always remain, and symptoms of poisoning are endemic in areas where people rely on these cassava varieties. Potatoes and other plants in the same family contain toxic alkaloids. When phytochemicals have obvious health effects, people usually know about them, but they may not recognize subtle effects.

Epidemiological studies and nutritional surveys have led researchers to identify dietary components that may have significant health benefits. Researchers have identified both antioxidants and phytoestrogens as potentially beneficial bioactive food components. For example, soy-rich diets (the so-called Asian diet) correlate with a low incidence of cardiovascular disease, osteoporosis, and estrogen-related cancers such as breast and endometrial cancer. Soy-supplemented diets relieve symptoms of menopause such as hot flashes. As a result of these findings, soy-based diets have become more popular in the United States (**Figure 7.7**). Soybeans contain several phytoestrogens (plant estrogens), some of which become biologically active only after microorganisms in the gut metabo-

Figure 7.7 Popularity of soy-based food products. Reports that phytoestrogens found in soybeans prevent certain types of cancers in women have fueled a boom in soy-based food products. *Source:* Courtesy of R. Morillon.

Box 7.3

Biologically Active Peptides and Proteins in Food

Protein quality depends on its digestibility and its protein score (presence of specific essential amino acids), but that is not all there is to it. Some whole proteins or short peptides formed in the human digestive tract in the digestion process also have biological activities. These short peptides, usually only 3 to 10 amino acids long, may act in the gastrointestinal (GI) tract or elsewhere in the body. They are formed through the action of digestive enzymes or through the action of enzymes released by bacteria in your GI tract. Intestinal flora plays an important but still little understood role in our health. Its composition (more or fewer lactobacilli, for example) changes depending on the foods you eat.

People have long known that breast milk provides newborn babies with immunoglobulins, lysozyme, and lactoferrin. The two latter proteins help protect babies against pathogenic bacteria. It isn't just the whole protein that is active; small peptides derived from these and other proteins have strong antibacterial properties and stimulate the immune system. Peptides derived from wheat protein (gluten) enhance the recovery of damaged intestinal cells. Peptides rich in the amino acid tryptophan can serve as precursors in your brain for serotonin, a signaling molecule that affects mood, stress, and sleep. On the whole people are still rather ignorant about the importance of the biologically active peptides released from the proteins in food.

One group of proteins whose action scientists somewhat understand is the plant lectins. These proteins, which are quite ubiquitous in plants, are very poorly digested and can be taken up intact into the bloodstream. Scientists believe they evolved as plant defense proteins. Many can bind to mammalian or insect gut cells where they may disrupt signaling pathways needed for these cells' normal functioning. Or they may be taken up intact and affect cells farther away. Some food sources, such as kidney beans and black beans, have so much lectin that people cannot eat them raw, but must cook them until the protein is denatured and loses its biological activity.

Once people understand the roles that specific small peptides or proteins play, they may be able to manipulate them, either by adding them to foods—food fortification—or by genetically manipulating plants to contain proteins from which these peptides can be readily released in the human GI tract.

---

lize them. Phytoestrogens can block the activity of normal estrogen by binding to estrogen receptors and may elevate the level of serum proteins that bind this sex hormone. Through this dual action they may lower the risk of certain cancers correlated with elevated plasma estrogen levels. High plasma estrogen levels occur in people who eat diets heavy in meat and dairy products, and oncologists think these higher levels may induce precancerous cells to become cancerous. However, studies of diets supplemented with specific phytoestrogens have not yet identified the active ingredients in the soy-rich diet. It is possible that soy products also contain other active ingredients, which interact with the phytoestrogens to provide the beneficial effect.

Human cells cannot live without oxygen, but its role in many oxidative processes causes the appearance of molecules called *reactive oxygen species,* highly active derivatives of oxygen that damage body molecules (DNA, RNA, protein, and membrane lipids). This damage accompanies pathological processes such as cancer and inflammation and is also an integral part of normal cellular aging. The cell's antioxidant defenses normally neutralize these highly reactive molecules, but oxidative stress occurs when those defenses are compromised. The interaction of reactive oxygen species with cellular molecules creates destructive free radicals. People can eliminate these if they eat food rich in antioxidants.

All the highly colored molecules (pigments) in human foods—the anthocyanins found in cherries and blueberries, the lipid-soluble pigments of carrots (alpha- and beta-carotene), tomatoes (lycopene), and deep green, leafy vegetables (lutein), act as antioxidants. People who eat diets rich in fruits and vegetables reduce free radical damage to

Figure 7.8 Fortified and functional foods. Fortified foods contain elevated levels of nutrients such as vitamins to ensure an adequate intake. Functional foods provide ingredients that may have health benefits but are not necessarily nutrients. In many processed foods, the two are combined to produce "health" foods, whose popularity is rising fast in developed countries. *Source:* Courtesy of R. Morillon.

their DNA (a risk factor for cancer development) and decrease their chance of getting certain cancers. However, supplementing diets with a variety of dietary antioxidants has not been shown to consistently reduce DNA damage or cancer risk. In addition to all these phytochemicals, our food sources contain proteins that have biological activity (not just nutritional value after breakdown into amino acids) and some of the large proteins are broken down into smaller peptides with biological activity (see **Box 7.3**).

The practice of fortifying foods with specific vitamins and minerals, and the discovery that foods contain chemicals with unexpected health benefits has given rise to the concept of functional foods (**Figure 7.8**). Functional foods are formulated with specific levels of certain ingredients that promote health beyond supplying essential nutrition. For example, the Netherlands Nutrition and Food Research Institute tested margarine fortified with high levels of plant sterols, to find out how this would affect serum cholesterol levels. The study found a 7–10% reduction in low-density lipoprotein (LDL) cholesterol and no effect on high-density lipoprotein (HDL) cholesterol, also known as "good" cholesterol (see Box 7.1). Because plant sterols are not nutrients we need, such margarine would be considered a functional food. Unfortunately, many functional foods and other nutraceuticals (a combination of nutrition and pharmaceutical) for which health claims are unsubstantiated, are appearing in the marketplace. In the United States, government agencies leave this area, like herbal medicine, largely unregulated.

7.9 The consequences of nutritional deficiencies are serious and often long-lasting.

Energy stores in the human body—glycogen and lipid deposits—are designed to tide a person over the inevitable situations where the demand for food exceeds dietary intake.

You use these stores between meals, for example, to keep the glucose level in your blood constant (glucose is the major energy source for your brain). A myriad of complex feedback loops, hormones, and so on, provides the mechanism for regulating and using these stores.

But when people must live on diets that provide only 25 g of plant protein and 1,100 kcal per day, the stores are soon exhausted. Only about one to two days' worth of energy stored in glycogen, and at most a month in stored fats, are available for use. After this, the only source of calories left is proteins, which can be broken down into amino acids and these in turn metabolized for energy. But as mentioned earlier, humans do not store proteins—all are used in some vital way. So breaking down an essential protein means it is then not available to carry out its role in the body.

Among the most accessible proteins for breakdown are those in blood, including antibodies, which are proteins the immune system makes to fight infections. As a result, people who are chronically undernourished are highly susceptible to infections. The poor sanitation of many regions where people are poorly nourished adds to the danger, creating a breeding ground for infectious agents. Children are especially vulnerable (**Table 7.4**). Many infectious diseases that kill children affect those whose immune systems are severely compromised by inadequate nutrition. What is more, poverty puts treatments (such as antibiotics) or prevention (vaccines) out of reach. A second impact of undernutrition, again most dramatically seen in the young, is growth retardation. During the first few years of life, when a child undergoes rapid growth, the body must have the necessary calories, proteins, and other nutrients with which to build more tissues.

Most importantly, nutrient deprivation harms brain development. The human brain grows most rapidly during later stages of prenatal development and the first year of life, and is most sensitive to undernutrition at these times. Before birth, the fetus obtains its food from the mother. As she must eat more to gain the necessary nutrition for both herself and the fetus, dietary deprivations at this time can harm both. Although the mother's nutritional status is sacrificed in favor of the fetus, both animal studies and human observations show that severe caloric restriction causes lower birth weight and later mental retardation. Undernutrition during the first year of life, even if later remedied, often results in a physically smaller brain, as reflected by a reduced head circumference. Children who suffered from malnutrition when they were younger often score lower on intelligence and adaptive behavior tests than their counterparts who have been adequately nourished and live in a similar environment. These studies suggest that poor nutrition of infants (especially during the first year) may permanently restrict their mental abilities.

Table 7.4	Infections that kill poorly nourished children
Diseases	**Estimated Deaths, 1998 (millions)**
Acute respiratory diseases (pneumonia)	2.2
Diarrhea	2.2
Malaria	0.7
Measles	0.8
Neonatal infections and meningitis	1.1
Sexually transmissible diseases, including HIV infection	1.5

Source: Data from Food and Agriculture Organization.

7.10 Millions of healthy vegetarians are living proof that animal products are not a necessary component of the human diet.

As pointed out in Chapter 3, a small number of plants—principally cereal grains, legumes, and root crops—supply most of the energy and nutrients in the human diet. In general, seed staples have a favorable protein content (8 to 15% of the dry weight), whereas roots and tubers have a much lower protein content (1 to 3%). Scientists often express the ratio between protein and carbohydrates as grams of protein per 100 kcal (see **Figure 7.9**). Because an average adult should have a daily food intake of at least 50 g protein and 2,500 kcal, or about 2 g protein per 100 kcal, staples that contain 2 or more grams protein per 100 kcal are therefore also good sources of protein. It is no accident that early human societies adopted staples that meet or exceed the protein-calorie ratio standard. Currently, more than half the human diet (55%) comes from the grain staples wheat, rice, and maize. Another 13% is provided by the protein-rich seeds of legumes (peas, chickpeas, beans, lentils, soybeans, cowpeas, and so on). Animal products supply 20% of human beings' dietary protein.

The lower protein–energy ratio of plant foods in comparison with animal foods and the low protein scores of plant proteins when compared with animal proteins (Table 7.1) also help partly explain the correlation between nutritional deficiencies and dependence on plant protein. Plant proteins are usually low in certain essential amino acids, and heavy reliance on a single staple can result in an inadequate intake of these. People can correct this deficiency by supplementing their diet with plant proteins from different sources (for example, by mixing cereal grains and legumes) (**Table 7.5**) or by including small amounts of animal proteins in their diet.

If animal proteins have a higher nutritional value for human beings than plant proteins because they have a higher protein score, why not convert more plant protein into animal protein by raising more livestock and producing more milk and cheese? In the past, production of animal protein depended largely on food sources that human beings could not eat. Herbivores such as cattle, sheep, and goats can live entirely on grass and other plants that human beings cannot digest. Animal protein produced in these ways is said to come "cheaply," because food sources of little value to human beings are being converted into foods valuable to humans.

Eating meat nearly every day is so culturally ingrained in developed countries that it is easy to forget that many vegetarians never eat any meat and vegans don't eat any animal

Figure 7.9 Protein–calorie ratios of several foods. The dashed line shows the approximate adult daily dietary requirement for protein. Note that the three major staples—wheat, rice, and maize—provide sufficient protein if calorie needs are met. However, a single staple does not provide a proper ratio of essential amino acids. *Source:* Data from Food and Agriculture Organization.

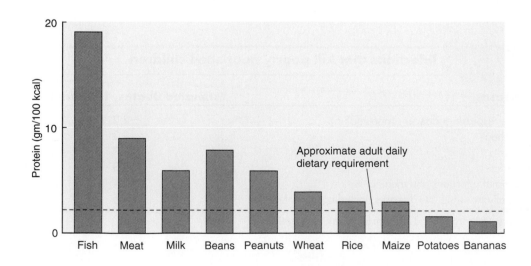

Approximate adult daily dietary requirement

Table 7.5	Protein score and protein content of various mixtures of staples		
Cereal	**Supplement[a]**	**Percentage of Protein**	**Protein Score**
Wheat	None	11.2%	62
	Groundnut	14.2	67
	Soybean	13.8	76
Maize	None	9.5	49
	Groundnut	12.6	58
	Soybean	12.2	67
Rice	None	6.7	69
	Groundnut	10.0	73
	Soybean	9.6	77

[a]Adding 8 g of protein concentrate to 100 g of staple.

Source: G. R. Jansen (1974), Amino–acid fortification of cereals, in A. A. Altschul and H. L. Wilcke, eds., *New Protein Foods* (New York: Academic Press), pp. 94–99.

products at all. Some do so out of choice, and others are vegetarians because of poverty. Can plants and plant products be an exclusive food source for humans? And can people remain in good health if they only eat plants? The answer to this question is "Yes, indeed"—if people carefully consider their intake of vitamin B_{12}, which neither plants nor animals can make. Only bacteria and other microorganisms such as yeast make this cobalt-containing vitamin. By carefully selecting a variety of plant proteins, vegetarians can obtain a diet with a high protein score by eating complementary foods, usually cereals and legumes. It is easier to ensure a proper amino acid balance by eating a small amount of animal protein to supplement a larger amount of plant protein. The large amounts of animal protein now consumed by most people in technologically advanced countries are nutritionally superfluous and may carry particular health disadvantages, such as types of cancer and heart disease.

Diets rich in plants may have particular health benefits. As already noted, phytoestrogens from soybeans and the antioxidants present in highly colored fruits and vegetables can prevent certain types of cancers (**Figure 7.10**). The nutritional recommendations from the USDA include eating five portions of fruits and vegetables per day.

Figure 7.10 Colored fruits and vegetables are an excellent source of antioxidants such as carotene, lycopene, and lutein. *Source:* Courtesy of R. Morillon.

7.11   Genetic modification of plants can permit more efficient use by animals.

Because monogastric (poultry, pigs, and fish) and ruminant (cattle, sheep, and goats) animals have very different digestive systems (see Section 7.4), the goals of the animal feed industry for genetically modifying animal feed are diverse. For ruminant animals the goals are to lower the lignin content of the pasture plants (to improve their digestibility) and to increase soluble, readily digestible components. For monogastric animals such as pigs, chickens, and fish that are fed primarily on seeds, the goals are to eliminate antinutrients. Antinutrients are chemicals that lower animals' absorption of nutrients from the feed.

The cell walls of many cereal grains are rich in arabinoxylans and glucans, which are polymers of xylose and glucose, respectively. These polymers are antinutritional in monogastric animals, particularly poultry, because they increase the viscosity of the gut content. The feed industry commonly pretreats cereal grains with the enzymes xylanase and glucanase from microbial cultures that break down these polymers to the corresponding monomers, which are not antinutritional. Researchers modified a fungal xylanase gene so that it would be expressed in seeds and introduced the gene into barley. The enzyme was highly expressed in the developing seeds and remained active during storage at 37°C for at least six months, making these seeds a great potential source of feed.

Similarly, researchers engineered a bacterial gene for a heat stable β-glucanase for high-level production of the enzyme in barley endosperm. They predict that this added enzyme will help break down the β-glucans in barley cell walls and thus increase the energy value of barley in poultry diets. Barley is well known for having low metabolic energy for poultry. The heat stability of the β-glucanase means that the enzyme can survive the pasteurization process needed during feed processing to prevent chick infection by salmonella.

Phytate (myo-inositol hexaphosphate) is the main storage form of phosphorus in cereal and legume feed grains. When the seeds germinate, they produce the enzyme phytase, which liberates the phosphate from inositol. However, neither the mature, ungerminated grain nor the digestive system of monogastric animals has a significant level of phytase activity, so animals use the phosphorus in grain very poorly. Phytate is also a known antinutritional factor because it binds zinc, iron, and other essential minerals. To overcome these two problems with phytate, manufacturers routinely add fungal phytate preparations to stock feed mixes. This greatly improves phosphorus use and increases availability of essential mineral elements. Researchers recently made genetically engineered pigs that secrete phytase in their saliva. These pigs could digest the phytate in their feed. This will obviate the need for separate treatment of feed mixes with the enzyme preparations isolated from fungi and will improve growth rates in animals.

Another research approach is to look for mutant seeds that store phosphorus as inorganic phosphate rather than as phytate. Researchers have identified such a line of low phytate maize, and animals and humans that eat this maize have an increased ability to absorb iron.

7.12   Food safety has become an international concern.

Is our food safe to eat? The popular press bombards us with messages about food safety. The food supply in developed countries is much safer than ever in the past, but warnings occur about pathogenic bacteria in poultry and eggs; hormones in beef, carcinogenic pesticide and fungicide residues on fruits and vegetables; myco-toxins (toxins produced by fungi) on maize, soybeans, and peanuts; the unknown effects of radiation to kill bacteria; and antibiotic residues in meat.

Although some pesticides and fungicides are carcinogenic, the residues that remain on fruits and vegetables are low enough so that they do not threaten the population. The main threat from pesticides and fungicides is to the workers who apply them, especially in developing countries, and—to a lesser extent—to adjacent ecosystems. Scientists estimate that over 95% of our carcinogen load in food comes from the natural chemicals in food plants. These chemicals are part of the defense mechanism plants use to ward off insects, bacteria, and fungi (discussed in Chapter 15).

Mycotoxins should be of greater concern, especially if seeds are stored for prolonged periods of time under conditions that permit the growth of fungi. If the seeds are also insect infested, as often in developing countries, then synergism arises between the insects and the fungi, because the inroads made by the insects promote the growth of fungi (**Figure 7.11**). One benefit of genetically engineered *Bt* maize is that the reduction in insect damage results in much less growth of the fungal pathogen *Fusarium*, which produces the potent mycotoxin fumonisin.

Food-borne illnesses associated with microbial pathogens are a serious health threat in developed as well as developing countries (**Table 7.6**). The problem is compounded by the emergence of antibiotic-resistant pathogens as a result of feeding antibiotics to poultry, pigs, and cattle to promote their growth. The Centers for Disease Control and Prevention report that in the United States alone, food-borne pathogens and contaminants account for 76 million cases of illness and 5,000 deaths annually. This is probably a substantial underestimate, because many cases of food poisoning are not reported. On a global scale, there are annually as many as 1.5 billion episodes of diarrhea, mostly in developing countries. The World Health Organization estimates that biologically contaminated food causes 70% of these. Providing clean food and clean drinking water is a major challenge for developing countries. These preventable problems put an unnecessary stress on the already overburdened health systems of these countries.

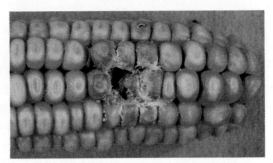

Figure 7.11 Insect damage and fungal growth are synergistic. An insect damaged this ear of maize, allowing the fungus *Fusarium* to grow. This fungus produces potent mycotoxins such as fumonisin. Insect control therefore is an important strategy to produce healthier food. *Source:* Courtesy of Gary Munkvold.

Table 7.6	**Food–related diseases**	
Disease Agent (and example)	**Where Found**	**Food Source**
Viruses (Hepatitis A)	Wide range includes shellfish, raw fruit, and vegetables	Poor hygiene, growing plants in area of raw sewage or organic wastes
Bacteria (Salmonella)	Raw and unprocessed foods, such as cereals and fish	Poor hygiene, carried by rodents, birds, humans
Molds (Aspergillus)	Nuts and cereals	Storage at high temperature, humidity
Protozoa (Amoeba)	Vegetables, fruits, raw milk	Contaminated water
Helminth worms (Ascaris)	Vegetables and undercooked meat and fish	Contaminated soil and water
Agricultural agents (pesticides)	All foods treated	Use of banned chemicals or excesses

Source: Adapted from Food and Agriculture Organization (1992), *Nutrition: The Global Challenge* (Rome: FAO), p. 24.

Irradiating food is undoubtedly the most effective way to prevent a substantial part of those deaths and illnesses. Although all major international public health organizations have endorsed the practice as safe and effective, it is not yet generally used because some consumer groups have organized opposition to it.

In the United States, consumers seem to have confidence in the government agencies that regulate food safety. In Europe, partly as a result of several food safety breakdowns in the late 1990s (such as the outbreak of mad cow disease in Great Britain and the contamination of animal feed with dioxin in Belgium), consumers have less faith in their regulatory agencies. The problem needs an international solution in part because international trade in food is continually increasing. For example, according to the U.S. Food and Drug Administration 38% of the fruits and 12% of the vegetables consumed in the United States are imported.

7.13 Genetically modified plants and the foods derived from them are the subject of safety concerns and special regulations.

Medicines produced by genetic modification techniques have been widely used since 1982, when regulatory authorities first approved the use of insulin isolated from genetically modified bacteria. Food processing aids, such as chymosin (rennet) used in cheese making, which was usually produced from calf stomach, can now be obtained by genetic engineering. Chymosin has been approved since 1990. Food, food ingredients, and feed produced from genetically modified plants are subject to a stringent regulatory process in the countries where they are produced or used. In the United States, for instance, three agencies—the Food and Drug Administration (FDA), the U.S. Department of Agriculture (USDA), and the Environmental Protection Agency (EPA)—have legal and regulatory power to ensure safety of feed and food and to set tolerances for pesticides. The products of genetically modified plants must be shown to be as safe as the products of plants derived by conventional plant breeding. These agencies can remove a food from the market if it poses a risk to public health.

Some consumers have voiced concern about the safety of food from genetically modified plants. They worry that scientists may introduce new allergens or toxins into the plants when they insert a new gene. Clearly, the same problems may arise in conventionally modified crop varieties, but these are not subject to regulation. Occasionally, new traditionally bred varieties have been withdrawn from the market by their producers because they contained unacceptable levels of toxins (glycoalkaloids in potatoes for example). Some people also worry that if the selectable marker gene used to develop the genetically modified plant is a gene conferring resistance to antibiotics, this gene could spread to pathogenic microbes and thus increase resistance to clinically important antibiotics. In fact, many pathogenic bacteria are already antibiotic resistant, but this situation arose from misuse of antibiotics by physicians and their gross overuse in the animal feed industry. No antibiotic resistance has been traced to the genetic engineering of crops.

These and other concerns about GM crops are addressed in depth during the evaluation of any new genetically modified plant for regulatory approval by the same three agencies (see above). The assessments are stringent, science-based and comprehensive. For example, regulatory boards pay special attention to the allergenicity of novel proteins. Some people develop allergic reactions when they consume certain foods—peanuts, soybeans or wheat, for example. These seeds contain allergens, proteins that may cause the body to synthesize a special class of antibody, which triggers severe inflammation. Scientists have traced such allergic responses to certain regions (short sequences of amino acids called *epitopes*) of the food proteins on the surface of the protein (**Figure 7.12**). People who know they are allergic to peanuts know not to eat them—but what happens if genes encoding peanut proteins show up in other plants? For example, researchers isolated a gene for a high-methionine protein from Brazil nuts and introduced it into soybean with the aim of improving the methionine status of soybean meal for stock feed. The Brazil nut protein was

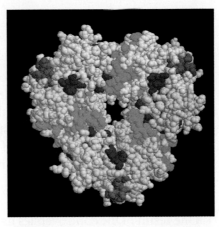

**Figure 7.12
Structure of a soybean protein that can cause allergies in people.** Shown here is a space-filling model of legumin, the most abundant protein in soybean seeds and all soy-derived products. Different epitopes—short stretches of amino acids that can evoke allergic responses in some people—are shown in different colors.
Source: Unpublished photo courtesy of Niels Nielsen, Purdue University.

successfully expressed in soybeans. However, it was well known that some people are highly allergic to Brazil nuts, although the actual allergen had not been identified. When scientists ran allergenicity tests on the methionine-rich protein, it proved to be a major culprit, so they did not develop this particular strategy for improving protein quality further. Given the widespread use of soy proteins in prepared foods, this modification was considered too great a risk to human health by the company that did the work. The result was that the project was discontinued long before the new soybeans reached commercial production.

The approval process for a single GM crop can take up to five years and involves numerous tests that potential distributors of the crop carry out. For instance, for maize plants carrying the gene for the insect toxin from *Bacillus thuringiensis* bacteria, these tests ranged from toxicity studies in animals, to digestibility of the toxin in humans, to toxicity in non-targeted insects. Government scientists and regulators then examine the results of such tests, often comprising thousands of pages. Each country has its own regulations and regulatory agencies, but confidence in such agencies varies. It is important that the procedures regulatory agencies use be transparent to the public and that all regulatory decisions be based on sound science. The perception of risk by the public is a complex phenomenon that social scientists have long studied and biotechnology companies must address in the future.

Despite assurances from government agencies, the public in some countries is concerned about the safety of GM foods. Such uneasiness is not limited to GM foods. Historically, the public has actively opposed some food-related health measures, such as pasteurization of milk in Great Britain (as noted earlier), and fluoridation of water in the United States. Government agencies use science (epidemiology or toxicology) to determine the risk associated with chemicals that occur naturally in foods or that are added during food processing. The public often takes a different view. This discrepancy caused Peter Sandman from Rutgers University in the United States to propose the following equation:

$$\text{Risk} = \text{Hazard} + \text{Outrage}$$

In this equation "Hazard" is the scientifically determined risk factor, or the probability that eating the food will have an adverse outcome. "Outrage" covers the nonquantitative, nonbiological properties of the food. The public sometimes pays little attention to hazard but is greatly influenced by outrage. Scientists and governmental advisory groups, in contrast, pay little attention to outrage and concentrate entirely on hazard. "Outrage" factors that Sandman identified include

- Whether the risk is voluntary
- Whether the risks and benefits are equitably distributed in society

- Whether the risk is from natural or synthetic sources
- Whether the risk is subject to individual control
- Whether the risk is familiar or not

For example, the public is not concerned about natural carcinogens in food, but very concerned about synthetic pesticide residues that might be carcinogenic. Physicians and cancer researchers estimate that of the one third of cancers caused by dietary factors, less than 1% are related to synthetic pesticides. Also, the benefits of GM foods have so far not been equitably distributed—most of the benefits have gone to companies and farmers, not to the public. Controversies about risks can generally not be solved by better communication of those risks to the public, although poor communication can exacerbate the problems.

One result of the controversies has been a move to label foods that are produced by GM technologies. A number of European Union countries have made such labeling mandatory. In the United States and Canada (as of 2002), labeling is not mandatory but is used rather, in reverse, to advertise that foods are "GMO-free" (GMO means "genetically modified organism") (**Figure 7.13**). This allows companies making those foods to develop a niche market, much as the organic food industry has. The U.S. government has so far rejected mandatory labeling because it takes the position that what matters most is the composition of the crop or the food prepared from it, not the method of producing it. Thus the nutritional content is reported on all types of packaged foods, but not how the crops were bred, grown, or processed.

Figure 7.13 Voluntary labeling of foods not containing GM crops. Uncertainty about the safety of foods derived from GM crops and opposition to the power of multinational corporations in agriculture has led to strong opposition to GM crops in some countries and to demands that the products be labeled. The United States has taken the approach that labeling should be voluntary and that companies can label their products as being GMO free (arrow).

CHAPTER
SUMMARY

Organisms can be classified as autotrophic (green plants and algae) or heterotrophic (animals and microbes) depending on their mode of nutrition. Humans are heterotrophs and need to eat energy-rich molecules (fats or carbohydrates), proteins, essential fatty acids, vitamins, and minerals. Human food sources contain all these in different proportions, and in addition they contain hundreds of chemicals that are not nutrients, but can positively or negatively affect our physiology and health. The roles of some of these chemicals, such as antioxidants and phytoestrogens, are just now being discovered.

Researchers estimate that some 800 million people, mostly in developing countries, are malnourished, receiving either too little food or inadequate supplies of certain nutrients (such as vitamins). In developed countries, poorly balanced diets or excess caloric intake and the resulting obesity are related to major diseases, causing premature death and considerable public health expenditures.

In developed countries people eat substantial quantities of meat and meat consumption is also rising sharply in developing countries. Meat consumption may be responsible for some of the diet-related diseases experienced in developing countries. People who do not eat meat (vegetarians) or any animal products (vegans) can easily remain in good health, showing that meat is not a necessary component of the diet. In some ecosystems ruminant animals perform the essential task of converting plant materials not directly useful to humans as food, into edible meat and animal products.

Public health measures have resulted in the fortification of widely distributed foods (such as milk and wheat flour) with specific nutrients. This practice has evolved in developed countries into the fortification of all kinds of foods by food manufacturers, giving rise to a new generation of functional foods. Genetic modification of crop plants with specific nutrients joins this trend. Such a development could help alleviate nutritional deficiencies in developing countries. Vitamin A-rich Golden Rice is a typical example.

The production of GM crops for human consumption is subject to strict government controls. Despite such controls, a substantial segment of the population in some countries opposes GM foods. This opposition is based in part on the perception that some risk may be associated with eating GM foods. Social scientists are beginning to study the public's perception of risk, which biotechnology companies must address.

Discussion Questions

1. What are the environmental and health advantages of a vegetarian diet? Of a vegan diet?

2. Discuss pastoralism—economies that depend entirely on the conversion of forages to animal products. Is this an inefficient use of resources to fill human needs? Explain.

3. Discuss food fortification. What examples are available in your local supermarket? Do they seem to meet real needs? Why or why not?

4. Research "functional foods"—a new development in food technology. Visit your local market, and list all the soy-based food products you find there. What problems could you speculate might be caused by reliance on functional foods?

5. Find examples of health claims made by the manufacturers of "health foods." Discuss the wording of these claims. What evidence is offered for such claims? What is the role of government agencies in labeling foods?

6. What are some connections between diet and disease? How are these connections established? Which types of diseases clearly appear diet related?

7. Some people who eat "unhealthy" diets live long and healthy lives. Why might this be so? Discuss the role of human genetics and lifestyle in our response to diet.

8. Meat consumption in developing countries is expected to increase dramatically in the next 50 years. Discuss the implications for health, disease, trade, research funding, and land use.

9. How do government agencies regulate genetically modified foods? Do you think these controls are adequate? Explain.

10. How do human fallibility and conflict of interest enter into the public's assessment of government, scientist, and manufacturer ability to promote and provide food safety? What about communication, control, and fear of the unknown? How would you handle fears of GM products, if you were in charge of mitigating them?

Further Reading

Ames, B. N., L. S. Gold, and W. C. Willett. 1995. The causes and prevention of cancer. *Proceedings of the National Academy of Science USA* 92:5258–5265.

Halbert, S. C. 1997. Diet and nutrition in primary care. From antioxidants to zinc. *Primary Care* 24:825–843.

Hester, R. E., and R. M. Harrison. 2001. *Food Safety and Food Quality Issues in Environmental Science and Technology*. Cambridge, U.K.: Royal Society of Chemistry.

Nelson, G. C. 2001. *Genetically Modified Organisms in Agriculture: Economics and Politics*. San Diego: Academic Press.

Nestle, M. 2002. *Food Politics*. Berkeley, CA: University of California Press.

Sanders, M. E. 1998. Overview of functional foods: Emphasis on probiotic bacteria. *International Dairy Journal* 8:341–347.

Shabert, J. 1996. Nutrition and women's health. *Current Problems in Obstetrics, Gynecology, and Fertility* 19:113–166.

Willett, W. C. 1994. Diet and health: What should we eat? *Science* 264:532–537.

———. 1999. Goals for nutrition in the year 2000. *CA: A Cancer Journal for Clinicians* 49:331–352.

Winter, C. K., and F. J. Francis. 1997. Assessing, managing and communicating chemical food risks. *Food Technology* 51:85–92.

U.S. Public Health Service. *Healthy People 2010*. Available at http://www.healthypeople.gov. Accessed August 22, 2001.

Internet sites for nutrition and health information:

http://www.discoveryhealth.com

http://www.HealthScout.com

http://www.mayohealth.org

http://www.NIH.gov

http://www.nysaes.cornell.edu/comm/gmo/

http://www.WebMD.com

The Genetic Basis of Growth and Development

Maarten J. Chrispeels
University of California San Diego

In Chapter 6 we discussed the molecular basis of crop improvement and noted that specific traits are encoded in genes, short segments of DNA that can be transferred from one organism to another. The expression of genes is highly regulated and changes dramatically during the plant's development and as a result of environmental signals. To understand how scientists might change plants to suit human needs as food sources, you first need to understand plants. That fertilized egg cell that constitutes the beginning of the life of every plant already contains all the 20,000 to 50,000 genes the plant will ever have! Their orderly expression controls development, which involves formation of organ systems such as roots and shoots. Such development requires integrated cell division, which gives rise to *more* cells, and cell differentiation, which gives rise to *different* cell types as new organs, each with multiple tissues, form. During this process many genes respond to environmental stimuli, changing the course of development. In addition, the plants establish defense mechanisms against pathogens, insects, and other pests. These mechanisms consist of proteins or other molecules, the information for which is encoded in the genome. Understanding how genes regulate plant development will help you gain insights into what scientists might accomplish by manipulating the expression of specific genes to transform crops into more efficient food producers.

Each type of cell in a tissue has its own specific set of proteins, required to carry out the unique functions of that cell, and each protein is encoded by a gene. Many so-called housekeeping genes are expressed in all cells and at all times, but thousands of genes are activated or up-regulated and down-regulated at specific stages of plant development. How the plant regulates the orderly expression of these thousands of genes greatly interests developmental biologists. When a plant develops from an embryo, a seed forms first; after seed germination, the vegetative body (roots and shoot) forms; finally, the appearance of flowers initiates the reproductive phase. Each stage requires the activation of new genes.

The expression of genes during development is influenced by such environmental signals as light, the availability of minerals, or a period of low temperature, and is coordinated by internal signaling molecules. These signals can be small molecules such as hormones that circulate within the plant, or proteins that pass from one cell to the next. Such signaling molecules help cells or organs communicate with one another and with meristems, the sites where cell division occurs and new organs are initiated. Cell differentiation is accompanied by the elaboration of the biosynthetic machineries that enable the plant to carry out photosynthesis, take up nutrients from the soil, transport sugars, and store protein and starch in special structures. The integrated, orderly expression of all this genetic information stored in the cell nucleus (and to a small extent in the chloroplasts) is truly awe-inspiring.

8.1 Plants are made up of cells, tissues, and organs.

The vegetative body of all plants, in all their apparent diversity, consists of two parts: a root system and a shoot system. The root system consists of a primary root with secondary and tertiary roots; the shoot system has a primary stem with branches, all of which usually carry leaves and may produce flowers. **Figure 8.1** shows the arrangement of these major organs (roots, stems, leaves) in young plants belonging to the two major groups, the monocots and the dicots. The main functions of the shoot system are photosynthesis and the formation of the reproductive organs. A stem consists of a number of segments, called *internodes,* connected by nodes. In a young plant a leaf is attached at each node. The functions of the root system are to take up water and minerals, to store starch, and to anchor the plant in soil.

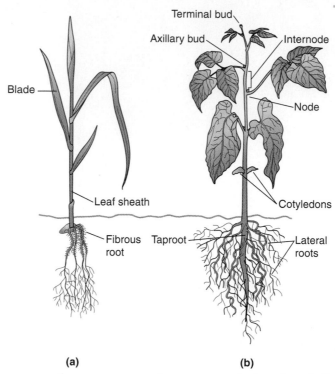

Figure 8.1 The organs of young vegetative monocot (wheat) and dicot (bean) plants.
(a) Monocots have elongated leaves with parallel veins (vascular bundles) and a system of numerous parallel roots; grasses, cereals, and palms are monocots. **(b)** Dicots have broad leaves with branched veins and a branched taproot with lateral roots. Most common plants, such as soybeans and other legumes, potatoes, tomatoes, and cotton, are dicots.

Plants have four reproductive organs, all contained in the flower: sepals, petals, stamens (male reproductive organs), and carpels (female reproductive organs). As in animals, each plant organ consists of several tissues and each tissue contains several cell types. Plant organs have three types of tissues—dermal tissues on their surface, conductive tissues, and ground tissues—and their locations in a dicot plant are shown in **Figure 8.2**. All together these tissues contain some 40 different cell types that are distinguishable on the basis of their morphology and function. The human body, in contrast, contains hundreds of different cell types indicating that plants are simpler organisms.

Cells are the basic building blocks of tissues, and the cells of all organisms contain more or less the same internal (subcellular) structures (organelles) (see **Box 8.1**). Compared to animal cells, plant cells have at least four unique structures: large chloroplasts for photosynthesis or other plastids in nongreen cells, a large central vacuole that occupies most of the volume of the cell, a cell wall made out of cellulose and hemicellulose that encases the entire protoplasm, and plasmodesmata, which are channels through the wall that connect one protoplast with another. In common with animal cells, plant cells have a nucleus surrounded by a nuclear envelope, endoplasmic reticulum and Golgi apparatus, plasma membrane, cytoskeleton consisting of tubules and fibrils, ribosomes for protein synthesis, mitochondria for respiration, and peroxisomes where reactions are carried out that produce hydrogen peroxide.

Each subcellular organelle has its own unique set of proteins that carries out the functions of that organelle. For example, chloroplasts have the enzymes necessary for

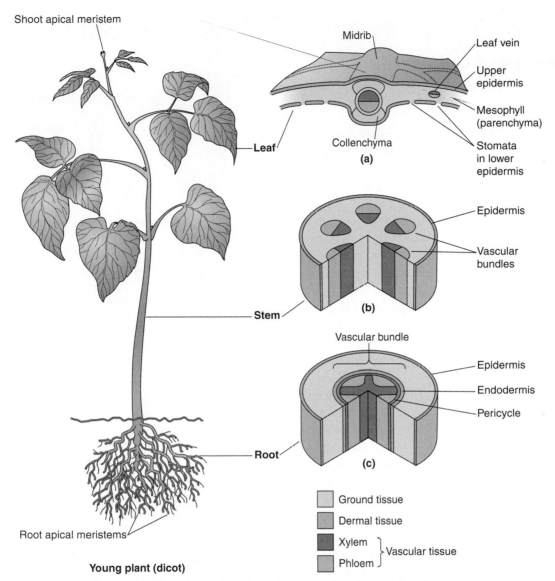

Figure 8.2 The body of a plant showing the organization of the three tissue systems in the organs. The cross sections through each organ shows the different organization of the three tissue systems in the leaf **(a)**, stem **(b)**, and root **(c)**. Each organ has vascular tissues (xylem and phloem), dermal tissue (epidermis), and ground tissue (mesophyll, cortex, and pith).

photosynthesizing and for assimilating CO_2. The same enzymes are not found in other organelles. One interesting question in cell biology is how proteins find their correct destination. Nearly all proteins are synthesized in the cytoplasm, yet they end up in about 25 different locations in the cell! Each protein contains its own address label, and the cell has machinery ("mail slots") that inserts the correct protein in the correct organelle.

Certain cells that are specialized for storage, as in seeds, roots, and tubers, also have unique organelles specifically to sequester protein (protein storage vacuoles or protein bodies), starch (amyloplasts), or oil (oil bodies). Such storage cells do not have a large central vacuole, and most of the cellular volume is taken up by these storage organelles.

Box 8.1

Plant Tissue Systems and Cell Types

The Three Tissue Systems of Plants

The meristems, where cell division takes place, give rise to three tissue systems with specialized functions. All organs contain the same three tissue systems: ground, dermal, and vascular tissues.

Dermal tissues: These form the plant's protective outer covering that is in contact with the environment. In young roots the epidermis facilitates ion and water uptake, and in leaves and stems specialized cells regulate gas exchange.

Vascular tissues: The phloem and the xylem form a continuous vascular system throughout the plant that conducts water, solutes (minerals and organic molecules), and some hormones. The vascular tissues provide mechanical support (as do bones in humans).

Ground tissues: These metabolic tissues make up the bulk of young plants and function in food manufacture in leaves, and storage in stems and roots.

Ground Tissues The ground tissue system contains three main cell types: parenchyma, collenchyma, and sclerenchyma.

Parenchyma cells are living cells that have a thin, flexible cell wall and that are found in all tissue systems. Given the right hormonal stimulus, cell division can be induced. In leaves they function in photosynthesis (mesophyll), in roots and stems they function primarily in starch and sucrose storage. Parenchyma cells form the bulk of most of the fruits and vegetables we eat. These cells have very large central vacuoles, and the cytoplasm forms a thin layer sandwiched between the wall and the vacuolar membrane.

In seeds, storage parenchyma cells are packed full of amyloplasts (starch grains), protein storage vacuoles, and oil bodies. Adjacent to the phloem sieve tubes are transfer cells, specialized parenchyma cells that help transfer organic molecules from the conductive tissues into the large parenchyma cells.

Collenchyma cells are also living cells, but with much thicker cell walls than parenchyma cells. They are elongated and occur in vertical files just underneath the epidermis where they provide mechanical support.

Sclerenchyma cells also have a supportive/protective role, but they are dead and have very thick cell walls with lignin. They form fibers that protect the phloem in the stem.

Dermal Tissues In young organs the epidermis is the main dermal tissue; in leaves and stems, some epidermal cells differentiate into guard cells; in all organs, epidermal cells can differentiate into hair cells. In older stems and roots, the periderm, which includes the cork layer, is the main dermal tissue. The periderm appears after roots and stems start to thicken and the epidermis is shed.

The **epidermis** is usually only a single cell layer. Leaf epidermis has a thick outer wall that is covered with a layer of wax, a polyester that is secreted by these cells. Root epidermis is not covered with wax.

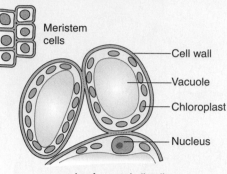

Meristem cells

Cell wall
Vacuole
Chloroplast
Nucleus

Leaf mesophyll cells

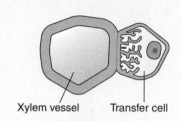

Xylem vessel   Transfer cell

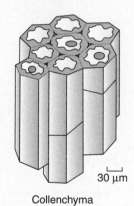

30 µm

Collenchyma

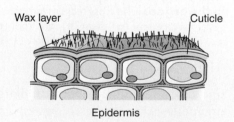

Wax layer   Cuticle

Epidermis

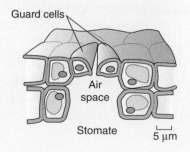

Guard cells

Air space

Stomate   5 µm

Stomates are small holes or pores in leaves and stems that regulate gas exchange. They are bordered by two kidney bean-shaped cells called **guard cells**. The diameter of the pore is regulated by the turgor pressure in the guard cells.

Hairs occur on all organs. On roots (root hairs), they are important for mineral and water uptake. On leaves and seeds, hairs (trichomes) sometimes produce important substances for defense against insects. The hair cells on cotton seeds are extremely long and represent the cotton fibers that can be spun into threads.

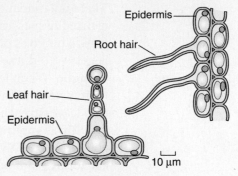

Vascular Tissues   Together, the phloem and the xylem form a continuous vascular system that extends through the entire plant. These complex tissues each have several cell types. In young plants, they form vascular bundles. In roots, there is a single vascular bundle at the center of the root, but in stems and leaves there are multiple vascular bundles. Each bundle has phloem and xylem cells and associated transfer cells. The thin-walled phloem is often flanked by sclerenchyma fibers for protection.

Phloem: The main phloem cells, called **sieve tube elements**, are vertically aligned to form sieve tubes. They transport organic solutes (sugars and amino acids) throughout the plant: from the leaves to the roots and to the developing flowers and fruits or from senescing organs to growing organs. They are living cells but have lost their nucleus, vacuole, and much of their cytoplasm. They rely on a companion cell for their maintenance, including the import of mRNAs and proteins. The elements are connected by sieve plates, areas of the cell wall that are completely perforated. Researchers have recently shown that the phloem also transports mRNAs.

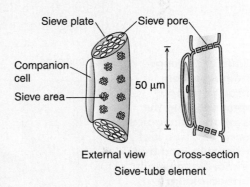

External view   Cross-section
Sieve-tube element

Xylem: The main xylem cells, called **vessel elements**, are vertically aligned to form vessels. The end walls between xylem elements are either heavily perforated or have been completely removed (dissolved) during the last phase of cell differentiations. This allows very long capillary tubes to form. The walls are thick, often reinforced by rings or spirals on the inside and completely encrusted with lignin, a hydrophobic polymer. Xylem elements are dead and have no cytoplasm. The xylem transports water and minerals from the roots to the shoot. The upward flow is caused by constant evaporation of water from the leaves. The water column in each vessel is thin and water molecules bind to one another. The xylem also transports the hormone abscisic acid from the roots to the shoots when the roots are experiencing water deficit or excess salt.

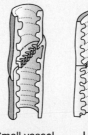
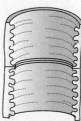

Small vessel element in root tip

Large, mature vessel element

Plant development is characterized by repetitive organ formation.

There is a fundamental difference in the way plants and animals develop. Plants have specialized tissues, called *meristems,* that can continuously give rise to new cells and new organs. This gives plants a rather simple way of solving the problem of aging: when an organ is getting old, the plant allows it to die and makes a new one. Leaf shedding is a typical example: When leaves are young, they photosynthesize vigorously, but as they get older their metabolic rate declines. Eventually they age and die. Perennial deciduous plants make new leaves in each growing season. When annual plants are crowded, as in a field of wheat, new leaves appear at the top of the plant while older leaves near the bottom are dying. This ability to repetitively make new organs also permits plants to recover from major damage to the plant body (as done by grazing animals or insects).

At the tip of each shoot is a shoot apical meristem (SAM) where the cells continuously divide (**Figure 8.3**). This is a very small, dome-shaped mass of cells. The SAM cells are the progenitors of all cells in the shoot. The tissue just below the SAM is also meristematic: The cells are dividing and contribute to the formation of new organs, especially the stem. Small protrusions of dividing cells, called *primordia,* arise on the surface of the SAM and just below it, and one by one these primordia grow out into small leaves. At the same time new meristems form, becoming axillary buds that remain quiescent until hormones trigger them into activity. Together, the apical meristem, the primordia, and the tiny leaves make up the apical bud you can find at the end of every twig. Apical buds also appear at the ends of other shoots, but are not always as easily identified as in twigs. The environment (temperature and day length), which causes changes in the cells' hormone level, regulates the activity of the SAM. These hormones in turn may activate gene cascades that ultimately control all the events in a meristem: rate of cell division, meristem size, exact position where a leaf primordium forms, and rate at which that primordium grows. (In addition to these apical meristems, cereals and other grasses also have growth zones called *intercalary meristems* just above each node in the stems. These growth zones explain why grass shoots up rapidly after being cut.) Manipulating these genes may allow scientists to change the course of plant development.

The root apical meristem (RAM) is present near the tip of every root, just behind the root cap, a small structure that protects the apical meristem as the root grows through the soil. The root cap has its own stem cells or initials, which continue to produce root cap cells. Root cap cells shed as the root grows, so the root cap never grows larger. The root itself also originates from just a few (three to six) initials or stem cells. As a root elongates, lateral roots form not from the RAM at the tip of the root, but from meristems that are newly initiated somewhat back from the tip and below the surface of the root, in a tissue called the *pericycle* (**Figure 8.4a**). The pericycle can initiate new meristems when the hormone

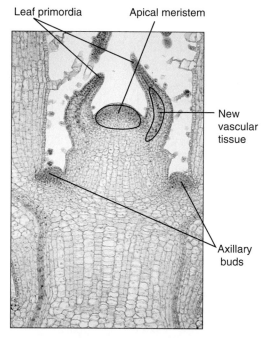

Figure 8.3 Longitudinal section through a shoot apex showing the location of the apical meristem, two leaf primordia, and bud primordia (partially formed axillary buds). *Source:* J. D. Mauseth (1998), *Botany: An Introduction to Plant Biology,* 2nd ed. (Boston: Jones and Bartlett), Figure 5.38(b).

Leaf primordia

Apical meristem

New vascular tissue

Axillary buds

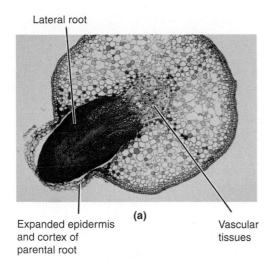

Lateral root

Expanded epidermis
and cortex of
parental root

(a)

Vascular
tissues

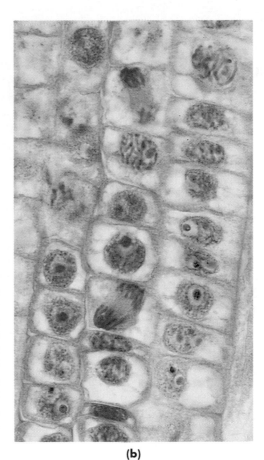

(b)

Figure 8.4 (a) Initiation of a lateral root in the pericycle of the main root. The hormone auxin triggers new cell divisions in the pericycle, and a meristem is formed. When the cells begin to enlarge, they push the meristem forward, breaking through the cortex and the epidermis. **(b) Portion of the root meristem of a hyacinth.** Here, two cells are in mitosis, the process in which the chromosomes duplicate and are partitioned to the two daughter cells. *Source:* J. D. Mauseth (1998), *Botany: An Introduction to Plant Biology,* 2nd ed. (Boston: Jones and Bartlett), Figures 7.17 and 4.13.

auxin triggers it to do so. As in stems, the pattern of root development is one of continued meristem and organ formation.

The SAM and RAM cells are small, about 100 to 1,000 times smaller than parenchyma cells (see **Box 8.2**), which make up the bulk of a young plant. Meristematic cells have a prominent nucleus, a dense cytoplasm full of ribosomes, and relatively undeveloped plastids (**Figure 8.4b**). The main function of these cells is to divide and give rise to more cells, and this they do with amazing regularity—once every 36 to 48 hours. Meristematic cells have a thin, flexible cell wall, and after each nuclear division a new cell wall forms separating the two new nuclei. At the periphery of the meristem, the cells begin to escape the signals that keep them continuously dividing. Now the cells begin to enlarge enormously, mostly because of the increased vacuolar volume. Enlargement is accompanied by differentiation and by the expression of thousands of new genes that are not, or only barely, expressed in meristematic cells.

The function of the SAM is to produce the cells that will make up the shoot system of leaves and stems. However, given the right environmental and/or hormonal stimuli, this apical meristem can be converted to a meristem that makes floral primordia. These floral primordia then grow out into flower organs (sepals, petals, stamens, and carpels). At that point, meristem activity ceases, and whatever cells remain stop dividing. The conversion of vegetative meristem to floral meristem is also controlled by a series of genes, as is the differentiation of the floral organs. In many plants, leaves, sepals, and petals have a very similar morphology yet are unique in function and in the proteins they contain. This uniqueness is caused by the expression of specific genes.

Box 8.2

The Cell Is the Basic Unit of Life

Nucleus: The control center of the cell that contains the chromosomes, which have all the nuclear DNA. As much as one meter of DNA is in a $20\mu m$ nucleus! All the machinery to duplicate DNA and to synthesize RNA is also in the nucleus. The nucleus is bordered by the nuclear envelope, which has large pores through which RNA and protein molecules can pass. The nuclear envelope disaggregates as a prelude to chromosome duplication and nuclear division, just before cell division.

Chloroplasts: Large, disk-shaped organelles with parallel internal green membranes (thylakoids) on which are located the proteins needed for the light reactions of photosynthesis. Chloroplasts contain about 100 minicircles of DNA (the chloroplast genome), which harbor about 100 genes. Chloroplasts also have ribosomes and their own protein synthesis machinery. Chloroplasts import numerous proteins made in the cytoplasm and encoded in nuclear genes, and are the site of starch biosynthesis and accumulation. In roots and seeds, the main function of the modified chloroplasts (called *amyloplasts*) is starch storage.

Plasma membrane: The membrane that surrounds the entire cytoplasm and controls (with special transport proteins) the movement of mineral ions, metabolites, and water in and out of the cell. It contains numerous receptors that have one part of the protein on the outside and one part on the inside of the membrane. These receptors transduce (relay) signals from the outside to the nucleus. The outer face of the plasma membrane is the site of cellulose synthesis.

Mitochondria: Small organelles with an outer and an inner (folded) membrane that are the site of all biochemical reactions of respiration.

Endoplasmic reticulum: An extensive network of intracellular double membranes (very flat sacs, really) in which the synthesis of many proteins takes place, especially those that are secreted out of the cell or accumulate in vacuoles.

Golgi apparatus: A series of short, flat sacs that are the site of synthesis for hemicelluloses destined for the cell wall. Transport vesicles bud off from this organelle and ferry proteins and polysaccharides to other destinations.

Vacuoles: Can be small or very large, filling almost the entire cell. They contain mostly mineral ions and some soluble metabolites (sugar and acids). Nearly always contain digestive enzymes that help dissolve cellular structures after they are taken into the vacuole. Each vacuole is surrounded by a membrane called the *tonoplast*. In some cells, storage vacuoles are specialized for the storage of proteins (in seeds), sucrose (in stems), or secondary metabolites and defense chemicals.

Ribosomes: Small, round particles consisting of half protein and half ribonucleic acid (RNA). Proteins are assembled on their surface.

Cytoskeleton: An extensive network of protein tubules and fibrils that extends through the entire cytoplasm. The tubules are important for the separation of chromosomes during cell division (the spindle). Small cytoplasmic structures can slide rapidly along the fibrils in a movement that is powered by motor proteins.

Cell wall: Cell walls are a characteristic feature of all plant cells. They can be thin and elastic, permitting the cell to grow in size or to become thick and hard in some cell types. They always contain cellulose microfibrils embedded in an amorphous matrix of hemicellulose polysaccharides. Many cell walls are totally encrusted with lignin, a hydrophobic polymer very resistant to degradation.

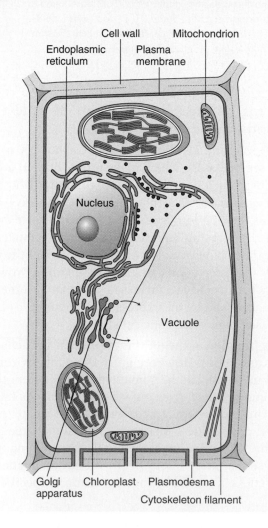

8.3   Environmental stimuli and hormones activate genetic programs that guide plant development.

Plant development—the orderly outgrowth of organs and the progression from seed to seedling to vegetative plant and finally to flowering plant—is profoundly affected by the environment. Indeed, a multitude of environmental stimuli or signals affect plant morphology and development (**Figure 8.5**). Many plant species heed environmental cues, such as changes in day length or a cold period, to proceed from one developmental stage to the next. The amount of light received, spectral quality of that light, relative lengths of day and night, and temperature regime during day and night all affect plant development. There is, first of all, a quantitative relationship both between light intensity and growth, and between temperature and growth. Up to a point, plants generally grow more vigorously if the temperature is higher and if there is more light. However, because plants have evolved in specific environments to which they are adapted, they may have quite different optimal light intensities and temperatures. Plants adapted to growing in dim light—on the floor of the tropical forest, for example—suffer from light stress and photooxidation if planted in full sun. Apart from these general effects of temperature and light, specific environmental regimes are often needed to let plants progress from one stage of development to the next.

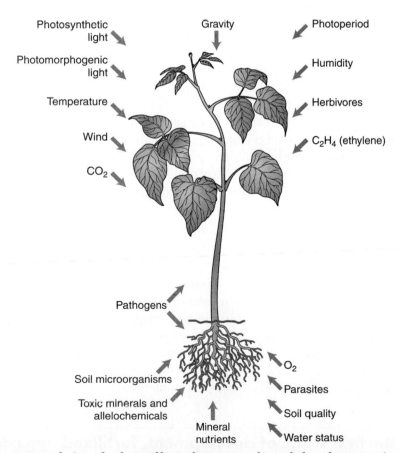

Figure 8.5 External signals that affect plant growth and development include many aspects of the plant's physical, chemical, and biological environments. Some external signals come from other plants. Apart from gravity, all other signals vary in intensity, often from minute to minute. *Source*: B. B. Buchanan, W. Gruissem, and R. L. Jones, eds. (2000), *The Biochemistry and Molecular Biology of Plants* (Rockville, MD: American Society of Plant Biologists), p. 931 (Figure 18.1).

Figure 8.6 Comparison of dicot seedlings grown in the dark and in the light. Dark-grown seedlings are yellowish (etiolated) and tall, with a long hypocotyl (young stem) and a clearly defined apical hook and unexpanded, unopened cotyledons. The dark-grown seedling is much shorter and dark green. The hook has straightened, and the cotyledons have expanded. The first true leaves are not yet visible at this stage.

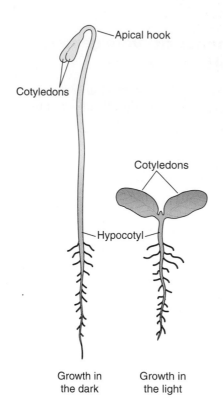

Apical hook

Cotyledons

Cotyledons

Hypocotyl

Growth in the dark

Growth in the light

Many species require light to break seed dormancy, and seeds buried too deeply in soil do not germinate. Moreover, seedlings need light to start greening. When seedlings emerge from soil, they are faintly yellow (etiolated); light causes them to turn green. In addition, when light touches the seedling the stem elongates less rapidly, and the internodes are much shorter (**Figure 8.6**). In dicots the leaves expand dramatically, and the plastids in leaf parenchyma cells develop into chloroplasts. In grasses, the leaves unroll and chloroplasts also develop. When the leaf parenchyma cells are exposed to light, they make the enzymes necessary to carry out photosynthesis; but if the leaves remain in darkness, they do not make these enzymes. If you think about this process in genetic terms, you see that light activates hundreds of genes that encode all the proteins needed for leaf development and photosynthetic function.

Reproductive growth or flowering can also depend on a particular environmental regime. As noted earlier, some biennial plants, including many varieties of winter wheat, absolutely require a period of cold weather before the vegetative apical meristem can become a reproductive meristem and give rise to flower primordia. This process is called *vernalization*. Winter wheats are normally planted in the fall and grow 10–20 cm tall before winter arrives. The wheat overwinters under a blanket of snow, and growth resumes in the spring. To come into flower and produce seed, winter wheat needs the cold period. Many other plants, including wild carrot, need to vernalize. The relative lengths of the day and the night determine whether flowering will occur. This phenomenon, called *photoperiodism*, is discussed later in Section 8.6 of this chapter. The photoperiod also has other effects on plant development. For example, the development of fleshy storage roots (such as carrots and beets) is governed by day length.

Scientists can mimic some of these environmental effects on plant development by applying plant hormones. During a developmental transition, the level of a particular hormone within the plant may change dramatically, and if the plant is sprayed with a solution of the same hormone, the transition occurs even in the absence of the correct environmental regime. For example, spinach normally requires long days and short nights to form flowers; scientists can circumvent this environmental requirement by spraying the plants with the hormone gibberellic acid. Analysis of unsprayed plants shows that levels of specific gibberellins rise as the plants make the transition from vegetative to reproductive growth. Such observations have led to the hypothesis that changes in the hormone level within the plant mediate the effects of the environment.

8.4 In the first stage of development, fertilized eggs develop into seeds.

The Plant Life Cycle. The plant life cycle (**Figure 8.7**) differs from the more familiar mammalian one in that male and female reproductive organs usually occur on the same plant and more often than not in the same flower. The cells of flowers are diploid, with two sets

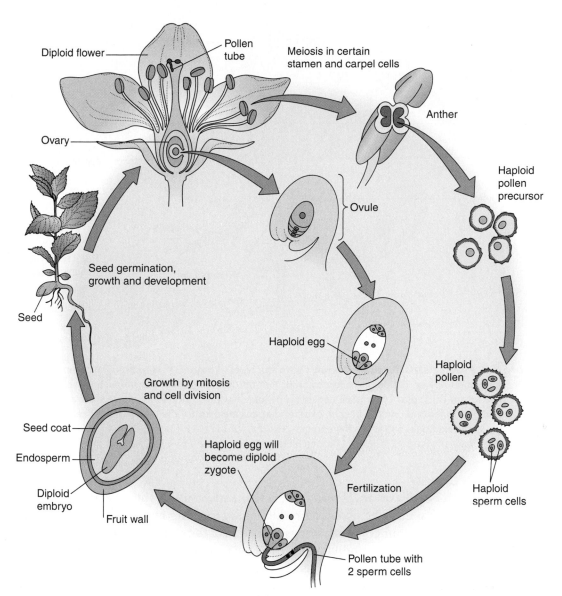

Figure 8.7 The life cycle of a plant showing the production of gametes (sex cells) and their union at fertilization. The flower, top left, is diploid and contains male (stamens) and female (carpels) reproductive organs. Specialized cells in each of these undergo meiosis to produce haploid egg and sperm cells. A single egg cell is formed in an ovule in the carpel. Two sperm cells are present in each pollen grain produced within the anther of the stamen. Fertilization results in the formation of a diploid zygote, which develops into an embryo as part of a seed. The seed germinates to form a vegetative plant, which eventually produces flowers. *Source:* J. D. Mauseth (1998), *Botany: An Introduction to Plant Biology,* 2nd ed. (Boston: Jones and Bartlett), Figure 9.5b.

of chromosomes. Some specialized cells in the reproductive organs undergo meiosis, a process that halves the chromosome number and produces haploid cells (see Chapter 6). These haploid cells undergo a few normal mitotic cell divisions, and some of their progeny then differentiate either into haploid egg cells or into haploid sperm cells. In animals, the haploid cells produced by meiosis differentiate directly into the sex cells (gametes) and there is no haploid developmental stage. The union of haploid egg and sperm produces a diploid zygote, which is the first cell of the new organism. Further cell divisions of this zygote result in the formation of the embryo, which forms a seed and then a new plant. Formation of the embryo and the seed occurs within the tissues (ovary) of the mother plant.

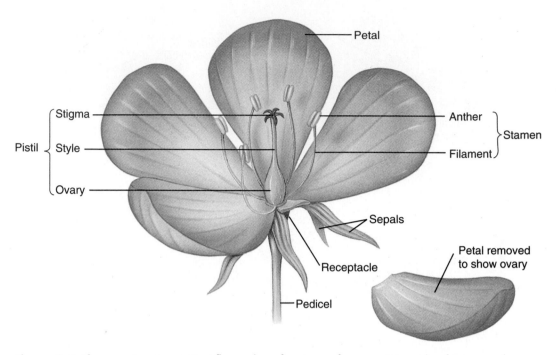

Figure 8.8 Flower structure. Most flowers have four types of organs: (1) sepals, often green, that protect the unopened flower, (2) petals, often colored, that attract pollinators, (3) stamens, which produce pollen with sperm cells, and (4) a pistil consisting of fused carpels (individual carpels are not shown). Carpels contain ovules and one cell in each ovule undergoes meiosis. Of the four daughter cells formed in meiosis, three die and only one forms the egg cell. *Source:* J. D. Mauseth (1998), *Botany: An Introduction to Plant Biology*, 2nd ed. (Boston: Jones and Bartlett), Figure 9.3a.

Flowers and Fertilization. We said earlier that conversion of the vegetative meristem to a flower meristem initiates sexual reproduction by forming reproductive organs. The flower organs occur in four concentric rings at the end of the stem: sepals, petals, stamens, and carpels (**Figure 8.8**). The individual carpels are usually fused to form a pistil. Sepals and petals often have the same shape as leaves, but stamens and carpels look quite different. A stamen consists of an anther, which produces the pollen, perched atop a long filament. The pistil has three parts: a swollen ovary with one or more ovules at the base of a style, and a stigma at the top. Pollen grains that land on the stigma germinate, and pollen tubes grow into the stigma and style to find the egg cell in the ovule. When the pollen tube reaches the ovule, it bursts and spills out two sperm cells that fuse, one with the egg cell and one with another cell that will produce the endosperm. This double fertilization requiring two sperm cells is unique to flowering plants.

Flower structures vary from plant to plant, but in general the flowers of the important cereal crops occur in clusters. In wheat and rice, small flowers are connected to a central stalk termed a *spike*; here, both male and female reproductive organs are in the same flower. For maize, in contrast, female flowers are located in lateral ears and male flowers in terminal tassels, at different locations on the same plant.

The union of egg and sperm that initiates development of a new plant occurs within the ovule. Depending on the species, pollen may come from the same plant (self-pollination, as in wheat) or from another plant of the same species (cross-pollination, as in maize). The type of fertilization that normally occurs for a particular species has important implications for plant breeding and for creating hybrid seed.

Embryogenesis. After fertilization, the zygote—the cell that results from the union of egg and sperm—undergoes many cell divisions and forms a multicellular embryo (see Chapter 9). The growth of the embryo occurs within the ovule in a rich nutrient medium called the *liquid endosperm*. At this stage, the new plant is growing completely heterotrophically, and the liquid endosperm provides the embryo with sugars, amino acids, vitamins, and the proper balance of hormones. The endosperm synthesizes these substances from sucrose and amino acids provided by the mother plant. As the embryo grows, it acquires more and more biosynthetic capacities; soon it can make all the complex molecules it needs, but it always continues to depend on a supply of sucrose, amino acids, and minerals from the mother plant.

An important aspect of embryo formation is the organization of the embryo axis: The two meristems, the SAM and the RAM, are located at opposite poles of the embryo. The asymmetric distribution of the hormone auxin plays a major role in this process. But that asymmetric distribution implies that all cells in a young embryo are not alike. So how soon do the cells become different from each other? Scientists have studied this process in greatest detail in *Arabidopsis thaliana*; at the 16-cell stage, the embryo is already divided into an upper half (the future SAM and cotyledons) and a lower half (the future RAM and stem of the seedling). By looking at the expression of specific genes at the 16-cell stage, researchers can find genes that are expressed only in some of the cells. Such observations suggest that cell differentiation in the embryo begins very early indeed.

Seed Formation. Once the embryo is formed, food reserves (proteins, oils, starch, and minerals) are synthesized and begin to accumulate in the storage parenchyma cells of the cotyledons or in other tissues closely associated with the embryo. In the majority of dicots, most of the reserves accumulate in the cotyledons, which themselves become very large. In cereals and other grasses, most of the food accumulates in the endosperm, a tissue located just next to the embryo (see Chapter 9). Food reserves accumulate in specialized organelles within the cytoplasm: starch in amyloplasts, a special type of plastid; protein in small protein storage vacuoles (**Figure 8.9**); and oil in very small oil bodies.

During the developmental period that stretches from fertilized egg to mature seed, three important genetic programs are activated: The first regulates cell division and the formation of tissues and organs in the embryo; the second regulates expansion of the storage organs and the massive biosynthetic effort to make reserves. Finally, the seed needs to prepare for survival in the dry state and for responding to environmental cues that will bring it out of dormancy. Survival in the dry state (15% water) depends on accumulating sugars, oligosaccharides, and special proteins that bind water and protect the cytoplasm components from damage when the cells dry out.

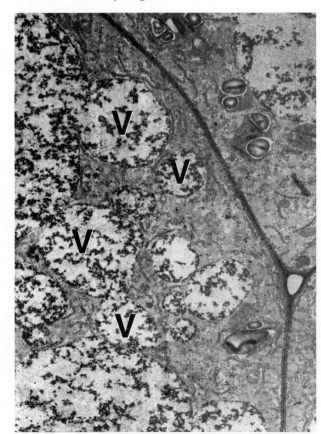

Figure 8.9 Legume seeds are a rich source of protein stored in vacuoles. In this electron micrograph of a soybean cell, the vacuoles (V) are not yet full and protein will continue to accumulate during the 40-day seed maturation period. *Source:* Courtesy of E. Herman, Agricultural Research Service, Beltsville, MD.

8.5 Formation of the vegetative body is the second phase of plant development.

With the appearance of conditions that break seed dormancy and allow germination, the seed restores its former water content and swells. This rehydration triggers the two main events of germination: embryo growth and use of the seed's food reserves. Embryo growth depends on using the reserves, because not until green leaves have formed can the seedling grow autotrophically. Its initial growth depends on the carbohydrates, fats, proteins, and minerals in the storage tissues of the seed. The growing embryo first breaks down the starch, fats, and proteins into their simple constituents (sugar, fatty acids, and amino acids), and then absorbs and uses them.

As the embryo absorbs the digested products of the seed's storage material, it uses them to build more cells and to expand cells once made. Within a few days, a recognizable root is growing down through the soil and the shoot (seedling stem and cotyledons) pushes above ground. Once above ground, light causes the stem to elongate less rapidly, and the leaves to expand. In the light, chlorophyll is synthesized and the plant begins to carry out photosynthesis and ceases to depend on seed reserves (see Figure 8.6).

Formation of the Root System. We noted earlier that cell growth involves three basic processes: division, enlargement, and differentiation. In the root tip these events are spatially separated (**Figure 8.10**). At the very tip, just behind the protective root cap, is the RAM, a region of continuous *cell division*. These new cells then become about 100 times larger; behind the region of cell division there is a region of *cell elongation*. Finally, as they are enlarging and after they have reached their full size, the cells specialize, depending on location, into the various cell types of the root; behind the enlargement region there is thus a region of *cell differentiation*. As the root grows through the soil, this sequence is always maintained.

Root growth is accompanied by the formation of lateral roots, which in turn form other lateral roots, until the plant has established a widely branched root system capable of taking up water and minerals from a large volume of soil. The size and shape of that root system are genetically determined, and some plants form very deep root systems, others very shallow ones. Like many other features of the plant, these represent ecological adaptations. There may be considerable variation in the root systems of wild relatives of crops that might be exploited by plant breeders. However, unlike shoot systems or flowering characteristics, root systems are difficult to study because people cannot see them. Moreover, given the differential growth of the roots in different soil types and under different climatic conditions, traditional plant breeding for better root systems is almost impossible. Nevertheless, molecular geneticists using genomic information will be able to determine in the near future the important genes that determine the characteristics of root systems. What now seems impossible to achieve with traditional breeding may be within reach using the tools of molecular breeding.

In addition to having an extensive root system, many plants greatly increase the area exposed to the soil by forming root hairs. These are tiny projections from individual epidermal cells. Roots also form symbiotic associations with soil fungi called *mycorrhizae*. These serve the same purpose as root hairs: The fungal threads take up minerals from the soil and transfer them to the root cells for transport into the plant (see Chapter 12). To provide the plant with water and mineral nutrients, the root system must grow continuously, and new root hairs are always being formed as the older ones die off.

Storage of food reserves is another important function of the root system. Some biennial plants, such as beets and carrots, develop a thick, fleshy taproot at the end of the first growing season. The reserves stored in this taproot are used at the beginning of the second growing season, and a new shoot produces flowers and seeds. Other plants, such as cassava, develop numerous fleshy, tuberous roots for food storage and vegetative reproduction (see later discussion).

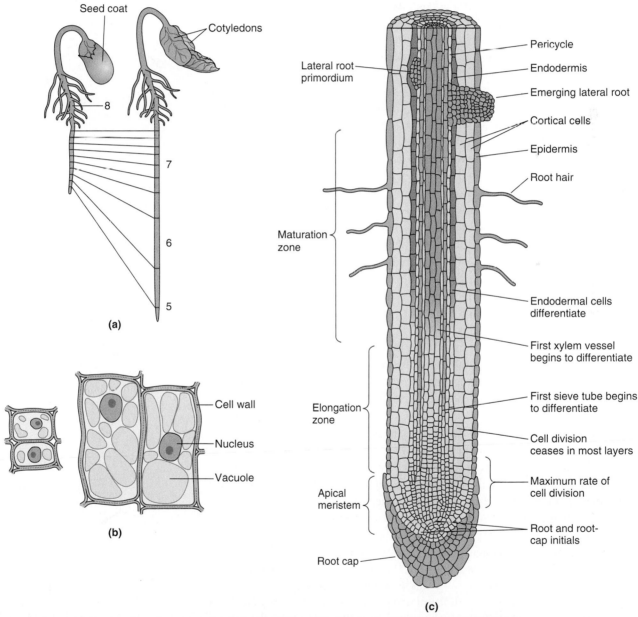

Figure 8.10 Diagrams showing root growth. (a) The distribution of growth zones in the root: 5 = apical meristem, 6 = zone of rapid elongation, 7 = elongation has ceased, 8 = mature portion; **(b)** cells are shown enlarged about 1,000 times; small cells have very small vacuoles and enlargement means a big increase in vacuolar volume; **(c)** detailed view of the different zones on the root: root cap, root apical meristem, elongation zone, maturation zone, and lateral root.

Box 8.3

Plant Hormones

Plants make at least seven kinds of hormones. The hormones are small molecules that are carried in the transpiration stream of the xylem (cytokinin, abscisic acid), are transported from cell to cell (gibberellin, auxin, brassinolide), or are released into the air (ethylene, methyljasmonate) like animal pheromones.

Auxin is produced primarily in the shoot apices, and it is transported toward the base of the stem, away from the SAMs. As auxin is transported, it is broken down, so that away from the meristem the auxin concentration gradually declines. The high levels of auxin that occur just below the meristem in the zone of cell elongation are necessary to maintain the high rate of cell elongation. Auxin is also involved in phototropism, the bending of stems toward the light, and gravitropism, the bending of the roots toward Earth and of stems away from Earth when plants are placed horizontally.

Gibberellins are a large group of related compounds (more than 100 common have been identified) defined not by their biological action but by their structure. One of their important roles in plant growth is that they are responsible for how tall plants grow. The stem of a tall plant contains more biologically active gibberellin than the stem of a short plant. As noted earlier, gibberellins are involved in the flowering process of certain plant species. In addition, they can break the dormancy of many seeds and obviate the need for a specific environmental regime for germination.

Cytokinins are a smaller group of related compounds that are produced in the roots and move in the transpiration stream to the leaves. They regulate leaf senescence. When an annual plant gets older and its root system stops growing vigorously, it produces less cytokinin. The older leaves receive less cytokinin in the transpiration stream, and this contributes to their loss of chlorophyll, loss of protein, and general senescence. Cytokinins regulate cell division, and the ratio of auxin to cytokinin in the stem determines the branching pattern of the stem. Beside a terminal bud, every stem also has quiescent buds, which do not grow out into shoots when the ratio of auxin to cytokinin is high. When this ratio falls because auxin levels go down or cytokinin levels go up, the quiescent buds grow into side shoots.

Abscisic acid is involved in seed and bud dormancy, as well as in the response of the plant to conditions of stress (such as not enough water, too much salt, or too cold). Water deficit and salt stress increase abscisic acid in the roots, and this abscisic acid moves to the leaves through the xylem. In the leaves, this hormone causes the stomates to close so that further water loss is prevented. At the same time, new proteins are made that help the cells cope with these stresses. Abscisic acid often acts antagonistically to gibberellins.

Brassinosteroids are a class of plant hormones that induce a broad spectrum of developmental responses, such as stem elongation, pollen tube growth, leaf bending and epinasty, root growth inhibition, and xylem differentiation. The brassinosteroids coordinate these multiple developmental programs with the presence of darkness or light.

Ethylene is a gas released by many different tissues, but especially by ripening fleshy fruits. Ethylene has numerous effects on plants, and these effects vary with the species. It regulates flowering in some plants, and flower aging in others. It promotes fruit ripening in many fleshy fruits and is commercially used to ripen bananas. Ethylene causes flowers and leaves to be shed. Chemicals that cause plants to produce ethylene are therefore used as defoliants.

Jasmonic acid is also a gas released in the form of methyljasmonate when plants are attacked by insects. It regulates the defense response of the plant to the insects. In addition, it regulates genes during normal plant development.

Indole–3–acetic acid (an auxin)

GA$_1$ (a gibberellin)

Zeatin (a cytokinin)

(S)-Abscisic acid

Ethylene

(—)-Jasmonic acid

Brassinolide (a brassinosteroid)

Formation of the Shoot. The processes involved in shoot growth are generally similar to those for root growth except that the three basic processes are not spatially separated. As noted earlier, a shoot consists of a stem and of leaves attached to the stem at points called *nodes*. Of particular interest to plant physiologists and plant breeders are the processes that control stem elongation in the shoot. Scientists know that at the cellular level hormones influence elongation. Auxin, gibberellins, and brassinolides (see **Box 8.3**) all affect stem elongation in different plant species and under different conditions. Cell elongation requires, among other processes, that the network of large molecules in the cell wall loosen; this loosening depends on a slight increase in acidity and on enzymes and other proteins secreted from cells into the wall. Hormones activate some of the genes that encode these enzymes.

Dwarf mutants with short stems in which the cells fail to elongate have been found for a number of plant species. Such mutants either have very low levels of a particular hormone (such as gibberellin) or are unresponsive to the hormone they synthesize. As part of the development of high-yielding wheat varieties for the Green Revolution (see Chapter 13), breeders purposefully selected dwarfs because shorter plants with large heads of grain do not fall over (lodge) so easily. Moreover, if the plant doesn't waste energy making stem tissue, it can channel more energy into seeds.

Vegetative Reproduction and Propagation. In addition to sexual reproduction, marked by seed formation, many plants can also reproduce asexually. In asexual (vegetative) reproduction, new plants form, not from seeds, but from specialized structures of the root, stem, or leaf (the vegetative organs of the plant). In nature, such vegetative reproduction often proceeds by means of horizontal stems or roots, which let the plant spread over a larger area. For example, the strawberry plant sends out runners (horizontal stems), which make new plants by sending down roots and forming leaf clusters where they touch the ground. In some plants, horizontal stems that grow underground may become thick and fleshy and form tubers, as in white potatoes. Each tuber can give rise to a new plant. People have learned to encourage the growth of such horizontal stems by mounding soil around the plant to cover the lower portion of the shoot (**Figure 8.11**). Tuber formation is also controlled by day length and mediated by genes that direct the growth of these structures and control starch accumulation. In other plants as well, the specialized structures involved in vegetative propagation are at the same time food storage organs and enable the plant to survive adverse conditions.

Human use of the natural process of vegetative reproduction is termed *vegetative propagation* or *cloning*. People use cloning to propagate potatoes, sweet potatoes, sugar cane, berries, nuts, and a variety of fruit trees and ornamental plants. The advantage of asexual reproduction is that all the offspring have exactly the same characteristics (genetic makeup) as the parent plant. This homogeneity is especially important when the parent is a genetic hybrid whose sexually produced offspring would not be the same as the parents. For example, an apple tree grown from a seed does not usually produce apples of the same quality as the parent plant. Vegetative propagation through grafting ensures uniform quality in such plants.

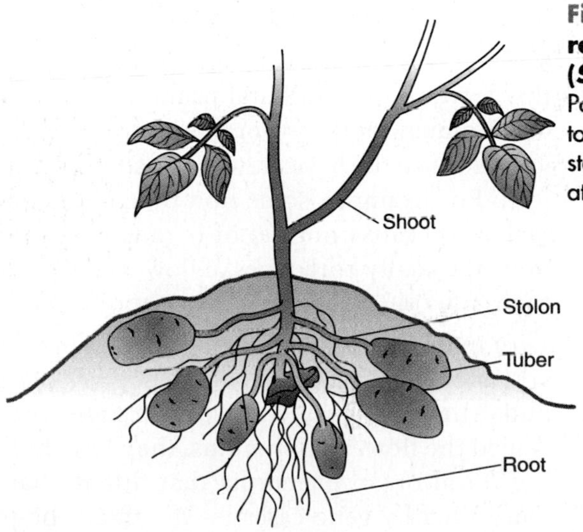

Figure 8.11 **Vegetative reproduction of potato (*Solanum tuberosum*).** Potato tubers are attached to underground horizontal stems or stolons that are attached to the main stem.

Shoot

Stolon

Tuber

Root

Plants that produce specialized structures for vegetative reproduction are easily propagated. Thus pieces of white potato tubers or of the tuberous sweet potato roots sprout and produce new plants when they are put in the soil. Other plants, such as sugar cane, pineapple, cassava, and many ornamental plants, are propagated by stem cuttings. In these species, pieces of stem produce roots spontaneously when placed in moist soil. The discovery that in many species the plant hormone auxin promotes the rooting of stem cuttings has greatly enlarged the list of plants that can be propagated this way. Many stem cuttings that would not ordinarily produce roots do so after the lower end of the stem is dipped in a solution of auxin. It is also possible to dissect out the meristem of a plant, cut it into small pieces, and regenerate new plants. This procedure, called *meristem culture*, must be done under sterile conditions. The goal of meristem culture can be genetic uniformity of superior strains, as with oil palms or orchids, or the rapid propagation of plants free of pathogenic viruses, as with strawberries.

8.6   Reproductive development involves the formation of flowers with their male and female reproductive organs.

We noted earlier that sex cells or gametes are produced in the reproductive organs of the flowers. Most dicot flowers consist of four concentric circles of organs (see Figure 8.8). The sepals, which are often small and green, completely enclose the flower bud before the flower opens. Next are the petals, which are often colored and showy, especially in plants pollinated by insects and birds. Inside the ring of petals are the stamens, and at the center is the pistil. Some of the most important crops—wheat, rice, maize, barley, and millet—are of the grass family (Poaceae), in which the flowers have a somewhat different structure. The inner two concentric rings of floral primordia also develop into female and male reproductive organs, but how the other organs of the flower correspond to sepals and petals is not clear. Grass flowers are called *spikelets*, and each floret has two leaflike organs, the lemma and the palea, that enclose the male stamens and female carpels (**Figure 8.12**). In wheat, barley, oats, and other grains, a long, slender awn is attached to the palea. When these flowers are mature, the lemma and palea separate slightly, exposing two stigmas and allowing the anthers to hang free. These wind-pollinated flowers have no need of showy petals to attract pollinators.

The elaborate floral organs, each with their own tissues and cell types, originate in the flower bud from floral primordia, just like leaves originate as leaf primordia. The development of these floral primordia is itself precisely regulated by genes, and many variations to the basic flower structure result from variations in the developmental pattern. For example, some flowers contain only male or only female sex organs—as in maize—because one set of organs fails to develop in either the male or the female flower. Usually both types of flowers are present on the same plant, but in some species each individual is genetically either male or female.

In many plants, a flowering signal or stimulus that originates in the leaves can convert a terminal vegetative bud that gives rise to leaves and stem tissues into a flowering bud. The chemical identity of this signal remains undiscovered. Once the SAM has received the flowering stimulus, the SAM changes into a floral meristem and starts forming floral organ primordia that differentiate into floral organs. These processes are controlled by gene cascades in which the protein product of one gene activates the next gene. A combination of genes controls the identity of each organ (see **Box 8.4**).

(a)   **(b)**

Figure 8.12 Structure of a grass flower. (a) Spike of wheat (*Triticum aestivum*). **(b)** Spikelets attached and detached from the rachis of wheat. The glume is thought to correspond to a bracht, a small leaflike structure found on many plants. The lemma and palea are sepal-like organs, based on the types of genes expressed there. The lodicule, at the base of the lemma and palea, swells up and causes the flower to open. The genes expressed in the lodicule identify it as equivalent to the petals of a dicot flower.

Reproductive development is often extremely sensitive to stress, and a single day of heat stress at the wrong time is often enough to block one or more aspects of this important phase of development.

As mentioned, when it comes to initiating flowers, many plant species respond to the photoperiod. Plants can be classified as "short day," "long day," or "day neutral" with respect to their flowering response to the photoperiod (**Figure 8.13**). Many plants are day neutral, meaning that the photoperiod does not influence them. Although they are called "short day" plants and "long day" plants, they are really long night plants or short night plants, respectively. Indeed, plants do not monitor day length, but rather night length. You can easily demonstrate the dependence of flowering on the length of the night (or the day) by turning on a light over such a plant in the middle of the night for a few minutes. This short light period alters the plant's response, even if the length of the day remains the same (Figure 8.13).

Because of their response to the photoperiod, certain plants flower only at a particular time of year. This selectivity may disadvantage the farmer, who thus can grow only one crop per year. Multiple cropping—which farmers practice in many tropical areas that have abundant water and sunlight—is only possible if the crop plants are insensitive to the photoperiod. The new strains of wheat and rice that allowed much of the increased grain production in India, China, and Mexico in the 1970s and 1980s, were selected to be photoperiod insensitive, so that farmers could grow several crops in a 12-month period.

Box 8.4

The Genes That Determine Floral Organ Identity

The genes involved in converting a vegetative meristem into a floral meristem and those that determine the identity of floral organs have been studied in greatest detail in *Arabidopsis,* by isolating mutants with unusual flowers. The names of the genes reflect the appearance of the plant or the mutant flowers. Thus, the *leafy* mutant of *Arabidopsis* has green flowers with whorls of sepal-like and carpel-like structures. Another mutant, called *apetala-2,* has, as its name implies, no petals. Its flower organs consist of whorls of carpels and stamens. The mutant *agamous* produces only sepals and petals and does not make carpels or stamens.

Scientists have found a large number of such mutants and have isolated the genes that are mutated (inactivated) in the mutant plants. The proteins encoded by such genes are almost invariably transcription factors that reside in the nucleus and play a role in the activity of specific genes. In some cases the protein made by one gene activates another gene; in other cases, the protein re-presses another gene. By studying a large number of these genes and determining in which cells each gene is expressed, scientists have been able to construct a model for how the identity of a floral organ is determined: the "ABC model for floral organ specification." This model proposes that the identity of the four concentric circles is determined by the activity of genes that function in three "fields" (see figure). Field A corresponds to the prospective sepals and petals, field B to the petals and stamens, and field C to the stamens and carpels. By looking at mutant flowers, scientists can determine in which field a gene is active. For example, the *agamous* gene must be active in field C, because when it is not working properly the flowers have no stamens and carpels. The model suggests that if only A function genes are active, the primordia become sepals. If A and B function genes are active, the primordia become petals. If B and C function genes are active, the primordia become stamens; and if only C function genes are active, the primordia become carpels. Together with some additional assumptions, this scheme explains how young primordia that look pretty much the same under the microscope, nevertheless all develop their own identity.

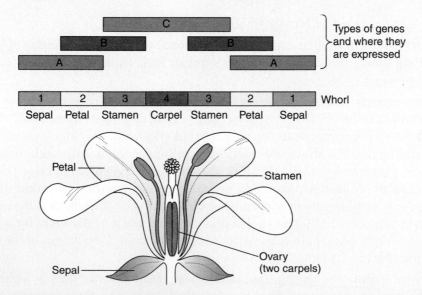

Control of organ identity by the overlapping expression of three classes (A, B, and C) of genes. Floral organs are organized in whorls, from outside to inside: sepals, petals, stamens, and carpels. Cells that express only A genes become sepals. Cells that express A and B genes become petals. Cells that express B and C genes become stamens, and cells that express only C genes become carpels. Plants in which all three types of genes are inactivated by mutation produce "flowers" in which all the organs look like small leaves.

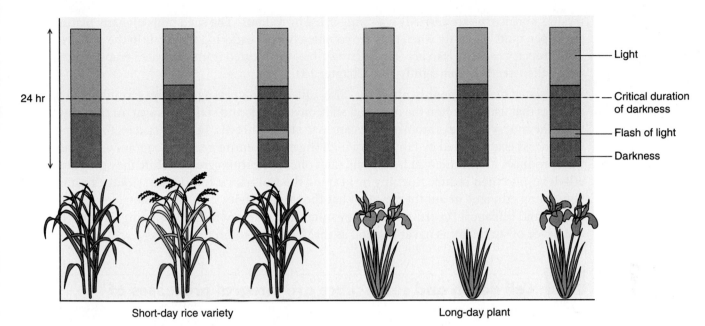

Figure 8.13 Photoperiodic regulation of flowering. The photoperiod is the number of hours of light and dark within a 24-hr period. Short-day plants (left) flower when the length of the night exceeds a critical number of hours. If the night is long but interrupted by a few minutes of light, they do not flower. Shown here is a short-day rice variety. (Many varieties of rice are day-neutral plants.) Long-day plants flower when the night falls below a critical number of hours (right). If the night is long but interrupted, they flower.

8.7   The formation of fruits aids seed dispersal.

Fertilization of the egg cell by the sperm cell is followed by embryo growth and seed formation. Along with seed formation, the ovary surrounding the growing embryo(s) forms a fruit. Depending on the species, the fruit may also incorporate other tissues in close proximity to the developing ovary. A peach is a "typical" fruit. The fleshy part of the peach is the greatly thickened wall of the ovary, and the seed is inside the stone.

The explanation just given should make clear that any structure containing seeds is a fruit, not a vegetable. Tomatoes, peppers, eggplants, zucchinis, cucumbers, okra, and green beans are all fruits. Each has a more or less thick, fleshy wall that originates in the ovary and contains seeds. So why do people call them *vegetables*? Well, the word *fruit* comes from the Latin *frui*, meaning to enjoy, to delight in, to have the use of something desirable. Sweetness is innately the most preferred taste sensation of mammals, and people call *fruits* those plant parts that are fleshy, sweet, and enjoyable. (So why is a lemon a "fruit"? Language and custom can be mysterious!)

The early development of the fruit-also called *fruit set* or *setting*—depends on hormones, especially auxin and gibberellin, produced by the growing embryo within the seed. These hormones stimulate the cells of the ovary to divide and expand. Spraying with synthetic auxin is sometimes used to promote fruit set and cause the development of fruits without fertilized seeds. In addition, plant breeders have found mutants in which fruit development does not depend on hormones produced by the seeds. Such mutants often produce seedless fruits. The fruit is the vehicle that aids seed distribution. Dandelion fruits (commonly called *seeds*, but to a botanist they are fruits) are carried away (for example) by the wind, berries are eaten by small mammals whose droppings are scattered over the countryside, and acorns are carried away and buried by rodents. The seed pods of many plants split open quite violently when they are completely dry, scattering the seeds in the process. This property, called *dehiscence* (or *shattering*), has disappeared from cultivated crops, which retain their seeds rather tightly (see Chapter 13).

The later development of the fruit, also called *fruit ripening*, involves a number of changes that usually soften the fruit wall and convert acids and starch to sugar. At the same time, the fruit synthesizes aromatic substances to attract animals. In many species the ripening process is controlled by ethylene, which triggers an entire genetic program when the fruit produces it. Commercial fruit companies often pick fruits green, before they produce ethylene, and then induce ripening by keeping the fruits for a few days in storage rooms containing ethylene or another gas that has the same activity. This procedure lets people handle and transport the fruit before they soften as a result of natural ripening. However, the flavor of such fruit is never as good as that left to ripen on the plants.

8.8 | Cell death and senescence are integral processes of development.

The death of individual cells, groups of cells, or entire organs is an integral part of development. The best known example of this phenomenon is the shedding of leaves by trees. Whether on deciduous trees or not, after leaves mature they gradually start losing their capacity for photosynthesis and undergo senescence. Such senescence is visible at the ultrastructural level: The internal membranes of the chloroplasts become disorganized, and little fat globules accumulate in the chloroplasts. Eventually the cells undergo the process of autophagy (self-eating) in which the major macromolecular components (proteins, nucleic acids, lipids, and starch) are digested and the resulting small molecules are exported to the growing part of the plant or to storage tissues. Thus, when a tree sheds yellowed leaves, these are not full of protein and minerals—everything has been digested and exported. In the final phase, the cells digest themselves and only the cell walls remain. Because the organism controls both the initiation and the execution of this senescence, the process is called *programmed cell death*. Programmed cell death may be triggered by environmental factors (annual leaf shedding is a photoperiodic response in many plants) and is controlled internally by hormones. Once again, the expression of specific genes underlies the whole process.

When you look at different aspects of plant development, you can see that programmed cell death occurs all through the plant and throughout its life. To begin with, digestion of reserves stored in the cotyledons of many dicots, or in the endosperm of cereals, is followed by cell death. In the aleurone cells of cereals, which secrete enzymes that digest stored reserves, cell death is accelerated by gibberellin and retarded by abscisic acid. As vascular tissues develop, precursors to vessel elements undergo programmed cell death and these elements do not become fully functional until the cells die. The sclereids in ground tissues also undergo programmed cell death. Another example of programmed cell death is senescence of sepals, petals, and stamens after fertilization.

When the root systems of some plants, such as maize, are flooded and thereby deprived of oxygen, they can develop internal channels through programmed cell death. Water has much less oxygen than does the soil atmosphere, and lack of oxygen triggers this developmental process. The channels permit faster diffusion of oxygen throughout root tissues. The channels develop in the cortex, and the cortex with its channels is now called the *aerenchyma*. The channels form when the cells are triggered to undergo programmed cell

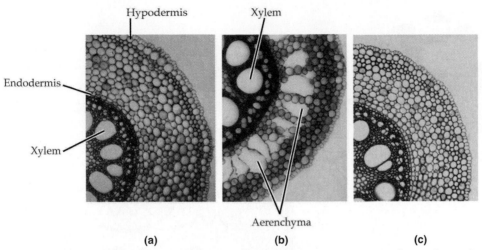

Hypodermis   Xylem

Endodermis

Xylem

Aerenchyma

(a)   (b)   (c)

Figure 8.14 Aerenchyma formation in maize roots. (a) Cross section of control root of 11-day-old plant. **(b)** Plant exposed to an unaerated nutrient solution for 4 days, depriving the roots of oxygen. **(c)** Inclusion of 50 micromolar EGTA in the nutrient solution. EGTA penetrates the root cells and binds up calcium, making it impossible for calcium to fulfill its signaling function. *Source:* B. B. Buchanan, W. Gruissem, and R. L. Jones, eds. (2000), *The Biochemistry and Molecular Biology of Plants* (Rockville, MD: American Society of Plant Biologists), p. 1051 (Figure 20.8) and p. 1086 (Figure 20.46A).

death and enzymes synthesized as part of aerenchyma formation digest the cell walls (**Figure 8.14**). This entire process is regulated by ethylene, the synthesis of which is induced by the absence of oxygen.

The cellular program leading to programmed cell death involves the fragmentation of nuclear DNA, caused by enzymes called *nucleases*; the biosynthesis of enzymes that digest proteins, lipids, and RNA; and the vacuole's engulfing of cytoplasmic structures that vacuolar enzymes then digest. The genes that encode all these enzymes are activated as part of the programmed cell death process. Scientists have known for some time that hormones are involved in different aspects of senescence, but they are only now exploring the genetic basis of programmed cell death.

8.9 Signal transduction networks relay environmental and hormonal signals to the nucleus to regulate gene activity.

The preceding description makes clear that hormones and the environment regulate many aspects of plant development. Because development requires the orderly expression of genes in specific genetic programs, the question arises, How does the genome "know" when to activate the correct genes? How is the environmental signal transduced to the genome? How does light turn genes on? How does ethylene turn on the genes that encode the enzymes needed for fruit ripening?

To perceive a light signal, there must be a receptor, a light-absorbing molecule that interacts with a protein to start the chain of events that leads to gene activation. To perceive a hormonal signal, there must be a receptor to which the hormone binds. The interaction of the signal (whether light or a hormone) with the receptor usually subtly changes the shape of the receptor, affecting its activity or its ability to interact with another protein.

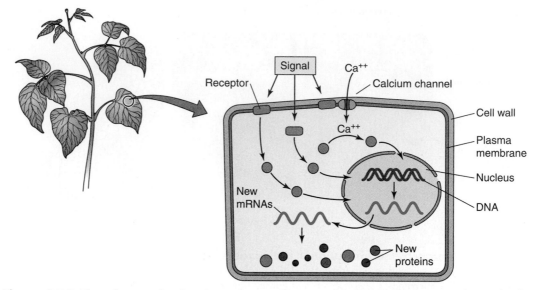

Figure 8.15 Signal transduction in a plant cell. A signal (light or a hormone or other molecule) reacts with a receptor (R), which is in the plasma membrane (PM) or in the cytoplasm. In one case, proximity to a calcium channel causes the channel to open, resulting in sudden influx of calcium. In each case, other proteins are modified down a "chain of command" that reaches the nucleus where genes are activated. New mRNAs are made and exported from the nucleus. In the cytoplasm, they direct the synthesis of new proteins.

This protein in turn passes the "signal" on to another protein, and so on, until genes in the nucleus activate (**Figure 8.15**).

In the case of light, the receptors are pigments that have a light-absorbing part and a protein part. Such pigments belong to a class of proteins called *chromoproteins*. (Hemoglobin is a typical chromoprotein.) Plants are green because they contain chlorophyll, but in addition to chlorophyll they contain blue chromoproteins, called *phytochromes*, that specifically absorb red light, and they contain cryptochromes, which are yellow chromoproteins that absorb blue light. Both phytochromes and cryptochromes are present in very small quantities, but these amounts are nevertheless large enough to register subtle changes in the light environment and to affect plant development. Plants use these pigments to measure the presence and absence of light, duration of the dark period (photoperiod), intensity of light, and spectral quality of light (in the shade underneath a tree canopy, there is relatively more red light).

The absorption of light energy by the light-absorbing portion of the molecule changes the shape of the protein. This change causes phytochrome to acquire a negatively charged phosphate group, which now permits phytochrome to enter the nucleus through pores in the nuclear envelope. It may well do so bound to other proteins. It has recently been shown that phytochrome's entry into the nucleus is followed by its binding to proteins responsible for DNA transcription. These DNA transcription complexes then bind to the promoters of genes whose activity is up- or down-regulated by the light signal.

Investigating signaling pathways is now a very active research area, not only to understand how they work, but because scientists may be able to alter the course of plant developmental processes if they interfere with the signaling pathways. For example, how does the plant perceive the gaseous hormone ethylene? In this case the receptor is a protein in the plasma membrane that has a domain (a portion of the protein) on each side of the membrane, with a short piece of the polypeptide spanning the membrane. Ethylene binds to this receptor, and binding turns the developmental switch that precipitates a phosphorylation cascade, meaning that phosphate groups become attached to a protein and are then transferred to the next protein, and on to the next. Many organisms use such cascades to sense what is going on in the environment and to respond. Remarkably, yeast cells use a very similar system to measure salt concentration in their environment. Eventually such a cascade ends with a change in activity of a number of transcription complexes, and genes are activated or silenced.

In many cases, the signal causes a rapid influx of calcium into a localized area of the cytoplasm and the calcium is equally rapidly brought back to its basal level. This is referred to as a *calcium spike*. The influx of calcium occurs because proteins that form calcium channels in certain membranes (for example, the plasma membrane or the tonoplast) are opened (Figure 8.15). Other proteins, called *calcium pumps*, in those same membranes bring the level of calcium down again. However, this brief elevation of cytoplasmic calcium permits certain enzymes that require calcium to be active for a short while. This activity propagates the signal in the cell. Figure 8.14c shows that the ethylene-induced formation of aerenchyma is inhibited by EGTA, a substance that binds cytoplasmic calcium and does not permit the calcium spike to occur. Such experiments suggest that a calcium signal mediates this particular effect of ethylene.

8.10 Genetic modification of crops can redirect developmental processes.

Because genes and their expression are at the basis of all developmental processes—from fertilization to senescence—it stands to reason that scientists can alter development by introducing new genes or by changing the expression of existing ones. Actually, many crop plants differ from their wild relatives by having bigger seeds or tubers, because specific developmental processes have been eliminated or altered. This modification of their development makes them more suitable as crops. Few crops could actually survive in the "natural" world, as their development has been too drastically modified. As a result, crops generally do not escape and become weeds. The following examples show some ways in which gene technologists are modifying crop development.

Engineering Male Sterility for Hybrid Seed Production. Hybrid plants produced by crossing two different lines often show increased yields because of hybrid vigor (see Chapter 14). However, large-scale and cost-effective production of hybrids requires the availability of male sterile lines *and* a way to restore male fertility in the next generation. For some crops, there are natural male sterile mutants and other mutant lines that restore male fertility. A combination of such plants can be used to produce hybrid seeds, if the two properties are bred into high-performing lines. This is how hybrid maize is produced. Male sterility can also be genetically engineered by the expression of a bacterial gene that encodes barnase, an enzyme with ribonuclease activity. If the expression of the gene is targeted specifically to the tapetal cells of the anther, pollen formation is prevented. The enzyme hydrolyzes the RNA in the tapetal cells, thereby killing them (**Figure 8.16**). Male fertility can be restored if in the same cells a protein is expressed that binds ribonuclease extremely tightly, so that there is no chance it will have any activity. Such a ribonuclease inhibitor, called *barstar*, is found in the same bacterial species from which the barnase gene was obtained. Scientists used these two genes to create lines of canola that can then be crossed to efficiently generate hybrid canola seeds. The hybrid seeds, now used in Canada, can raise canola yields by 20% above those from nonhybrid seeds.

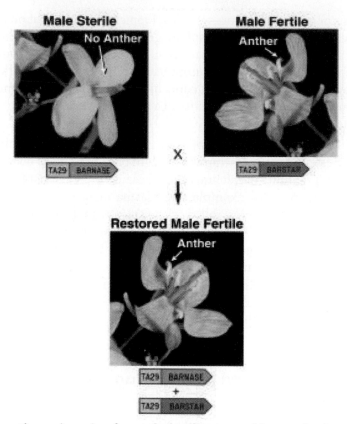

Figure 8.16 Genetic engineering for male fertility control in canola. Canola was transformed with a gene that encodes a ribonuclease enzyme (barnase) or a ribonuclease inhibitor (barstar). In each case, the regulatory region (promoter) used to drive gene expression was TA29, which directs gene expression specifically to the tapetal layer in the anther. In the plants expressing barnase, the flowers have no anthers and they are male sterile. In the plants expressing barstar and in the barnase x barstar hybrid, the flowers have normal anthers. *Source:* Courtesy of R. Goldberg, University of California, Los Angeles.

Modification of Fruit Ripening. We said earlier that ethylene regulates fruit ripening of many fleshy fruits and that the process can be speeded up by exposing fruit to ethylene. Ripening can also be slowed down by interfering with ethylene production or with ethylene perception by the cells of the fruit. Biosynthesizing ethylene requires two enzymes, and eliminating either one, by suppressing the expression of these genes, results in fruits that make very little ethylene. Such fruits ripen very slowly, an advantage to farmers and customers: Such fruits can be left on the plant longer—it is no longer necessary to pick them while they are green—and can often be sold in markets that are farther away (**Figure 8.17**).

Modification of Senescence with an Autoregulated Gene for Cytokinin Biosynthesis. In annual plants, seed formation is accompanied by senescence of the vegetative body, and nitrogen and other minerals stored in the leaves are used to make the seeds. So what would happen if the plant stayed green longer? Would it make bigger seeds, because it retains the capacity for photosynthesis? Plants such as lettuce and broccoli generally begin to senesce as soon as they are harvested. Could their senescence be delayed? Scientists had the idea of hooking the gene that encodes an enzyme for cytokinin biosynthesis to a promoter that is induced when cytokinin levels fall. Thus, as soon as the leaf begins to senesce, the gene for cytokinin synthesis is induced, and cytokinin synthesis ac-

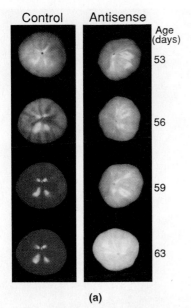

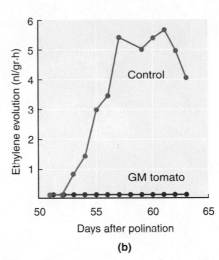

(a)   (b)

Figure 8.17 Genetic engineering for ripening control in tomato. Ethylene production in tomato has been reduced by genetically modifying expression of the gene that encodes one of the enzymes responsible for the ethylene biosynthesis. **(a)** The control tomatoes ripen normally between 53 and 59 days. The genetically modified (GM) tomatoes ripen extremely slowly during this same period. **(b)** Graph showing ethylene production in control and GM tomatoes. *Source:* Courtesy of A. Theologis, USDA Plant Gene Expression Center.

celerates, delaying senescence. The technology is being applied in Taiwan, with the goal of delaying senescence in broccoli after harvest.

Delaying Seed Pod Splitting to Avoid Seed Scatter. Most plants have evolved mechanisms to help them scatter seeds over a wide area. When the seed pods of a number of plants dry out, they split open quite abruptly—even explosively—scattering the seeds. This process, dehiscence, has disappeared from crops such as beans and peas because plant domestication selected against this trait. Dehiscence remains a serious problem in canola, where up to 20% of the seeds can be lost during harvest. Seeds scattered in this way germinate in the field the next year and pose a problem for the farmer who must now treat them as weeds. Scientists identified the genes responsible for dehiscence in *Arabidopsis*, a close relative of canola, and were able to generate mutant plants in which the pods do not split open (see Chapter 13). Commercial application of this discovery to canola is now being considered.

Genes That Accelerate Flowering. As noted earlier, flower formation depends on conversion of a vegetative meristem to a flowering meristem, a process controlled by the environment and by hormones that activate specific genes. Overexpression of such genes converts the vegetative meristem to a floral meristem much earlier in the plant's life than is normal. One application of this modification is in tree breeding. Many trees don't flower until they are 3 to 10 years old, making tree breeding a rather slow process. Tree breeding could be accelerated if the trees flowered after a year of growth (**Figure 8.18**).

These examples show that, in the future, genetic engineering of plants will move beyond introducing a single new characteristic with an unusual gene—as in the case of herbicide resistance—but will modify the development of the plant in subtle ways for the benefit of farming and food production. The new science of genomics will greatly help us identify the important genes.

Figure 8.18 Genetic engineering of flowering time in citrus. Flowering in citrus and other trees does not occur until the trees are several years old. The six-month-old plants on the right, which are flowering, have been transformed with a meristem identity gene. Such genes determine when a vegetative meristem converts to a floral meristem. The control plant on the left is not yet flowering. *Source:* Courtesy of José Martinez Zapater, University of Valencia, Spain.

CHAPTER SUMMARY

Plants are multicellular organisms consisting of different organ systems (roots, stems, leaves, flowers), each with a number of tissues and cell types. Plants grow by continuously forming new organs as a result of cell division activity in meristems. Meristematic cells are small and relatively undifferentiated, but they greatly enlarge and differentiate as they escape the meristematic zone. Apical shoot meristems form leaf and stem tissues. A hormonal trigger converts such a vegetative meristem into a floral meristem, initiating the formation of floral organs (sepals, petals, stamens, and carpels). Roots grow as a result of apical meristem activity, but lateral roots arise in an internal tissue, the pericycle.

Sexual reproduction requires the production of pollen by the stamens and egg cells in the carpels. Fertilization—the union of sperm cells in the pollen and egg cells—produces a zygote, which gives rise to a seed and then to a plant.

Many types of signals (internal and external) modify plant development by prompting the expression of specific genes. Signals interact with receptors, and this message is then relayed to the nucleus via a signal transduction pathway. Because development has a genetic basis, GM technology can be used to change developmental processes with the goal of crop improvement.

Discussion Questions

1. How are plant and animal development alike, and how do they differ? Consider that plants have meristems.

2. Discuss how the structure of different cell types relates to their function. How is differential gene expression related to cell differentiation? What other mechanisms exist for having cells that contain different complements of proteins?

3. Humans and plants have about the same number of genes, yet plants are said to be simpler. How could higher levels of complexity arise from the same number of genes?

4. A single hormone has many effects on plants, and many processes are affected by the same hormones or signals. List the five major hormones and which plant processes they affect. You may have to consult other sources to do this.

5. Discuss the difference between signal transduction pathways and networks.

6. The last section of this chapter gave several examples of modifications of plant development by GM technology. Which other aspects of development could be modified to benefit crops?

7. Discuss the difference between the life cycle of plants and animals as it relates to haploid and diploid cells.

8. Cell enlargement that depends on vacuoles is a phenomenon not found in animals. What is its significance in evolution?

9. Discuss the ecological functions of phytochrome and how it allows plants to adapt to different climatic zones. What is the relationship to the rapid movement of wheat out of its center of domestication in the Middle East (east and west migration) and the slow movement of maize out of its center of domestication in southern Mexico (north and south migration)?

Further Reading

Bewley, J. D., and M. Black. 1994. *Seeds: Physiology of Development and Germination*. New York: Plenum Press.

Buchanan, B. B., W. Gruissem, and R. L. Jones, eds. 2000. *The Biochemistry and Molecular Biology of Plants*. Rockville, MD: American Society of Plant Biologists.

Howell, S. H. 1998. *Molecular Genetics of Plant Development*. Cambridge, U.K.: Cambridge University Press.

Raven, P. H., R. F. Evert, and J. E. Eichhorn. 1999. *Biology of Plants*, 6th ed. New York: W. H. Freeman and Co.

Taiz, L., and E. Zeiger. 1998. *Plant Physiology*, 2nd ed. Sunderland, MA: Sinauer Associates.

Seeds: Biology, Technology, and Role in Agriculture

Kent J. Bradford
University of California Davis

J. Derek Bewley
University of Guelph

Seeds play a central role in agriculture, serving as the means to propagate plants from one generation to the next as human food, as animal feed, and as an important commodity in international trade. Both people and domesticated animals ultimately rely on plants as their food source, which means both depend on seeds. Seeds are a concentrated source of carbohydrates, proteins, and fats for humans and livestock alike, and are a significant source of minerals, vitamins, and fiber. They also provide raw materials for industrial products (see Chapter 19). Cereals are the main source of food, followed by legumes and oilseeds, with soybean being both a legume and an oilseed crop (**Table 9.1**).

Seeds are important for other reasons. Most weeds sprout from seeds buried in the soil, and new weed seeds sprout each time the soil is cultivated. Because seeds are dry (6–15% water), they can be easily stored for long periods of time (**Figure 9.1**) and shipped in vast quantities across the oceans. As propagules, seeds are the means by which genetically improved crops, whether achieved by conventional breeding or as a result of biotechnology, can be mass produced for use by farmers. Agricultural biotechnology companies purchased seed companies in the 1990s, because genes specifying valuable traits can be sold to

Table 9.1	World production (in millions of tons) of 15 major seed crops, 2001				
Cereal Crops	**Production**	**Legume Seed Crops**	**Production**	**Other Oilseed Crops**	**Production**
Wheat	518	Soybean	176	Rape/canola	36
Maize	605	Peanut	31	Sunflower	21
Rice	590	Common bean	17		
Barley	139	Dry peas	12		
Sorghum	58	Chickpea	8		
Oats	27				
Millet	29				
Rye	22				

farmers as seeds. Finally, seeds are the means by which the genetic diversity in domesticated and wild crop species can be conserved in specialized facilities called *gene banks*.

The seeds of today are the products of thousands of years of domestication and decades of controlled breeding (see Chapters 13 and 14). All genetic improvements are passed from one generation to the next via the seeds. Domestication eliminated certain characteristics that benefit the survival of the species but are detrimental to the farmer. Elimination of seed dispersal mechanisms and seed dormancy occurred quite early during domestication. Other selected traits were an increase in seed size, with its accompanying increase in stored reserves, and the production of more seeds per flowering head (inflorescence). For example, teosinte, the plant that is the closest living relative of maize and that probably resembles

Figure 9.1 Seed storage in Africa. Because seeds are dry, unlike roots and tubers, they can be stored for long periods of time. Nevertheless, they must be protected from predators and pests, especially weevils, whose larvae burrow into the seeds and eat the reserves. African farmers build elaborate granaries to protect their harvest. *Source:* Courtesy of Larry Murdock, Purdue University.

its wild progenitor, yields about 100 kg of seed per hectare. In contrast, modern maize hybrids can produce 8,000 kg of seed per hectare (also see Chapter 13).

9.1 Seeds, the products of sexual reproduction, accumulate nutritional reserves to support the growing seedling.

As discussed in Chapter 8, pollination followed by the fusion of a haploid sperm cell and a haploid egg cell, creates a diploid zygote. This process initiates seed formation. To understand the process, it is useful to first look at the end result, the mature seed. All seeds contain three essential parts: an embryonic root (radicle) and shoot (plumule), food reserve tissues or organs (cotyledons or endosperm), and a covering structure, the seed coat (testa). After germination, food stored in seeds supports seedling establishment until the plant becomes self-sufficient through photosynthesis. Because of selection and breeding, seeds of food crop species now contain far more reserves than are required for seedling growth. The mature seeds of cereals and legumes are distinctly different in structure. In cereals, most food reserves are in the endosperm (**Figure 9.2a**), a tissue that is not part of the embryo, whereas in legumes such as soybean or the common bean, reserves are in the seed leaves or cotyledons (**Figure 9.2b**), which are part of the embryo.

Seed development can be divided into three major phases, each with its own genetic program (**Figure 9.3a**). In the first phase, all the structures and tissues of the embryo form as a result of cell division and cell differentiation; in the second phase, the seed grows larger and the genetic program for synthesizing seed reserves is activated. Finally, during seed maturation the seed dries out and prepares to survive in the dry state. This phase requires the expression of a whole new set of genes. Maturation can also entail the expression of genes that create dormancy, a condition lost from most crop species.

The seed forms from the ovule after the egg cell within the ovule has been fertilized by a sperm cell from the pollen. As shown in detail for canola or oilseed rape (**Figure 9.3b**),

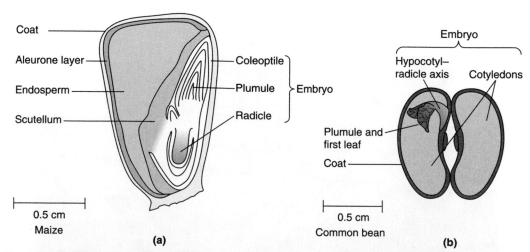

Figure 9.2 The structure of a cereal and a legume seed, using maize (*Zea mays*) and common bean (*Phaseolus vulgaris*) as examples. (a) In maize, the major site of stored reserves is in the endosperm, which is nonliving at maturity, except for the peripheral aleurone layer. **(b)** In bean, the cotyledons are the site of stored reserves, and the cells remain alive at maturity.

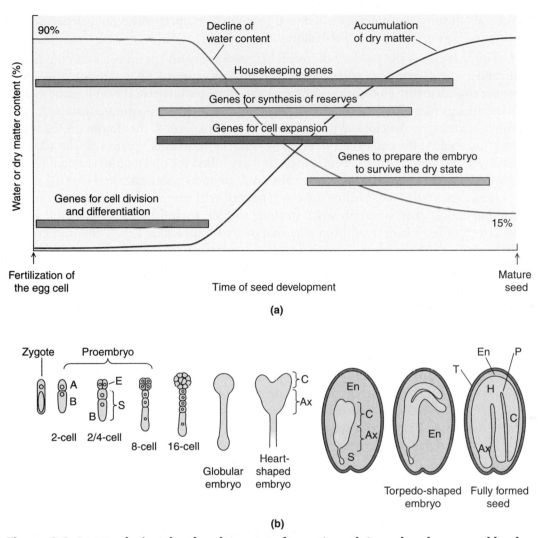

(a)

(b)

Figure 9.3 Events during the development of rape (canola) seed embryos and barley endosperm. (a) General scheme of the major events occurring during seed development. Dry weight increases as the storage reserves (protein, starch, and oil) are laid down, displacing water in the storage tissues. A final decline in water content to 12–15% occurs as the seed undergoes maturation drying. During development of the seed, different sets of genes come into play, as indicated by the horizontal bars. **(b)** Rape seed embryogenesis. After fertilization to form the zygote, mitosis occurs to produce an apical (A) and a basal (B) cell. The latter forms the suspensor (S), which facilitates development of the embryo (E) by being a conduit for nutrients but degenerates during the later stages of seed maturation. The apical cell undergoes extensive cell division and tissue formation until it forms a torpedo-shaped embryo embedded in a nutritive endosperm. Reserve deposition then occurs, with the endosperm and parent plant providing nutrients for the formation of oils and proteins in the cotyledons. The supply of nutrients from the parent plant for reserve formation arrives via the phloem to the seed coat, from where the nutrients diffuse inward to the developing embryo. The endosperm is completely reabsorbed by the time the embryo has completed reserve accumulation. **A:** apical cell; **Ax:** axis; **B:** basal cell; **C:** cotyledons; **E:** embryo; **En:** endosperm; **H:** hypocotyl; **P:** plumule (shoot apex); **S:** suspensor; **T:** testa (seed coat).

the resulting diploid zygote initially undergoes a single mitotic cell division to produce two different cells: a small apical cell and a large basal cell. The smaller apical cell undergoes a series of cell divisions and differentiates into the embryo. The larger basal cell forms a short temporary structure, the suspensor, that serves as a conduit for nutrients into the growing embryo. In the dicot legumes and oilseeds such as canola, the two cotyledons are the prominent structures in the mature embryo (Figures 9.2b and 9.3b), but in the monocot cereals the single cotyledon is reduced to a shield-shaped embryonic structure, the scutellum (Figure 9.2a), which also stores oils.

The endosperm is the product of fusion between a second sperm cell from the pollen grain and a special ovule cell with two nuclei. This fusion product has a triploid (3n) chromosome complement. This unique cell forms the entire endosperm through many cell divisions. In legumes and oilseeds the endosperm develops but is often consumed again during the embryo's development. In cereals, the endosperm is the dominant feature of the mature seed. At the end of cereal seed development, most endosperm cells die and only a thin layer of starch-free cells around the periphery called the *aleurone layer* remains alive (Figure 9.2a). After germination, these cells produce and secrete enzymes that digest the stored reserves to support seedling growth. The seed coat surrounds and protects the embryo (and endosperm, when present). In some species, including the cereals, the outermost covering layer is derived from maternal ovary tissue and is a fruit coat rather than a seed coat.

Deposition of the insoluble stored reserves occurs in storage parenchyma cells and begins after formation of the embryo and endosperm (Figure 9.3a). The storage tissues expand greatly, and the dry weight of the seed increases as reserves replace water in the cells. In cereals, starch and protein in the endosperm account for about 75% and 12% of the mature dry weight of the seed, respectively, whereas the oil deposited in the scutellum is usually less than 5% of the dry weight. There are predominantly starch-storing legumes (dry beans, peas, chickpeas) and oil-storing (soybean, peanut, or groundnut) legumes; mature soybean seeds contain approximately equal amounts of starch and oil. Protein content of legume seeds is usually 25–40% of dry weight, with starch comprising about 40–55%, when present, or oil at 25% (soybean) to 50% (peanut). Sunflower and rape seeds contain 40–45% oil on a dry weight basis, and about 25% protein.

Starch is synthesized and stored as starch granules contained within amyloplasts, special starch-storing plastids. In certain mutants of maize, there are defects in the biosynthetic pathway for starch, leading to the accumulation of sucrose and accounting for the sweetness of their seeds, commonly known as *sweet corn*. Protein is deposited within discrete protein bodies in the storage cells (see Figure 8.9). In cereals, the proteins (prolamins) are nutritionally deficient in the essential amino acid lysine (see Chapter 7). Oats, in contrast, are rich in other proteins and are a better source of dietary protein than other cereals. Legume seeds contain proteins (globulins), which are deficient in the essential sulfur-containing amino acids, cysteine and methionine.

Seed oils, or triacylglycerols, are deposited within small oil bodies. Species differ in the types of fatty acids (length, and degree of unsaturation) contained in their oils. For cooking and human health, the presence of unsaturated fatty acids (those containing one or more double bonds in the carbon chain, such as oleic and linoleic acids) is a desirable feature (see Chapter 7).

Seeds may also store antinutritional compounds. Phytic acid, an insoluble phosphate derivative, frequently is a minor component in seeds (1% dry weight or less), but it can bind essential dietary minerals in the intestines, making them unavailable for absorption. Processing and preparation of many foods for humans from seed may remove most of the phytic acid, but its presence in animal feed slows livestock growth. In addition, the phosphorus contained in phytic acid is not readily digested, so animal feeds are supplemented with phosphate, which can end up polluting waterways.

Some seeds, particularly those of legumes, may also contain enzyme inhibitors to protect them from predation by insects; these inhibitors can reduce digestibility of uncooked seeds. These and other proteins can evoke severe allergenic responses in susceptible people; for example, gliadins and glutens in wheat, globulins in legumes (see Figure 7.12), and an albumin storage protein in peanut.

Many variables affect seed set (the initial formation of the embryo and first growth of the seed), seed growth rate, and duration of storage reserve accumulation. Environmental stress at the time of pollination and fertilization can cause many seeds to abort, reducing the potential yield. Inherent genetic differences can regulate the number of cells in cotyledon or endosperm storage tissues, but the environment during seed development can modify the supply of assimilates available for storage product synthesis. Water deficit, temperature, light, plant competition, and nutrient availability are all variables affecting seed yield. High temperatures during seed filling can shorten the filling period, reducing seed size and yield. Water deficit can induce the same effects, and in some species it accelerates leaf senescence, prematurely terminating seed filling.

9.2 Seed maturation and entry into quiescence are important aspects of seed development.

The third phase of seed development is maturation (Figure 9.3a) and entry into quiescence. During the seed-filling phase, storage reserves replace water in the cell vacuoles, reducing seed water content to around 40–50%. During maturation, the seed synthesizes special sugars such as raffinose and a number of unusual proteins that bind water molecules. These may help protect cellular membranes and proteins and enable the cells to survive the final desiccation phase, when seed water content drops rapidly to less than 15%. At this low water content, biochemical processes essentially stop, allowing the seeds to persist for long periods in the dry state. Many maturation processes are controlled by the hormone abscisic acid (ABA), and mutant plants that do not make ABA or have a defect in gene regulation by ABA are viviparous: They germinate while still attached to the parent plant. After seeds are shed and fall to the ground, they imbibe water and, if dormant, they can survive in the soil for years (see **Box 9.1**).

Maturation and loss of water are essential in most crops, to let seeds be harvested and stored in the dry state. However, maintenance of seeds in the dry state does not deter insects from completing their life cycle in the stored seeds, a problem that is especially acute in developing countries, where people often store seeds in granaries (see Figure 9.1) rather than in hermetically closed silos where they can be fumigated with insecticidal gases. The insects that damage stored seeds and their products (such as flour) are referred to as *stored product pests*. Prominent among these are the bruchids, beetles that lay eggs on pods in the field or on dry harvested seeds. The young larvae then burrow into the seed and continue to develop as they consume the seed's contents. When the adults emerge after about a month, they leave holes in the seeds. They then mate, and each female lays about 75 eggs on the seeds, thus starting a new life cycle. Low-resource farmers use various approaches to try to kill as many larvae as possible, such as spreading the seeds out in the hot sun (**Figure 9.4**).

Seed drying also plays an important physiological role, because it signals to the seed that developmental processes are concluded. Thereafter, when the dry seed takes up water, its metabolism is now geared exclusively toward germination and seedling growth.

In many wild plant species, seed maturation is accompanied by the imposition of dormancy, a condition that also has a genetic basis. The term *dormancy* refers to the ability

Box 9.1

Many Weeds Propagate by Seeds

The presence of weeds in cultivated crop fields is a scourge for the farmer, often requiring expensive remedies such as hand or mechanical hoeing or the application of herbicides. In contrast to cultivated species, seeds of wild species have inherent dormancy mechanisms that allow them to remain in the soil in vast numbers for many years in a persistent soil seed bank. In arable land, from 12,000 to 650,000 seeds per square meter have been recorded in the autumn in the top 20 cm of soil, of which about 90% probably perish or germinate during the first four years (see figure). However, in controlled seed burial experiments many weed species were able to survive 15–20 years, and some seeds of *Verbascum* were still viable after 100 years.

Soil disturbances enhance germination of the dormant seeds in the seed bank, bringing them to the surface where they are exposed to conditions that promote germination, such as exposure to light, improved oxygen supply, and warmer or cycling temperatures. Seeds that germinate when buried too deeply in the soil exhaust their reserves before they can emerge and become photosynthetic, and thus die. Fallowing of land for one or more years (with cultivation to prevent weed growth) both reduces seed production and stimulates the germination of buried seed, consequently reducing the persistent soil seed bank.

Some weeds, such as dodder, *Orobanche*, and *Striga*, are parasitic and must attach to a host plant for survival. Seeds of these species have a unique dormancy mechanism, allowing them to germinate only when stimulated by compounds released by the roots of a host plant growing nearby (see Chapter 12).

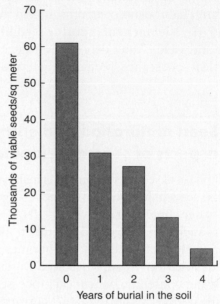

Survival of natural populations of weed seeds buried in soil. The histogram shows the survival of seeds of the grass *Poa annua* in undisturbed arable soil in England. *Source:* Data of H. A. Roberts and P. M. Feast, redrawn by C. C. Baskin and J. M. Baskin (1998), *Seeds: Ecology, Biogeography and Evolution of Dormancy and Germination* (San Diego, CA: Academic Press).

of seeds to delay germination until the environmental conditions are appropriate for survival of the seedlings. Dormancy is beneficial for wild species by timing seedling emergence to occur in the proper season, distributing seed germination over many years, and reducing competition among offspring. In crop species, seed dormancy is usually undesirable, because it leads to unpredictable germination, uneven crop growth, and variable yields. When early farmers began to harvest and store seeds for replanting, seeds that germinated readily after the rainy season and grew to maturity during the dry season were collected preferentially, resulting in a strong selection against dormancy. Thus, most domesticated crops show relatively little seed dormancy.

Nevertheless, a limited amount of dormancy is desirable in some crops. If rain and high humidity occur prior to harvest, cereal grains, especially wheat and barley grown

Figure 9.4 Protecting seeds from damage caused by the larvae of seed weevils. When seeds are harvested they may contain larvae of seed weevils, insects that propagate when the seeds are in storage. Spreading the seeds out in the hot sun before storage in granaries can kill most larvae (see Figure 9.1). *Source:* Courtesy of Nathan Russell, Centro Internacional de Agricultura Tropical.

in temperate climates, are prone to germinate while still on the parent plant, a phenomenon known as *pre-harvest sprouting*. The result is poor-quality grain with unacceptable milling and malting properties and poor baking quality of the flour (**Figure 9.5**), causing losses to farmers running into the hundreds of millions of dollars. The retention of limited dormancy is therefore desirable in these cereals, preferably with dormancy being lost from mature dry grain while it is stored in preparation for planting the following spring. Loss of dormancy during dry storage (called *after-ripening*) is common for seeds of many species.

Figure 9.5 Precocious germination or pre-harvest sprouting of wheat. When wheat is left on the field too long in a wet summer, it may begin to sprout (arrowheads) before it is harvested. Sprouting occurs because domesticated wheat has lost its dormancy, still found in wild grasses. Flour made from such sprouted wheat produces poor quality bread and flour mills do not pay full price for such grain if they accept it at all. *Source:* Photo courtesy of Daryl Mares, Adelaide, Australia.

9.3

Seed germination, seedling establishment, and seed treatments are important agronomic variables.

When farmers talk about seed germination, they mean the emergence of the seedling above the soil, but botanically speaking, germination starts when the dry seed begins to take up water and is completed when the embryo elongates and the embryonic root pushes its way through the seed or fruit coat. Imbibition, or the uptake of water, is a strictly physical phenomenon, much like water uptake by a sponge. Emergence of the seedlings, also called *stand establishment,* is caused by the growth of the embryo, supported by the reserves stored in the storage organs. After imbibition, metabolism resumes very quickly, initially using enzymes and reserves synthesized during development and conserved in the dry seed. Soon new genes are transcribed, and enzymes and new cellular components are made (**Figure 9.6**).

Respiration during germination uses stored sucrose and a series of modified sucrose compounds that contain galactose and are called the *raffinose series oligosaccharides (RSOs).* RSOs, which occur at high levels in seeds of legumes, have attracted some interest in relation to human and monogastric farm animal nutrition because they escape digestion in the upper intestinal tract. Instead, they degrade and ferment in the colon, leading to the production of H_2 and CO_2, causing flatulence. The common bean is a notorious source of flatulence. Eliminating RSOs from soybeans improves efficiency of energy conversion when farmers use such soybean lines as feed for monogastric animals. Genetic engineers are now seeking ways to eliminate RSOs by disrupting their biosynthesis.

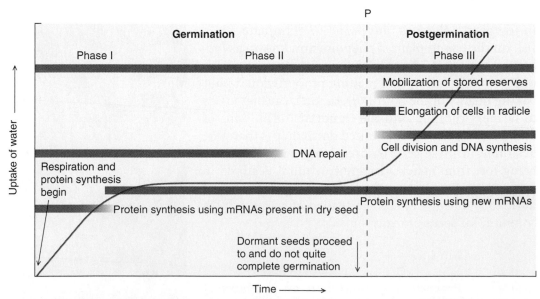

Figure 9.6 Some of the major events that occur during seed germination and subsequent postgerminative growth. Events are completed within hours or after many weeks, depending upon the plant species and germination conditions. Dormant seeds complete all events up to the time of radicle emergence. The graphed line shows the increase in fresh weight (water uptake). The extent to which germination proceeds during hydropriming, osmopriming, and matrix priming is marked with vertical line P. *Source:* Modified from J. D. Bewley (1997), Seed germination and dormancy, *The Plant Cell* 9:1057, Figure 1.

Postgermination events, such as the utilization of the stored polysaccharides, proteins, and oils and their conversion to small molecules such as sucrose and amino acids that can be transported, are quite well understood. In cereal grains, the living aleurone layer produces the digestive enzymes for starch in response to the hormone gibberellic acid released from the embryo. These enzymes are then secreted into the storage cells of the starchy endosperm, which the aleurone layer surrounds. Malting of barley for producing beer involves acceleration of this process so that the sugars released through the action of digestive enzymes are available for subsequent fermentation (alcohol production by yeast) in the brewing process. The success of malting requires maximizing endosperm reserve modification while minimizing seedling growth, a process that would consume the sugars.

Mobilizing reserves in the cotyledons of legumes and oilseeds during seedling establishment involves enzymes produced within these tissues. As with the carbohydrates, the oils are converted to sucrose, which is transported to the growing seedling. The proteins are hydrolyzed into their 20 constituent amino acids, which are then converted to the amino acids asparagine and glutamine for transport.

Germination of (nondormant) crop seeds is affected mostly by soil temperature and moisture availability. Although many species and varieties can achieve satisfactory stand establishment in a wide range of environments, excessively low or high temperatures or water deficit can severely reduce the percentage of seeds that develop into plants. For some species seed priming can enhance seed performance. This technology, which controls water uptake by seeds, allows many early germination processes—but not radicle emergence—to be completed before sowing (**Figure 9.7**). Seed priming can thus lead to increased germination percentage and rate, germination under a broader range of environments, and improved seedling vigor and growth.

Several priming techniques can be used:

- *Osmopriming* soaks seeds in aerated solutions of certain dissolved substances such as polyethylene glycol, potassium nitrate, or mannitol to control imbibition.

- *Hydropriming* or *drum priming* introduces liquid water to seeds in controlled, precise amounts to achieve a desired level of hydration.

- *Matripriming* mimics the natural process of imbibition from soil particles by mixing seeds with solid materials (such as diatomaceous earth, clay, or vermiculite products) containing water to control water uptake by seeds.

After priming, seeds retain tolerance to desiccation and can be dried for packaging, transportation, and marketing. However, storage life of primed seeds is generally reduced, so they must be handled with more care than untreated seeds. Seed priming has been applied most successfully to vegetable and ornamental species—such as lettuce, tomatoes, peppers, carrots, onions, pansies, and impatiens, as well as some turfgrass species—where seed quantities are relatively small and seed value is high. For field crops such as maize, wheat, cotton, or soybean, where seed quantities are high and seed costs are comparatively low, the expense of priming and the difficulties of processing huge quantities of seed has thus far outweighed its benefits; it is simply cheaper to sow more seed to assure satisfactory plant stands in the field.

Other technologies can also enhance seed performance and improve crop establishment. Inoculating legume seeds with bacterial *Rhizobium* species has long been used to help establish a symbiotic nitrogen-fixing association with the growing plant (see Chapter 12). Seeds are often treated with fungicides or insecticides to protect seedlings against soil-borne

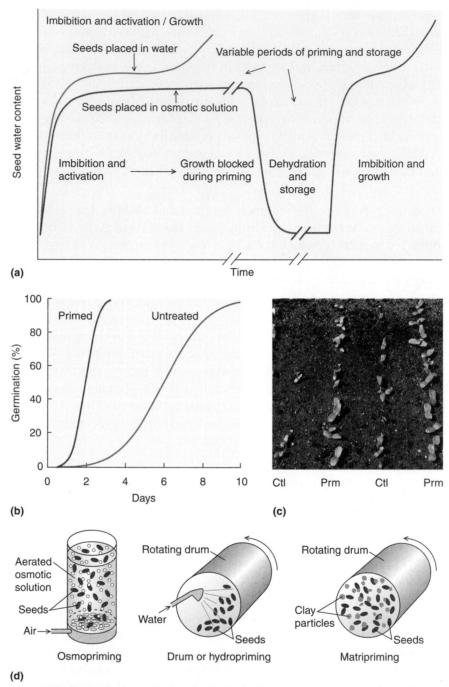

Figure 9.7 Seed priming can enhance seed germination. Seed priming involves allowing seeds to absorb enough water to initiate metabolic processes, but insufficient water to complete germination **(a)**. After the priming period, the seeds can be dried, stored, and transported to the farmer. When the seeds are imbibed again, the lag period before radicle emergence occurs is considerably reduced, improving the rate and uniformity of germination **(b)**. For example, the emergence of muskmelon seedlings at a cool temperature (15°C) is improved by priming **(c)**; **Ctl** = untreated control, **Prm** = primed. Priming can be achieved in several ways **(d)**, including imbibition in an osmotic solution that controls water uptake by the seeds (osmopriming), slow addition of measured amounts of water to bring the seeds to a specific water content (drum or hydropriming) or combining seeds, clay particles, and water to allow the seeds to imbibe to a specific water content (matripriming).

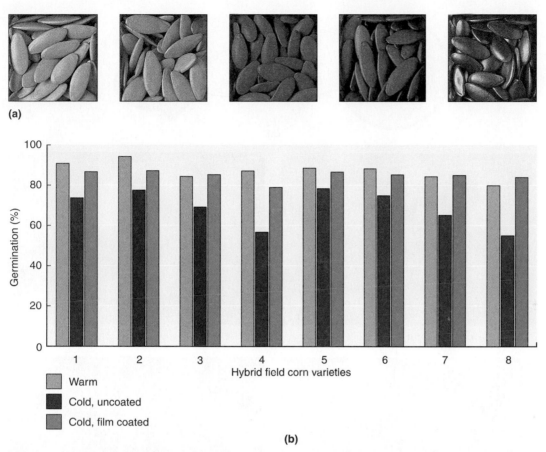

Figure 9.8 Film coating can be used to improve seed performance. Film coating applies a thin layer of a polymeric material to the outside of the seed **(a)**. The polymer is applied as a mist as seeds fall though a special treatment chamber, then is quickly dried. Protective chemicals such as fungicides included in the polymer are bound to the seed, allowing the use of less fungicide and reducing exposure of workers to the chemicals. Seeds treated with pesticides are required to be colored to prevent their accidental use as foods. Colors can also be used to distinguish between varieties or treatments, as illustrated here with cucumber seeds **(a)**. Some coating materials, such as an organic starch-based polymer, also slow the rate of water uptake by seeds and prevent damage that can occur when dry seeds are planted in cool, wet soils **(b)**. *Sources:* (a) Courtesy of Celpril; (b) courtesy of SeedBiotics.

diseases and insects. Recently, application of such chemicals to seeds is shifting from a dusting or slurry treatment to incorporating the chemicals into a polymeric film coating that adheres tightly to the seed (**Figure 9.8**). This prevents loss of the applied material from the coated seeds, reducing the amount of chemical required and improving safety for people handling the seeds. Some coating materials, including a biodegradable polymeric starch product, can slow water uptake by the seed and prevent the cellular damage that rapid imbibition can cause in cold soils (Figure 9.8b). Other polymers have temperature-dependent water-permeability properties that prevent imbibition until a particular temperature is exceeded. This may allow coated seeds to be planted early in the fall or spring, because they undergo imbibition to start germination only when favorable temperature and moisture conditions have developed in the soil. Coating or pelleting are also used to make irregularly shaped seeds more uniform, making them more amenable for mechanical planting (for example, lettuce or sugar beet seeds; see **Figure 9.9**).

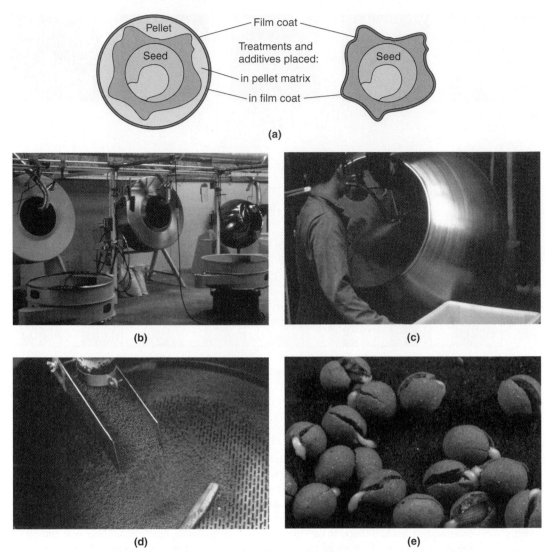

Figure 9.9 Seed pelleting allows irregularly shaped seeds to be planted by machines. Some seeds, as illustrated here by sugar beet **(a)**, are irregularly shaped or are too small to be easily planted with mechanical planters. In these cases the seeds can be made more uniform in shape and increased in size by encasing them inside pellets of clay and other materials. Pelleting is done in large rotating pans **(b)** in which the seeds are tumbling and being moistened with water or a binder. The pelleting material is introduced a little at a time and gradually starts to form a ball around each seed **(c)**. As the pellets near the desired size, they are screened and those that are too small are returned to the pan for additional coating **(d)**. The finished pellets split open when the seeds imbibe, allowing the seedling to emerge, as illustrated here for lettuce seeds **(e)**. Film coats can be applied directly to seeds or to pelleted seeds (see a). *Source:* (a) P. Halmer (2000), in M. Black and J. D. Bewley, eds., *Seed Technology and Its Biological Basis* (Sheffield, U.K.: Sheffield Academic Press).

9.4 Seed production is often distinct from crop production.

The production of seeds for planting can be essentially the same as producing the crop, or it can be a completely different commercial operation. For a number of the world's major crops, including wheat, rice, soybeans, and cotton, seeds are produced primarily using a self-pollinated or pure-line system (see Chapter 14). These crops do not outcross with other plants of the same species. Rather, each flower pollinates itself, resulting in seeds genetically identical to the parent plant. This has the advantage that once a superior variety has been developed, it can be propagated indefinitely by simply harvesting the seeds each year and replanting them. Random mutations and a low incidence of outcrossing do occur, however, so vigilance is needed to maintain the integrity of the original variety.

For self-pollinated crops, quality assurance systems based on certification of planting stock pedigrees and field inspections for any off-type plants are efficient and effective (see Section 9.5). Because it is also possible for farmers to save their own seed of these types of crops for subsequent planting, people can widely distribute superior varieties quite easily. This system of developing superior varieties by breeding and then releasing the seeds to farmers through government-sponsored programs drove the Green Revolution advances in producing wheat and rice.

For some crops that normally outcross, or that can be readily manipulated to make crosses between parent lines, hybrid seeds are widely produced by seed companies and sold to farmers. To produce hybrid seeds, two inbred parent lines, each possessing beneficial traits, are developed by repeated self-pollination. These parent lines are then crossed in a controlled way, with one line (male) pollinating the other (female) to produce F_1 hybrid seeds (see Chapter 14). Seeds produced on the female plants are then sold to farmers, who plant them to produce the commercial crop. The primary advantage of hybrid varieties is that when two distinct inbred (or genetically homozygous) parents are crossed, the offspring often grow much larger and more vigorously than either of the parents. This phenomenon, known as *hybrid vigor* or *heterosis*, has been exploited in many crops to improve yield and quality. Maize is the oldest example, but also sorghum, sunflowers, and in developed countries many vegetables, including tomatoes, peppers, carrots, cabbage, broccoli, melons, and onions, are grown almost exclusively from F_1 hybrid seeds.

Given the yield advantages of hybridization, efforts are under way toward developing hybrids in additional crops in which it is currently difficult to produce them. In many vegetable and flower species, pollination can be accomplished by hand, because the value of the resulting seeds justifies the additional expense. This labor-intensive production of hybrid seeds occurs primarily in China, Thailand, Chile, Central America, Kenya, and other countries where labor costs are comparatively low. In self-pollinated crops such as wheat, rice, cotton, or soybeans, hand pollination is not feasible and other methods must be employed. Methods to hybridize wheat or cotton by the use of chemicals to induce male sterility (failure of plants to produce viable pollen) have been developed but are not widely commercialized. However, much progress has been made in China with development of hybrid varieties of rice using a combination of genetic male sterility and labor-intensive methods. Hybrid varieties now represent 50% of the rice crop area in China and yield 15–20% more per hectare than conventional varieties. In addition, genetic engineers have developed new methods to control the fertility and pollination of crops to produce hybrid seeds, and are now introducing these into the market in crops such as the canola and maize (see also Figure 8.16). For many crops—including hybrids of all types, most vegetables, alfalfa, grasses, and other forages—seed production is a completely separate and specialized operation distinct from production of the marketed commodity. For example, field production of hybrid seeds requires interplanting two to eight rows of the female parent alternately with one or two rows of the male parent (**Figure 9.10**). The female parent must be male sterile so that it can only be fertilized by pollen from the adjacent male rows. In maize, male sterility can easily be assured by removing the pollen-producing tassel (the male flower) at the top of the plant and leaving the ears (the female flowers) below.

(a)   (b)

Figure 9.10 Production fields for F₁ hybrid sunflower and maize seeds. (a) The two rows of multiflowered sunflower plants in the center are the male parents, and the single-flowered female parent plants are in the adjacent rows. The female flowers are male sterile (do not produce viable pollen) and are dependent on pollen carried by bees from the male plants for pollination and fertilization. After pollination, the male plants are removed, and only the female plants are harvested for the F₁ seeds. These will then be sold to farmers to produce the sunflower oilseed crop. In this case, the male parent also carries a gene that can overcome the sterility of the female parent, making the F₁ generation of plants fertile and allowing seed set. **(b)** F₁ hybrid seeds of maize are produced in a similar manner, except that the female rows are emasculated by removing the tassels (male flowers) by hand before they emerge. The single row in the center with the darker leaves is of the male parent, which soon produces tassels and sheds pollen to pollinate the ears (female flowers) on the three adjacent female rows. In both sunflower and maize, note the distinctive plant types of the two parents. *Sources:* (a) Courtesy of K. J. Bradford; (b) courtesy of Mariano Battista.

Producing hybrid seeds is a challenge, because to succeed the producer must control many factors. For example, for pollen to be available when the female flowers are open and receptive, the two parent lines must flower at the same time. Management of pollination is also crucial for successful seed production. If the male flowers are not releasing pollen when the female flowers are receptive, then fertilization will not occur and no seeds will be produced. This difficulty is exacerbated by the fact that the best hybrids result when the parents are quite distinct and may have very different morphologies and growth rates (Figure 9.10).

In maize, wind carries the pollen to the adjacent female rows, so no special operations are required. In rice, however, even though genetically male-sterile female lines have been developed, the male flowers normally self-pollinate and do not readily release pollen to fertilize the female flowers. In China, the plant growth stimulator gibberellin is sprayed on the male plants to make them taller at the time pollen is shed, and workers drag ropes across the male rows to disperse the pollen so that the wind can carry it to the adjacent female rows. Many plants rely on insects to carry pollen among flowers. This is crucial in hybrid seed production, and large numbers of beehives are moved into seed fields during the flowering period to encourage pollination. However, bees forage widely (as much as 5 km) in search of nectar and pollen, and if crops of the same species are in the area, bees can carry pollen from other fields and contaminate the seed crop. Thus seed fields of insect-pollinated crops must be isolated by distances of 1–5 km from any crops of the same species to prevent genetic mixing by pollen transfer.

These examples illustrate that seed production of modern agricultural varieties is a highly specialized operation conducted primarily by commercial breeding and seed production companies. In the developed countries, only 25–50% of a few crops (such as wheat and soybeans) are planted each year from farmer-saved seeds; farmers plant most crops from seeds they buy from seed companies each year. However, on a global basis, more than 85% of crops are planted from farmer-saved seeds. These seeds have the advantage that the plants are adapted to local conditions, but the disadvantage that they may not produce yields as high as those obtained with elite lines grown with the best management practices.

Expected yield increases are the primary reason why farmers pay companies for hybrid seeds. Farmers buy new seeds for every crop planting because self-harvested seeds will not "breed true": The plants from such seeds are not uniform in characteristics and yields. This follows from the laws of genetics and the independent assortment of chromosomes (see Chapter 14).

Hybridization is therefore a way to automatically protect the variety for the seed company, requiring the farmer to buy seeds from the supplier each year. Thus, in addition to yield advantages, the protection from unauthorized propagation that F_1 hybrid seeds offer has been a significant impetus to their commercial development.

New developments in biotechnology, however, may dramatically change this situation in the future. In some plants, seed development can occur without pollination and fertilization. In these plants, seeds naturally develop directly from cells of the ovary, so that the genetic composition of the seeds is identical to that of the parent. This phenomenon, known as *apomixis*, is essentially the same as when a potato is vegetatively propagated by a tuber or a fruit tree by grafting. If apomixis could be introduced in a controlled way into rice, for example, the F_1 hybrid would need to be created only once, and then that exact genetic combination could be transmitted to all subsequent generations.

9.5 Seed certification programs preserve seed quality.

After the development of improved varieties by plant breeding, seeds must be maintained and propagated for distribution to farmers. If care is not exercised throughout the seed production process, genetic changes can occur over generations that reduce the quality of the variety. Random mutations, outcrossing with other varieties, or inadvertent mixtures with seeds of other varieties can result in contamination. This was often the fate of varieties released before the development of seed certification programs. These programs maintain a pedigree system that tracks each bag of seeds back to its parents over a limited number of generations (**Figure 9.11**). Breeder's seed is produced by the breeder and is the purest possible sample of seed representing the variety. This small amount of seed is then used to grow foundation seed, which in turn is used to produce registered seed, which is used to produce certified seed. The certified seed is sold to the farmer for producing a commercial crop. In this way, farmers are assured that the seed they purchase is only a few generations away from the pure variety. The pedigree system allows the small amount of breeder's seed to be multiplied for commercial planting while maintaining the genetic purity of the variety.

In addition to maintaining the pedigree system, seed certification staff also inspect crops during growth for abnormal plants, diseases, weeds, and isolation (a minimum distance from plants of the same species that could outcross with plants in the seed field). Seeds from fields that do not meet the standards are rejected and cannot be sold as certified. Certified seeds must also meet standards of purity (freedom from weeds and foreign matter) and germination (viability). Together, these standards increase the benefits from improved varieties by maintaining genetic purity and providing high-quality seed for planting. Seed certification is most common in field crops such as cereals and legumes, although in many countries vegetable seed is also certified.

Figure 9.11 Flow chart for variety development and certified seed production. This diagram illustrates the essential features of a seed certification program from variety development through several multiplication steps before distribution to farmers. Quality standards (such as for isolation distances, occurrence of off-types, and absence of weeds) are highest in the earlier generations and become somewhat less stringent in later generations.

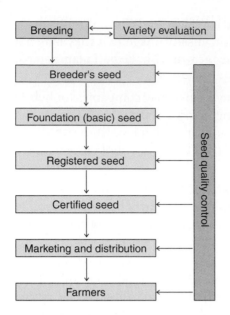

In developed and a number of developing countries, cultivar improvement, seed production, and seed certification programs are well established, and competition among commercial seed companies further ensures high standards for seed quality. However, many developing countries lack basic seed production, certification, and distribution systems, and most seeds are saved by farmers themselves or purchased locally. Without certification programs to maintain varietal purity, the advantages of improved varieties introduced under these conditions can quickly disappear. Establishing seed production and certification programs is a high priority for international agricultural development, because most benefits of improved seed quality do not depend on farm size or level of technological sophistication.

Another important function of seed programs is to prevent the spread of diseases through contaminated seed. Some plant diseases are seed-borne, or capable of being carried on or in the seed. In many cases, even a very small percentage of contaminated seed is enough to cause a disease epidemic in the production field. For example, lettuce seed lots in California are rejected if only one seed in 30,000 carries lettuce mosaic virus. Because there is a global trade in agricultural seed, contaminated seed can introduce a new disease into a region where it has not existed before, creating destructive outbreaks and reducing crop yields. Many countries have phytosanitary regulations that strictly monitor the presence of diseases in seed production fields or on the seed, and obtaining a phytosanitary certificate is often required for the export or import of seed. The detection and monitoring of seed-borne diseases has been greatly aided by the development of laboratory-based tests in which the pathogens are cultured from seed and identified or where they are detected by antibodies or DNA amplification techniques.

9.6   Seed banks preserve genetic diversity for the future.

In addition to their roles in crop propagation and as foods, seeds are also a primary means of preserving crop germplasm. Most seeds can survive dehydration and remain viable for long periods if kept dry. This desiccation tolerance was an important evolutionary development for plants, allowing them to survive periods of adverse conditions such as a dry season or freezing and to re-establish a plant population again when conditions were favorable. This cycle happens annually in many environments, or at much longer intervals in deserts, for example, where germination, plant growth, and seed dispersal may occur only once every few years or even decades in response to rare rainfall events. The ability to harvest, dry, and transport seeds that people could use to establish the next year's crop was fundamental to their adopting and expanding agriculture.

To sustain the cropping cycle, seeds generally only need to be stored from one harvest to the next planting season—a period of nine months or less, in most cases. However, the natural ability of dry seeds to remain viable for much longer periods under proper storage conditions has allowed people to develop seed banks as a way to preserve the diversity of plant types important in agriculture. Early examples of seed banks were the All-Union Institute of Plant Industry in Russia, established by Nikolai Vavilov in the 1920s, and four regional plant introduction stations established in the United States in the 1940s. Currently, the U.S. National Plant Germplasm System consists of the National Center for Genetic Research Preservation in Fort Collins, Colorado, and 21 regional plant introduction stations where seeds and plant materials are preserved and propagated. There are now about 440,000 accessions, representing more than 38,000 species of plants, cataloged in the U.S. Germplasm Resources Information Network (GRIN), an electronic database of all genetic resources preserved by the system (www.ars-grin.gov/npgs/). Each year, the U.S. germplasm system alone distributes over 150,000 packets of seeds and other plant materials to more than 100 countries for use in crop improvement programs.

The introduction of high-yielding varieties of wheat and rice in many countries during the 1960s led to concern that traditional varieties, local landraces, and wild relatives of crop species would be lost if they were not collected and preserved. Landraces are not genetically uniform, a property reflected in the many colors of seeds found in local markets of developing countries (**Figure 9.12**). These resources are crucial for maintaining the genetic diversity plant breeders require to continue to improve crop varieties (see Chapter 14). International initiatives to establish a global network of gene banks resulted in the establishment in 1974 of the International Board for Plant Genetic Resources (IBPGR, subsequently renamed IPGRI, International Plant Genetic Resources Institute) under the auspices of the Consultative Group on International Agricultural Research (CGIAR). This network has facilitated the collection and conservation of crop genetic resources and the establishment of germplasm repositories in developing countries. These efforts have focused primarily on crop plants, with the majority of the accessions coming from the top 20 of the world's major crop species and their relatives. Because seeds are much easier and cheaper to store, of over 2.3 million crop accessions held in banks around the world 97% are seeds and only 3% are plants or clonal materials.

More recently, attention has turned toward the remaining 99.99% of the world's plant species that are not crops and the growing concern about loss of global biodiversity. By the end of 2000 the Convention on Biological Diversity (CBD), developed following the

Figure 9.12 Landraces of cowpeas collected in a local market in Nigeria. The genetic diversity of the cowpeas grown by different farmers is reflected in the seed coat colors. *Source:* Courtesy of Larry Murdock, Purdue University.

Rio Earth Summit in 1992, had been ratified by 179 countries (see Chapter 2). The objectives of the CBD are the conservation of biological diversity, the sustainable use of biological resources, and the fair and equitable sharing of benefits arising from the use of genetic resources. The third objective recognizes that the vast majority of plant genetic resources originate in less developed countries, whereas these resources in crop improvement have been used primarily in the developed countries. Lobbying by the U.S. agricultural biotechnology industry because of concerns about limitations on its ability to continue to use these resources and retain ownership of varieties produced with those public resources has thus far prevented the United States government from ratifying the CBD. Other specific conservation initiatives include the Millennium Seed Bank Project of the Royal Botanic Gardens in Great Britain, which has the goal of collecting and preserving seeds of all of the United Kingdom's native seed-bearing plants and an additional 25,000 plant species from the tropical drylands by the year 2010.

Preserving seeds for the long term is not as simple as you might expect. The initial quality of the seeds before storage is important, and storage conditions are crucial for extending storage life. Most seed banks store seeds in freezers at –20°C (**Figure 9.13**), but at the U.S. National Seed Storage Laboratory, seeds are increasingly being stored in insulated tanks over liquid nitrogen (approximately –150°C). Depending on the species, seeds properly dried and stored at –20°C may retain viability for 20 to 50 years or more, whereas researchers expect those stored at –150°C to remain viable much longer. Researchers do not know how

(a)   (b)

Figure 9.13 Storage of seeds at low temperatures in seed banks. At the National Center for Genetic Resources Preservation in Fort Collins, Colorado, USA, seeds are **(a)** dried and sealed in foil packets, then stored in large rooms held at –20°C, or **(b)** placed in plastic tubes and stored in insulated vessels over liquid nitrogen at –150°C. Seeds are expected to remain viable much longer at the lower temperature, reducing the frequency at which they need to be grown into plants to produce new seeds. *Source:* (a)Courtesy of the U.S. Department of Agriculture; (b) Courtesy of Loren Wiesner of the National Center for Genetic Resources Preservation.

long, as experiments have been under way for only 30 years, and viability has changed little over this time. Experimental research on seed longevity in storage is a very slow process.

Collecting germplasm as seeds and preserving them in seed banks is *ex situ* conservation, in contrast to *in situ* conservation, which emphasizes preserving habitats and maintaining living populations of endangered species. Although *in situ* preservation might be preferable, because it maintains species or varieties as living plants that continue to adapt and create genetic diversity, in many cases it is not practical. Loss of habitat and changing crop production practices threaten many plants and crop varieties with extinction, and *ex situ* conservation may be the only way of preserving this germplasm. In addition, *ex situ* conservation complements *in situ* conservation by serving as an insurance policy against loss and by providing propagation material for *in situ* conservation programs.

However, *ex situ* conservation is not simply putting seeds in storage. Because all seeds lose viability over time in storage, at some point the seeds must be planted to produce more seeds. Because the objective is to preserve as much genetic diversity as possible, regeneration is desirable when a high percentage of the seeds are still viable, rather than when only a few seeds are still able to germinate. Otherwise, the genetic contributions of all the dead seeds in the population will be lost. You can appreciate the magnitude of the challenge by considering that in the United States alone, over 400,000 seed accessions are in the collections. If the average lifetime in storage is 40 years (an optimistic assumption under standard conditions), then more than 10,000 accessions must be grown each year just to replenish the existing stocks.

Thus far, we have considered only seeds that exhibit desiccation tolerance and can therefore be dried for storage. Such *orthodox* seeds represent about 90% of flowering plant species. In contrast, so-called *recalcitrant* seeds cannot tolerate dehydration and can generally be stored only for relatively short periods without either germinating or losing viability. Many tropical plants produce recalcitrant seeds, as the temperature and moisture conditions are generally conducive to germination soon after the seeds are shed. However, many temperate trees and other plants also produce recalcitrant seeds. A number of important crops in tropical regions either produce recalcitrant seeds (rubber, oil palm, cocoa, mango, citrus, coconut, jackfruit, avocado) or do not produce seeds and are propagated vegetatively from cuttings or roots (bananas, cassava), making germplasm conservation much more labor intensive and costly (see Figure 13.14).

9.7 Seeds are the major delivery system for plant biotechnology.

Most crops are propagated by seeds, and seeds are therefore the primary delivery system by which breeders transfer the products of plant biotechnology to the farmer and then to the consumer. The unique traits adding value to a crop are sold to the farmer via the seeds, and the higher potential value is reflected in higher seed prices that enable the commercial developer to recoup the costs of creating the variety. Public institutions may not need to recoup their development costs when they release genetically engineered lines. The farmer expects a higher yield, a lower cost of production, or a higher price for the product because of the characteristics passed on through the genetically enhanced seeds. Thus genetic improvements increase seed value throughout the agricultural production and marketing chain.

"Value-added" traits developed through biotechnology (such as pest and disease resistance or improved nutritional quality) are most valuable if incorporated into elite varieties already highly adapted for yield and quality. The relatively limited number of sources for the best crop germplasm led biotechnology companies who developed the first genetically engineered seeds to acquire the leading seed companies, such as Du Pont's purchase of Pioneer Hi-Bred International (the world's largest seed company) and Monsanto's purchase of Holden Foundation Seeds and DeKalb Seeds. These and other companies in developed countries have also acquired seed companies in developing countries. Consolidation of seed companies by the corporations active in plant biotechnology research occurred in the mid-1990s as the first products resulting from biotechnology began to reach the market. The high prices paid for these seed companies reflected the importance of introducing value-added traits into varieties that were already competitive in the marketplace.

In addition, as inputs such as protection from pests are increasingly delivered via improved seeds rather than as chemical pesticides, the value of seeds in the overall production system increases (**Figure 9.14**). A significant transfer of income from companies that make chemicals for crop protection (such as insecticides, fungicides, and so forth) to seed/biotechnology companies has already occurred in the United States, Argentina, and other countries where crops engineered to resist pests have been introduced (although in many cases these represent separate divisions of the same company). For agronomic crops, where seeds have traditionally been a relatively low-cost component of the total production system, seed cost may increase considerably in the developed countries. As long as this is compensated by even greater savings in other agronomic inputs, such as reduced insecticide or weed control costs, or by increases in market price of the product commodity, higher seed costs will be accepted. As has already occurred in high-value vegetable and ornamental crops, however, high seed costs lead to additional demands by the companies that create them for intellectual property protection and by the farmers for enhanced seed performance in crop establishment.

Figure 9.14 Transgenic seeds ready for planting. GM seeds are more expensive than non-GM seeds because they contain value-added traits that save the farmer money. GM seeds can reduce the number of pesticide applications needed (for example, GM *Bt* cotton) or make weed control simpler and less expensive (for example, herbicide-resistant soybeans).

Legal mechanisms are in place protecting the intellectual property represented in genetically enhanced varieties. In the United States, the Plant Variety Protection Act (PVPA) provides a mechanism for registering seed-propagated varieties and restricting their sale: They can be sold only by the breeder or the company that developed them (**Figure 9.15**). Similar mechanisms are internationally recognized (for example, UPOV, the International Convention for the Protection of New Varieties of Plants). In addition, general utility patents can be awarded for novel plant varieties. These PVPA and patent rights give the developer of an improved crop variety the same type of protection that other inventors can receive for novel machines, pharmaceuticals, or processes. As discussed earlier, in self-pollinated crops it is difficult for a breeder to prevent unauthorized propagation and sale of a variety once it has been released. The situation is analogous to that in the publishing or music industries, where pirated copies of books or recordings are often sold in places where copyright laws do not exist or are not enforced. Because farmers can buy the improved variety once and then propagate their own seed thereafter, opportunities for economic return to seed companies are reduced and the companies have therefore been reluctant to invest heavily in breeding and genetic improvement of self-pollinated crops. In these crops, seeds have traditionally been a relatively low-cost input into the production system, with many varieties being developed by university and government-sponsored breeding programs and released to the agricultural industry with little or no royalty or patent protection. However, with the array of new traits that can be genetically incorporated and marketed through seeds, both public and private breeders developing those traits expect to receive a higher price for the seed and will enforce their intellectual property rights. Thus, higher prices for seeds can be anticipated in these crops, and the issue of farmer-saved seed will need to be addressed.

This problem has already become an international issue, for example, with herbicide-tolerant soybeans. In the United States, farmers were required to sign "technology agreements" not to save or sell seed from crops planted with these varieties, and some companies selling these seeds aggressively pursued violators of this agreement through lawsuits. At the same time, similar varieties were being sold in Argentina, where patent protection laws do not give the seed companies the ability to require the farmer to sign a restrictive technology agreement. Farmer-to-farmer sales quickly distributed seeds of herbicide-tolerant (Roundup Ready®) varieties, and within only a few years over 90% of the soybeans grown in Argentina carried this trait. As might be expected, U.S. farmers were angry that Argentine farmers were able to obtain the same varieties of soybeans with desired traits without paying the additional fees associated with the technology agreements, and the seed company's returns were reduced by the so-called brown-bag sales (off-market sales of seeds among farmers). With annual global seed exports estimated at US$3.6 billion and internal commercial seed markets totaling over six times this amount, considerable opportunity exists for profiting from avoiding laws protecting plant varieties. Inability to control propagation of self-pollinated crops deters seed companies from marketing improved varieties in countries where legal protection is not available or not enforced.

At the same time, some developments are so valuable to society that such constraints should not limit widespread distribution. On the one hand, for example, should rice hav-

Figure 9.15 Mickylee watermelon variety developed by the commercial vegetable breeding company Seminis Vegetable Seeds Inc. and protected by the Plant Variety Protection Act. This act protected plant varieties and the companies that created them before the more recent court rulings started protecting genetically improved plants using general utility patents. *Source:* Copyright Seminis Vegetable Seeds, Inc. Used with permission.

ing increased iron and β-carotene content (*Golden* Rice™) be distributed only to those who can pay a licensing fee, when people who most need it probably cannot afford the fee? Fortunately, seeds of this nutritionally enhanced rice will be further developed and released through the International Rice Research Institute to make it freely available to people who need it. On the other hand, companies that have invested millions of dollars in developing improvements that benefit farmers or consumers of agricultural products also expect a reasonable return on investment. A two-tiered system, where royalties are waived for countries where average farm incomes are below a given threshold, has been negotiated for *Golden* Rice™, and has been proposed for general international adoption.

9.8 The conflict over "terminator gene technology" brought the importance of seeds in delivering technology into sharp focus.

The conflict over control of seed distribution was sharply focused by the development and patenting of a method to genetically engineer seeds so that they would not be viable as propagules for producing a next generation. Developed by researchers at the U.S. Department of Agriculture (USDA) in collaboration with the Delta & Pine Land Company, a major U.S. cottonseed company, this genetic "Technology Protection System" allows the distribution of seeds for planting that will germinate to establish the crop, but the seeds produced on those plants will not be viable for planting to produce a subsequent crop (**Box 9.2**). This system was promptly termed the "terminator gene" by groups opposed to genetic engineering of crops, and a worldwide controversy erupted over it, even though it was and remains in the research stage and has not been transferred to commercial crops. Some countries prohibited its use, and various groups, including the Rockefeller Foundation, a major sponsor of the international agricultural research centers, criticized its intent. People expressed fear that poor farmers would be forced to buy expensive seeds each year and that multinational companies would hold a monopoly on crop varieties. There was also concern about the potential spread of the nonviability trait to other nearby crops, making them useless as seed. The outcry was sufficient to cause Monsanto (which was in the process of acquiring Delta & Pine Land Company; that acquisition was not completed, however) to announce that it would not commercialize this technology. However, the USDA, Delta & Pine Land Company, and other agricultural biotechnology companies are continuing research on this and other methods to prevent unauthorized propagation of self-pollinated varieties.

The clear intent of this technology was to provide a mechanism for releasing self-pollinated varieties while affording the same protection from unauthorized propagation that is currently available only in hybrids. This would maintain profitable sales without competition from farmer-saved seed. However, the system would also prevent escape of genetically engineered varieties into the environment, because any seeds carrying the trait would not be viable. In addition, farmers in developed countries generally do not save seeds of most crops, either because they plant hybrids or because producing high-quality seed is difficult. In these cases, seed production has already been primarily in the hands of private companies for the past 50 years or more. Nonetheless, active competition still exists among seed companies selling those crops, including maize, sunflowers, and many vegetables, suggesting that limiting the ability of farmers to save their own seed does not automatically result in monopolistic seed companies. The reverse may be true: The opportunity for economic return increases investment and competition in crop breeding, as has been the case for hybrid varieties.

Box 9.2

A Biological System to Prevent Unauthorized Propagation of Seeds

In the "technology protection system" developed by the USDA and Delta & Pine Land Company, also dubbed the "terminator gene," the parent plants are engineered to contain a gene that would kill the seeds after they had completed development in the crop production generation. This mechanism requires several components: (1) a DNA recombinase, an enzyme that can recognize a specific piece of DNA, excise it, and rejoin the two cut ends; (2) an inhibitor protein that can block the action of the recombinase; (3) a genetic promoter that allows an adjacent gene to be expressed late in seed development; (4) a piece of "blocking" DNA that the recombinase enzyme recognizes; and (5) a gene that makes a product that when expressed, kills the cell. The blocking DNA is inserted between the promoter and the lethal gene, preventing the latter from being expressed. When these components are transformed into a plant, both the recombinase and the inhibitor are expressed, so essentially nothing happens and viable seeds are produced. To trigger the system, a chemical is applied to the seeds that prevents production of the inhibitor protein. This frees the recombinase to excise the blocking DNA and join the promoter directly to the lethal gene. Still, nothing happens immediately because the promoter is only active late in seed development. The treated seeds can germinate normally, grow to maturity, and produce seeds. Only then, after seed development is essentially completed, does the promoter become active, causing the lethal gene to be expressed, and resulting in the loss of seed viability. This system was designed to prevent unauthorized propagation of seeds; many other applications could be conceived where the ability to control the generation in which a specific trait is expressed would be useful.

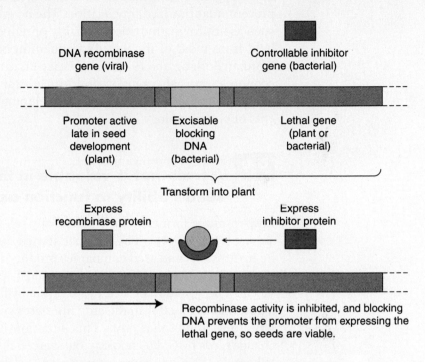

DNA recombinase gene (viral)

Controllable inhibitor gene (bacterial)

Promoter active late in seed development (plant)

Excisable blocking DNA (bacterial)

Lethal gene (plant or bacterial)

Transform into plant

Express recombinase protein

Express inhibitor protein

Recombinase activity is inhibited, and blocking DNA prevents the promoter from expressing the lethal gene, so seeds are viable.

To induce expression of the "terminator" trait, seeds are treated with a triggering compound that blocks the inhibitor.

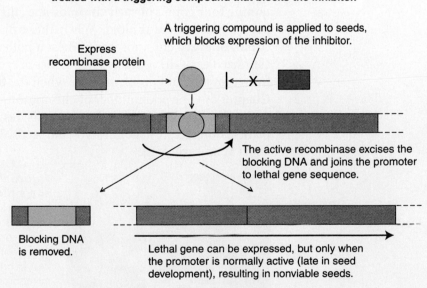

Express recombinase protein

A triggering compound is applied to seeds, which blocks expression of the inhibitor.

The active recombinase excises the blocking DNA and joins the promoter to lethal gene sequence.

Blocking DNA is removed.

Lethal gene can be expressed, but only when the promoter is normally active (late in seed development), resulting in nonviable seeds.

In countries where agricultural technology is less advanced, in contrast, the situation is quite different, because most farmers save their own seed or rely on government-sponsored seed distribution systems to provide seeds at low cost. The CGIAR, which sponsors a network of 16 international agricultural research centers, announced it would not release any germplasm incorporating technologies that prevent farmers from saving and replanting seeds. However, private companies developing genetically enhanced varieties may choose not to release them in countries lacking mechanisms (legal or biological) to prevent unauthorized propagation. The potential benefits of agricultural biotechnology, such as improving nutritional quality or reducing pesticide use, are too great to be withheld from most of the world's population because of intellectual property issues. As Norman Borlaug, the 1970 Nobel Peace Prize recipient noted, "This issue goes far beyond economics; it is also a matter for deep ethical consideration. Fundamentally, the issue is whether small-scale farmers of the developing world also have a right to share in the benefits of biotechnology."

9.9   A trait that is valuable in the final product can harm the seed's ability to function as a propagule.

Low-resource farmers have traditionally selected their own seeds for planting the next season (**Figure 9.16**). This practice started disappearing in developed countries as public institutions and seed companies produced elite varieties that were either hybrids, as with maize, or were resistant to the most recently evolved pathogen strains, as with wheat. The situation of seed propagation by farmers becomes even more complex when a trait that is valuable in the crop interferes with the seed's ability to function as a propagule to establish that crop. This situation is arising more and more often as people change seeds by GM technology, but it is certainly not new. Consider seedless watermelons. Where do the seeds come from to plant a field of seedless watermelons? In this case researchers use a specific strategy of hybridization. The female parent is a tetraploid, containing four copies of each chromosome (instead of the usual two copies), and the male parent is a normal diploid. When these plants make gametes, each egg cell has two copies of every chromosome and each pollen grain has one copy. The resulting zygote (fertilized egg cell) contains three copies of each chromosome. This does not stop the plant from growing, but problems arise when the flowers are forming and meiosis occurs (see Chapter 6). The odd number of chromosomes prevents pairing and proper distribution of the chromosomes to the daughter cells, resulting later in the fertilized egg cells in those plants failing to produce embryos and seeds. The fruits continue to develop, but the seeds abort at an early stage.

Figure 9.16 Collecting maize seeds for next season's planting in Colombia. Low-resource farmers collect their own seeds, but in both developed and developing countries commercial farmers obtain seeds from public institutions or seed companies for planting. The reason is that they want elite varieties with the best resistance genes or the best hybrids. Plants grown from seeds produced by hybrids never produce the same high yield as the hybrids themselves. *Source:* Courtesy of Nathan Russell, Centro Internacional de Agricultura Tropical.

Some traits incorporated into varieties for their value in the end product can also impact how seeds are propagated and marketed. One example is the use of certain naturally occurring mutations that increase sugar content in maize (sweet corn). Some of these mutations (such as *shrunken2* or *sh2*) limit the maize seed's ability to synthesize starch; instead, the seeds accumulate large amounts of sucrose. This is a valuable trait for sweet corn, because consumers prize the flavor, and the sweetness lasts longer after harvesting. However, because of the low starch content the seeds shrivel up when they dry out, hence the name "shrunken." The high sugar content makes the seeds attractive to insects and pathogens. These properties make it hard for the farmer to establish a crop in the field. Similarly, researchers have modified canola seeds to produce different types of oils, some of which reduce the ability of the seeds to germinate at lower temperatures. In soybean, eliminating raffinose to improve digestibility by animals may decrease seed longevity.

One application of the "terminator technology" discussed earlier is to make seeds nonviable and prevent unauthorized propagation. However, at a more basic level it represents the ability to control the seed generation in which a specific trait is expressed—or remains silent until a specific trigger activates it. Many potential applications of this technology to crops need not involve the protection of intellectual property. For example, in sweet corn, it would be possible to genetically engineer a normal *Sh2* gene into *sh2* mutant plants during the seed production generation. This would allow starch synthesis during seed development, resulting in higher-quality seed for the farmers. This normal *Sh2* gene could then be silenced in the next generation, which produces the sweet corn for the consumer. Thus, although many condemned the terminator technology, people will find other useful applications. As is often the case in science, discoveries are neutral, but what people do with them is not.

CHAPTER SUMMARY

Seeds are both the starting point and the primary product for most agricultural commodities. Unique properties of seeds, including accumulation of storage reserves and ability to dehydrate for transportation and storage, have played a major role in the development and spread of agriculture. During domestication, some seed traits have been reduced or eliminated, such as dispersal mechanisms (shattering) and dormancy; selection and gradual genetic modification have increased others, including seed size and number. These changes have greatly increased seed yield from domesticated crops. Seeds accumulate reserves of carbohydrates, oils, and proteins for use by the seedling after germination. These same reserves explain why seeds are the basis of food production and nutrition for livestock and humans.

Seed germination is initiated by water absorption and concludes with the emergence of the embryonic tissues from the seed. Storage reserves are then mobilized and transported to the growing seedling. The timing and percentage of germination and seedling emergence are important components of crop production. Both physical (coating and pelleting) and physiological (priming) techniques can be used to improve seed performance in establishing crop plants. At the same time, the abundance and persistence of weed seeds in the soil is a major factor in their success as competitors to crop plants. Preventing seed production and dispersal is an important method for controlling weeds.

Producing seeds for planting is often a distinct agronomic operation from producing the crop commodity. Specialized procedures are used, for example, to produce F_1 hybrid seeds that exhibit more vigorous growth and higher yields than their parents. Special care must also be taken to prevent contamination of seed crops by off-types or by pollen transferred from nearby crops. Seed certification programs are important in distributing improved crop varieties by maintaining their genetic purity through a pedigree and inspection system. Similarly, phytosanitary regulations are crucial in preventing the spread of plant diseases via the international seed trade. In addition to their use as propagules, seeds are also the primary means for storing plant germplasm in gene banks. Nonetheless, the 2.3 million accessions of seeds and plant materials conserved in gene banks around the world represent only a minute fraction of the total biodiversity present in the plant species of Earth.

Seeds are and will remain the primary system for delivering improvements from plant breeding and biotechnology into agriculture. Most enhanced traits are only valuable when incorporated into a high-yielding, high-quality, and disease-resistant variety, placing a premium on the value of elite crop germplasm and breeding lines. This focus has led to mergers between major crop biotechnology and seed companies. The higher value of seeds, in turn, spurs greater emphasis on protecting the research investment they represent through legal mechanisms such as plant breeders' rights and utility patents. Biological mechanisms have also been developed that could potentially prevent unauthorized propagation of crop seeds. However, legal and economic mechanisms are needed to ensure that the benefits of biotechnology are not withheld from developing countries by intellectual property issues. Seeds of improved varieties remain the most efficient and equitable means of increasing the productivity of the world's farmers.

Discussion Questions

1. How well would a cultivated cereal or legume species survive if it were to grow in the wild again? In what ways would it be more or less competitive than the natural wild species with which it would have to compete?

2. Assuming you had the appropriate genetic engineering tools, what would you design into a seed to optimize its nutritional content, both quantitatively and qualitatively?

3. Given that the population of the world is increasing, while agricultural land is decreasing, what properties of a seed would you genetically enhance to increase yield?

4. Why do weed seeds often exhibit strong and complex dormancy mechanisms? What seed characteristics might you expect to be different for a weed adapted to rice fields versus a weed adapted to sunflower or maize fields?

5. In what ways does producing a crop for use as seeds differ from producing a crop for food? What are some specific procedures for seed production that might not be necessary for producing the crop?

6. What features of F_1 hybrids make them valuable for crop production? Why can seeds produced from F_1 hybrid plants not be planted to produce a subsequent crop of the same quality?

7. Why would apomixis be a useful trait to incorporate into crop varieties? What are some potential problems that could arise if this were accomplished?

8. What characteristics of seeds make them particularly well adapted for germplasm preservation? What is the difference between *ex situ* and *in situ* conservation of germplasm, and what are the advantages of each method?

9. Discuss legal and biological mechanisms for preventing unauthorized propagation of crop varieties. What additional issues does each method raise?

10. How would you reconcile the dilemma of providing access to improved crop varieties to the developing countries while allowing the breeding and biotechnology companies to receive an equitable return on their investment in research and development? Explain your approach.

Further Reading

Baenziger, P. S., R. A. Kleese, and R. F. Barnes. 1993. *Intellectual Property Rights: Protection of Plant Materials.* CSSA Special Publication Number 21. Madison, WI: Crop Science Society of America.

Baskin, C. C., and J. M. Baskin. 1998. *Seeds: Ecology, Biogeography and Evolution of Dormancy and Germination.* San Diego, CA: Academic Press.

Basra, A. S., ed. 2000. *Heterosis and Hybrid Seed Production in Agronomic Crops.* New York: Food Products Press.

———. 2000. *Hybrid Seed Production in Vegetable Crops.* New York: Food Products Press.

Bewley, J. D., and M. Black. 1994. *Seeds: Physiology of Development and Germination,* 2nd ed. New York: Plenum Press.

Black, M., and J. D. Bewley, eds. 2000. *Seed Technology and Its Biological Basis.* Sheffield, U.K.: Sheffield Academic Press.

Copeland, L. O., and M. B. McDonald. 2001. *Principles of Seed Science and Technology,* 4th ed. Norwell, MA: Kluwer Academic Publishers.

Derera, N. F., ed. 1989. *Preharvest Field Sprouting in Cereals.* Boca Raton, FL: CRC Press.

Desai, B. B., P. M. Kotecha, and D. K. Salunkhe. 1997. *Seeds Handbook. Biology, Production, Processing, and Storage.* New York: Marcel Dekker.

Egli, D. B. 1998. *Seed Biology and the Yield of Grain Crops.* Wallingford, U.K.: CAB International.

James, C. 2000. *Global Status of Commercialized Transgenic Crops: 1999.* ISAAA Briefs No. 17. Ithaca, NY: International Service for the Acquisition of Agri-biotech Applications.

Kelly, A. F., and R. A. T. George, eds. 1998. *Encyclopaedia of Seed Production of World Crops.* New York: Wiley.

Kigel, J., and G. Galili, eds. 1995. *Seed Development and Germination.* New York: Marcel Dekker.

Larkins, B. A., and I. K. Vasil, eds. 1997. *Cellular and Molecular Biology of Plant Seed Development.* Boston: Kluwer Academic.

Converting Solar Energy into Crop Production

Donald R. Ort
University of Illinois

Stephen P. Long
University of Illinois

The energy contained in the food that people eat is ultimately derived from sunlight through photosynthesis. Agriculture and crop production are basically all about capturing solar energy and converting it into food and fiber with the highest possible efficiency in a sustainable manner. The amount of solar energy received at the site where a crop is grown sets the upper limit on potential photosynthetic production. The solar energy at a given location on Earth on a clear day may be predicted for any minute of the day and time of the year from Sun–Earth geometry (**Figure 10.1a**). At 10°N latitude of the equator, as with all sites within the tropics (0–23.5°) there is only modest annual variation in irradiance (sunlight) contrasting with 70°N, the northern limit of any crop production, where there is strong seasonal variation. At the polar latitudes (66.5–90°), there is no sunlight for a period of days in midwinter, but in midsummer total solar radiation can exceed that of the tropics because of very long summer days. When the sky is clear, solar energy over the course of a single day shows symmetrical variation around midday, but during intermittent cloud cover there are frequent variations (**Figure 10.1b**). Both clouds and atmospheric pollution can significantly lower the actual amount of solar radiation reaching Earth's surface.

Whereas solar energy input sets the upper limit on the energy that can be transformed into crop yield, the actual yield of energy in food (Y) depends on the product of solar input (S) and the efficiency with which the solar energy is transformed into the harvested product. This efficiency is expressed as a number between 0 and 1 that shows what proportion of the solar energy reaching the field is transformed into food. This overall efficiency is the product of the efficiencies of the interception of sunlight energy by the leaves (ε_i), the conversion of the intercepted energy into plant matter (ε_c), and the partitioning of this plant matter into the harvested food product (ε_p):

$$Y = S \cdot \varepsilon_i \cdot \varepsilon_c \cdot \varepsilon_p \tag{10.1}$$

Box 10.1 gives a worked example of this equation and shows that for a wheat field in England 5,000 megajoules (MJ) of solar energy was converted into 21 MJ of wheat grain.

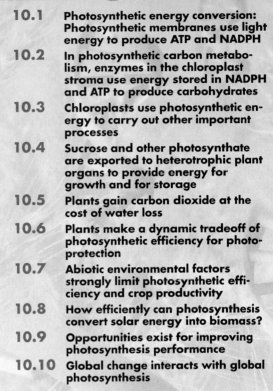

The last 50 years have seen large increases in crop yields worldwide as a result of the joint efforts of plant breeders and agronomists (see Chapters 3 and 14). Using wheat and rice—the world's two most important food sources—as examples, we can see that two factors have contributed to these yield increases. First, increased use of fertilizers and pesticides has resulted in larger and longer-lived leaves, improving ε_i. Second, plant breeders have selected genotypes that in-

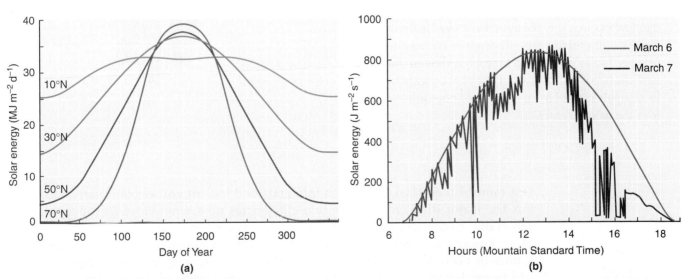

Figure 10.1 Annual and diurnal solar energy profiles. (a) Annual variation in daily solar energy received at Earth's surface under clear skies at four latitudes: 10°N (for example, Caracas, Venezuela), 30°N (for example, Austin, Texas, United States), 50°N (for example, Vancouver, Canada), and 70°N (for example, Tromsø, Norway). **(b)** The daily course of solar energy for two separate days over a wheat crop at Maricopa, Arizona (33°N latitude), in early March, showing the effect of intermittent cloud cover.

Box 10.1

Efficiency of Food Production from Solar Energy to People

Cereal grains represent the most important source of energy and protein for the world's population. A square meter of land in southern England receives about 5,000 MJ of solar energy annually. A winter wheat crop intercepts this light with an efficiency (ε_i) of 0.7. This absorbed light is converted in crop biomass with an efficiency (ε_c) of 0.01. In wheat about 0.6 of the biomass is partitioned into the grain (ε_p). Substituting into Equation 10.1,

$$Y = 5,000 \text{ MJ} \times 0.7 \times 0.01 \times 0.6 = 21 \text{ MJ}$$

Therefore the input of 5,000 mJ results in just 21 mJ of energy stored in the grain. The fate of energy that is lost is illustrated in the figure. This seemingly low efficiency of solar energy conversion by crop plants is among the best achieved by farmers worldwide.

Grain is converted into energy stores in the bodies and of humans and other animals at an efficiency of about 10%. Therefore, about 10 times as much grain is required if people obtain their energy by consuming a farm animal rather than directly consuming the grain. Despite a growing world population, per capita increase in cereal production grew each year until the 1990s, when for the first time it declined. A reasonable solution to feeding an increasing world population would be to decrease meat production, but in reality the trend has moved in the opposite direction. Increasing the efficiency of conversion of intercepted solar energy into biomass is thus an important new course of action to meet the 33% increase in world food supply needed to meet projected population increase over the next 20 years.

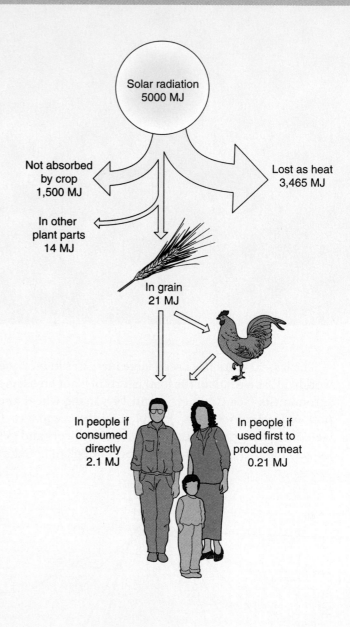

Solar radiation 5000 MJ

Not absorbed by crop 1,500 MJ

Lost as heat 3,465 MJ

In other plant parts 14 MJ

In grain 21 MJ

In people if consumed directly 2.1 MJ

In people if used first to produce meat 0.21 MJ

vest more of their total production into grain and less into other plant parts, raising ε_p. For the major food crops, these two efficiencies now seem close to the maximums achievable. In contrast, there has been very little change in ε_c, the efficiency of conversion of solar energy intercepted by leaves into crop biomass through photosynthesis. Improving ε_c must be a central focus in meeting the challenge of doubling global food production that will be needed to feed the human population by the middle of this century. The following sections explain photosynthesis and explore the prospects for improving its efficiency in crops.

10.1 Photosynthetic energy conversion: Photosynthetic membranes use light energy to produce ATP and NADPH.

Photosynthesis is a two-stage process that occurs in chloroplasts. In the first stage (**Figure 10.2**), the plant converts light energy into chemical energy in the form of two high-energy compounds called ATP and NADPH. These reactions are intimately associated with the photosynthetic membranes. In the second stage, the plant uses these energy-rich biochemicals to convert carbon dioxide (CO_2) and water into simple sugars. These two processes occur simultaneously in the chloroplast stroma, and simple sugars are exported from the chloroplast into the cytoplasm, where they are used to make sucrose.

An Antenna System Containing Chlorophyll and Carotenoid Pigments Captures Light Energy. The action spectrum of higher plant photosynthesis—that is, the colors of light that can energize photosynthesis—includes wavelengths from about 400 nm (violet) to 700 nm (red). About 60% of the sunlight incident on Earth's surface falls within this photosynthetically active range. The light that penetrates the leaf is captured by light antennae containing two classes of pigments, chlorophylls and carotenoids, that are responsible for the absorption of light that energizes photosynthesis in higher plants. Chlorophyll is the dominant pigment and strongly absorbs red and blue light, while scattering most of the green light that falls on the leaves (which is why most plants appear green).

To convert the transient energy of a photon—photons are elementary light energy packets—into stable chemical energy, the photosynthetic apparatus performs a series of energy-transforming reactions. The process is initiated by absorption of a photon by a chlorophyll or carotenoid molecule that converts light energy to an excited electronic state of the pigment molecule. These pigment molecules are arranged in the photosynthetic membrane of chloroplasts in groups of 250–300 anchored there by specialized proteins, which provide a scaffolding for the precise arrangement of each molecule within the antenna. Because of the proximity of other unexcited pigment molecules with the same or similar energy states, the excited state is rapidly transferred over the antenna system. Under optimum conditions, over 90% of the photons captured by this antenna are successfully used to drive the next step in photosynthesis: primary charge separation.

The Energy from Absorbed Light Is Used to Separate Positive and Negative Charge. Photosystem II (PSII) and photosystem I (PSI) of plants are the sites of the primary photochemical reactions of photosynthesis, which involves separation of a positively charged molecule from a negatively charged one. Each photosynthetic membrane in a chloroplast contains thousands of these two photosystems, which are themselves complexes of multiple proteins embedded in the membranes. In PSII the excited state energy is transferred to a specialized group of six chlorophyll molecules associated with proteins at the core of PSII. In the primary photochemical reaction, a negatively charged electron is transferred from an excited chlorophyll molecule to an electron-accepting pigment, creating positively and negatively charged molecules that are adjacent to each other. The negatively charged electron cannot flow back to the positively charged molecule, and this separation of charge "captures" the energy and drives the subsequent electron transfer reactions in the photosynthetic membrane (Figure 10.2). In PSII the energy is used to remove electrons from water and to add electrons, as well as protons, to plastoquinone (PQ).

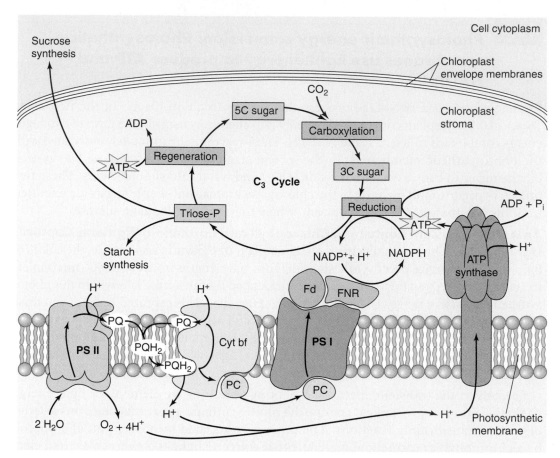

Figure 10.2 Schematic drawing of a chloroplast. The photosynthetic membrane contains the major protein complexes and pigments responsible for light absorption and photosynthetic electron and proton transfer. The reactions of the photosynthetic membrane drive the C_3 photosynthetic carbon reduction cycle that takes place in the chloroplast stroma. Illustrated is the concept of light-driven electron transport coupled to the accumulation of protons in the membrane vesicle lumen that is in turn used to drive ATP formation by ATP synthase. In addition to the energy stored in ATP formation, energy derived from absorbed light is also stored in the formation of NADPH. Photosynthetic carbon reduction is shown as a three-stage cycle. **Carboxylation**: A molecule of CO_2 is covalently linked to a five-carbon sugar. **Reduction**: Energy in the form of NADPH and ATP is used to form the simple carbohydrate triose-P. **Regeneration**: Energy in the form of ATP is used to regenerate the 5-carbon sugar for carboxylation. (PSII, photosystem II; PSI, photosystem I, PQ and PQH_2, plastoquinone and reduced plastoquinone; cyt, cytochrome; FeS, Rieske iron-sulfur protein; PC, plastocyanin; Fd, ferredoxin; FNR, ferredoxin-NADP reductase.) *Source:* Modified from D. R. Ort (1994), Photosynthesis: The chloroplast, in C. J. Arntzen, ed., *Encyclopedia of Agricultural Science* (San Diego, CA: Academic Press), pp. 187–195, Figure 3.

In PSI, the energy captured in the primary charge separation drives the oxidation of plastocyanin, a soluble copper-containing protein located in the lumen, and the reduction of ferredoxin, a soluble iron-containing protein located in the stroma (Figure 10.2). As with PSII, the reactions of PSI produce an electric potential across the photosynthetic membrane and generate a strong reductant. PSI differs from PSII in that it does not deposit protons into the lumen but uses its energy to reduce NADP-producing NADPH.

Electron Transport between PSI and PSII Creates a pH Difference Plus an Electric Potential Across the Photosynthetic Membrane. The light-driven electron and proton transfer reactions of the two photosystems are interconnected through the activity of another membrane protein complex (cytochrome bf complex, abbreviated cyt bf), which catalyzes the energetically downhill reaction of reducing plastocyanin. Plastoquinone, a special membrane lipid, serves as a mobile hydrogen carrier, transporting hydrogen (that is, a proton plus its electron) from PSII to cyt bf, and plastocyanin serves as a mobile electron carrier (that is, does not carry a proton), transporting electrons from cyt bf to PSI. In addition to linking the activity of PSII and PSI, the cyt bf complex plays a central role in energy transformation and storage by converting energy available in reduced plastoquinone (PQH_2) into a transmembrane pH difference as well as an electric potential difference. As illustrated in Figure 10.2, this feat is accomplished as cyt bf oxidizes PQH_2 at a site near the inside (lumenal side) of the membrane vesicle, resulting in the release of protons into the lumen. Because plastoquinone is reduced near the outside (stromal side) of the membrane, the protons for its reduction are taken up from the stroma (Figure 10.2). In addition to oxidizing plastoquinone, cyt bf also reduces plastoquinone at a second site that is near the stromal side of the membrane. The net result of these reactions is the transfer of protons from the stroma to the lumen, creating a transmembrane pH difference and an electric potential difference. The energy stored in the pH difference and electrical potential is used for the energy-requiring reaction of converting ADP to ATP by the addition of a phosphate group. This conversion is done by the enzyme ATP-synthase located in the photosynthetic membrane (Figure 10.2).

10.2   In photosynthetic carbon metabolism, enzymes in the chloroplast stroma use energy stored in NADPH and ATP to produce carbohydrates.

Although the energy stored in ATP and NADPH is chemically stable, plants do not accumulate high levels of these chemicals. The ATP and NADPH are rapidly used in the biosynthesis of carbohydrates from CO_2 and water. This intricate biosynthetic pathway, known as the C_3 *photosynthetic carbon reduction cycle* (C_3 *cycle*), takes place in the stroma of chloroplasts, and involves more than a dozen different enzymes.

The C_3 Photosynthetic Carbon Reduction Cycle Fixes CO_2 and Produces Sugar. The C_3 cycle begins with a carboxylation reaction (Figure 10.2) in which CO_2 is attached to the five-carbon acceptor molecule ribulose bisphosphate, abbreviated as RuBP. This reaction is catalyzed by Rubisco (ribulose bisphosphate carboxylase/oxygenase), an exceptionally abundant enzyme in leaves, often accounting for 40% or more of the soluble leaf protein. The resulting 6-carbon compound is not stable and immediately splits into two 3-carbon molecules, which accounts for the C_3 name of the cycle. The next stage of the C_3 cycle requires energy in the form of both ATP and NADPH to form triose phosphate (glyceraldehyde-3-phosphate). Figure 10.2 illustrates that triose phosphate is the principal branch point within the C_3 cycle. Some of the triose phosphate is transported out of the chloroplast and used in synthesizing sucrose. Triose phosphate is used within the chloroplast for starch synthesis and is reinvested into the C_3 cycle to regenerate the five-carbon CO_2 acceptor RuBP, thereby completing the photosynthetic carbon reduction cycle.

After Rubisco Fixes O_2 Rather than CO_2, the C_2 Photorespiration Cycle Retrieves Some of the Lost Carbon, but at a Cost. The fossil record indicates that photosynthetic organisms have been producing oxygen for over 3 billion years, although stable molecular oxygen did not appear in Earth's atmosphere until about 2 billion years ago. This means that photosynthesis evolved in an oxygen-free or oxygen-poor atmosphere. This fact probably explains a curious feature about Rubisco that is the origin of a major inefficiency in photosynthesis. In addition to the carboxylation of RuBP by CO_2, Rubisco also catalyzes its oxidation by atmospheric O_2 (oxygenation) to yield one molecule of a three-carbon compound (PGA) and a molecule of a two-carbon compound. This oxygenation reaction creates a significant inefficiency in the photosynthetic process, because the two-carbon compound cannot enter the C_3 cycle. In a typical C_3 crop (such as wheat), the rate of Rubisco-catalyzed oxygenation is about 20% of the rate of Rubisco-catalyzed CO_2 fixation. The inability of Rubisco to distinguish between molecular O_2 and CO_2 appears to be an unavoidable consequence of having evolved in an oxygen-free atmosphere.

To compensate for the oxygenation of RuBP by Rubisco, a metabolic pathway evolved in plants that recovers the two-carbon skeletons that are diverted from the C_3 cycle by the oxygenation reaction. However, this scavenging operation, known as *photorespiration,* is energetically expensive. In the biochemical reactions of photorespiration, two of the two-carbon molecules are converted into a single three-carbon molecule with the release of one CO_2 and at the expense of ATP and reducing power. The cell thus succeeds in returning 75% of photorespiratory carbon to the C_3 cycle, with the remainder released as CO_2.

C_4 Photosynthetic Metabolism Defeats Photorespiration but at a Considerable Energy Cost. The photorespiration reactions just discussed make the best of a wasteful situation caused by Rubisco's oxygenase activity. An alternative "strategy" taken by plants in the course of evolution is to prevent or greatly reduce Rubisco's oxygenase activity by exploiting the competition between CO_2 and O_2 as substrates of Rubisco. Plants such as maize, sorghum, and sugar cane greatly suppress or completely inhibit the oxygenation reaction of Rubisco by concentrating CO_2 in specialized leaf cells that contain Rubisco. These species are known as *C_4 plants* because the initial carboxylation reaction produces a four-carbon acid. They have a unique leaf anatomy with two distinct photosynthetic cell types in which chloroplast-containing mesophyll cells surround chloroplast-containing bundle sheath cells, which in turn encircle the vascular bundles of leaf (**Figure 10.3a**).

The basic pathway of C_4 photosynthesis and the interplay of the two photosynthetic cell types are shown in **Figure 10.3b**. A key feature of C_4 photosynthesis is that the initial fixation of CO_2 takes place in mesophyll cells and involves the carboxylation of the three-carbon molecule phospho(enol)pyruvate (PEP) by an enzyme called *PEP carboxylase.* Unlike Rubisco, PEP carboxylase does not bind O_2 in competition with CO_2. The four-carbon product of this reaction is transported to the bundle sheath cell and decarboxylated to release CO_2. The decarboxylation of the four carbon compounds significantly elevates the CO_2 concentration in the bundle sheath cell chloroplast where Rubisco and the other enzymes of the C_3 cycle are localized. The elevated concentration of CO_2 effectively competes with oxygen, virtually eliminating photorespiration. The three-carbon product of the decarboxylation reaction is transported back to the mesophyll cell so that the CO_2 acceptor PEP can be regenerated.

Although the C_4 photosynthetic pathway effectively suppresses photorespiration, it is important to recognize that there are substantial energetic costs (from ATP). In effect the C_4 cycle is a light-driven (that is, energy-driven) CO_2 pump, concentrating CO_2 in the bundle sheath cells, the site of the C_3 cycle in C_4 plants. Researchers estimate that only about 5% of all terrestrial higher plant species are C_4, indicating that the pathway with its higher energetic costs has an advantage in relatively few habitats. However, the much higher

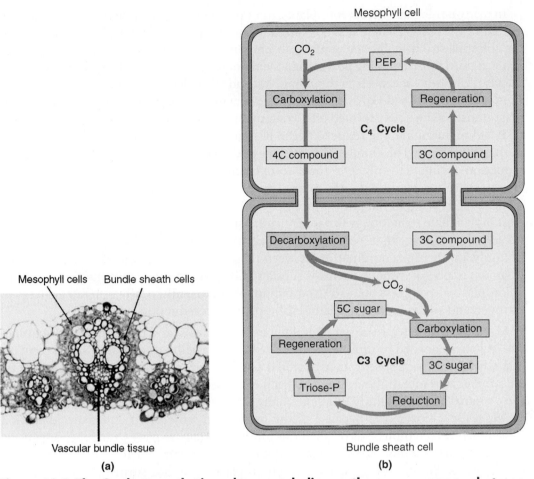

Mesophyll cell

Bundle sheath cell

Mesophyll cells   Bundle sheath cells

Vascular bundle tissue

(a)

(b)

Figure 10.3 The C$_4$ photosynthetic carbon metabolism pathway suppresses photorespiration by concentrating CO$_2$ at the site of carboxylation by Rubisco. (b) The C$_4$ pathway involves two different cell types, mesophyll cells and bundle sheath cells. Shown in the C$_4$ cycle is **carboxylation** of CO$_2$ into a 4-carbon molecule in mesophyll cells; transport of the C$_4$ compound into bundle sheath cells; **decarboxylation** of the C$_4$ compound, producing a high concentration of CO$_2$ within bundle sheath cells where the C$_3$ cycle produces carbohydrate; transport of the resulting three-carbon acid back to the mesophyll cell; and the **regeneration** of the CO$_2$ acceptor, phosphoenolpyruvate (PEP). **(a)** Cross section of sugar cane leaf showing the C$_4$ placement of mesophyll and bundle sheath cells. *Source:* (a) David Webb, University of Hawaii at Manoa. (b) Modified from D. R. Ort (1994), Photosynthesis: The chloroplast, in C. J. Arntzen, ed., *Encyclopedia of Agricultural Science* (San Diego, CA: Academic Press), Figure 5.

representation of C$_4$ species in hot/dry climates is well documented and probably reflects the higher water use efficiency associated with C$_4$ metabolism (see later discussion) as well as the increasing affinity of Rubisco for O$_2$ with increasing temperature.

10.3 Chloroplasts use photosynthetic energy to carry out other important processes.

Although photosynthesis is usually considered in the context of CO$_2$ reduction to form sugars, significant amounts of reducing power and ATP are used within the chloroplasts for other processes essential to plant growth and development.

In many nonlegume herbaceous plants, most of the nitrate taken up by the roots is transported to the leaves, where it is converted in the cytoplasm to nitrite and imported into the chloroplast. The reduction of nitrite to ammonia in the chloroplast consumes, per molecule, a third more reducing power than the reduction of CO_2 to carbohydrate. Nitrite reductase is located exclusively in the chloroplast and takes electrons directly from ferredoxin, thus competing with NADPH formation. The enzyme glutamine synthetase in the chloroplast transfers the newly formed NH_4^+ to glutamate forming glutamine at the expense of ATP. From glutamine, the reduced N can be channeled into a host of other amino acids in the chloroplast. The amino acids cysteine and methionine contain sulfur (S), and the reduction of sulfate to sulfide in the chloroplast requires, on a per mole basis, 50% more energy than does carbon assimilation. This multistep process also uses electrons that come directly from ferredoxin to produce sulfide, which is incorporated into cysteine. Although sulfate reduction is very energy intensive, the fact that the ratio of reduced sulfur to reduced nitrogen in plants is maintained within a narrow range near 1:20 implies not only an intricate coordination between these processes but also that the overall investment of photosynthetic energy is substantially greater for nitrate reduction. Typically, chloroplasts expend less than 10% the amount of photosynthetically generated energy on nitrogen and sulfur reduction than is spent on carbon reduction.

Many biosynthetic processes are unique to chloroplasts, and those that are not unique use enzymes encoded by unique genes. For example, a cell may contain two different but related enzymes for a specific step in amino acid biosynthesis: one enzyme located in the chloroplast and one in the cytoplasm. Each enzyme is encoded by a different gene, and the two enzymes are slightly different. Many herbicides specifically inhibit biochemical processes in chloroplasts because of the manner in which herbicides are identified. Researchers screen thousands of chemicals for their toxicity to plants and other organisms. Chemicals that are more toxic to plants than to other organisms are candidates for herbicide status. Because only plants have chloroplasts, it is not surprising that when researchers discover such chemicals' mode of action, they often find that the herbicides inhibit a biochemical process specific to the chloroplast.

10.4 Sucrose and other photosynthate are exported to heterotrophic plant organs to provide energy for growth and for storage.

Photosynthetic organs such as leaves are fully autotrophic and are considered to be *source* organs because they are sources of photosynthetic products such as sucrose. But plants also have heterotrophic tissues and organs that import the products of photosynthesis (photosynthate). Such *sink* organs fall into two categories (1) storage sinks such as developing tubers or seeds that accumulate carbohydrates, lipids, and proteins; and (2) metabolic sinks such as developing leaves or meristems that require the import of metabolites to sustain growth (**Figure 10.4**). Although seeds and tubers function as an important source of energy and metabolites during seedling growth or tuber sprouting, mature green leaves are the fundamental source organs in the life cycle of plants. The function of a particular plant tissue as a source or a sink is dynamic, depending on its stage of development. Thus an expanding leaf changes from a sink to a source organ during maturation, but a mature leaf can later revert to a sink organ if it becomes heavily shaded by the leaves above. A po-

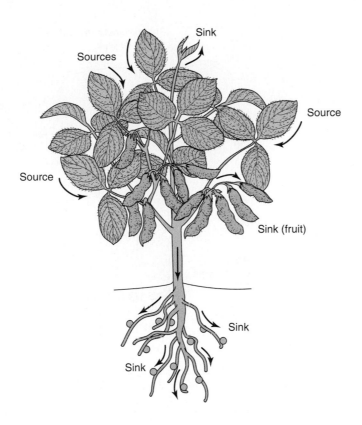

Figure 10.4 Diagram of soybean plant showing patterns of photoassimilate distribution. Movement of photoassimilate in the phloem is from sites of production in the mature leaves (source tissue) to different sites of use (sinks) within the plant. Sinks for photosynthate include developing leaves, roots, and developing seeds. The direction of flow from source leaves can be up or down, depending on the location of sinks on the plant.

tato tuber is a sink organ when it develops, but a source when the tuber sprouts. Such transitions are genetically programmed (see also Chapter 8).

Long-Distance Transport from Source Leaves to Sink Organs Requires the Input of Energy to Drive Mass Flow. Long-distance transport of the products of photosynthesis is carried out by the phloem, a vascular tissue with two important cell types: sieve tubes and companion cells. Products of photosynthesis, such as sucrose or certain oligosaccharides and amino acids produced in the mesophyll cells of mature leaves, move by diffusion via plasmodesmata (channels that connect cells, see Chapter 8) toward the phloem, where loading into the sieve tubes for long-distance transport takes place. In most herbaceous crop plants, sucrose is exported from source cells near the phloem through the plasma membrane and into the cell walls. This space outside the cells is called the *apoplast* (**Figure 10.5**). Such export is from an area of high concentration in the cytoplasm to a much lower concentration in the apoplast and therefore does not require the input of energy. However, the subsequent loading of photoassimilates into the phloem is a different story, because the concentration of sucrose in the phloem sap of a source leaf is quite high. Transporting sucrose from the leaf cell apoplast into the phloem against this high concentration gradient requires the input of significant energy resources in the form of ATP. In other species, such as squash, that transport raffinose oligosaccharides rather than sucrose, these compounds never enter the apoplast. Transport occurs via numerous plasmodesmata between the source mesophyll cells and the phloem. In these plants the concentration of photosynthate needed to drive mass flow in the phloem is achieved through the ATP-requiring biosynthesis of the oligosaccharides from sucrose.

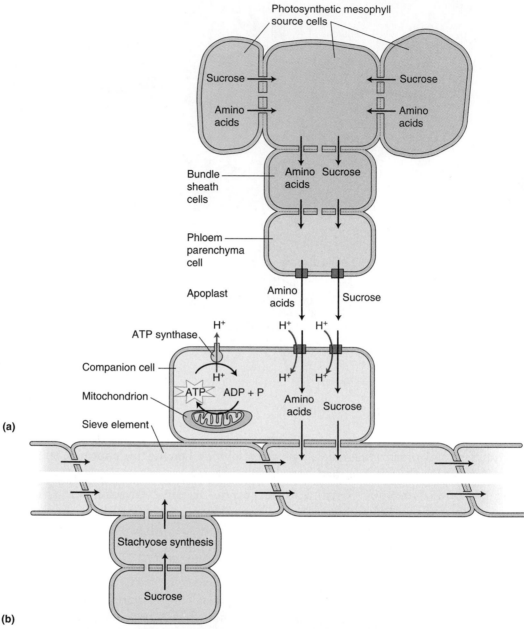

Figure 10.5 Phloem loading. (a) Apoplastic phloem-loading pathway for sucrose and amino acids. The transport of photoassimilates from the photosynthetic mesophyll cells in source leaves includes their export into the extracellular apoplastic space in the near vicinity of minor phloem elements. The accumulation of sucrose and amino acids in the phloem is driven by sucrose and amino acid carriers, which couple photoassimilate and proton transport in opposite directions. This is energized by the maintenance of a pH difference across the companion cell-plasma membrane by a proton-pumping ATPase-consuming ATP. The high concentrations created in the phloem of source leaves coupled drive long-distance mass flow in the phloem toward sites of unloading in sink tissues. **(b)** Symplastic phloem-loading pathway. Certain plant species transport larger soluble sugars (such as stachyose) that are synthesized from sucrose. In these species, phloem loading occurs via direct plasmodesmata connections without passage through the apoplast.

During periods of active photosynthesis, phloem transport velocities of 30–150 cm h^{-1} are typical. As shown in Figure 10.4, the direction of the mass flow is determined by site of consumption; up the aerial portion of a plant to a developing leaf, or underground to provide the energy roots need to function.

Sink Strength Regulates Photosynthetic Capacity of Source Leaves. To maximize growth, biomass production, and reproductive capacity, it is essential that photosynthetic production be coordinately balanced with mobilization, allocation, and use of photosynthate. A balance between the supply and demand for photosynthate in response to changing environmental conditions may be maintained in the short term by the modulation of metabolic factors and in the longer term by changes in gene expression that, for example, alter the capacity for photosynthate production.

The rapid fine-tuning of photosynthetic carbon metabolism to match changes in the demand for photosynthate can be achieved through changes in enzyme activity. Balance between production and use of triose phosphate is at the center of such metabolic control and operates to modulate carbon flux into sucrose. Thus when the sink demand for photosynthate declines, sucrose accumulates in mesophyll cells of source leaves, in turn elevating triose phosphate and diverting a larger portion of carbon flow from sucrose production to starch production.

Sucrose plays important roles in plant metabolism and as a transportable form of energy in most plants. In addition, it serves a central regulatory role in maintaining the balance between the supply and the demand for photosynthate. Many genes, whose products are involved at key points in photosynthetic metabolism, are either induced or repressed depending on the level of sucrose or other simple sugars in the source leaf mesophyll cytoplasm. For example, within hours of moving a plant from a condition of normal to elevated atmospheric CO_2, the genes of a number of proteins involved in photosynthesis (such as Rubisco) are repressed, as a direct response to increased soluble sugar levels in the cytoplasm. Signaling by sugar interconnects with signaling by hormones and stress, indicating that the signal transduction networks in plants (see Chapter 8) are very complex.

10.5 Plants gain carbon dioxide at the cost of water loss.

The photosynthetic assimilation of CO_2, an atmospheric gas, and its conversion into carbohydrates, takes place in the chloroplast-containing tissue inside the leaf, termed the *mesophyll*. Therefore, CO_2 must enter the leaves from the atmosphere. Furthermore, as with all biological reactions, photosynthesis takes place in water. This means that CO_2 from the atmosphere must dissolve into wet surfaces inside the leaf. The cost of exposing wet surfaces to the atmosphere to allow CO_2 to enter is the loss of water, yet water often limits plant production and survival on land. To understand how much water is lost to obtain CO_2 for photosynthesis and how the plant can regulate this loss, consider the physical processes that determine gas movement between surfaces and the atmosphere. The rate of water loss from a wet surface is determined by the law of diffusion, which states that the net flux of a gas (F_{gas}) is equal to the ratio of the concentration gradient (ΔC_{gas}) and the resistance (r) to transfer of that gas. The symbol F is used to indicate a flux, the symbol C to indicate concentration, and Δ to show the concentration difference between two points.

Fick's law of diffusion:

$$F_{gas} = \Delta C_{gas}/r \qquad\qquad (10.2)$$

Using Fick's law of diffusion, you can predict the rate of loss of water from a wet surface. The air molecules immediately above a wet surface will be saturated with water vapor. At 25°C, air saturated with water vapor contains 23 grams of water vapor per cubic meter (g/m^{-3}). Such air is said to have a relative humidity (RH) of 100%. The air outside the leaf is almost always less saturated with water vapor and can be very dry. This means the water vapor concentration gradient between the inside of the leaf and the outside is usually quite steep. If the outside air is moisture laden, then the concentration gradient is less steep. Molecular diffusion is the movement of molecules from a high to low concentration. In still air, the resistance to diffusion is proportional to distance and the rate of molecular diffusion. Wind accelerates movement, and decreases resistance. By assuming some reasonable values for the relative humidity (around 50%) and the resistance to diffusion, you can calculate that a square meter of surface would lose about 500 mg of water per second, or 1.8 kg (4 pounds) per hour. If a leaf really lost water at this rate, it would need to completely replace its water once every three minutes.

How can plants expose wet surfaces to the atmosphere to obtain carbon dioxide and conserve their water? They do this by maintaining a barrier to water loss, the leaf epidermis (see Chapter 8). A waterproof, waxy cuticle covers the outer surfaces of the epidermal cells. Within the epidermis there are stomates, small pores bordered by a pair of specialized epidermal cells, called *guard cells* (**Figure 10.6**). These cells alter their width to open or close the pore. When open, CO_2 can diffuse from the atmosphere to the wet surfaces of the photosynthetic cells of the mesophyll below. Typically, stomates open during the day, when photosynthesis requires CO_2, but close at night. Water vapor evaporates from the wet surfaces of the mesophyll cells surrounding the cavity below the stomate and escapes to the atmosphere through the open stomate.

How much water is lost to gain CO_2? Let's return to the example of evaporation from a wet surface, but this time, consider the wet surface of the mesophyll below a stomate. For a well-watered plant in full sunlight, when the stomates are open stomatal resistance to water vapor diffusion is 10 times greater than from an open surface (see preceding discussion). Water loss is therefore 10 times less, or about 50 mg per square meter per second. Furthermore, this is the minimum stomate resistance; at night the stomates will close and water loss will cease.

How much water is actually lost for each milligram of CO_2 used in photosynthesis? Photosynthesis within the mesophyll removes CO_2, creating a lower concentration inside the leaf. CO_2 then diffuses down this gradient from the higher concentration outside the leaf. Typically the CO_2 concentration below the stomate pore of a C_3 leaf during daylight will be about 70% of that in the atmosphere, where it is about 360 parts per million of air. Knowing the concentration gradient of CO_2, you can calculate that the rate of CO_2 diffusion over this same pathway is around 0.56 mg per square meter of leaf surface per second. Thus, the ratio of the mass of water lost per unit mass of CO_2 used is 89 in this example (50 mg m^{-2} s^{-1} divided by 0.56 mg m^{-2} s^{-1} = 89).

Plants using the C_4 photosynthetic pathway fix CO_2 with PEP carboxylase, which is more efficient than Rubisco. As a result, the CO_2 concentration below the pore inside the leaf is much lower in C_4 plants than in C_3 plants, and the diffusion gradient for CO_2 twice as steep. Thus C_4 plants can double the stomatal resistance and still take up CO_2 at the same rate. However, doubling the resistance halves water vapor loss, and C_4 plants only require about half the water needed by C_3 plants to assimilate the same amount of CO_2; that is, C_4 plants have greater water use efficiency than C_3 plants. This advantage is reflected at the whole plant level (**Table 10.1**): C_4 crops may produce twice as much biomass per unit of water transpired over their lifetime compared to C_3 crops growing in similar environments.

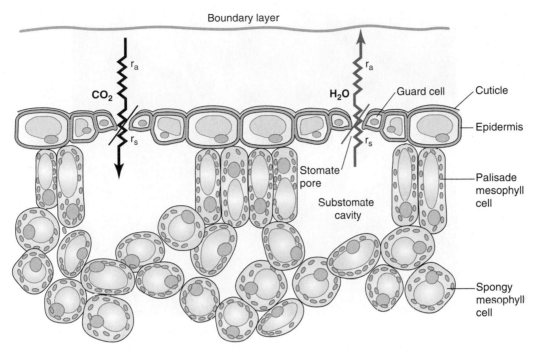

Figure 10.6 Cross section of a leaf showing the path of diffusion of CO_2 and H_2O in and out of leaves. The boundary layer is a very thin layer of unstirred air next to leaf surface that presents a diffusive resistance (r_a) to the movement of water vapor and CO_2. Guard cells regulate the aperture and therefore the resistance of stomates (r_s) to the flux of CO_2 and water vapor in and out of the leaf.

The amount of water vapor that air can hold increases exponentially with temperature. The potential for evaporation of water similarly increases with temperature. At 40°C the mass of water required to assimilate 1 mg of CO_2 would rise from 89 mg at 25°C, in our example, to 300 mg at 45°C. This presents a particular problem to plants of hot semiarid regions, where daytime leaf surface temperatures may exceed 45°C and where both relative humidity and soil moisture are also typically very low. However, in these areas of the world nighttime temperatures may be quite low. One group of plants have evolved a way to take advantage of this marked difference in potential evaporation between night and day. These crassulacean acid metabolism (CAM) plants include the cacti (**Figure 10.7**).

Table 10.1	Amounts of biomass produced per liter of water over the growing seasons for the five major cereal grain crops. Crops with C_4 photosynthetic metabolism are substantially more water efficient than C_3 crops		
Zea mays, corn		C_4	0.0033
Sorghum vulgare, sorghum		C_4	0.0036
Triticum aestivum, bread wheat		C_3	0.0020
Hordeum sativum, barley		C_3	0.0019
Oryza sativa, rice		C_3	0.0015

Sources: Calculated from G. R. Squire (1990), *The Physiology of Tropical Crop Production* (Wallingford, U.K.: CAB International); A. H. Fitter and R. K. M. Hay (1987), *Environmental Physiology of Plants,* 2nd ed. (London: Academic Press); R. K. M. Hay and A. J. Walker (1989), *An Introduction to the Physiology of Crop Yield* (Harlow, U.K.: Longman).

Figure 10.7 Pineapple, a commercially cultivated CAM plant. A field of pineapple in Hawaii. *Source:* Courtesy of K. G. Rohrbach.

They open their stomates at night and use PEP carboxylase and respiratory energy to absorb CO_2 by making organic acids that are stored in cell vacuoles. In the daytime the stomates close to prevent water loss and the CO_2 is released from the acids to allow C_3 photosynthesis. If the average temperature at night is 10°C but 45°C during the day, the nocturnal opening achieves a water saving of 95%. This explains why CAM plants are the dominant perennial plants of the hot semideserts of the world.

10.6 Plants make a dynamic tradeoff of photosynthetic efficiency for photoprotection.

All crops are inherently capable of photosynthetic efficiencies for carbon reduction that closely approach theoretical limits. With two photosystems operating in tandem (see Figure 10.2), this high efficiency is possible only because the amount of light absorbed by the antennae serving the two photosystems is closely balanced. However, actual average CO_2 assimilation efficiency is much lower and the degree of inefficiency is often much greater than can be accounted for just by photorespiration, one of the major factors that limits efficiency (see discussion earlier).

Early in the day, the light levels are low and photosynthetic carbon reduction responds linearly to increasing light intensity (**Figure 10.8**); however, crop plants frequently encounter light intensities that exceed their photosynthetic capacity (Figure 10.8; see red shaded region). Exactly what constitutes excess light for a leaf depends on its instantaneous environmental conditions and can vary over quite a wide range of light levels. For example, irrigated field-grown sunflower is typical of C_3 crop plants, exhibiting maximum photosynthetic capacity during midmorning, with photosynthesis declining throughout the afternoon as stomates partially close in response to declining leaf water status. Thus, even under conditions that may not generally be considered stressful, partial stomate closure can

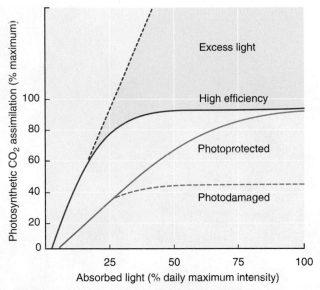

Figure 10.8 A schematic of the light response of photosynthesis in a leaf in the high-efficiency state (blue curve), in the photoprotected state (red curve), and the photo-damaged state (red dashed curve). In the high-efficiency state, the quantum requirement of CO_2 assimilation approaches the theoretical limit in the linear portion of the light response curve. As irradiance is increased further, the rates of the component photosynthetic reactions become limiting, causing photosynthesis to light saturate and light to be in excess (dark green shaded area). Excess light induces a photoprotective response that dramatically lowers or even eliminates the threat of damage by excess light via a thermal dissipation process. If the aggregate photoprotective capacity is exceeded under severe circumstances, photodamage to the photosynthetic apparatus can cause a persistent and sometime permanent inefficiency. *Source:* Modified from C. B. Osmond (1994), What is photoinhibition? Some insights from comparisons of sun and shade plants, in N. R. Baker and J. R. Bowyer, eds., *Photoinhibition in Photosynthesis—From Molecular Mechanisms to the Field* (Oxford, U.K.: Bios Scientific Publishers), p. 4, Figure 1.2.

substantially restrict CO_2 entry into leaves, causing even moderate light levels in the top layers of leaves of a crop to exceed the photosynthetic capacity for CO_2 reduction.

A Dynamic Process Enabling Leaves to Regulate Thermal Dissipation of Excess Absorbed Light Trades Photosynthetic Efficiency for Photoprotection. When environmental conditions prevent photosynthetic and photorespiratory carbon metabolism from using all the light absorbed by a leaf, the likelihood of biologically damaging oxidants forming through photosynthetic processes increases dramatically. Although some plants can reduce the amount of incident light they absorb, through strategic leaf and chloroplast movements, such strategies for rapidly reducing light absorption appear to play only a minor role in the challenge of coping with excess light. Instead, plants rely on a dynamic tradeoff between photosynthetic efficiency and photoprotection. Under circumstances that occur essentially on a daily basis, over half the light absorbed by photosystem II (PSII) chlorophylls in healthy, fully functional leaves can be redirected by a process, operating within the antenna of PSII, that harmlessly discharges excess light energy as heat.

The blue curve of Figure 10.8 illustrates that plants achieve the highest efficiency of photosynthesis at low light when the component reactions of electron transfer, ATP formation, and carbon metabolism are least limiting. Photosynthetic efficiency drops at higher light levels as these reactions become increasingly limiting and as the amount of excess light increases,

Figure 10.9 Photodamage in a mango orchard. A commercial mango orchard in Israel following four sequential cold (0.5–2°C) nights in February. Leaf chlorosis and necrosis is primarily localized on branches exposed to highest light levels (south facing) on the days following the cold nights. *Source:* Courtesy of Yosepha Shahak, Volcani Institute, Bet Dagan, Israel.

portending an ever greater *threat* of photodamage. By greatly reducing the amount of excess excitation within the antenna, thermal energy dissipation decreases photosynthesis efficiency throughout most of the light intensity profile without lowering the photosynthesis rate at the very highest intensities (Figure 10.8, see red curve). However, the photoprotected state is dynamic and within minutes of a reduction in excess light there is a compensatory reduction in thermal protection, restoring efficiency at low irradiance to nearly maximum values. In contrast, when photoprotective measures are inadequate to prevent photodamage (Figure 10.8, red dashed curve) both the efficiency at lower intensities and the maximum rate of photosynthesis at light saturation are decreased, and these inhibitions do not rapidly recover as the excessive light condition recedes.

How Effective Is the Tradeoff of Efficiency for Photoprotection? Although photodamage has been documented in crops grown outside of their ancestral geographic range (**Figure 10.9**), the vast majority of plants in native habitats and even most crops under cultivation deal successfully with excess light, avoiding photodamage even under daunting environmental challenges. Clearly, photosynthesis is a balancing act in which plants trade photoprotection for photosynthetic efficiency. Apparently evolution has refined the photosynthetic apparatus by emphasizing high efficiency at limiting light with a failsafe that ensures that the plant can endure high intensities without accumulating photodamage. Under intense light (for example, midday at the top of a crop canopy) other factors, such as maintenance of water status, take physiological precedence over maximizing photosynthesis. Although the tradeoff between efficiency and photoprotection is clear, it is less apparent how well the dynamic range of the tradeoff is suited for agricultural environments and productivity goals. Genetic variation in the ability of crop plant varieties to maintain photosynthetic efficiency at somewhat higher irradiances may prove an important factor in the search to improve photosynthetic productivity of crops.

10.7 Abiotic environmental factors strongly limit photosynthetic efficiency and crop productivity.

Plants frequently face periods of environmental stress sufficient to limit their growth and reproductive capacity. In natural populations, these periods represent a major force leading to genetic change, and those individuals with superior characteristics leave more progeny in succeeding generations. In agriculture similar forces are at work, but because of human intervention, successful genetic traits are not so closely linked to reproductive success, and so the meaningful selection criterion becomes production on a land area basis. However, because the plant's reproductive structures are often the sought-after agricultural product, there are clearly parallels between how environmental factors affect natural plant communities and how they affect crops.

A comparison of average and record yields reveals the large impact of unfavorable environments on crop production. The maximum production potential of a particular crop cultivar is called its *genetic potential,* and is set by the genetic makeup of the plant that can be expressed in optimum, nonlimiting growing environments. Although environmental constraints can never be entirely eliminated, a useful benchmark for estimating the genetic potential of a crop is the highest yield ever been attained under conditions of minimal unfavorable environmental conditions. A comparison of the world record yields versus the average yields for eight major crops of the temperate and tropical regions of the world show a stunning disparity. Even for rice, where the disparity is smallest, average yields are only 40% of record levels in temperate regions and below 30% in the tropics (**Table 10.2**). These comparisons illustrate not only that the vast majority of agricultural environments greatly reduce yields but in addition that the largest shortfalls from the genetic potential of crops occur in the tropics and subtropics.

Table 10.2 Average and world record economic yields of crops from temperate and tropical areas

Crop	Temperate Yields (t/Ha)[a]		Tropical Yields (t/Ha)	
	Average	Record	Average	Record
Potato	18.1	126.0 (U.S.)	8.7	60.0 (Central America)[b]
Sweet potato	13.6	65.0 (U.S.)	6.9	—
Rice	4.1	10.5 (Japan)	2.0	7.4 (Asia)[b]
Corn	4.0	22.2 (U.S.)	1.4	12.9 (Zimbabwe)
Wheat	3.0	14.1 (U.S.)	1.4	10.3 (Asia, Zimbabwe, Central America)
Sorghum	2.3	20.1 (U.S.)	1.2	10.3 (Asia)[b]
Groundnut	1.7	8.6 (U.S.)	1.0	9.6 (Zimbabwe)
Soybean	1.6	7.3 (Japan)	1.0	4.8 (Zimbabwe)
Average % of record yield	20.8	100	15.7	100

[a]Subtropical and temperate zones are considered temperate, whereas the zones between 23.5° north latitude and 23.5° south latitude (excluding dry areas) are considered tropical.
[b]Asia is represented by the Philippines, Thailand, and India. Central America is represented by Mexico and Colombia.

Source: Modified from J. S. Boyer (1987), in W. R. Jordan, ed., *Water and Water Policy in World Food Supplies* (College Station: Texas A&M University Press), p. 234, Table 1.

Table 10.3	Record yields, average yields, and yield losses resulting from pests and unfavorable abiotic conditions for major U.S. crops			
	Yields (t/Ha)		**Average Losses (t/Ha)**	
Crop	**Record**	**Average**	**Pests[a]**	**Abiotic Stress[b]**
Corn	19.3	4.6	2.5	12.2
Wheat	14.5	1.9	0.9	11.7
Soybeans	7.4	1.6	0.8	5.0
Sorghum	20.1	2.8	1.3	16.0
Oats	10.6	1.7	1.2	7.7
Barley	11.4	2.1	0.9	8.4
Potatoes	94.1	28.2	15.9	50.0
Sugar beets	121.0	42.6	24.0	54.4
Mean % of record yield	100.0	21.5	11.6	66.9

[a]Cumulative losses due to diseases, insects, and weeds.
[b]Calculated as record yield minus (average yield plus pests loss).

Source: J. S. Boyer (1987), in W. R. Jordan, ed., *Water and Water Policy in World Food Supplies* (College Station: Texas A&M University Press), p. 234, Table 2.

But which environmental limitations are the most important? For eight major U.S. crops, completely eliminating pests (insects, weeds, and disease) would bring the average collective yields up to about one third of the genetic potential of these crops (**Table 10.3**), implying that abiotic (physical) stresses cause the largest portion of the yield suppression. Abiotic factors that diminish yields are numerous, but the most significant ones can be categorized into two broad areas: unfavorable soils and unfavorable climates. Researchers estimate that 12% of the land surface in the United States and perhaps as little as 10% worldwide can provide a soil environment for plants that does not normally limit production. The impact of unfavorable climates is also large and widespread. Inadequate water, excessive water, and cold extract the largest toll in the United States, as judged from insurance payouts by the U.S. government to farmers for crop losses (**Table 10.4**). Inadequate water availability is without question the single most important factor in limiting crop production throughout the world (see Chapter 11).

Abiotic Stresses Limit Photosynthetic Efficiency. Because plant growth is the result of many integrated and regulated physiological processes, it is seldom possible to assign limitations of plant growth to a single process. Certainly in the field a number of limiting processes control plant growth simultaneously and to varying extents. However, among the diversity of overlapping factors are likely to be some central physiological processes that are highly sensitive/responsive to the environment and are, therefore, dominant in determining plant response to stress. In many instances, the dominant process is photosynthesis. Plant growth as biomass production is in fact a measure of net photosynthesis integrated over time, so factors limiting plant growth are unavoidably the same factors that limit net photosynthesis.

When soils are too dry to replace water loss from the leaves, plants respond by partially or fully closing their stomates, thereby lowering the rate of water loss and improving water use efficiency, but also lowering the rate at which CO_2 can enter the leaf. Strong photoprotective measures, most importantly the thermal dissipation of energy within the light-harvesting antennas, trade off photosynthetic efficiency (productivity) for photoprotection. However, as shown in Figure 10.8, very severe or prolonged stresses can exceed the photoprotective capacity of plants and damage to the photosynthetic apparatus can occur. This creates a persistent decrease in photosynthetic efficiency (that is, the plant does not rapidly recover when the stress is removed). For example, when sunflower plants grown under well-watered conditions are suddenly deprived of water, photosynthesis decreases dramatically, as shown in Figure 10.10. At atmospheric CO_2 levels, light-saturated photosynthesis was inhibited 75% in this experiment (compare points A and B in **Figure 10.10**). This decrease is not caused simply by closure of the stomates, and in such cases photosynthesis never fully recovers. Only new leaves that emerge after the stress recedes are again capable of high-efficiency photosynthesis.

Certain plants have a significant capacity for adapting to stress. If such plants are exposed to a mild stress, they can better withstand a subsequent more severe stress. This phenomenon is called *acclimation*. Thus, sunflower plants grown under moderately dry conditions are able to acclimate so that they photosynthesize more rapidly under subsequent drought conditions (see **Figure 10.11**). Understanding how plants cope with stressful environments will help breeders improve crop species more quickly. Drought stress very rapidly inhibits cell elongation, preventing growth. Cell elongation can be sustained if the growing cells contain more soluble products of photosynthesis. An important aspect of acclimation to drought is the accumulation of photosynthate (soluble sugars and amino acids) in the regions of cell expansion. Thus the response of photosynthesis to drought is an important aspect of acclimation.

Table 10.4	Causes for crop insurance payments in the United States, 1939 to 1978

Cause of Crop Loss	Proportion of Payments (%)
Drought	40.8
Excess water	16.1
Cold	13.8
Hail	11.3
Wind	7.0
Insect	4.5
Disease	2.7
Flood	2.1
Other	1.5

Source: J. S. Boyer (1987), in W. R. Jordan, ed., *Water and Water Policy in World Food Supplies* (College Station: Texas A&M University Press), p. 235, Table 6.

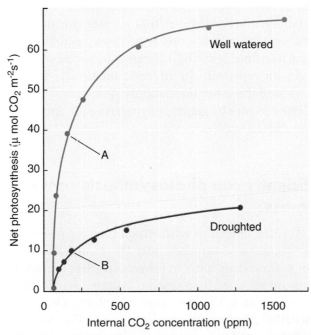

Figure 10.10 The dependence of photosynthesis in non-drought–acclimated sunflower on the intercellular CO_2 concentration when measured in a well-watered and in a drought condition. Point A shows the intercellular CO_2 concentration under normal atmospheric ambient conditions of 350 ppm in a well-watered condition. Point B shows that although the intercellular CO_2 concentration was actually a bit higher in the drought plant, photosynthesis was nonetheless significantly inhibited. *Source:* Modified from Figure 3 in D. R. Ort and J. S. Boyer (1985), Plant productivity, photosynthesis and environmental stress, in B. G. Atkinson and D. B. Walden, eds., *Changes in Eukaryotic Gene Expression in Response to Environmental Stress* (Orlando, FL: Academic Press), p. 288.

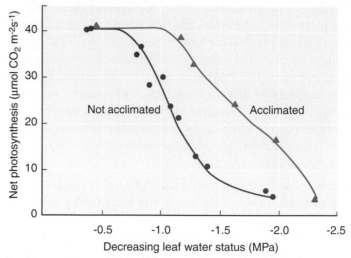

Figure 10.11 The dependence of photosynthesis on leaf water status in drought-acclimated (▲) and nonacclimated sunflower (●). Acclimated plants were subject to a moderate drought for two weeks prior to the measurements. *Source:* Modified from M. A. Matthews and J. S. Boyer (1984), Acclimation of photosynthesis to low leaf water potentials. *Plant Physiology* 74, 161–166.

Unfortunately, not all crop plants can measurably acclimate to all types of stresses. Thus although many wheat varieties can harden to withstand temperatures well below freezing, other important crop plants such as maize, soybean, and tomato have rather little ability to acclimate even to cool (above freezing) temperatures. These chilling-sensitive species were imported into temperate regions from warmer climates where they originated. Thus although genetic variability in osmotic regulation capabilities is providing powerful sources of drought tolerance, chilling tolerance in maize is not likely to be found within its ancestors' genetic repertoire. In this case, the need to understand the underlying biological mechanism of the lower temperature intolerance is even more crucial, because scientists need to introduce by genetic engineering the appropriate suite of genetic traits from sources outside the species.

10.8 How efficiently can photosynthesis convert solar energy into biomass?

As we noted in Box 10.1, the efficiency with which plants convert intercepted solar energy into plant matter seems to be very low; just 1% in our example. This value was an average over the life of a wheat crop. Early in the season when crops are growing most rapidly, higher efficiencies are achieved for a few days. For C_3 crops, the highest efficiencies are about 3.5% and for C_4 about 4.3%. This section explores why even these record numbers are so low and whether photosynthesis in crops is really as inefficient as you might at first assume from such numbers.

Only about 50% of solar energy can be used in crop photosynthesis, because the other half is in the near infrared part of the spectrum and is not energetic enough to drive photosynthesis. Leaves reflect some of the photosynthetically active light. In C_3 plants, the minimum number of light photons required to fix one molecule of CO_2 is 8, regardless of wavelength within the photosynthetically active spectrum; that is, a red photon has the same effect as a violet photon. However, a violet photon contains about 70% more energy than a red photon. The additional energy of the violet photon is lost instantaneously on

Figure 10.12 Inactive absorption by nonphotosynthetic pigments. The purple segments in this young maize plant grown with insufficient phosphorous are caused by the accumulation of high levels of anthocyanin pigments within the leaf cell vacuoles. Plants frequently respond to a variety of stress conditions by synthesizing high levels of anthocyanins and other water-soluble nonphotosynthetic pigments. These are the same compounds that impart fall color in the leaves of deciduous trees. *Source:* Courtesy of Howard Woodward, South Dakota State University.

absorption as heat, representing an intrinsic photochemical inefficiency. Other pigments, such as the anthocyanins, may absorb light but cannot pass the energy on to photosynthetically active pigments (**Figure 10.12**). This type of light absorption is called *inactive absorption*. In common with all syntheses, transfer of energy from the short-term energy stores of the chloroplast, ATP and NADPH, to the long-term energy store of most plants, carbohydrate, proceeds at the cost of energy. In C_3 plants, the efficiency of this conversion is about 35%. Because the C_4 pathway requires more ATP, carbohydrate synthesis in these plants has a lower efficiency. However, this difference is offset in C_3 plants by photorespiration, which converts a portion of this carbohydrate back to CO_2. Finally, mitochondrial respiration—necessary for synthesizing new tissues and maintaining existing tissues in all plants—uses about 40% of the remaining energy. All these losses encountered during the energy transformations of photosynthesis are summarized in **Table 10.5,**

Table 10.5 Energy losses during conversion of intercepted solar energy into biomass

	% Remaining Lost at Each Stage		Remaining (kJ)	
Intercepted solar energy			1,000	
Outside the photosynthetically active waveband	50		500	
Reflected	10		450	
Inactive absorption	4		432	
Photochemical inefficiency	20		346	
	C_3	C_4	C_3	C_4
Inefficiency of carbohydrate synthesis	66	72	118	97
Photorespiration	30	0	83	97
Dark respiration	40	40	50	58
% Remaining			**5%**	**5.8%**

The processes are in sequence of the stages from light interception through to the final cause of loss, dark respiration. What remains is the energy in biomass. The first column shows the percentage of the remaining energy that is lost at that stage, the second shows how much of 1,000 kJ of intercepted solar energy remains after each process. For example, after we have taken account of losses from energy outside of the photosynthetically active waveband, reflection and inactive absorption, 432 kJ of the original 1,000 kJ of intercepted solar energy remains, photochemical inefficiency results in a loss of 20% of this remaining energy, decreasing the remainder to 346 kJ. The energy required for carbohydrate synthesis and lost during photorespiration differ between C_3 and C_4 photosynthesis, so the columns are subdivided accordingly.

Source: Adapted from C. L. Beadle, S. P. Long, S. K. Imbamba, D. O. Hall, and R. J. Olembo (1985), *Photosynthesis in Relation to Plant Production in Terrestrial Environments* (Oxford, U.K.: United Nations Environment Programme/Tycooly International), p. 156.

which illustrates that in C_3 plants a maximum of about 5% of the total incoming solar energy can in theory remain as biomass, and in C_4 plants, about 5.8%.

So why do plants in the absence of pests, diseases, and environmental stress, fail even to achieve these modest efficiencies? Even under optimal growing conditions it is impossible to remove all environmental limitations. For example, even when soil moisture is high, excessive evapotranspiration can outpace the vascular system's ability to deliver enough water for leaves to sustain fully open stomates. In addition, photosynthesis must operate over a wide range of solar energy input each day (Figure 10.1b). It operates at maximum efficiency only at low light. In the next section we discuss how redesigning the arrangement of crop leaves improves the efficiency with which plants of some crops use photosynthetically active light within the canopy.

10.9 Opportunities exist for improving photosynthesis performance.

Earlier (Equation 10.1) we showed that the major opportunity that remains for increasing maximum yields of major crops is increasing conversion efficiency (ε_i). Stressful growing environments can and do substantially decrease conversion efficiency. Does this represent an important opportunity to make improvements, and will this translate into higher crop yield? Another major cause of inefficiency is that leaf photosynthesis responds nonlinearly to increased solar energy. In C_3 crops, leaf photosynthesis is saturated at solar energy levels of about one quarter of maximum full sunlight ((Figure 10.8), so any solar energy above this level is wasted. In this section we look at different potential approaches to overcoming these constraints and achieving a higher ε_i.

The Distribution of Light Among Leaves Can Be Altered. A mature, healthy crop may have three or more layers of leaves; that is, above each square meter of soil may be 3 square meters of leaves. This ratio is described as a leaf area index of 3. If the leaves are roughly horizontal (**Figure 10.13a**, Plant X), the uppermost layer will intercept most of the direct light, about 10% may penetrate to the next layer, and 1% to the layer below that. With the sun overhead, the photosynthetically active energy intercepted per unit leaf area by a horizontal leaf at the top of a plant canopy would be $900\,J\,m^{-2}\,s^{-1}$, or about three times the amount required to saturate photosynthesis (**Figure 10.13b**). Thus at least two thirds of the energy intercepted by the upper leaves is wasted. A better arrangement for an agricultural situation would be for the upper leaf layer to intercept a smaller fraction of the light, allowing more to reach the lower layers. This is achieved when the upper leaves are more vertical and lowermost leaves are horizontal, as in the example of Plant Y (Figure 10.13b). For a leaf at a 75° angle with the horizontal, the amount of light energy intercepted per unit leaf area would be $300\,J\,m^{-2}\,s^{-1}$, just enough to saturate photosynthesis, but the remaining direct light ($600\,J\,m^{-2}\,s^{-1}$) would penetrate to the lower layers of the canopy. By distributing the energy among leaves in this way, in full sunlight Plant Y would have over double the efficiency of solar energy use that Plant A has (**Figure 10.13c**). Researchers have developed mathematical models to design optimum distributions of leaves for maximizing efficiency, which have been used as guides for selecting improved crops. This approach has been a major factor in improving rice productivity. Older varieties with more horizontal leaves such as Plant X have been replaced by newer varieties that have been bred to have more vertical leaves in the top layer, such as Plant Y (**Figure 10.14**).

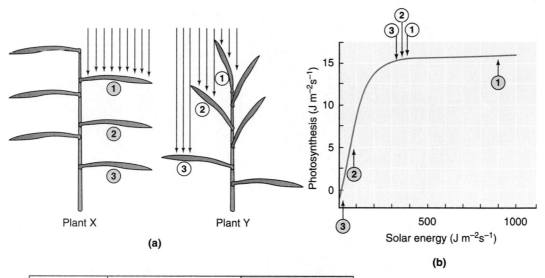

Plant X   Plant Y

(a)

(b)

	Solar energy		Photosynthesis	
	$J\,m^{-2}\,s^{-1}$		$J\,m^{-2}\,s^{-1}$	
Plant Leaf layer	X	Y	X	Y
①	900	370	16	16
②	90	330	8	15
③	10	300	1	14
Total	**1,000**	**1,000**	**25**	**45**
Efficiency			*0.025*	*0.045*

(c)

Figure 10.13 Light distribution in a crop canopy. (a) Plant X has horizontal leaves, such that the upper layer ① will intercept most on the incoming solar energy, shading the lower layers ② and ③. Plant Y has vertical leaves at the top, becoming more horizontal near the bottom. This arrangement spreads the solar energy more evenly between layers. **(b)** Photosynthesis for a leaf, in terms of energy trapped in stored carbohydrate, is plotted against solar energy. Arrows below the curve indicate the average amounts of solar energy at the three leaf layers of Plant X and arrows above the three leaf layers of Plant Y. **(c)** From the graph of **(b)**, the amount of solar energy and the photosynthesis for each leaf layer, and their totals, for the two plants are given. Note that by spreading the same amount of solar energy more evenly among its leaves, Plant Y can achieve almost double the rate of photosynthesis of Plant X.

Can Rubisco Be Altered? In considering how to redesign plant canopies, we noted that amounts of solar energy well below full sunlight saturate photosynthesis at the leaf level (Figure 10.13b). Referring back to Figure 10.1b you see that for most of a sunny day solar energy exceeds the amount needed to saturate leaf photosynthesis. Are there other approaches to using this excess energy? The response of photosynthesis to solar energy is hyperbolic, rising rapidly with increases in light at low levels, but saturating at about 25% of full sunlight. Why does this saturation occur? Several analyses suggest that a major limitation is the amount of Rubisco in the leaf. So why not just increase the amount of Rubisco per unit of leaf area? Rubisco is already the most abundant protein in crop leaves. It is doubtful that leaves can profitably divert much more of their resources into making this one enzyme. An alternative would be to increase the efficiency of this enzymatic

Figure 10.14 Modern rice cultivars have been selected for vertical leaf orientation for the uppermost leaves (a), thereby improving light penetration into the canopy. This change in leaf orientation, compared to the more horizontal leaf position of older varieties **(b)**, has been a major factor in improving the productivity of rice. *Source:* Courtesy of Dr. Shannon Pinson, USDA-ARS, Rice Research Unit, Beaumont, Texas.

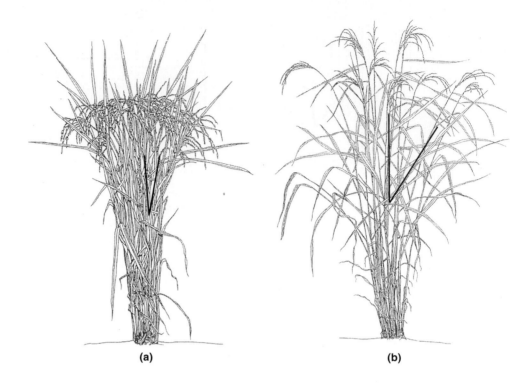

(a)   (b)

reaction. Build a better Rubisco! This would allow higher maximum light-saturated rates of photosynthesis without any additional investment into Rubisco. Researchers have shown that forms of Rubisco in some photosynthetic bacteria have catalytic rates four times faster than those from land plants. Modifying the catalytic site of crop Rubisco to that of the bacterial form is a possible approach to increasing efficiency of light use.

Can Photorespiration Be Suppressed? About 30% of the carbohydrate formed in C_3 photosynthesis is lost via photorespiration. The amount increases with temperature, so photorespiration is a particularly important inefficiency for C_3 crops in tropical climates and during hot summer weather in temperate climates. As you have seen, photorespiration results from the apparently unavoidable oxygenation reaction of RuBP by Rubisco. Beyond this point the purpose of photorespiratory metabolism is to recover the carbon diverted into this pathway. Blocking photorespiratory metabolism downstream of Rubisco simply results in this carbon entering a dead-end metabolic pathway. Indeed, mutants that lack any photorespiratory enzymes die, unless they are grown at low oxygen or at very high CO_2 to inhibit oxygenation of RuBP. The only remaining prospect for decreasing photorespiration, then, is decreased oxygenation. Forms of Rubisco from different plant species differ in their tendency to catalyze the oxygenation reaction. The Rubiscos of some red algae carry out the oxygenation reaction only half as often as the Rubiscos of crops. If the genes encoding such algal Rubisco proteins could be transferred successfully to crops, or if crop Rubisco could be redesigned to mimic the red algal Rubisco, then crop photorespiration might be more than halved.

Commercial growers of greenhouse crops take advantage of the properties of Rubisco to suppress photorespiration and obtain higher yields. In this closed environment, the CO_2 concentration is raised by injection of CO_2 to three or four times the outside concentration. This inhibits the oxygenation reaction of Rubisco, increasing photosynthetic efficiency and final yield. At the present time the global concentration of CO_2 is rising and this too may diminish photorespiration, but atmospheric change also includes many potentially negative effects for crops and natural ecosystems, which we examine in the next section.

10.10 Global change interacts with global photosynthesis.

Monitoring of atmospheric CO_2 concentrations at sites far away from any industrial sources of CO_2, such as the mountains of the Big Island in Hawaii and in Antarctica, has shown a steady increase since the late 1950s. Bubbles of air trapped in the ice of glaciers in Antarctica and Greenland can be dated and show that this increase began in the early 1800s, the start of the industrial era, and has accelerated dramatically over the last three decades (**Figure 10.15**). The increase is proportional to the amounts of CO_2 humans have released into the global atmosphere by burning fossil fuels and forests. In 1800, the concentration of CO_2 in the atmosphere was about 250 parts per million parts of air (ppm), in 2001 it reached 370 ppm, and by 2100 some researchers expect it to be between 600 and 700 ppm. If rising CO_2 were the only change occurring in the atmosphere, this might be expected to benefit crops. A doubling of CO_2 concentration would roughly halve photorespiratory losses in C_3 crops. Photosynthesis would be further increased, because the present CO_2 concentration is insufficient to saturate Rubisco, with its surprisingly low affinity for CO_2. Finally, an increase in the atmospheric CO_2 concentration would allow plants to maintain the same photosynthetic rate, with a higher stomatal resistance, so less water would be lost for each CO_2 molecule assimilated by the plant. Elevating CO_2 concentration in greenhouses has been shown to increase yields of the major crops. However, enclosing crops such as wheat in a greenhouse affects growth, production, and appearance. Such plants may respond very differently to environmental treatments from plants in the field. But unlike nitrogen fertilizers or pesticides, CO_2 added in the field simply blows away. So how can researchers discover the effects of the future atmospheric CO_2 concentration on field crops, without using an enclosure?

A new series of experiments being conducted at various locations around the world take advantage of natural air movements to enrich crops with CO_2 in the open air in a precise,

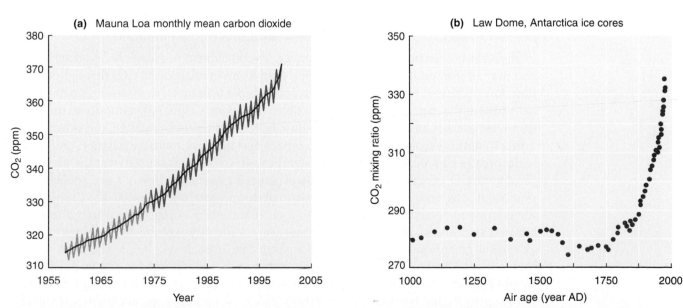

Figure 10.15 Human activities initiated a rapid increase in the atmospheric CO_2 concentration. (a) The concentration of CO_2 in the air on Mauna Loa mountain, Hawaii. The year-by-year increase at a remote location far from any source of major industry, first alerted the world to the fact that fossil fuel burning and tropical deforestation were affecting the composition of Earth's atmosphere. **(b)** Ice formation on glaciers traps bubbles of atmosphere; as ice layers are added each year, they preserve a record of past atmospheric composition. Analysis of bubbles from deep layers of ice at Antarctica now show that the change in Earth's atmosphere began about 1750, corresponding with the start of the Industrial Revolution, and has greatly accelerated since 1950.

Figure 10.16 One of four 20-meter diameter Free-Air Carbon dioxide Enrichment (FACE) rings in a soybean field at the University of Illinois, Urbana. Computer feedback control releases CO_2 from the vertical pipes into the wind so that the concentration within the ring is held at a constant elevated CO_2 concentration 50% higher than today, simulating the concentration of the global atmosphere of the future.

controlled manner. These experiments use a new technology: Free-Air Carbon dioxide Enrichment (FACE). FACE consists of rings of pipes that release CO_2 into the air (**Figure 10.16**). A computer continuously measures wind speed and direction, and the CO_2 concentration within the ring determines which pipes release CO_2 and how much they release. The CO_2 is released into the naturally moving air so that the concentration within the ring is elevated to a constant level. The advantage is that the crop is not enclosed and microclimate is unaltered, simulating the atmosphere of the future without otherwise altering the environment. The world's first such system, in Arizona, examined wheat crops over four years. Increase of CO_2 concentration to 550 ppm, the level possible for the second half of this century, resulted in a 28% increase in rates of photosynthesis; however, grain yield only increased about 10%. One urgent task will be to select cultivars of crops that can translate the increased photosynthesis allowed by rising CO_2 into increased yield. But atmospheric change may have adverse effects on crop production.

Rising atmospheric CO_2 concentration has several indirect negative effects on agriculture and may alter natural plant communities. Crops grown in elevated CO_2 throughout their life show decreased nitrogen and protein contents per unit mass, so increased quantity may be gained at the expense of nutritional value. Also, plants differ substantially in their ability to use the additional CO_2—some show large production increases, and others show none. In natural ecosystems, such differences may alter competitive balance, and lead to widespread extinctions. Because photosynthesis in C_4 plants is saturated at the present atmospheric CO_2 concentration, this group of plants is expected to lose out to C_3 plants where photosynthesis has the potential for substantial stimulation by increased CO_2. Carbon dioxide is a potent "greenhouse gas," that is, a gas that traps the long-wave infrared radiation emitted by Earth's surface. This trapping warms the atmosphere in the same way the glass of a greenhouse warms the atmosphere that it encloses, hence the term "greenhouse effect." As a result, average global temperatures are expected to rise by 2–6 degrees C, and probably more, in this century. This may allow crop production at higher latitudes than at present, but is expected to depress yields in warm and tropical climates. It will also alter rainfall patterns, portending increased drought in some areas of the globe.

Is photosynthesis providing some protection against atmospheric change? **Figure 10.17** shows that Earth's atmosphere contains about 775 gigatons (Gt) of carbon (1 gigaton = 1 billion metric tons or 10^{15} g). Each year photosynthesis removes about 101.5 Gt of this carbon to the oceans and to land, and each year the respiration of plants, animals, and microorganisms releases about 100 Gt back into the atmosphere. This balance was stable for centuries, until humans started to release significant additional carbon to the atmosphere from fossil fuel use. Today fossil fuel combustion annually adds another 5.5 Gt and forest destruction 1.5 Gt, to the atmosphere. Because this anthropogenic emission of 7 Gt of carbon is about 1% of the total atmospheric concentration, the atmospheric concentration would be expected to rise at about 1% per year. However, the actual rise is about 0.5% per year, therefore half of the CO_2 that people add to the atmosphere is being absorbed somewhere—this somewhere has been termed the "missing sink." Photosynthesis is apparently halving the rate of rise in CO_2 that would otherwise occur, and at present is protecting Earth from a more rapid rate of atmospheric and climate change. It is important to know where this photosynthesis occurs, to assess how long it may be sustained and to determine priorities for protecting the ecosystems that are providing this critical environmental benefit. How can people measure the photosynthesis and carbon dioxide exchange of landscapes, large tracts of ocean, and the globe? Two techniques are now becoming widely used within international networks.

The first technique measures vertical wind speed and CO_2 concentration differences close to Earth's surface, to determine the amount of CO_2 being absorbed by surface vegetation. In the second technique, chlorophyll fluorescence–measuring devices lowered from research ships or from automated buoys in the ocean, are providing a detailed picture of the dynamics of photosynthesis in the oceans. This research into global photosynthesis is beginning to reveal where some of the additional carbon is and is not being consumed. Originally scientists assumed that most of the additional carbon was deposited into the

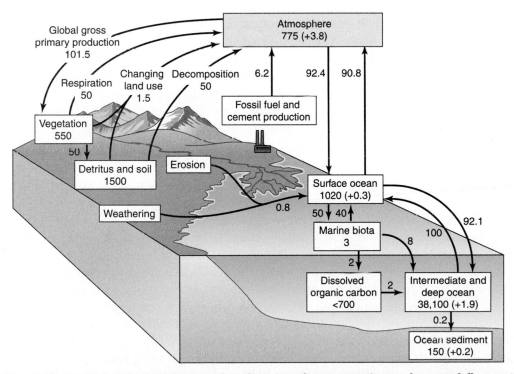

Figure 10.17 The global carbon cycle, showing the reservoirs and annual fluxes (in gigatons of carbon) averaged over the period 1992–1997. Annual photosynthetic carbon uptake by terrestrial vegetation is 101.5 Gt (labeled Global gross primary production) and 50 Gt by aquatic vegetation, primarily oceanic phytoplankton. By 2001 the reservoir in the atmosphere had risen to 786 Gt. *Source:* Figure adapted and updated by Dr. W. M. Post, Oak Ridge National Laboratory, from Houghton et al. (1995), *Climate Change 1995: The Scientific Assessment. The Intergovernmental Panel on Climate Change* (Cambridge, U.K.: Cambridge University Press).

oceans. However, it is now clear that about half of the "missing sink" is actually on land. Areas of the temperate zone that people have allowed to revert from cropland to forest during the last few decades, appear to be a major sink for this additional carbon. Although tropical deforestation is a source of CO_2, remaining forest and regrowth secondary forest is another strong sink. In contrast, the tropical Pacific Ocean has very low photosynthetic rates despite warm temperatures and a good supply of nitrogen and phosphorus for plankton growth. The major limitation to growth of photosynthetic algae in this region seems to be iron deficiency. Exact identification and monitoring of these sinks will be vital to understanding the areas of the globe in critical need of protection. In addition, there is an urgent need to discover how long these sinks will last and continue to offset a significant portion of humanity's release of carbon dioxide into the atmosphere.

CHAPTER
SUMMARY

All the energy contained in the food people eat is ultimately derived from sunlight through photosynthesis. Whereas solar energy input sets the upper limit on energy that can be transformed into crop yield, the actual yield of energy in food depends on the product of solar input and the efficiency with which the plant transforms solar energy into the harvested product. To convert the transient energy of a photon into stable chemical energy, the photosynthetic apparatus performs a series of energy transforming reactions. The photosynthetic membranes capture light energy and convert it into stable chemical energy in the form of ATP and NADPH. Chloroplasts then use this energy to assimilate atmospheric CO_2 to sugars while extracting electrons and protons from water to liberate oxygen. Sucrose and other photosynthate formed in leaves are exported to heterotrophic plant organs to provide energy for growth and to be stored. In addition to being the basic energy currency of plants, sucrose and other photosynthate also act as regulatory compounds able to signal the balance between demand and photosynthetic production. A significant amount of reducing power and ATP can be diverted from CO_2 reduction to other processes within the chloroplast that are essential to the growth and development of plants, such as nitrate and sulfate assimilation, the regeneration of photoprotective compounds, and the biosynthesis of proteins and lipids.

Crop plants daily encounter light energy that exceeds their photosynthetic capacity. Plants have evolved a sophisticated set of regulatory photoprotective measures that safeguard plants but at significant cost to photosynthetic efficiency. Significant improvements in production of some crops may be possible by introducing steeper leaf angles, thereby reducing the amount of excess light at the top of canopies and allowing greater light penetration to otherwise shaded leaves. Plants also frequently encounter a variety of environmental stresses that exacerbate the need to engage photoprotective measures as well as diminish photosynthetic efficiency in other ways. For eight major U.S. crops, these stress conditions conspire to reduce yields >30% below their genetic potential. Water availability is the single greatest constraint on agricultural production because stomates frequently must close to conserve water but in so doing starve photosynthesis of CO_2. There are compelling reasons to believe that understanding how plants cope with stressful environments will accelerate crop improvement.

The increasing CO_2 content of the atmosphere since 1800 has had both positive and negative impacts on photosynthesis and agricultural production, and these impacts can be expected to intensify. Photosynthesis in turn has had an impact on atmospheric change, roughly halving the rate of rise in CO_2 caused by human activities at the present time.

Discussion Questions

1. Discuss the different ways that biomass production could be increased. What would you need to know to calculate if biomass production could ever replace fossil fuels in the United States?

2. Discuss genetic engineering strategies to make photosynthesis and biomass production more efficient.

3. Discuss the evolutionary tradeoff between having stomates wide open to take up carbon dioxide and losing water as a result.

4. Discuss the difference between global warming and climate change. What is the relationship of global warming to anthropogenic atmospheric gases?

5. Discuss the scientific and political aspects of reducing carbon emissions and sequestering carbon (see also Chapter 4).

6. Plants as a whole are autotrophic, but most parts of the plant are not. By combining what you learned about development, discuss the respective roles of respiration and photosynthesis in the plant. Discuss how organs other than leaves assimilate nitrate and sulfate.

7. Discuss the sequence of energy conversions in photosynthesis that are necessary for transforming light energy into the chemical energy of carbohydrate.

8. Discuss the relationship between photoprotection and photosynthetic efficiency.

9. Discuss the possible identities of photosynthetic sinks for carbon dioxide released into the atmosphere by mankind's activities. What are the prospects that each of these sinks will sustain itself and continue to act against an even more rapid rise in atmosphere carbon dioxide content?

10. Discuss the benefit of increased light penetration into a crop canopy. Why might increased light penetration be a disadvantage to some species within a natural plant community?

Further Reading

Allen, D. J., and D. R. Ort. 2001. Impacts of chilling temperatures on photosynthesis in warm-climate plants. *Trends in Plant Science* 6:36–42.

Baker, N. R. 1996. *Photosynthesis and the Environment.* Dordrecht: Kluwer.

Beadle, C. L., and S. P. Long. 1985. Photosynthesis—is it limiting to biomass production? *Biomass* 8:119–168.

Boyer, J. S. 1982. Plant productivity and environment. *Science* 218:443-448.

Evans, L. T. 1998. *Feeding the Ten Billion: Plants and Population Growth.* Cambridge, U.K.: Cambridge University Press.

Intergovernmental Panel on Climate Change. 1996. *Global Climate Change 1995: Economic and Social Dimensions of Climate Change.* Cambridge, U.K.: Cambridge University Press.

Leegood, R. C., T. D. Sharkey, and S. von Caemmerer. 2000. *Photosynthesis: Physiology and Metabolism.* Dordrecht: Kluwer.

McLeod, A. R., and S. P. Long. 1999. Free-air carbon dioxide enrichment (FACE) in global climate change research: A review. *Advances in Ecological Research* 28:1–55.

Niyogi, K. K. 1999. Photoprotection revisited: Genetic and molecular approaches. *Annual Review of Plant Physiology and Plant Molecular Biology* 50:333–339.

Ort, D. R., and C. F. Yocum. 1996. *Oxygenic Photosynthesis: The Light Reactions.* Dordrecht: Kluwer.

Plant Nutrition and Crop Improvement in Adverse Soil Conditions

Idupulapati M. Rao
International Center for Tropical Agriculture

Grant R. Cramer
University of Nevada Reno

Plant growth requires not only carbon dioxide and oxygen from the air but also water and mineral nutrients from the soil. Soil has been called the "placenta of life," because it supplies essential nutrients to all land plants, and the plants in turn feed all the terrestrial ecosystems. Throughout history, humanity's standard of living has depended on the fertility and productivity of the soil. The Fertile Crescent of the Middle East, one of the areas in the world where agriculture originated (see Chapter 13), is considerably less productive today than it was 10,000 years ago. This decrease in productivity is caused, in part, by changes in weather patterns—possibly as a result of deforestation of the Mediterranean basin—and, in part, by failure of the inhabitants to maintain soil productivity. Soil erosion and salinization are accelerated by poor agronomic practices. Mismanagement and neglect of soil can ruin the arable land, which is a fragile and precious resource. The Harappan civilization in western India, Mesopotamia in Asia Minor, and the Mayan culture in Central America all collapsed partly because of soil degradation. Maintaining productive soils should be one of society's important goals.

11.1 Soil is a vital, living, and finite resource.

The formation of soil is a long and complex process involving breakdown of the parent rock into small mineral particles, chemical modification of these particles, and finally continuous addition and decomposition of organic residues from plants, animals, and microorganisms. Solid rock is continually broken down into small particles, a process termed *weathering*. Heating and cooling of rocks, freezing and thawing of water that seeps into cracks, running water, the scouring action of winds carrying small particles, glaciers that creep over the rocks, grinding them into small particles, all contribute to the weathering process.

Physical forces break up rocks into smaller particles, and chemical forces can change their chemical properties. The more a rock is fractured by physical weathering, the faster chemical weathering will occur. Earth's crust contains more than 90 different elements that

exist in certain combinations called *minerals* (**Table 11.1**). These minerals usually form small crystalline grains that become cemented together to form rocks. Many minerals are specific combinations of silicon, aluminum, iron, and oxygen, because these four elements are by far the most abundant in Earth's crust.

Once the *parent rock* has been broken down into the individual *mineral grains*, these grains are differentially affected by temperature, rainfall, and vegetation. Some, such as quartz grains, are not changed, but are slowly broken into even smaller particles, whereas others, such as feldspar or mica, are actually modified. Feldspar reacts with the carbonic acid in rainwater and forms soluble potassium carbonate and kaolinite, a clay mineral. Mica is even further degraded. It falls apart into potassium carbonate, oxides of iron and aluminum, and clay. This complete dissolving of certain components of the mineral particles is an important aspect of weathering. Indeed, the minerals must be dissolved in the soil solution before plants can take them up. Once the minerals have been dissolved, they can remain in solution, bind to the outside surface of the soil particles, or react with other dissolved minerals to form insoluble compounds.

Soil formation requires the accumulation of soil particles. Both water and wind can carry soil particles from their site of formation to other regions. For example, some of the soils that are the most agriculturally important today formed 10,000 years ago, when winds deposited enormous amounts of clay and silt particles in certain areas. But just as these factors help to form soils, so too they can remove them, in the process of *erosion*. Plants play an important role in preventing erosion, because their roots hold the soil together—a lesson painfully learned in the United States, in Oklahoma and Texas, during the 1930s "dust bowls." Whenever the soil is not covered by plants, winds can whip up huge dust storms, carrying away the soil and depositing it elsewhere.

Table 11.1	Some primary minerals
Mineral Group	**Typical Composition**
Mica	$KH_2Al_3(SiO_4)_3$
Feldspar	$KAlSi_3O_8$
Quartz	SiO_2
Iron oxide	Fe_2O_3
Carbonate	$CaCO_3$

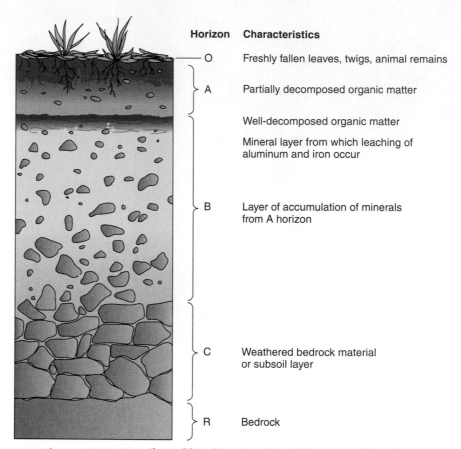

Horizon	Characteristics
O	Freshly fallen leaves, twigs, animal remains
A	Partially decomposed organic matter
	Well-decomposed organic matter
	Mineral layer from which leaching of aluminum and iron occur
B	Layer of accumulation of minerals from A horizon
C	Weathered bedrock material or subsoil layer
R	Bedrock

Figure 11.1 A soil profile, showing the different soil horizons.

A layer of soil particles may form rapidly or slowly, depending on the nature of the parent rock, weather, vegetation, and topography. Level terrain with forest vegetation, with parent rock that is easily broken down, and with a warm, humid climate, all favor rapid soil formation. High rainfall and high temperatures promote rapid chemical weathering, because the downward movement of water carries dissolved mineral components to lower layers of the soil and also because chemical reactions go faster at higher temperatures. The percolating water also leaches acids out of the decaying plant residues. Decaying needles of conifers produce more acid than leaf litter of deciduous forests, which in turn are more acidic than the decay products of prairie grasses. Thus, the type of vegetation affects chemical weathering.

When rain percolates into the soil, there is a slow downward movement of dissolved minerals, bits of organic matter, and the smallest mineral particles. Once formed, a mature soil usually shows at least three distinct layers or horizons (see **Figure 11.1**). The A horizon, or topsoil, is the layer richest in the decaying organic matter and dissolved minerals that sustain plant growth. On the average, it is no more than 20 to 30 cm thick. Below the topsoil is the B horizon, or subsoil. Minerals leached out of the A horizon accumulate here. Most of a plant's roots are in the topsoil, but good farming practice often involves breaking up the subsoil to let plants find additional water and nutrients. The third layer, or C horizon, consists of parent rock in the process of being broken up.

So far we have discussed only the nonliving part of the soil, but soil is a dynamic, living ecosystem that contains thousands of living species. This aspect of the soil is discussed in Chapter 12. From the microorganisms to the large animals and plants living in and on the soil, all affect the fertility, aeration, and water-holding capacity of soils. Thus soils represent a dynamic and vital niche in Earth's environment.

Only about a third (4.5 billion of the 13.2 billion hectares) of Earth's land surface is used for crop cultivation and animal husbandry. Equal portions are covered with forests or are too rocky, steep, or dry (deserts) for agriculture. The arable soils—those suitable for crop agriculture—represent only 11% of the total land surface. However, about three quarters of the arable soils have poor fertility, and about one half of the arable soils is constrained by steep terrain and subject to erosion.

11.2   A soil consists of mineral particles, organic matter, nutrients, air, water, and living organisms.

The soil particles formed by weathering are not uniform in size, and their size distribution and the way in which they aggregate greatly affect the water- and air-holding capacities of the soil and therefore its ability to support plant life.

Soil Texture and Structure. Agronomists classify soil particles according to size as sand, silt, and clay. Sand particles range in diameter from 2 to 0.02 mm, and silt particles from 0.02 to 0.002 mm; particles smaller than 0.002 mm are considered clay. Soil scientists classify soils according to the proportion of sand, silt, and clay particles in each (see **Table 11.2**). These proportions, generally called the *soil texture*, are determined by parent material and by extent of weathering. Sandy soils are often formed on sandstone, whereas limestone gives rise to loam soils, and shale results in clay-rich soils.

Both water and nutrients are *adsorbed* (bound) on the surface of the soil particles. Obviously, the more finely divided the particles, the greater their surface area per unit mass, and the greater their capacity to bind water and nutrients. Clay particles have often weathered so much that they have become porous and thus present an even greater binding surface. Sandy soils, having relatively more large particles, do not bind much water or plant nutrients in comparison to clay soils.

The individual soil particles generally clump together to form aggregates of varying size and shape. This is especially common with silt and clay particles. The "glue" that sticks the particles together is a combination of lime, iron hydroxides, and humus and other decaying organic matter. The degree of aggregation of the particles is generally referred to as the *soil structure*. Although soil structure is difficult to define, you can readily recognize it from the size and shape of the lumps formed when a soil is crumbled.

Table 11.2   **Composition of three typical soils classified according to size of mineral particles**

Soil Type	Type of Particle		
	Sand	Silt	Clay
Sandy loam	85%	6%	9%
Loam	59%	21%	20%
Clay	10%	22%	68%

Another important property of an agricultural soil is its tilth. When a soil is tilled (plowed, leveled, raked, or rolled), the original structure is partially disturbed and new air spaces are created. The new pores may occupy up to half the soil's volume. The tilth of a soil determines how well water percolates into it and whether seedlings can easily push through the surface layer. With time and as a result of gravity and rain, the tilth will change and become less favorable. In soils that are never tilled—as in no-till agriculture—the continuous channels created by earthworms and roots that permit water percolation and gas exchange are very important. Tilling disturbs these natural continuous channels, but creates new channels for water percolation and gas exchange.

Air and Water. The size of the individual pores is related to the size of the particles or aggregates. Small pores normally occur between small particles, whereas larger pores exist between large particles or aggregates. The small pores, called *capillaries*, are usually filled with water, whereas the larger ones are filled with air. A good, fertile soil should have half the pore space filled with water and the other half with air. Such a distribution provides a good balance among aeration, water percolation, and water storage capacity.

The importance of air for plant roots and most other soil organisms is commonly underestimated. Plant roots need energy to grow and to take up minerals, and they obtain this energy by respiring sugars made in the leaves. This respiration requires oxygen and causes CO_2 to be given off. To maintain the proper balance of oxygen and CO_2 in the soil, gases must move continuously in and out of the soil. Oxygen from the atmosphere must move into the soil, and CO_2 must escape or it will build up to toxic levels. This gas movement requires the presence of continuous air channels. Such channels are made by small burrowing animals, such as worms, or are formed when dead plant roots decay. Gas exchange between the atmosphere and the soil readily occurs when the soil has good structure and good tilth.

What happens to dry soil during a prolonged slow rainfall or irrigation? As soon as the water touches the soil particles and organic matter, it binds to them. A soil thus saturated with water is said to be at field capacity: The small pores are filled with water, and the large pores are filled with air (**Figure 11.2**). If still more water is added, it will fill not only the small pores but also the large ones, expelling the air. The soil is now waterlogged. A waterlogged soil literally suffocates most plants and soil organisms because they die from lack

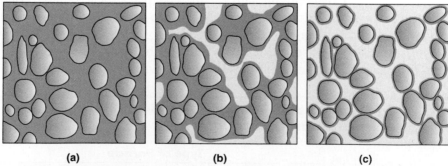

(a)   (b)   (c)

Figure 11.2 Distribution of water and air in a structured soil. In **(a)** the soil is waterlogged and all spaces are filled with water. In **(b)** the soil is at field capacity; there is air in the air spaces and water in the capillary spaces. In **(c)** the soil is beginning to dry out; most of the capillary water is gone and only a thin layer of water surrounds each particle. *Source:* Adapted from P. M. Ray, T. A. Steeves, and S. A. Fultz (1983), *Botany* (Philadelphia: Saunders), p. 183.

of oxygen. Some plants, such as rice, are specially adapted to thrive in waterlogged soils (see Chapter 12).

The texture, structure, and tilth of a soil together determine how well it can store water and permit water percolation. A clay soil, with its preponderance of small particles, has many small pores and a large water storage capacity. There may be so few large pores, however, that water cannot percolate downward. Rainwater then gathers on the surface of the soil and eventually runs off, often taking topsoil with it. Adding organic matter to such soils causing aggregates to form, creating larger pores and allowing better percolation. Maintaining plant cover on the land thus helps prevent erosion in three different ways: (1) The roots of the plants stabilize the topsoil, keeping it from being carried away; (2) decaying plants also supply organic matter necessary for aggregate formation, allowing more rapid percolation; and (3) root penetration of the soil offers ways for water to enter after these roots die and decay.

The decomposition of organic matter by microorganisms, called *mineralization,* produces organic particles that bind together small soil particles into larger aggregates. The addition of organic matter may increase the water-holding capacity of sandy soils and improve the drainage of heavy, clay soils. Besides being a source of energy for soil microorganisms, organic matter also acts as slow-release fertilizer to provide mineral nutrients for plants. Thus organic matter improves the physical structure of soils, releases acids for chemical weathering, and feeds the soil ecosystem.

11.3 The acidity, alkalinity, and salinity of soils are important determinants of productivity.

Because soil acidity influences the physical properties, the availability of certain plant nutrients, and the biological activity of the soil, it greatly affects plant growth. A soil's degree of acidity depends on the concentration of hydrogen ions (H^+) dissolved in the soil water. In a neutral soil, the H^+ concentration is about 1 part per billion parts of water. An acid soil may have a concentration of H^+ that is 100 to 1,000 times higher, whereas an alkaline soil has a lower H^+ concentration. The acidity or alkalinity of a solution is expressed by a single measurement called the *pH.* A pH of 7 is neutral, a pH of 5.0 (100 times more hydrogen ions) is acidic, and a pH of 9 (100 times fewer hydrogen ions) is basic or alkaline. Neither extreme acidity nor extreme alkalinity is suitable for plant growth or for most other soil organisms. Such conditions also upset soil weathering and the availability of nutrients. Although some plants can grow in strongly acidic or alkaline soils, most crop plants grow best in neutral or slightly acidic soils. Just over a quarter (26%) of the world's arable land is classified as acidic (**Figure 11.3**). In the tropics the percentage is even greater (43%). Acidic soils account for 68% of tropical America, 38% of tropical Asia, and 27% of tropical Africa.

We have already noted that acidity depends on vegetation. It also varies as a result of the types of fertilizers that are used (long-term use of ammonium or urea fertilizers cause acidification), and as a result of rainfall that is loaded with acids. The pH of rain is normally just below neutral, but pollution from power plants and cars create strong acids (nitrous acid and sulfurous acid) when they dissolve in the rain. When such acid rain falls on the land, it lowers the pH of the soil. The solubility of aluminum ions from the soil minerals increases when soil pH is 5.5 or lower, and can lead to aluminum toxicity in plants.

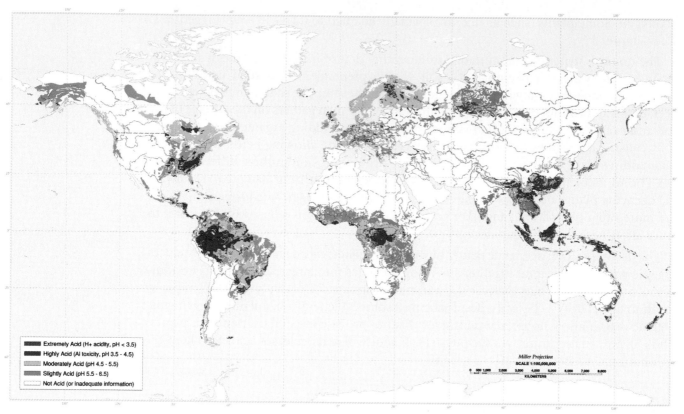

Figure 11.3 Global distribution of acid soils. *Source:* U.S. Department of Agriculture, Natural Resources Conservation Service, Soils Division, World Soil Resources.

The direct effects of air pollutants on plants and the indirect effects caused by acid rain and aluminum toxicity are a major cause of decline in forests in large areas of central Europe and North America.

The minerals that are dissolved by chemical weathering sometimes interact with each other to form new, insoluble complexes. A soil's acidity plays an important role in this process, and can greatly reduce the availability of certain nutrients to the plants (**Figure 11.4**). For example, phosphate, an essential nutrient, can form a variety of insoluble complexes with other ions in the soil solution. If the soil solution is too alkaline, phosphate readily combines with calcium to form insoluble calcium phosphate. When the soil solution is too acidic, phosphate combines with iron, aluminum, and manganese to form insoluble products. Thus phosphate is most readily available to plants when the soil solution is neutral or slightly acidic.

A soil's degree of acidity can be adjusted to make it more nearly neutral by adding lime. Acidic fertilizers, such as ammonium sulfate, can neutralize excess alkalinity with time. The uptake of ammonium stimulates acid excretion from the roots. The most effective method to acidify the soil is to use elemental sulfur. When microorganisms oxidize elemental sulfur, they release H^+ into the soil. Farmers may also use such treatments to optimize the growth of certain crops. For example, potatoes grow best in a somewhat acidic soil, whereas alfalfa thrives in a soil that is very slightly alkaline, but grows very poorly on acidic soil (**Figure 11.5**).

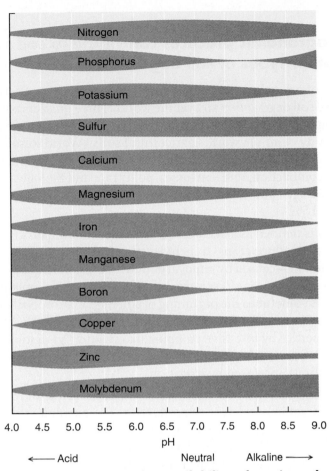

Figure 11.4 Influence of soil pH on the availability of nutrient elements in organic soils. The width of the shaded areas indicates the degree of nutrient availability to the plant root. All these nutrients are available in the pH range of 5.5 to 6.5. *Source:* R. E. Lucas and J. F. Davis (1961), Relationships between pH values of organic soils and availabilities of 12 plant nutrients, *Soil Science* 92:177–182.

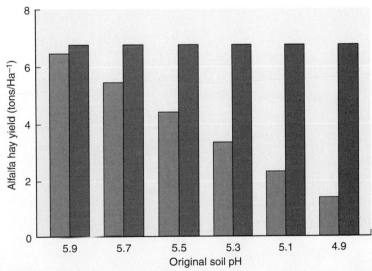

Figure 11.5 Yields of alfalfa as a function of original soil pH, with or without lime. Notice that as the original pH drops from left to right, the yield of alfalfa also drops dramatically (green bars). When lime is added to the soil to neutralize the acidity (low pH), all the yields are the same (brown bars). The final pH after adding lime is not shown. *Source:* Agriculture Canada (1974), Publication no. 1521 (Ottawa: Agriculture Canada).

11.4 Poor agronomic practices contribute to soil salinization and other types of soil degradation.

Poor management of natural resources—deforestation and misuse of agricultural lands—has led to extensive soil degradation all over the world. Three quarters of the area degraded by inappropriate agricultural practices, overgrazing, and deforestation are in the developing world. A recent study of all the world's agricultural soils (World Resources 2000–2001) indicates that about 67% of agricultural soils have been and are being degraded by erosion, salinization, compaction, nutrient losses, pollution, and biological deterioration. As much as 40% of the agricultural soils are seriously degraded, reducing world crop productivity by about 16%.

Soil degradation is defined as a decline in soil quality that impairs the soil's current or potential capacity to produce crops. It includes physical, chemical, and biological deterioration. Soil depletion—caused by removal of nutrients by water, or by cropping and removing harvested produce—is a less severe problem because people can remedy it by fertilization, but it is sometimes the beginning of soil degradation. Soil degradation takes many forms: erosion (removal) of the topsoil by wind or water; compacting by traffic; alteration of soil properties by intensive cropping; laterization by deforestation and raised temperature of tropical soils; decline in organic matter and biological activity of soil; and soil salinity or alkalinity increased by irrigation.

Although land degradation is widespread, it is extremely difficult to predict either the short-term or long-term impacts on food production. For example, although soil erosion is a problem in the United States, with 5 billion tons of topsoil being lost in some years (about 18 tons per hectare per year for arable land with seasonal crops), productivity in the United States has nevertheless increased steadily. These increases have resulted from adding more fertilizer, and some people argue that present agronomic practices allow U.S. farmers to ignore the steady decline in soil fertility from soil erosion. If the present rate of soil erosion continues, when will the United States have to start paying the price for these poor agronomic practices? Good agronomic practices such as contour plowing (**Figure 11.6**) and

Figure 11.6 Contour plowing helps prevent soil erosion. *Source:* U.S. Department of Agriculture.

Table 11.3	**Effect of soil removal depth on cassava yield from a tropical alfisol**	
Depth of Soil Removed (cm)	**Cassava Tuber Yield (t/ha)**	
	With Fertilizer	**Without Fertilizer**
Control	36.0	39.5
10	21.4	12.7
20	17.1	7.8

Source: Unpublished data of R. Lal (1989), cited in D. Pimentel and C. W. Hall, eds., *Food and Natural Resources* (New York: Academic Press), p. 103.

reduced tillage can minimize, but not eliminate soil erosion. *Ridge tilling*, a specialized type of land cultivation, can reduce water runoff to about 40% and soil loss to 10% of that observed with conventional tilling practices.

In tropical regions with abundant seasonal rainfall, erosion rates of 50–100 tons per hectare per year are not uncommon and often affect 30–50% of a country's arable land. Soil erosion generally does not greatly reduce yields on deep soils, especially if farmers use fertilizers to compensate for losing nutrients, but can be a real problem on shallow soils. An example for a shallow tropical soil in southern Nigeria is shown in **Table 11.3**. The loss of 20 cm of topsoil caused a massive decline in cassava yield, even with supplemental fertilizer.

A different type of soil degradation called *desertification* occurs in many arid and semi-arid regions. Some 50 years ago, observers noted gradual changes in the vegetation of the Sahel, a vegetation zone in West Africa, bordered to the north by the Sahara and to the south by dry savanna. The changes were described as the advance of the Sahara, because along its northern edge the Sahel was becoming like the Sahara, and along its southern edge the vegetation that characterized the Sahel was moving into the dry savanna. The changes in vegetation were caused by a combination of climatic changes and poor land management resulting from population pressures. Overgrazing irreversibly changed the land by denuding it of protective cover.

Overgrazing and the resulting desertification are not confined to the Sahel. Other semi-arid regions such as Patagonia, Iran, Syria, India, and Kenya also show desertification, often because pastoralists grazed three to five times more grazing animals than the meager vegetation could support.

When people irrigate land with water that contains dissolved salts (all groundwater and river water contains dissolved salts, but rainwater does not), there is always the danger that salinization, another form of land degradation, will occur. Salinization, or the accumulation of salts in topsoil, happens in areas where mean annual potential evapotranspiration (evaporation from soil plus transpiration by plants) exceeds precipitation by a factor of 1.3. The salts that accumulate can come from irrigation water, but salts can also rise to the surface in groundwater (**Figure 11.7**). In coastal areas, the intrusion of seawater below ground level can carry salts into topsoil.

Salts accumulate because there is not enough rain to carry them out of topsoil into groundwater and/or rivers. If water is abundantly available, people can sometimes leach salts out of a salinized soil to be carried away into streams. However, this procedure may create problems downstream. Thus a greater abundance of saline soils occur in semiarid and arid regions than in tropical regions.

Figure 11.7 The effect of soil salinity on the growth of wheat. The soil area in the foreground has a higher salt load because of a rising groundwater table. *Source:* Dr. Rana Munns, CSIRO, Canberra, Australia.

The salinity of a soil solution is expressed as molarity. A solution containing 58 g per liter of sodium chloride (sea salt) is said to be 1 molar (M) or 1,000 millimolar (mM). Saline soils are defined as soils that contain enough dissolved salts to osmotically stress plants and reduce growth. This is equivalent to approximately 40 mM NaCl (seawater contains approximately 500 mM NaCl or 29 g per liter). Some soils have reached salt concentrations twice that of seawater! Plants that are very sensitive to salinity may be affected by concentrations as low as 15 to 20 mM salts. The dissolved salts in saline soils are made up of many ions (just as in seawater); often Na^+ and Cl^- are the predominant ions but not always. Sulfates and carbonates may also significantly contribute to the saline solution.

Saline soils are increasing around the world, threatening the most productive, irrigated lands. Saline soils make up about 23% (340 million ha) of the world's cultivated land and are increasing by about 10 million ha per year. For example, two thirds of the saline soils are located in Asia, where two thirds of the world population depends on rice, a salt-sensitive crop produced primarily in low-lying areas. With global warming, the seas are expected to rise, which increases the threat of seawater intrusion into the low-elevation coastal areas.

11.5 In the transpiration stream, water lost from the leaves must be replaced from the soil.

Water is the medium in which soil minerals dissolve, and mineral nutrients must dissolve in soil water to enter the roots of plants. This important role of soil water, however, cannot explain the extremely high water requirements of many crops. As discussed in

Chapter 10, the requirement that CO_2 needed for photosynthesis must be dissolved in water has resulted in a two-way gas exchange: The cost to the plant of allowing the CO_2 to enter the leaf is that water vapor escapes through the stomates. Water vapor in intercellular spaces is quickly replenished by evaporation of liquid water inside the leaf. However, liquid water within the leaf must be replenished. Therefore leaves obtain water from the conductive tissues, which in turn get their water from the soil via the roots. Thus during the daytime, when the stomates are open, there is a continuous stream of water through the plant (**Figure 11.8**). This process is called *transpiration*. At night, when the stomates are closed, the transpiration stream stops, although the plant may continue to take up water from the soil until it has completely rehydrated.

Transpiration dissipates heat in plants, similar to perspiration in humans. It also helps transport minerals and organic molecules from the root to the shoot. If the rate of water uptake from soil is inadequate to replenish the water lost from leaves by transpiration, leaves will lack water, and if the deficit is really severe the stomates will partially or fully close. Partial closure of stomates reduces further water loss, but it also reduces photosynthesis, because CO_2 movement into the leaf is reduced. On a hot summer day, plants may experience transient wilting because water loss from leaves is more rapid than water uptake into roots, even though there is enough water in the soil. This transient wilting can severely depress photosynthesis in the middle of the day. Thus provision of adequate soil water is crucial for optimal plant productivity.

Water is lost not only by transpiration but also by evaporation from the soil surface. The sun warms the soil and causes some of the water to escape as water vapor. The term *evapotranspiration* describes the total amount of water lost from a plant–soil system. The total productivity of an ecosystem is very closely tied to the amount of water lost each year

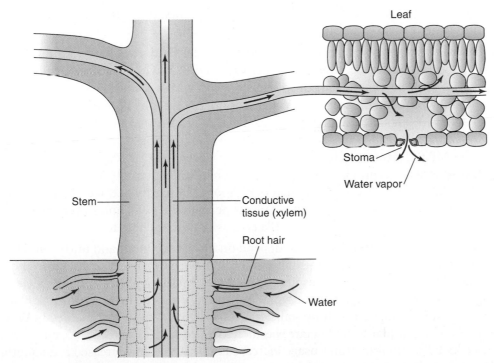

Figure 11.8 The transpiration stream through a plant. Arrows indicate the direction of movement of water or water vapor.

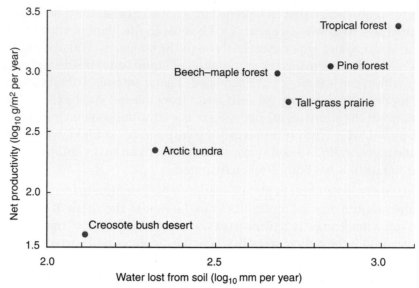

Figure 11.9 The relationship between the water lost from the soil (by evaporation and transpiration) and plant productivity. In general, it is possible to predict productivity from data on temperature and water availability. *Source:* Adapted from M. Rosenzweig (1968), Net primary productivity of terrestrial communities: Prediction from climatological data, *American Naturalist* 102:67–74.

through evapotranspiration (**Figure 11.9**). Availability of water in soil determines plant growth more than any other factor. Thus it is not possible to "make the desert bloom" without irrigation. Although farmers can obtain small yields with drought-adapted plants, large yields always depend on supplying additional water.

On a worldwide scale, water availability is a major limitation to crop production. Only a tiny fraction of the world's water is actually used by plants, but plants transpire daily an amount of water nearly equal to their total water content.

Water infiltrates the pores between soil particles and is held there with varying degrees of tenacity. Soil water can be measured directly by weight loss of the soil on drying, or by using special devices. The tenacity or soil tension with which soil particles hold water increases as soil water content decreases. Water tension in soil at any moment controls movement of soil water in soil and its use by plants. This water tension (a negative pressure) is expressed in units called megapascals or MPa. When tension is low—between −0.01 and −0.03 MPa (1 atmosphere is approximately equal to 0.1 MPa)—water moves to lower layers because of gravitational pull. Also, when soil water tension is −1.5 MPa or below the adhesive force is so strong that plant roots can hardly extract water from soil. At approximately this water tension, most plants permanently wilt and stop growing. Soil water between about −0.01 and −1.5 MPa is considered available to plants.

The water requirements of any plant depend on its environment and nutrition. Plants are grown in nutrient-rich soil are usually large and have extensive root systems. The total amount of water they lose by transpiration increases, but they use water more efficiently. Placing plants closer together increases total transpiration, because there are more plants, but decreases direct evaporation from soil, because sunlight does not hit soil directly. Where the number of maize plants per hectare was increased from 20,000 to 40,000, maize yield increased by 65%, but total water usage increased only 20%. The increased transpiration was largely compensated for by decreased evaporation from soil.

In Chapter 10 we noted that plants differ about twofold in their ability to produce dry matter per 1,000 kg of water used, depending on the type of photosynthesis, C_3 or C_4. Other properties of plants also affect their water use efficiency (kg of water used per kg of dry matter produced), as shown in **Table 11.4**. On this list only maize and sorghum have C_4 photosynthesis. The others are all C_3 plants. Water use efficiencies are usually measured under ideal growth conditions, and do not predict how plants will perform with a normal (or abnormal) rainfall regime. Crop yield (usually seed production) is the bottom line, not dry matter production.

Table 11.4	Water use efficiencies
Plant	**Kg Water Used per Kg Dry Matter Produced**
Alfalfa	850
Soybeans	650
Oats, potatoes	580
Wheat	550
Sugar beets	380
Maize	350
Sorghum	300

11.6 Plants require six minerals in large amounts and eight others in small amounts.

The fact that plants need to take up minerals from the soil was shown for the first time in 1699 by the British botanist John Woodward. He grew small cuttings of mint in rainwater, river water, and water to which he had added some soil. When he weighed the plants some time later, he concluded that growth was related to the amount of dissolved substances in the water. But which chemical elements in the soil water are required for growth? Earth's crust consists of over 90 different chemical elements. An analysis of plant ash, the residue that remains after plants are burned, reveals that plants may take up as many as 50 or 60 different elements from the soil. Are all these elements essential, or does the plant take up whatever it happens to find in the soil? This question was first investigated in a systematic way in 1860 by the German plant physiologist Julius Sachs, who grew plants with their roots immersed in solutions of minerals (hydroponics). He found that many plants could grow satisfactorily in solutions containing only three mineral salts: calcium nitrate, potassium phosphate, and magnesium sulfate.

These salts provided the plants with six elements: calcium, potassium, magnesium, nitrogen, sulfur, and phosphorus—termed the *major nutrient elements* or *macronutrients*. If any one of these elements was omitted from the culture solution, the plants did not grow well, so Sachs concluded that these six elements were essential for the plant. Later he also found iron to be essential for growth. These experiments showed that plants did not require any organic substances (such as vitamins), a question hotly debated at the time. Using hydroponics, it is relatively easy to demonstrate that these major nutrients are required for healthy plant growth (**Figure 11.10**).

Modern research suggests that the minerals used by Sachs were "contaminated" with small amounts of other minerals. Advances in chemistry made it possible to obtain much purer chemicals, and biologists later showed that plants also require trace amounts of seven other minerals. These seven, termed the *minor nutrient elements* or *micronutrients,* are boron, copper, chlorine, manganese, molybdenum, nickel, and zinc. Iron, which Sachs already knew to be essential, is also on the list of micronutrients. Together, these 14 nutrient elements are known as the *essential plant nutrients.* A mineral nutrient is said to be essential if without it the plant cannot complete its normal life cycle (flower and set seed).

A few other elements are apparently needed by only some plants. Sodium is required by some plants that prefer salty environments for growth. Diatoms (a major component

Figure 11.10 Tomato plants after several weeks' growth in solutions containing all the essential elements (complete) or in media lacking N, P, or K, respectively. *Source:* Courtesy of Grant Cramer.

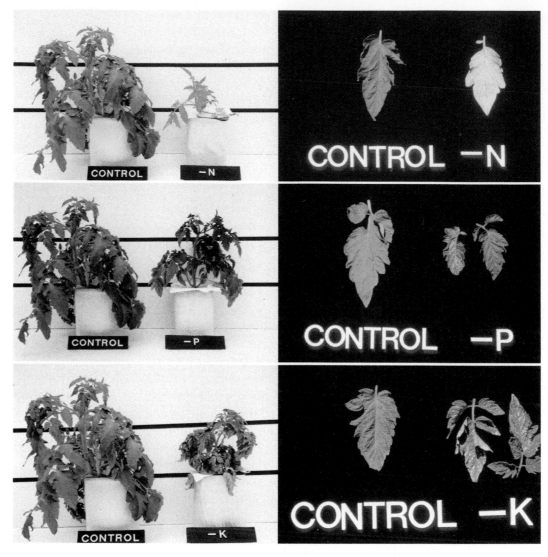

of oceanic phytoplankton) and some cereals, such as rice, need silicon. The microorganisms that live in the roots of leguminous plants and fix nitrogen need cobalt.

Plants take up from the soil the other 40 or 50 elements present in plant ash even though they may not require these chemicals for growth. This somewhat indiscriminate uptake of elements may benefit animals that eat the plants. For example, animals require sodium and iodine, two elements plants don't need. Plants may also accumulate elements that are toxic to animals. Certain plants in the genus *Astragalus*, called *locoweeds*, accumulate the element selenium. Selenium apparently does the plants no harm, but it kills the sheep, cattle, and horses that eat locoweeds.

11.7 To be taken up by roots, minerals must dissolve in the soil solution.

The 14 essential elements or nutrients just discussed must dissolve in soil water before plants can take them up. When minerals dissolve in water, they break up into smaller electrically

To be Taken up by Roots, Minerals Must Dissolve in the Soil Solution **285**

charged particles called *ions*: positively charged *cations* and negatively charged *anions*. To understand the importance of this charge for the plant nutrients, consider again the surface properties of the soil particles, both mineral and organic, which determine not only the soil's ability to bind water but also its ability to bind many plant nutrients.

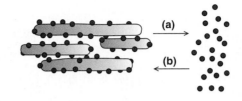

Figure 11.11 Equilibrium between cations in the soil solution and those bound to the surface of clay particles. (a) The roots' removal of nutrients from the soil solution causes more nutrients to leave the surface of the particles. (b) Adding nutrients, in the form of fertilizer or from decay of organic matter, causes more nutrients to be bound to the clay particles.

Mineral and organic soil particles have an overall negative charge. Because opposite charges attract, cations bind electrostatically to these particles. As a result, most cations in a soil are more or less firmly bound to the particles, although some remain in the soil solution. The anions, in contrast, remain in the soil water. Because of their negative charge, they do not bind to soil particles. As rainwater percolates through the soil, it carries anions and cations with it down to the groundwater and sometimes into streams and waterways. As noted earlier in this chapter, this process is called *leaching*. Because the anions are not bound to the soil particles, they are more easily lost by leaching than are the cations. However, phosphate is not leached, because it is rapidly tied up in the soil as insoluble precipitate. As soon as cations are removed from the soil solution, either by leaching or because they are taken up by plants, they are replaced by others released from particle surfaces. Thus an equilibrium is kept between cations in soil water and cations bound to particles, as illustrated in **Figure 11.11**.

Because the surface area available for binding nutrients depends on soil texture, a soil's texture greatly influences its fertility. The greater the surface area of the particles, as in clay soils, the more plant nutrients they can bind. As water percolates down through the soil, it carries some nutrients down with it, especially unbound anions such as nitrate. In some agricultural areas, nitrate coming from fertilizers or produced by the decay of organic matter sometimes pervades groundwater at such high concentrations that the water is not suitable for use by babies.

To take up nutrients from the soil solution, most plants must develop an extensive root system. The root system provides the plant with the surface area it needs to exploit a large volume of soil. In most plants, this surface area is increased even more either by root hairs—filamentous extensions of root epidermal cells—close to the tips of the roots, or by symbiotic fungi called *mycorrhizae* (see Chapter 12). Ions the root absorbs from soil come in contact with the root surface, either because they have moved through the soil with soil water (some nutrients move more readily than others), or because the root has grown into a previously unexploited area of soil. If the soil is poor in mineral nutrients, the plant develops a much more extensive root system than if the soil is nutrient rich. However, this root system growth will be at the cost of shoot system, which usually produces the crop.

Special proteins in the plasma membranes are needed for ion transport, forming either *ion channels* or *ion pumps*. The uptake of most ions into root cells requires energy because the concentration of the nutrient ions in the soil solution is much lower than within the cell. In terms of energy, this uphill process indirectly requires ATP. ATP maintains the plasma membrane in an "energized" state through the action of H^+-ATPases (proton or hydrogen ion pumps). The pumping of H^+ through the plasma membrane to the outside of the cell sets up a H^+ gradient: a greater concentration of H^+ and therefore greater acidity and lower pH on the outside than on the inside. This proton gradient powers the uptake of sugars, amino acids, and mineral nutrients such as nitrate through specific carrier proteins (**Figure 11.12**). The excess of negatively charged ions left in the cell creates a membrane potential that is a source of energy to move positively charged nutrient ions such

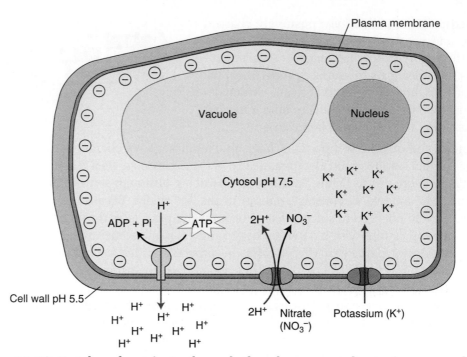

Figure 11.12 Uptake of nutrients through the plasma membrane is powered indirectly by ATP. ATP provides the necessary energy for a protein called a *proton* (H$^+$) *pump* to carry protons across the plasma membrane so that there are more H$^+$ outside than inside. This proton gradient now provides the energy for other proteins to transport sugars and nitrate into the cytoplasm. The excess negative charge left inside the cell creates a membrane potential that powers the uptake of positively charged ions such as potassium through channel proteins. Similar transport systems operate at the level of the vacuolar membrane, but much less is known about them.

as potassium to the inside. Once the mineral nutrients are inside the root cells, they pass from cell to cell until they reach the conductive xylem tissues in the center of the root. Once in the xylem, dissolved minerals can be carried upward with the transpiration stream.

The various nutrients have specific roles in plants that are often similar to their roles in animals. For example, both plants and animals require nitrogen and sulfur to synthesize amino acids, and they need phosphorus to make phospholipids and nucleic acids. Many minor nutrients, needed only in trace amounts, help enzymes carry out biochemical reactions. Some nutrients have unique roles in plants, such as calcium, which participates in maintaining cell wall integrity, and magnesium, which is part of the chlorophyll molecule.

Nutrients, especially potassium, also play an important role in maintaining the turgidity of the plant's cells and consequently of the organs as a whole. Potassium ions (and other solutes) become concentrated within the cell, and this drives water uptake into root cells. The influx of water causes cells to swell. Because cell walls resist stretching, pressure builds within each cell. Water pressure gives additional rigidity to cells and plant organs (analogous to air in a tire). When water loss by evaporation exceeds water uptake, pressure inside cells drops. As a result, plant organs (leaves, in many plants) do not retain their shape but collapse, and the plant wilts. Water pressure also drives cell enlargement in the plant's growing zones, thus playing a key role in plant growth.

11.8 Mineral deficiencies are prevented by applying organic or inorganic fertilizers.

If the soil is deficient in even one plant nutrient, plant growth will be retarded and crop yield may be diminished. If the deficiency is severe, the plants will develop visible symptoms that are diagnostic of the lacking element (Figure 11.10). Treatment usually involves fertilizing the soil with the nutrient. Unless this is done at an early stage in plant growth, severe crop damage may occur. Furthermore, nutrient-deficient plants are more susceptible to pathogens.

The appearance of symptoms such as stunted growth, leaf yellowing, "burned" leaf margins, or death of the terminal bud usually indicates a rather severe deficiency in one or more plant nutrients. Plants do not develop these signs when the nutrients are only marginally deficient. However, marginal deficiencies may also reduce crop yield. This "hidden hunger" is difficult to diagnose. It can sometimes be uncovered by measuring amounts of plant nutrients in soil or in plant tissues; however, neither method is completely satisfactory.

Although total amounts of particular nutrients present in soil can be measured, chemical tests cannot tell what proportion of the nutrients is available to the plants. Availability can only be measured by experiments with plants and measuring how crop yield responds to addition of nutrients (fertilizers) to a specific soil. After yield responses are known, plants can also be analyzed to measure how much of the nutrients the plants took up from the soil.

Harvesting crops removes large amounts of the major nutrients (see **Table 11.5**). These must be restored if soil fertility is to be maintained, and failure to restore them gradually reduces soil fertility and crop productivity. This is adequately demonstrated by the low yields obtained on experimental plots that have not been fertilized for many years. Plants may display nutrient deficiency symptoms even when a nutrient is present in the soil but unavailable to the plants, as when soil is too acid or too alkaline.

There are many ways to increase or restore the fertility of the soil so that it will support crop production. The oldest method of restoring soil fertility—allowing the land to lie fallow—was discussed in Chapter 4. A second widely practiced method is to *incorporate organic residues* and wastes into the soil. A survey of agricultural practices around the world reveals that farmers incorporate all kinds of organic residues into the soil: manure, fish wastes, algae, human excrement, crop residues, sawdust, and composted kitchen

Table 11.5 **Nutrients contained in the total above-ground plant material in a hectare of corn yielding 10,000 kg of grain**

Nutrient	Kg/Ha	Nutrient	Kg/Ha
Nitrogen	200	Iron	2.3
Phosphorus	42	Manganese	0.4
Potassium	205	Copper	0.1
Calcium	41	Zinc	0.42
Magnesium	48	Boron	0.19
Sulfur	24	Molybdenum	0.01
Chlorine	86.0		

Source: S. Barber and R. Olson (1968), *Changing Patterns of Fertilizer Use* (Madison, WI: Soil Science of America), by permission of the American Society of Agronomy.

Figure 11.13 Yield responses of three different crops to phosphorus fertilizers on a low-phosphorus silt loam soil. Different crops respond differently to a particular fertilizer. *Source:* U.S. Department of Agriculture.

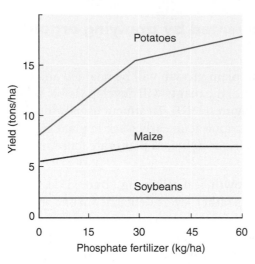

scraps, among others. These organic fertilizers decompose in the soil, releasing their nutrients. Organic fertilizers provide a steady supply of nutrients but may not release enough nutrients during the period of rapid vegetative growth, when demand is greatest. Slower growth in the spring and early summer often means a reduced crop in the fall.

A third method to restore the fertility of the soil is to add *inorganic fertilizers* (also called "chemical" or "synthetic" fertilizers). Chemical fertilizers release nutrients rapidly and can be made available when the plant needs them most. A disadvantage is that some plant nutrients, especially nitrate, may leach out of the root zone before plants can use them. This problem is greatest if the fertilizers are applied just before or at planting time. The seedlings take several weeks to develop a root system large enough to take full advantage of the fertilizer. Minor nutrient elements are sometimes applied directly to the leaves as a spray. Weak solutions of iron, zinc, or copper are commonly used on crops such as pineapple, citrus fruit, or avocado.

The three plant nutrients used most widely in inorganic fertilizers are nitrogen, phosphorus, and potassium. Adding fertilizer to the soil tends to stimulate overall plant growth and hence crop production. However, not all crop plants respond in the same way to adding a given amount of fertilizer to a certain soil. **Figure 11.13** shows how the yields of three different crops responded to adding phosphate fertilizer to a particular soil. Even within a given species there can be much variation from line to line, depending on genetic makeup.

Integrated nutrient management (INM) is an approach to farming that uses both organic and inorganic plant nutrients for correcting nutrient imbalances in soil, to attain higher productivity, prevent soil degradation, and safeguard future food supplies. It relies on nutrient conservation and application, new technologies to increase nutrient availability to plants, and transfer of knowledge between farmers and researchers. Resource-poor farmers in the tropics can benefit from INM because it helps maintain or adjust soil fertility and plant nutrient supply to an optimum level for sustaining productivity. The appropriate combination of mineral fertilizers, organic manures, crop residues, compost, or N_2-fixing crops varies according to land use system and ecological, social, and economical conditions.

11.9 Adaptations to water deficit and to adverse soil conditions are of great interest to plant breeders.

Of the total land in the world, only 11% of the land area is arable land and even on this land crops do not reach their genetic potential because of abiotic stresses (see Chapter 10). Climate (drought, cold, heat) and adverse soil conditions such as lack of nutrients, shallowness of the soil, and excess water limit crop productivity on the other 89% of the world's land. In the course of evolution, plants have adopted different strategies to cope with water deficit (drought) and adverse soil conditions such as salinity or acidity. These are of great interest to plant breeders and crop physiologists. The crops that provide most food for humans are not necessarily adapted to drought and adverse soil conditions. Plant breeders have genetically modified these crops to better adapt them to stressful environments. The improved varieties fall into three major categories: those with uniform superiority in all environments, those that perform relatively better in poor environments, and those that grow relatively better in favored environments. For a successful breeding result, a plant breeder will consider the following:

- Are screening techniques available for identifying tolerance to stress?
- Is there a reasonable range of variability within the species to be bred?
- Is the character heritable, and is some estimate of heritability available?
- Are there any strong, undesirable genetic correlations with stress tolerance?
- Can an estimate be made of improvement to stress tolerance in the field?

With this information in hand a breeding program can be undertaken.

Drought. Drought is a complex syndrome involving three main factors—timing, intensity, and duration of water deficit—all of which vary widely in nature. The extreme year-to-year and place-to-place variability in these three factors makes it difficult to define plant traits required for improved performance under all drought situations. Impact of drought on crop yield is a function of duration, crop growth stage, crop species or variety-within-species, and soil and management practices. A drought of about two weeks during flowering can result in complete loss of grain yields.

Droughts are an inevitable and recurring feature of world agriculture and, despite improved ability to predict their onset and modify their impact, drought remains the single most important factor affecting world food security. The problem of drought is most widespread in the developing world (**Figure 11.14**), where it has hastened the collapse of many fragile and unstable food production systems.

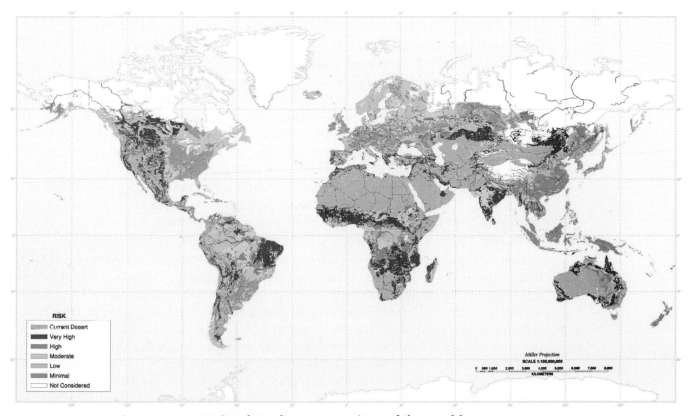

RISK
- Current Desert
- Very High
- High
- Moderate
- Low
- Minimal
- Not Considered

Miller Projection
SCALE 1:100,000,000

Figure 11.14 Major drought-prone regions of the world. *Source:* UNESCO, 1977.

Water deficits cause morphological and physiological changes in plants. Morphological responses to water deficits consist of leaf shedding, leaf rolling, leaf angle changes, and an increase in the root-to-shoot ratio. Physiological responses caused by water deficits include changes in leaf wax thickness, transpiration, respiration, stomate behavior, photosynthesis, translocation, mineral nutrition, and protein loss.

When the plant lacks adequate water, the stomates will close partially and eventually completely so that the plant does not dry out. This process is regulated by the hormone abscisic acid (ABA), which is transported by the sap from the roots to the leaves. The ABA triggers not only closure of the stomates, but a host of other processes that allow the leaves to survive this period of water deficit.

Plants have three ways of coping with drought: escape, avoidance, and tolerance. Drought escape allows a plant to grow and complete its life cycle before soil moisture becomes limiting. Drought avoidance is an alternate mechanism by which plants can maintain full hydration of their leaves and stems even under limited soil moisture conditions. They do so by decreasing water loss from the shoot or by more efficiently extracting moisture from the soil. The architecture of the plant is adapted to dry conditions, so the cells never experience water deficit. True drought tolerance, in contrast, functions at the tissue or cellular level. When the cells begin to dry out, they respond by synthesizing substances (sugars, proteins, and so on) that permit the cells to function normally even under water deficit.

Adverse Soil Conditions. Adverse soil conditions such as salinity, acidity, and nutrient deficiency and toxicity are problems for crop productivity throughout the world. In the past, farmers have amended such soils to meet plant needs. This approach requires farmers to purchase inputs: fertilizer, lime, irrigation water. These practices may no longer be economically feasible or environmentally sound. An alternative strategy is to select or improve plants for production on problem soils with fewer purchased inputs.

In many instances, adaptation to adverse soil conditions requires tolerance to excessive levels of mineral elements such as Al and Mn in acid soils, Mn and Fe in waterlogged soils, and sodium chloride in saline soils. Thus, tolerance to multiple stresses is often necessary for adaptation. In the case of rice, scientists from IRRI, the International Rice Research Institute in the Philippines have made impressive progress. In a large-scale screening and breeding program, they screened about 200,000 rice cultivars and breeding lines for tolerance to adverse soil conditions. They have developed adapted cultivars yielding up to 2,700 kg per hectare more than traditional cultivars in farmers' fields on unamended soils with adverse soil chemical conditions (**Table 11.6**). In hybridization programs, plant breeders have successfully used as parents materials identified as tolerant. This has allowed marginal land to be brought into rice production without the need to take costly land reclamation measures.

Poor crop productivity and soil fertility in acid soils are mainly caused by a combination of Al and Mn toxicity and nutrient deficiencies (mainly P, Ca, and Mg). Among these problems, Al toxicity and P deficiency have been identified as the most important constraints to crop production. These two factors limit production of maize, sorghum, and rice in developing countries located in tropical areas of Africa, Asia, and Latin America. The most easily recognized symptom of Al toxicity is inhibited root growth. Because cell elongation is inhibited, root cells become shorter and wider. As a consequence, root elongation is impaired and the roots have a "stubby" appearance when grown in the presence of Al in solution (**Figure 11.15**); this has become a widely accepted measure of crop sensitivity to Al. There are two major mechanisms of plant resistance to toxic levels of Al in soil. First, the

| Table 11.6 | Yield advantage of tolerant modern rice cultivars on problem soils in farmer's field in the Philippines |

	Mean Grain Yield (kg/ha)		
Soil Stress Factor	Landraces	Modern Cultivars	Yield Advantage (kg/ha)
Salinity	1,400	4,400	2,000
Alkalinity	800	3,400	2,600
Phosphorus deficiency	2,200	4,900	2,700
Zinc deficiency	1,800	4,400	2,600
Iron deficiency	900	2,800	1,900
Iron toxicity	2,200	4,100	1,900
Boron toxicity	1,100	3,000	1,900
Aluminum/manganese toxicity	1,200	3,000	1,800

Source: Data from S. K. De Datta, H. U. Neue, D. Senadhira, and C. Quijano (1994), International Sorghum and Millet Collaborative Research Support Program Publication No. 94-2, p. 249, available at INTSORMIL.unl.edu/index.htm (accessed October 23, 2001).

plant may exclude Al from being taken up by secreting organic acids that bind Al or by modifying the pH of the rhizosphere. Second, the plant may tolerate toxic concentrations of Al within the cell (for example, by binding Al internally after taking it up).

Salinity. Salinity can injure plants in three different ways: It can alter water relations, cause ion toxicity, and cause ion imbalance and nutrient deficiency within the plant. How each of these factors influences a plant depends on many environmental variables, such as humidity, temperature, soil structure and aeration, light intensity, composition and concentration of ion species in solution, and duration of exposure to salinity. Also, genetic and developmental variables influence the plant's sensitivity to salinity. Not surprisingly,

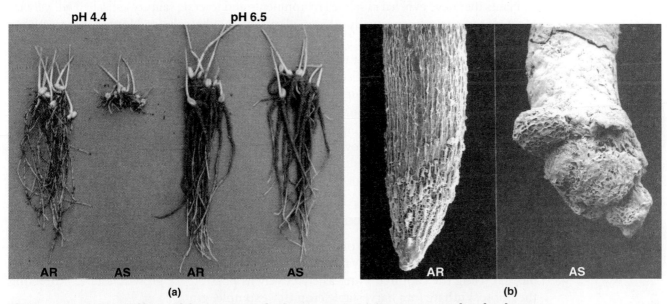

Figure 11.15 The effect of low pH and aluminum toxicity on growth of wheat. **(a)** Seedlings of two lines of wheat, one aluminum resistant (AR) and one aluminum sensitive (AS) were grown in an acid soil (pH 4.4) or a near-neutral soil (pH 6.5). The AS line does not form roots at pH 4.4 because this low pH solubilizes aluminum from the soil particles. **(b)** Scanning electron microscopy of the root tips of the AR (left) and the AS (right) lines.

Figure 11.16 The effect of salinity on the growth of different species of plants.
Source: H. Greenway and R. Munns (1980), Mechanisms of salt tolerance in nonhalophytes, *Annual Review of Plant Physiology* 31:149–190.

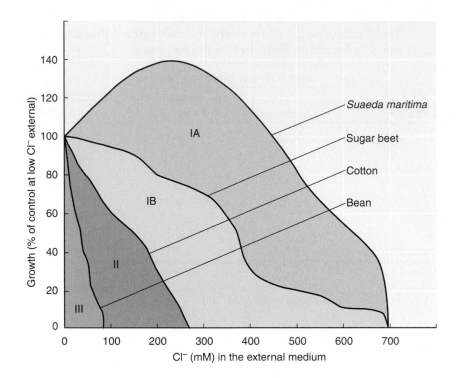

therefore, plant responses to salinity can be very complex and there is a wide range of responses, which vary considerably with different genotypes. Scientists who study responses and adaptations of plants to salinity want to understand the underlying physiological mechanisms and the genetic characteristics.

Plants that have evolved in saline environments and tolerate salinity are called *halophytes*. Most plants, including virtually all crop plants, are *nonhalophytes*. Nonhalophytic plants are not adapted to saline environments, although some (such as sugar beets) may have halophytic ancestry. Based on their growth response, plants have been classified into three groups (**Figure 11.16**). Group I consists of halophytes, Group II represents a group of halophytes and nonhalophytes whose growth response to salinity overlaps, and Group III represents very sensitive nonhalophytes. The growth of most plants (Groups IB, II, and III), including many of the halophytes, is inhibited as the salinity levels in the soil solution increase. However, the growth of some halophytes (Group IA) can be stimulated by saline conditions.

11.10 Crop resistance to water deficit can be improved.

Improving drought resistance of crops is an important aim of plant breeders. Three factors are involved in effective water use by plants: (1) capturing as much water as possible; (2) using the captured water as effectively as possible for growth; and (3) directing the photosynthate into harvestable crop (for example, grain).

Four basic approaches to the breeding for drought resistance have been used.

1. Breed for high yields under optimal conditions—that is, breed for yield potential—assuming this will provide a yield advantage under suboptimal conditions.

2. Breed for maximum yield in the target environment.

3. Select and incorporate morphological and physiological mechanisms of drought resistance into traditional breeding programs.

4. Do not use multiple physiological selection criteria, but establish without doubt that a single drought-resistance character will benefit yield under water-limited conditions, and then incorporate the character into an existing yield-breeding program.

As mentioned, drought escape means completing the entire life cycle from seedling to seed when water is available in the soil. Twenty years of drought in the Sahel have resulted in numerous crop failures, but plant breeders have now adapted cowpea varieties that produce acceptable yields even under conditions of low rainfall. Most cowpea varieties grown in the Sahel prior to 1980 required a growing season of 80–120 days. Analysis of rainfall patterns indicated that plants requiring only 65 days to produce a crop would be more effective in the drier areas of the Sahel. Crop breeders from the University of California crossed a variety that flowers early under the hot long-day conditions prevalent in the Imperial Valley of California with a day-neutral variety from Senegal, and selected from the progeny a variety that does well under conditions of low rainfall (short rainy season).

When rainfall was sufficient (452 mm), the new variety did not perform better than the parents or the local variety (**Table 11.7**). But as the rainfall decreased and yields dropped dramatically, the new variety produced four times more bean than the local variety and 50% more than the parent from Senegal. At intermediate rainfall, the new variety outperformed both parents. It is important to note that when rainfall is low, crop production is also low, even for a drought-adapted variety.

Scientists from Pioneer Hi-Bred International Inc. Ltd. of Iowa, in the United States, used 500 field sites to evaluate maize lines and tested whether plants selected to have improved yield under water-limited conditions also have improved yield under optimum conditions. Their experiments provide the strongest possible evidence that improvement under water-limited conditions need not sacrifice yield under favorable conditions. The number of replications, genotypes, and field sites was so large that this conclusion now seems beyond doubt.

To develop drought-resistant maize, researchers at CIMMYT, the International Center for Maize and Wheat Improvement in Mexico, began selective breeding with the *Tuxpeño* variety that was already well adapted to tropical lowlands. They bred this drought-tolerant *Tuxpeño* maize under drought conditions, selecting at flowering the plants with silks (female flowers) that appeared soon after the male flower emerged. When tassel and

Table 11.7 Comparison of yields of different varieties of cowpea under different conditions of rainfall in Senegal

Cowpea Variety	Bean Yields (kg/Ha)		
	452 mm Rain	315 mm Rain	135 mm Rain
Local variety	2,260	1,300	50
Senegalese parent	Not tested	1,070	130
U.S. parent	2,350	1,350	195
New variety	2,400	1,800	200

Figure 11.17 Mutants that resist water deficit. The two plants on the right are mutants of *Arabidopsis thaliana* that remain green much longer than the control on the left when the soil was allowed to dry out by withholding water. Identification of the mutated genes helps us understand how water loss is regulated. *Source:* Courtesy of N. Crawford.

silk development most nearly coincided, grain production was highest. The selected plants were then bred further over a decade for even greater resistance to drought. Scientists found that such plants allocate more carbohydrates, or energy, to the ear, which allows them to produce more grain with less water. This selective breeding led to the *Tuxpeño sequia* variety and others like it, which can be grown under a wide range of tropical conditions. The new breeds also show up to 10% increase in yield during non-drought seasons. In times of severe midseason droughts, these drought-resistant new maize varieties produce 2,800 kg per hectare, a 40% increase over regular maize yields under similar drought conditions.

Using molecular techniques, scientists have identified several classes of genes that confer resistance to water deficit (**Figure 11.17**). Some of these genes could be used to engineer plants for drought resistance and better crop yield under drought conditions. First they found enzymes that synthesize osmoprotectants, small molecules that accumulate in the cytoplasm of drought-stressed plants. Plants genetically engineered with the genes encoding these enzymes are more drought tolerant. Second, they found genes that encode transcription factors (see Chapter 6) that regulate entire metabolic pathways leading to drought adaptation. By incorporating such genes into crops, scientists hope to ensure that plants respond rapidly and efficiently to any water deficit and continue all their developmental processes. Yet another approach may be to make stomates more responsive to ABA in plants so that they close more rapidly as soon as the plant senses water deficit in soil. Genes encoding proteins that function in the ABA signal transduction pathway may have this desired effect. Overexpression of these proteins results in plants that remain green and active longer in a drying soil because they use water more slowly (Figure 11.17).

Conventional (see **Box 11.1**) and GM breeding are complementary approaches and can be expected to enhance the drought resistance and yield of crops. People have entered a new era in which enhanced knowledge of both the physiology of yield accumulation and the physiological basis of genetic variation in drought resistance traits offer the potential for improving breeding efficiency for major food crops in different target environments. Using physiological knowledge and powerful tools of molecular genetic analysis requires a systems approach, with agronomists, physiologists, breeders, and biotechnologists working together with farmers to raise crop yields and farmer income in dry areas, particularly in developing countries.

Box 11.1

Scientists at ICARDA, the International Center for Agricultural Research in Dry Areas, have developed an efficient approach to breeding improved barley cultivars. It involves direct selection of the germ plasm in the target environments. Scientists from National Agricultural Research Systems (NARS) and farmers actively participate in the selection process, and farmers provide a substantial number of landraces and farmer varieties, with specific adaptation to both biotic and abiotic stresses. As a result, more than 100 cultivars have now been released in 35 countries in West Asia, Africa, Latin America, South Asia, the Far East, and Ethiopia through collaborative work with national programs in both developing and developed countries.

The decentralized and participatory approach to barley breeding for dry regions at ICARDA, which involves national scientists and farmers as equal partners, is receiving widespread recognition and support because of its achievements in producing new varieties that increase production even in the harshest areas without relying on

Farmer Participatory Breeding for Improving Barley Yields in Dry Areas

external inputs. The results are sustainable, environmentally friendly, and of immediate benefit to the poorest farmers. This approach has led NARSS (national agricultural research systems) and international institutions to reassess the value of locally adapted germ plasm (landraces and farmer varieties) and wild relatives in breeding programs. A number of NARSS have changed their barley-breeding strategies as a result of cooperative research with ICARDA. Landraces are now a major component of breeding programs in Eritrea, Ethiopia, Nepal, and Syria. Participatory barley breeding has been adopted by Ecuador, Morocco, Syria, and Tunisia. An impact study of a newly released barley variety, Rihane 03, in Iraq indicated an increase of 79% in average net revenue per hectare or US$1,000 per farm, at international exchange rates.

11.11 Crops can be improved for efficient use of soil nutrients.

To achieve adequate levels of crop yields on various types of soils, nutrients must be applied. In addition, plants must take up these nutrients, which are present in very low concentrations in the soil solution, with high efficiency. Shortages and the high cost of fertilizer and other amendments, in association with environmental problems (soil and water quality), have intensified the need for more nutrient-efficient plants. Improved efficiency in recovering applied nutrients is becoming a prerequisite for lowering production costs, protecting the environment, and improving crop yield.

Nutrient efficiency is the ability of a crop system to convert nutrient inputs into desired outputs. Supply, availability, or amount of a mineral nutrient is the input, and plant growth, physiological activity, or yield is the output. Efficiency is the relationship of output to input. This can be expressed as a simple ratio, such as kg of yield per kg of fertilizer, or kg of dry matter per g of nutrient. In many soils, the efficiency of plants in recovering applied fertilizer is low.

There are two major approaches to improving nutrient efficiency of crops. First, developing appropriate nutrient management and supply strategies. Second, breeding for improved efficiency in acquiring and using nutrients. Nutrient uptake efficiency obviously depends on the characteristics of the root system, such as uptake rate of nutrients per unit root length, ability of a root system to quickly expand into a fertile zone of soil, ability of roots to modify the rhizosphere, and ability of the plant to grow symbiotically with mycor-

Table 11.8	Genetic control of nutrient efficiency in crops
Trait	**Gene Action**
Phosphorus efficiency	Multigenic (additive, dominance, epistasis, 3 chromosomes)
Phosphorus uptake	Multigenic
Potassium efficiency	Predominantly additive, but dominance and epistasis important
	Single gene recessive
Iron efficiency	Single locus dominant, polygenic, additive, two complementary
	dominant genes, quantitative. Major gene governs uptake.
Boron efficiency	Single gene
Copper efficiency	Single gene

Source: Data from R. R. Duncan (1994), in R. E. Wilkinson, ed., *Plant Environment Interaction* (New York: Marcel Dekker), pp. 1–38.

rhizal fungi. All these properties of the root system are determined by numerous genes. Certain properties of the shoot, such as ability to translocate nutrients from the roots or to redistribute nutrients from senescing leaves, are also important for efficient nutrient use.

Growing nutrient-efficient genotypes of crop plants on soils of low nutrient availability represents an environmentally friendly approach that would reduce land degradation by reducing use of machinery and by minimizing application of chemical fertilizers on agricultural land. The danger of exhausting soil nutrient resources ("land mining") would be negligible, at least for phosphate and micronutrients. The total supply of phosphate and micronutrients in soils is enough for hundreds of years of sustainable cropping by new, efficient genotypes that can gain access to nutrient pools generally considered plant-unavailable.

Growing nutrient-efficient plants on soils low in plant-available nutrients represents a strategy of "tailoring the plant to fit the soil," in contrast to the traditional practice of "tailoring the soil to fit the plant." The approach of developing nutrient-efficient crops is very significant because out of all agricultural innovations, farmers most readily accept new cultivars. This is mainly because new cultivars can yield better, demand less fertilizer input, and need minimum or no changes in agricultural practices. However, relatively slow progress in defining the genetic, physiological, and biochemical basis of nutrient efficiency has hampered development of superior nutrient-efficient cultivars through conscious breeding efforts geared specifically toward that purpose (**Table 11.8**).

In alkaline soils, phosphate is much less available to plants because it becomes insoluble (see Figure 11.3). Recently, scientists from Mexico expressed a bacterial gene that encodes the enzyme citric acid synthase in plants and found that the genetically engineered plants were better able to solubilize the insoluble phosphate present in alkaline soils. This work shows promise for enhancing the use of scarce nutrient resources by crops.

11.12 Crops can be improved for better performance on saline soils.

Historically, farmers dealt with soil salinity by replacing salt-sensitive crops with more salt-tolerant ones; for example, barley replaced wheat thousands of years ago in Ethiopia. Thus people probably used crop substitution as a method of dealing with salinity long before they developed technologies to leach salts from soils and to avoid salinity problems using various management strategies. In all the saline growing areas of the world, farmers substitute salt-tolerant crops for sensitive crops. Crops such as sugar beet, barley, cotton, asparagus, sugar cane, and dates are very salt tolerant.

Improved salt tolerance in both sensitive and tolerant crops would allow more extensive use of brackish water supplies—an especially important consideration where water costs are high or water availability is low. Salt tolerance is a complex, quantitative, genetic trait controlled by many genes. Recently, scientists have identified a few genes that provide information useful in screening and selection programs for salt tolerance.

Scientists study the salt tolerance of plants with the goal of developing more salt-tolerant crops. Four major strategies have been proposed to develop salt-tolerant crops.

1. Gradually improve the salt tolerance of crops through conventional breeding and selection.

2. Introduce traits for salt tolerance from wild relatives into the crops by the process of backcrossing, as explained in Chapter 14.

3. Domesticate wild species that currently inhabit saline environments (halophytes) by breeding and selecting for improved agronomic characteristics.

4. Use molecular techniques to identify genes associated with salt tolerance, and enhance their expression in the crop species or transfer the genes from a noncrop species to a crop.

One example of breeding for salt tolerance is the development of salt tolerance in rice, a salt-sensitive species. Pokkali rice is grown in the state of Kerala in the southern tip of India. The region where this rice is grown has many inland waterways subject to tidal flow from the nearby sea. Because of seawater intrusion, the land is too salty to grow rice. However, just before the monsoon season, the farmers plant rice seeds that have been primed for fast germination. The seeds are planted into mounds, and when the monsoons come, rain washes some salt out of the soil, allowing germination. After a month, farmers transplant the seedlings throughout the field. Then the rice fields are flooded with freshwater and the rice grows as usual through the rest of the monsoon season. Most rice varieties would still not grow well in these salty soils even during the monsoon season, but over many generations local farmers have selected rice that will grow in these soils. After the monsoon season when farmers harvest the rice, they flood the rice stubble with salty river water and these conditions provide a very rich habitat for shrimp production. IRRI, the International Rice Research Institute in the Philippines, has used the salt-tolerant Pokkali rice extensively in its breeding program to develop salt tolerance in other, more desirable rice genotypes.

A promising approach attempted for tomato, barley, and wheat is to introduce traits of salt tolerance into these crops from their wild relatives through interspecific hybridization. This requires extensive backcrossing (see Chapter 14) and may be difficult because a trait such as salt tolerance may be determined by several genes located on different chromosomes. Moreover, in the wild relatives these traits are favorable mainly for survival under natural conditions. For crop production in dry saline soils, high efficiency of water use (kg of dry matter produced per kg of water) is of key importance. Present wheat and barley cultivars have a higher water use efficiency and total biomass production on saline soils than most of their wild relatives that thrive in such environments. So far this approach has not yielded successful new crop strains.

Recent progress in genetic engineering has led to the development of salt tolerance in plants. Two notable examples are (1) increased expression of a gene that regulates stress

Figure 11.18 **Wild-type (left) and transgenic tomato overexpressing a vacuolar Na⁺/H⁺ antiport (right) grown in a solution containing 200 mM NaCl.** Seedlings were germinated in soil and after germination (6 leaves) were transferred to a hydroponics growth system and grown in a Hoagland growth medium supplemented with 5 mM KCl and 200 mM NaCl. The solution was continuously aerated, and the growth medium was replaced weekly. *Source:* Courtesy of Hong-Xia Zhang and Eduardo Blumwald.

responses in plants and (2) increased expression of a gene that encodes a sodium (Na⁺) transporter in the tonoplast membrane of the vacuole.

On the molecular front, scientists are identifying the genes involved in sensing salt in the environment (signal transduction genes), transcription factor genes that turn on batteries of other genes that prepare cells to withstand a higher rate of salt influx, and genes that are part of the plant's adaptation to the presence of salt. An example of the latter category is the gene that encodes a vacuolar membrane sodium pump. Plants that can turn this gene on rapidly when the cells are exposed to salt, will be able to transport the salt from the cytoplasm and into the vacuole, thereby detoxifying the cytoplasm. Scientists in Canada expressed such a gene with a strong general promoter in tomato plants and found that the plants could be grown in 500 mM sodium chloride. The tomato plants apparently pump the leaf vacuoles full of sodium chloride and then allow those leaves to die, while at the same time producing new leaves (**Figure 11.18**). The plants produced an abundant crop of tomatoes in the greenhouse, but need to be field-tested.

Groups of Japanese and Belgian scientists independently discovered transcription factors that are specifically expressed when a plant is exposed to dehydration. The gene is turned on early in the drought response, and the protein activates expression of a large suite of genes involved in adaptation to stress. When the scientists engineered plants so that this gene was quickly expressed in response to dehydration, the genetically modified plants were more tolerant to water deficit in the laboratory. In addition, these plants were more tolerant to salinity and cold stress. All three stresses (drought, salinity, and cold) cause osmotic imbalances to which the plant must respond.

11.13 Crops can be improved for greater tolerance to acid soils.

Strategies for maintaining production on acid soils include applying lime to raise soil pH (see Figure 11.5) using lime from limestone or dolomite, and breeding for tolerance to acid soils. Liming increases nutrient uptake by plants and enhances organic matter decomposition, thus improving the conditions for plant growth. Concurrent application of lime and balanced nutrient replenishment is usually necessary to ensure continued long-term soil fertility. A problem associated with acid soils is the low availability of phosphate at low pH (see Figure 11.4). However, the main problem on acid soils is aluminum toxicity, because at low pH the clay minerals are solubilized and aluminum oxide is released in the soil solution. Soluble aluminum inhibits root growth, causing the crops to be more easily water stressed, because they cannot mine a sufficient volume of soil for available water. The sensitivity of different crops to Al is shown in **Table 11.9**. Crops that are highly sensitive to toxic levels of Al in acid soils are maize, mung bean, cotton, and cocoa.

Breeding improved varieties for acid, infertile soils starts by identifying productive lines with general adaptation to biotic and climatic conditions of soils in the target environments. Scientists compare these varieties under conditions where soil fertility, but not pH, has been modified with modest quantities of plant nutrients. The immediate benefit for the farmer of cultivars adapted to acid soil is obvious and has been experimentally demonstrated for important crops such as maize (see **Box 11.2**), sorghum, and upland rice. For example, in South America the Colombian National Program released four Al-tolerant cultivars of sorghum that, in field tests, showed a two- to three-fold yield advantage over other varieties.

Also in Colombia, CIAT scientists successfully developed acid soil–adapted upland rice lines that were released to the farmers in the early 1990s. These improved rice varieties permit commercial rice commercial production on the acid soil savannas of Colombia.

Research has shown that certain plants that can tolerate high levels of soluble Al secrete organic acids into the rhizosphere. Researchers in Mexico transformed papaya plants with a bacterial gene that encodes the enzyme citric acid synthase and showed that the roots of these plants secreted citric acid and that the plants tolerated higher concentrations of soluble Al in their growth medium. Field evaluation of these plants is presently in progress.

Table 11.9 Sensitivity of major crops to aluminum

Sensitivity to Aluminum	Crops
Low	Cassava, brachiaria, andropogon
Low to moderate	Millet, rice, cowpea, mucuna, rubber, oil palm
Moderate	Groundnut, centrosema, stylosanthes, kudzu
Moderate to high	Soybean, sorghum, bean, wheat, panicum
High	Maize, mung bean, cotton, crotalaria, cocoa

Source: N. Caudle (1991), *Managing Soil Acidity, Groundworks 1,* Tropsoils Publication (Raleigh: North Carolina State University), p. 17.

Box 11.2

Improving Maize Yields on Acid Soils

Nearly 8 million hectares of maize are planted in acidic soils: 3 million hectares in South America; 2.5 million hectares in Asia; 1.5 million hectares in Africa; and about 1 million hectares in Mexico, Central America, and the Caribbean. Traditional maize varieties yield as little as half a ton per hectare of grain on moderately acidic soils, as compared with two tons per hectare, the average for developing countries.

During the late 1970s, through the South American Regional Program based at CIAT, CIMMYT began developing maize varieties with acid and aluminum resistance—demonstrating the possibility of developing maize varieties with genetic adaptation to acid soils. During the 1980s, researchers identified and collated maize characteristics with high yield potential, acid soil tolerance, and resistance or tolerance to other key constraints. With help from research and breeding programs in each nation, researchers then conducted recurrent selection and crossings of thousands of superior varieties with each population at a number of sites in Colombia, as well as on acid soils in Brazil, Indonesia, Peru, the Philippines, Thailand, Venezuela, and Vietnam. In the early 1990s, experimental varieties from these new populations were evaluated on acidic and normal soils in Latin America, Africa, and Asia, and compared with local control varieties. The CIMMYT genotypes out-yielded control varieties by an average of 33%, with one experimental variety delivering the highest yields across all environments. In subsequent tests, products of these experiments yielded as much as 700 kg per hectare more than a Brazilian hybrid under nonacid conditions in Colombia, showing that the acid soil–adapted maize is also productive in normal soils.

The most important outcome of this research has been the new maize variety, ICA-Sikuani V-110, developed by the Colombian Agricultural Corporation (CORPOICA) using acid soil–adapted maize generated through the collaborative research project. The variety is already sown on thousands of hectares in its native Colombia and is being tested for use in neighboring countries. In trials conducted on farmers' fields in acid soil areas of Ecuador and Peru, Sikuani consistently outyielded the best local varieties both under optimal and farmer management. More recently, acid soil–adapted hybrids derived from CIMMYT's research have produced as much as 70% more grain than Sikuani, and should prove valuable in Brazil, Colombia, and Venezuela, where many maize farmers sow hybrids. These new varieties and hybrids would permit sustainable maize cropping systems to be established on acidic savanna and would reduce the pressure to farm marginal forest and hillside lands. All this would help reduce deterioration of fragile agricultural lands and ease the pressure to cut down tropical rainforests to obtain additional farmland.

CHAPTER SUMMARY

Soil is a dynamic, living resource whose condition is vital for producing food and fiber and to the proper functioning of the global ecosystem. Indeed, it is essential to the sustainability of life on Earth. Plants need carbon dioxide and sunlight, as well as water and minerals for growth. Soil supplies the latter two. If the plant is to take up CO_2 for photosynthesis, the stomates must be open. As a result, water moves through the plant in a continuous transpiration stream. Thus water must be taken up rapidly by the roots.

Water is supplied by rain or by irrigation, but continuous irrigation of land with river water may cause salinization (buildup of salts). Salinization is one aspect of the general soil degradation now occurring all over the world as a result of poor soil management practices.

Nutrients become available to plants when the mineral particles derived from rocks slowly weather and dissolve. Plants need some nutrients in large amounts; others, in very small amounts. Harvesting removes substantial amounts of nutrients from the agricultural ecosystem, and these must be replaced by adding either chemical fertilizers or organic material.

During evolution, plants have adapted to certain rainfall regimes and soil conditions, and not all crops are adapted enough to drought and adverse soil conditions to produce acceptable yield. The old solution to the problem of "not enough water and nutrients" was to irrigate and fertilize. A new possible solution is to try to understand which genes offer adaptation to drought and adverse soil conditions and to try to breed plants to overcome these major climatic and soil constraints.

Selection processes used by scientists to evaluate crop varieties for adaptation to drought and adverse soil conditions are similar. Such traits are often determined by multiple genes located on different chromosomes, making improvement through breeding slow (see Chapter 14).

Conventional and biotechnological breeding are complementary approaches and can be expected to enhance the efficiency of breeding for stress resistance and yield. The use of physiological knowledge and powerful tools of molecular genetic analysis requires a systems approach, with agronomists, physiologists, breeders, and biotechnologists working together with farmers to raise crop yields and farmer income in stressful environments, particularly in marginal soils of the tropics.

Discussion Questions

1. Can people "make the desert bloom"? How? At what cost?

2. Would reductions in air pollution solve agricultural problems with soil acidification? Explain.

3. How will global warming and higher CO_2 concentrations affect the response of plants to stressful conditions?

4. Could the abundant nutrients and water in the oceans be used more effectively for food production? At what cost? Explain.

5. What kinds of strategies have people employed to produce more stress-resistant plants?

6. Will the development of stress-resistant plants contribute to the continued degradation of soils? Why or why not?

7. Is the addition of organic matter to the soil really necessary? What are its functions?

8. Is soil really necessary for growing plants? What are the costs of hydroponic plant growth? For which crops is it suitable?

9. Because there is a finite supply of arable lands, could people expand their ability to grow crops in other areas? How? What would be the costs?

10. How does raising plant productivity on arable land by plant breeders impact efforts by conservationists to preserve biodiversity?

Further Reading

Baker, F. W. G., ed. 1989. *Drought Resistance in Cereals.* London: CAB International.

Blum, A. 1988. Plant Breeding for Stress Environments. Boca Raton, FL: CRC Press.

Evans, L. T. 1993. *Crop Evolution, Adaptation and Yield.* Cambridge, U.K.: Cambridge University Press.

Follett, R. F., and B. A. Stewart. 1985. *Soil Erosion and Crop Productivity.* Madison, WI: American Society of Agronomy.

Kramer, P. J., and J. S. Boyer. 1995. *Water Relations of Plants and Soils.* San Diego: Academic Press.

Lal, R., and P. Sanchez, eds. 1992. *Myths and Science of Soils of the Tropics.* Madison, WI: Soil Science Society of America.

Marschner, H. 1995. *Mineral Nutrition of Higher Plants*, 2nd ed. London: Academic Press.

Peskin, H. M. 1986. Cropland sources of water pollution. *Environment* 28:30–36.

Pessarakli, M., ed. 1999. *Handbook of Plant and Crop Stress*, 2nd ed. New York: Dekker.

Rendig, V., and H. M. Taylor. 1989. *Principles of Soil–Plant Interrelationships.* New York: McGraw-Hill.

Rengel, Z. 1999. *Mineral Nutrition of Crops. Fundamental Mechanisms and Implications.* New York: Haworth Press.

Turner, N. C., and J. B. Passioura, eds. 1986. *Plant growth, Drought and Salinity.* Melbourne, Australia: CSIRO.

Tate, R. 1992. *Soil Organic Matter: Biological and Ecological Effects.* Melbourne, Australia: Krieger.

Tisdale, S., W. Nelson, J. Havlin, and J. Beaton. 1992. *Soil Fertility and Fertilizers.* New York: Macmillan.

Troeh, F., J. Hobbs, and R. Donohue. 1991. *Soil and Water Conservation.* Upper Saddle River, NJ: Prentice Hall.

United Nations Development Programme (UNDP), the United Nations Environment Programme (UNEP), the World Bank, and the World Resources Institute. 2000. *World Resources—People and Ecosystems: The Fraying Web of Life.* Amsterdam Elsevier Science.

Life Together in the Underground

Maarten J. Chrispeels

University of California, San Diego

Although many terrestrial ecosystems have been investigated quite thoroughly, most of these scientific inquiries stop where the plants disappear into the soil! As a result, scientists know very little about the ecological relationships between the thousands of species in the top meter (3 feet) of soil, the layer where plant roots find most of their nutrition. People usually think of roots as having three major functions: They anchor plants to the ground, they take up water to replace what is lost by transpiration, and they extract mineral nutrients from the soil solution. But this short list omits a major newly discovered function of roots: the creation and feeding of an entire and distinct soil ecosystem that can profoundly influence the growth of the plant. Plants transfer as much as 50% of the products of photosynthesis below ground. A large portion of this underground ecosystem inhabits the rhizosphere: the outer cell layers of the root and the layers of soil particles adjacent to the root. Beneficial, neutral, and harmful bacteria, fungi, algae, protozoa, nematodes, and microarthropods live in the rhizosphere in an intricate food web. The health of this underground ecosystem can be dramatically affected by agricultural practices, tipping the balance in favor of beneficial or harmful organisms. The abundance of beneficial organisms can be manipulated by practices such as crop rotations, by chemicals that kill harmful microbes, or by inoculation with beneficial ones. Some of the organisms in this underworld benefit the plants directly, others may keep plant pathogens at bay. In this chapter we explore these complex interorganism relationships.

12.1 Fungi, bacteria, protozoa, and plant roots interact in the rhizosphere.

People have known for a long time that plant roots are colonized by a variety of microorganisms, but scientists have only recently studied the beneficial and detrimental relationships between plants and microorganisms. The organisms that interact with the roots may live within the roots themselves, as do the nitrogen-fixing *Rhizobium* bacteria, or may

live in the soil in immediate proximity of the root. Certain fungi grow within the outside cell layers of the root, but their hyphae, the long cellular strands that make up the fungal body, extend into the soil. The entire space, consisting of the outer tissues of the root (cortex and epidermis), the root surface, and the soil around the root, is the rhizosphere. Plants stimulate this microbial colonization by secreting complex carbohydrates that the microorganisms can use as food. In return, the plants benefit from some associations that make mineral nutrients more available. In many grasses and cereal crops, such as maize, the roots and the soil particles intimately interact so that when the plants are gently removed from the soil, a layer of soil particles, embedded in mucilage secreted by the root cap, covers the terminal portion of the roots like a sheath (**Figure 12.1**). Bacteria and fungi extensively colonize these sheaths. The thickness of the sheath is the same as the length of the root hairs, which penetrate it.

A number of the microorganisms in the rhizosphere fulfill specific functions that benefit plants, such as fixing nitrogen or taking up phosphate; others simply help decompose the organic matter. Organic matter contains nutrients, such as nitrogen or

Figure 12.1 Root sheaths on a young bromegrass plant. The plant was gently removed from the soil and shaken to allow all the loose soil to fall away. The soil particles that remain are connected to the main roots by root hairs, and fungal hyphae, and stick together by slime secreted by the root caps. *Source:* Courtesy of M. McCully.

Figure 12.2 Soybeans resistant to root rot caused by the fungus *Phytophtora sojae.* Growth of this soil-borne pathogen is favored by wet soil conditions. This may happen during a wet spring or summer. The most important method of control is to identify resistant soybean varieties. In this figure, a resistant variety is on the left, a susceptible variety on the right. *Source:* Courtesy of Brian Hudelson, University of Wisconsin.

phosphate, that are tied up in larger molecules and cannot be used directly by the roots. The secretion of mucilage by the growing roots stimulates the growth of bacteria that break down this organic matter and use the nutrients for themselves. Protozoa feed on these bacteria and excrete ammonia that the plants can use. Fungi growing in the same environment can also break down organic matter and release ammonia; soil bacteria usually convert soil ammonia rapidly to nitrate, and for most plants nitrate is the principal source of nitrogen. Other nutrients, such as sulfur and phosphate, are also released by these biological processes and made available for plant growth. Because mineralization depends entirely on the growth and activities of microorganisms, it occurs most rapidly in late spring and summer when soils are moist and temperatures are high. Microorganisms that benefit plants by simply breaking down the organic matter and releasing nutrients are called *saprophytes.*

So far, we have emphasized the beneficial organisms that live in the soil. However, many harmful organisms lurk there as well, including bacterial and fungal pathogens and nematodes. For example, *Phytophtera sojae*, an organism that causes root rot in soybean, is favored by wet soil conditions. Its spores can survive for many years in soil even when no soybeans are grown, so crop rotation is not an effective control. Finding a resistant genotype and breeding in this resistance is the only solution (**Figure 12.2**).

The numbers of pathogenic microorganisms are often held in check by yet other, and often uncharacterized, microorganisms. An early example of such interactions between microorganisms in the soil was uncovered by microbiologist A. W. Henry. Seventy years ago, he studied foot rot infection, by the fungus *Helminthosporium sativum*, of wheat seedlings. When seedlings inoculated with the fungus were planted in sterile soil, 50% of the plants became infected. Henry observed that adding nonsterile soil greatly limited the infection and showed that soil fungi in the nonsterile soil were primarily responsible.

Similar experiments were occasionally reported by other scientists, but progress in this field was slow not only because the interactions between microorganisms in the soil are complex, but also because science is still relatively ignorant about the tens of thousands of soil-dwelling microorganism species that exist. Progress has accelerated in the past 20 years, and the study of the rhizosphere has become an exciting new branch of biology. Opportunities for improving crop productivity by manipulating organisms in the rhizosphere are challenging more biologists to enter this field. Much remains to learn about life in the underground (see **Box 12.1**).

12.2 Rhizodeposition feeds the underground ecosystem.

By allowing plants growing in the field to carry out photosynthesis in the presence of radioactive CO_2, scientists can follow the path of the radioactive carbon and determine where it ends up. Cereal crops send 20–30% of it below ground and pastures send as much as 50% below ground. In most plants the radioactive carbon moves in the form of sucrose (see Chapter 10); the plant respires some of this sucrose for energy; uses some to build the root system; uses some to synthesize mucilage that is secreted by the root cap or to synthesize organic acids and phenolics, small molecules that can be massively secreted

Box 12.1

ow little people really know about life underground is underscored by an article in the weekly scientific journal *Nature* in spring 1992, in which Canadian researchers reported the existence of a 40-acre fungus that lives beneath a forest floor. This fungus is about 1,500 years old and weighs 100 tons. No one has, of course, seen the entire organism, but its size was confirmed by taking samples over the entire site, isolating fragments from different locations. By using the same gene-mapping technique now widely used in forensic medicine to determine the genetic identity of sperm in rape victims, the researchers established that all the samples taken over the 40-acre site were genetically identical. If these samples had come from another fungus, or from the sexual offspring of the same fungus, the genetic fingerprints would have been different. The researchers deduced the age of this fungus from a measurement that shows the fungus expanding outward from its site of ori-

Is the World's Biggest Organism a Fungus?

gin at a rate of 20 cm per year. Assuming a constant growth rate, that means it is about 1,500 years old. However, unfavorable growth conditions or diseases may have slowed it down at times. The species of fungus, called *Armillaria bulbesa*, lives underground and produces edible fruiting bodies (mushrooms).

But is such a fungus really one organism, or is it the vegetative offspring of one fungal spore? That is a matter of semantics. Such a fungal "spread" should probably be likened to a stand of bamboo that covers a large area and is the result of vegetative propagation starting from a single plant. But whatever name is attached to these organisms, their discovery shows that people still know very little about the mysterious life of the underground.

---

by the root cells; and transfers some molecules directly to mycorrhizal fungi (see later discussion) that live partially inside the root cells. Soil biologists use the term *rhizodeposition* to describe this transfer of molecules from roots to soil (**Figure 12.3**).

Every root tip is covered by a root cap, a specialized tissue that forms a caplike structure at the end of the root tip and originates from a meristem in the root cap (see Chapter 8). Cell divisions in this meristem continuously produce root cap cells. As the cells mature, they pass through the cap and are shed or abraded from the cap surface as the root grows between soil particles. The cap of a single maize root produces up to 20,000 cells per day that are left behind in the soil. Keep in mind that a single maize plant may have hundreds of small side roots, each with its own cap. The root cap cells are differentiated cells whose main function is to produce mucilage or slime.

The mucilage lubricates the root tip as it pushes forward between soil particles. The mucilage binds water, protecting the root against drying out and ensuring that a continuous water film covers the soil particles and the root. At night, the root may secrete water to keep its surroundings moist. The mucilage, which is left behind as the root grows, also binds soil particles into larger aggregates, and serves as a source of food for bacteria that grow in the rhizosphere. Experiments show that a root sheath has many more and diverse microorganisms than the soil just

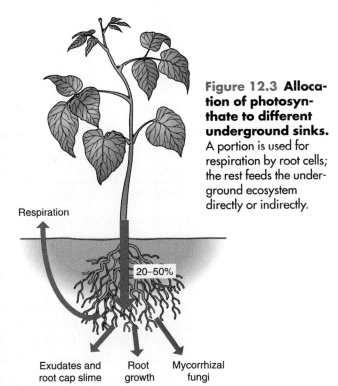

Figure 12.3 Allocation of photosynthate to different underground sinks. A portion is used for respiration by root cells; the rest feeds the underground ecosystem directly or indirectly.

Respiration

20–50%

Exudates and root cap slime

Root growth

Mycorrhizal fungi

beyond it. The sheath represents that portion of the soil that the root is mining for mineral nutrients.

Mucilage is not the only source of energy for root microbes and fauna. As roots grow through soil, root hairs die, break off, and leave behind root cap cells; in addition, fungal hyphae that derive nutrients directly from plants are fair game for microarthropods, nematodes, and worms. It is a veritable feeding frenzy down there. Eat and be eaten!

This rhizodeposition alters the rhizosphere in other ways. For example, root cell secretion of organic acids such as malic acid can tie up toxic aluminum ions that would otherwise inhibit root growth (see Figure 11.15).

12.3   Mineralization, the slow decay of soil organic matter, provides a steady stream of mineral nutrients for plants.

Although rhizodeposition is an important contributor to soil organic matter, residues of plants and animals that live in and on soil also make substantial contributions. Leaf litter accumulates on the soil surface in many forests, and crop wastes are left on the fields after the crops are harvested; as they decay, they are gradually mixed with the top layer of soil (earthworms are especially important in this process) and carried downward by percolating water. In this way, organic matter becomes distributed throughout the topsoil. Organic residues are also contributed by the roots of plants (see **Table 12.1**) and by the soil organisms. The organic matter in soil often gives it a characteristic brown color. Sandy soils and many tropical soils are usually light-colored, because they contain very little organic matter (1 to 2%), whereas heavy clays can vary from dark brown to black because they contain much more organic matter (5 to 10%).

Table 12.1	The organic residues in the roots of crops grown in central Ohio
Crop	**Residue (kg/Ha)**
Soybeans	600
Wheat	830
Corn	1,270
Alfalfa	3,850
Kentucky bluegrass	5,000

Source: Data from U.S. Department of Agriculture.

The amount of organic matter in a soil depends on the rate it is added to the soil and the rate it is broken down by soil organisms. Rate of decay is influenced by prevailing temperature, oxygen availability, and soil acidity. The first phase of the decay process is carried out partly by earthworms and other soil animals, which eat large amounts of leaf litter and dead roots. These materials are their food, and the animals excrete a black organic residue called *humus*. This transformation of organic residues into humus is aided by soil microorganisms that use organic matter as a source of energy and minerals, especially phosphorus, sulfur, and nitrogen. As the microorganisms multiply, they are in turn eaten by grazers (amoebas, nematodes, or microarthropods) that serve as food for yet larger soil organisms. Thus, the continuous process of "eat and be eaten," in which the whole soil ecosystem is involved, gradually converts fresh organic matter into humus and also results in the slow decay of humus (**Figure 12.4**). Humus has three main roles in the soil. It binds together small soil particles into larger aggregates, providing more pores for gas exchange; binds mineral nutrients loosely (by ionic interactions) so they can be used by plants; and improves the water-holding capacity of soil.

Mineral nutrients taken up and therefore immobilized by one group of organisms are released again when these organisms are eaten by organisms on the next trophic level.

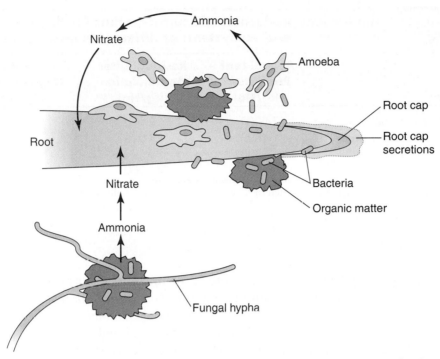

Figure 12.4 Model of proposed interactions in the rhizosphere and in the bulk soil.
A root is growing in the soil from left to right. Using root-derived carbohydrates as a source of energy, bacteria grow and mineralize nitrogen from the organic matter, which is immediately immobilized in the bacterial biomass. Amoebas attracted to the site consume bacteria. When digesting the bacteria, the protozoa release part of the bacterial nitrogen or ammonium on the root surface, where the root can take it up. Below the root, in the bulk soil, a fungal hypha is decomposing organic material. Ammonium released as a waste product can diffuse toward the root as ammonium or, after nitrification, as NO_3^-. *Source:* From a model proposed by M. Clarholm, Swedish University of Agricultural Science, Uppsala.

This complex process is called *mineralization*. For example, nitrogen in organic molecules is slowly released as ammonia (NH_3), which other soil bacteria then rapidly convert into nitrate (NO_3^-). The same types of microorganisms that degrade and mineralize organic matter can also degrade "biodegradable" herbicides and pesticides. However, not all such chemicals are readily biodegradable and those that are not may harm the soil ecosystem.

Decay is accomplished by organisms that require oxygen for their respiration. As a result, it usually proceeds much more rapidly in well-aerated soils. The soil temperature and degree of soil acidity also affect the decay of organic matter because they influence microbial activity. In temperate or cold climates, acidic soils are usually rich in organic matter, because neither earthworms nor most bacteria can thrive in such soils, so organic matter decomposes slowly. However, the acidic tropical soils are poor in organic matter, as are most tropical soils, because high temperature and rainfall speed decay despite the low pH (see **Table 12.2**).

To understand how decomposition works, examine what happens when fresh organic matter, such as a crop residue or a rich farm manure, is added to the soil. Soil microorganisms immediately start working on this organic matter, and because the microbes have a large source of food, they multiply rapidly.

Table 12.2	Annual leaf production and turnover time for the organic matter in several ecosystems at different temperatures		
System	Annual Leaf Production (kg/Ha)	Residual Litter Accumulation (kg/Ha)	Time for Decay of Organic Matter (years)
Rainforest (tropical)	14,000	9,000	1.7
Deciduous forest (temperate)	4,500	14,000	4.0
Conifer forest (northern)	2,700	40,000	14.0
Tundra (Arctic)	900	45,000	50.0

Source: Data from various sources, quoted by C. Kucera (1973), *The Challenge of Ecology* (St. Louis: Mosby), p. 64.

If the organic matter has more nitrogen than the microorganisms need for their own growth, the decay processes release nitrogen into the soil. Because farmyard manure contains nitrogen-rich organic matter (animal wastes) and many mineral nutrients, its decay provides the plants with a steady supply of nutrients. Straw and leaf litter are poor in nitrogen and minerals, so their decay does not help plant growth as efficiently. Farmers use compost to make organic matter relatively richer in nutrients. Compost is organic matter that microorganisms have partially decomposed. Basically, the first rounds of "eat and be eaten" have already taken place when farmers use compost as fertilizer. The enrichment in nutrients is relative, of course. There is no absolute increase in nutrients, only a decrease in carbon (from polysaccharides and lignin), which is released as carbon dioxide into the atmosphere as a result of respiratory activities by microbes. Adding compost to soil tends to boost plant growth more quickly than adding fresh organic matter, which must still decompose to release the minerals.

Converting a natural system to an agricultural one usually decreases the amount of organic matter in soil, for several reasons. Tilling soil and mixing crop residues into topsoil increases microbial activity because it increases topsoil aeration. Thus cultivation accelerates decay of organic matter by initially increasing microbial activity. Maintaining a healthy soil ecosystem therefore requires continuous addition of fresh organic matter to maintain the level of humus in soil and so a vigorous population of microbial decomposers can release nutrients. Organic matter improves soil structure, aeration, and water percolation, and these all contribute to healthy root systems that are not so readily invaded by the many pathogens present in the soil.

12.4   In flooded soils, the absence of oxygen creates a different environment.

So far we have discussed biological processes in well-aerated soils, conditions we usually associate with agriculture. But humanity's most important crop, rice, does not grow in such an environment. Rice is a semiaquatic plant that evolved in the floodplains and along the riverbeds of Southeast Asia. Although rice can be grown on dry land (so-called upland cultivation), lowland paddy rice produces higher yields. Rice is adapted to growing in water, and the soil and rhizosphere conditions in a rice paddy are very different from those discussed so far.

Gas exchange between soil and atmosphere is greatly inhibited when water covers soil, because the diffusion of gases through water is 10,000 times slower than through the air channels in a normal well-structured soil. Thus respiration by roots soon depletes the oxygen, and the soil environment becomes anaerobic. This has major consequences for both the plants and the microbes. Oxygen-requiring (aerobic) bacteria and other microorganisms die off, and anaerobic bacteria proliferate. They convert insoluble forms of iron and manganese into toxic and soluble forms of iron and manganese that the plant readily absorbs. They also convert sulfate into hydrogen sulfide, which is toxic (and can give swamps the smell of rotting eggs). In addition, they convert nitrate into nitrogen gas (denitrification). Thus the chemistry of a swamp or paddy soil is very different from that of a typical well-aerated soil.

Plants that evolved in this environment have special adaptations. They develop air channels (aerenchyma, see Chapter 8) in their stems and roots to conduct oxygen-rich air to the submerged roots (and sometimes stems and leaves). The roots modify the rhizosphere by releasing some oxygen, which oxidizes the toxic forms of iron and manganese to nontoxic forms. Oxygen diffuses out of the roots because the oxygen concentration on the outside is much lower than on the inside. The roots of rice and other flood-tolerant species are usually covered with red-brown deposits, which are precipitates of iron oxides resulting from this oxidation process (**Figure 12.5**). Oxygen released by the roots also allows aerobic bacteria to grow in the rhizosphere. Anaerobic conditions prevail elsewhere in the rice paddy, and the decomposing bacteria produce methane (CH_4) rather than carbon dioxide (CO_2). Methane is an important greenhouse gas, and the enormous expansion of paddy rice cultivation in the second half of the 20th century has resulted in massive increases in worldwide methane production.

Figure 12.5 Rice roots are red. Paddy rice grows in water-covered soil, and the roots of paddy rice are covered with red-brown deposits, which are precipitates of iron oxides resulting from oxygen being excreted by the roots into the oxygen-deprived paddy soil. *Source:* Courtesy of IRRI, Los Baños, The Philippines.

The oxygen level in the roots is not high enough to maintain normal respiration, and the cells switch to fermentation to derive energy from sugar. Fermentation does not release as much energy in the form of ATP as respiration does, but flooding-adapted plants can maintain a much higher level of fermentation than other plants by activating genes that encode enzymes needed for fermentation.

Another adaptation of floodplain plants such as rice is that their stems elongate faster, especially when submerged. This is especially important for deepwater rice, a type of rice particularly well adapted to areas where severe flooding occurs during the monsoon season. Certain types of rice, so-called floating rice varieties, can grow at rates of 20–25 cm per day when partially submerged. In this way they keep up with rising floodwaters. This rapid elongation is triggered by lack of oxygen, which stimulates synthesis of the hormone ethylene.

12.5 Signaling between organisms in the underground is a complex process.

So far we have portrayed plants as somewhat passive participants in the underground ecosystem, which secrete photosynthate to encourage growth of the microorganisms and benefit from the mineralization of organic matter. However, plants are constantly communicating with a multitude of organisms in the soil and actively modify the structure of the rhizosphere community by releasing chemical compounds. For example, plants release large quantities of phenolic compounds into the soil—as much as 100 kg/hectare can be added to the soil by a grassland community each year—that strongly affect neighboring plants and microbes. Other signals released from plants have more subtle effects and are aimed at attracting or repelling specific colonizers. These molecules tell other organisms that a particular species of plant is growing nearby. In the course of evolution, organisms have learned to use these molecules as signals to establish mutually beneficial symbioses, or to become pathogens or parasites. When you signal your friends, you also signal your potential enemies!

In the following sections we discuss two beneficial root symbioses: root–fungal associations called *mycorrhizal roots* for enhanced phosphate uptake, and root–bacterial associations for nitrogen fixation. Establishing such symbioses involves signaling by both partners. However, the soil also harbors numerous plant pathogens that are alerted to the presence of specific plants by the presence of certain chemicals in the soil.

In addition to microbial pathogens, plants may be parasitized by other plants. At least 3,000 species of flowering plants are parasitic on other plants, and this type of parasitism seems to have often evolved numerous times independently. Some parasites invade the roots, and others the shoots of plants. Some are true parasites—they have no chlorophyll—and others are hemiparasites or facultative parasites. Facultative parasites can survive without parasitizing a host, although they typically grow more vigorously when attached to a host. Establishing an active parasitism generally involves active signaling between the two plant species. The signals include (1) germination stimulants that enhance germination of the parasite seeds in the host's vicinity and (2) chemicals called *haustoria-induction factors* that promote growth of invasive haustoria on the roots of some parasitic plants.

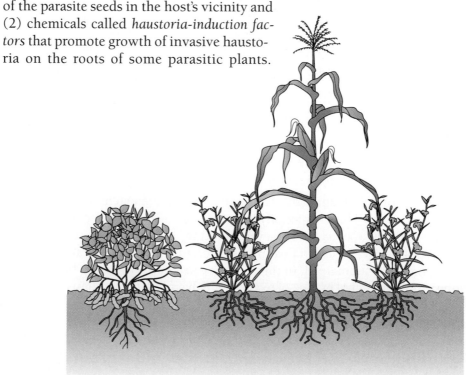

Figure 12.6 Plant-on-plant parasitism. Two *Striga* plants parasitize a field-grown maize plant. The same species of *Striga* does not parasitize the groundnuts growing nearby. Signaling is necessary to establish parasitism. The host (maize) secretes chemicals that cause the *Striga* seeds to germinate.

Haustoria are projections from fungal hyphae that function as organs that penetrate plant roots and transfer nutrients between plant and fungus in both directions.

The parasitic plants that have the greatest impact on worldwide food production are the many different species of *Striga* (witchweed) and Orobanche (broomrape) (see Figure 5.8). Together they parasitize the roots of important cereals, such as maize, sorghum, millets and rice, legumes, sunflowers, and vegetables (**Figure 12.6**). Different *Striga* species infect two thirds of the 75 million hectares of cereals and legumes in sub-Saharan Africa, with different species parasitizing different plants. Crop losses resulting from *Striga* invasion can reach 100%, and fields often must be abandoned. At this time farmers have no effective, low-input way to combat these infestations.

12.6 Mycorrhizae are root–fungus associations that help plants take up phosphate.

The majority of higher plants have fungal hyphae closely associated with their actively growing roots. This type of association, called a *mycorrhiza*, was described more than a hundred years ago, but its beneficial effects on plant growth were discovered relatively recently. Scientists now realize that a mycorrhiza is a true symbiotic relationship. The flow of photosynthetic products from the shoot to the roots benefits the fungus, which uses these products as a source of food; the plant benefits from growing in association with the fungus, especially if the soil is poor in nutrients. When plants are growing in phosphate-poor soils, or under other types of nutrient stress, mycorrhizal plants grow faster and more vigorously than nonmycorrhizal plants because the fungal hyphae take up phosphate from this phosphate-poor soil and

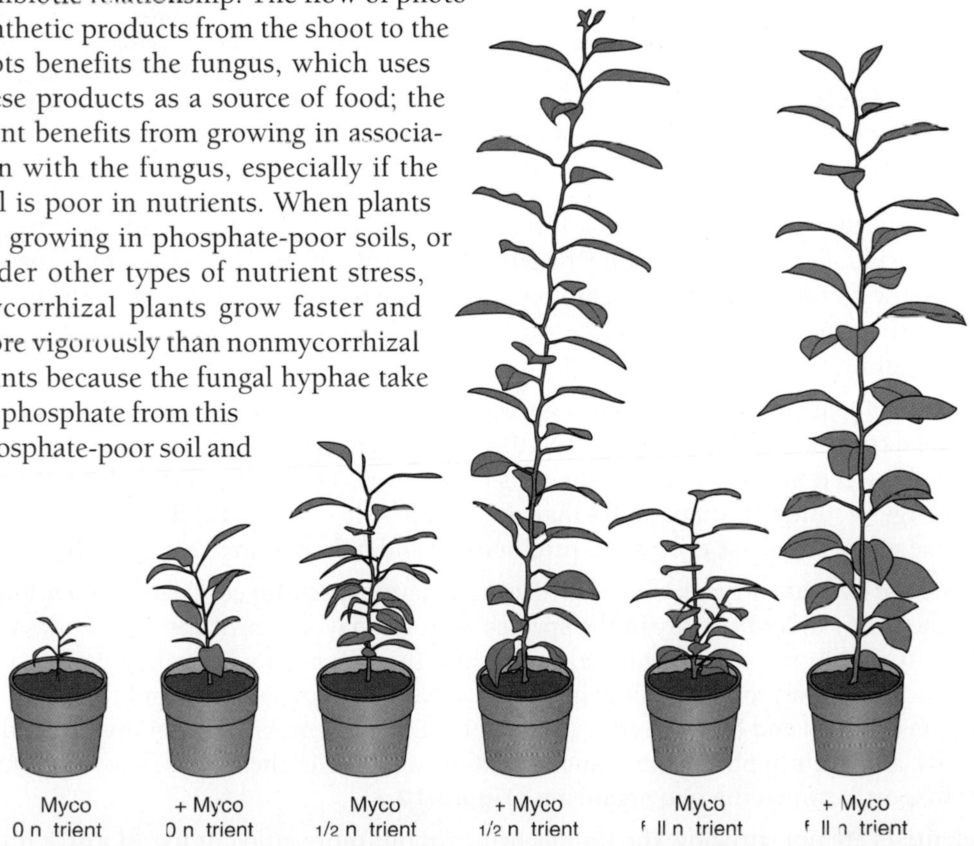

Myco	+ Myco	Myco	+ Myco	Myco	+ Myco
0 n trient	0 n trient	1/2 n trient	1/2 n trient	f ll n trient	f ll n trient

Figure 12.7 Cultivated citrus species show a marked dependency on an endomycorrhizal association for adequate growth. Shown here are rough lemon seedlings after six months, with and without mycorrhizae, and with weekly applications of no, half, or full-strength nutrient solution without phosphate. The presence of mycorrhizal fungi greatly helps the citrus seedlings obtain phosphate from the soil when other nutrients are supplied.

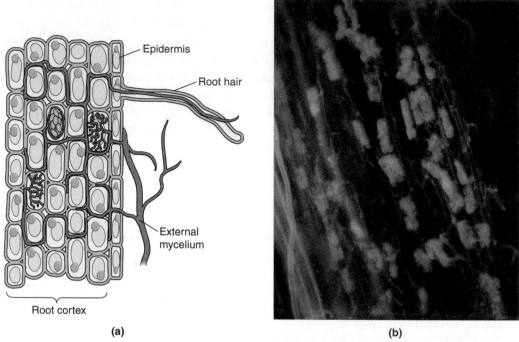

Figure 12.8 Mycorrhizal fungus penetrates a plant root. (a) Schematic drawing of the fungal hyphae. After penetration, the fungus forms treelike structures (arbuscules) in the cortex cells. **(b)** Image of *Medicago truncatula* roots colonized with an arbuscular mycorrhizal fungus, *Glomus versiforme*. The roots were stained with acid fuchsin and viewed by fluorescence microscopy. The root was squashed to enable visualization of the fungal arbuscules and intercellular hyphae within the root cortex, where the yellow masses represent the arbuscules. *Source:* Courtesy of Maria Harrison.

transfer it to the plant (**Figure 12.7**). If these same plants grow in rich soils, the mycorrhizal associations are not measurably beneficial. Some plants, such as certain pine species, cannot grow without their fungal symbionts. When such pines are transplanted to an area in which they are not native, it is necessary to inoculate the soil with soil fungi from the area of origin.

The fungal partner of a mycorrhiza grows on the root surface, but in addition its hyphae can extend far into the soil between the soil particles, as well as into the root cortex growing between the cells and into the cells themselves. The hyphae can draw nutrients from a larger volume of soil particles than a root can by itself. In natural ecosystems, mycorrhizae greatly increase ecosystem productivity and help maintain biodiversity.

There are two major types of mycorrhizal associations: ectomycorrhizae and endomycorrhizae. They differ not only in the species of fungi that can form the different associations but also in the way they invade roots and the extent to which they modify root morphology. Ectomycorrhizal fungi grow as a sheath on root surfaces, and their hyphae penetrate into soil and between root cortex cells. Endomycorrhizal fungi invade epidermis cells, and their hyphae form branched extensions within these cells, allowing nutrient transport between the two organisms (**Figure 12.8**).

Scientists are not sure how the fungus brings about more efficient use of nutrients in nutrient-poor soil. For example, roots with mycorrhizae of beech trees take up phosphate five times faster than nonmycorrhizal roots. Microorganisms generally secrete enzymes that help mineralize organic matter. Secretion of the enzyme phosphatase helps phosphate solubilization and uptake by breaking down larger molecules that contain phosphate.

Whether mycorrhizae can also solubilize the insoluble mineral phosphates is not entirely clear. Another way that mycorrhizae may obtain scarce nutrients is by secreting special organic molecules. After such an organic carrier molecule is secreted, it forms a complex with a single nutrient ion and is then taken up again by the organism that produced it. Although this mechanism has not yet been demonstrated for mycorrhizae, it is reasonable to hypothesize that mycorrhizal fungi acquire scarce minerals in this manner.

12.7 Cereals and most other plants use soil nitrate, but legumes have symbiotic bacteria that can use atmospheric nitrogen.

All but one of the inorganic nutrients that plants obtain from the soil are made available to the plants by the weathering of soil particles. The one exception is nitrogen, which plants normally take up from the soil solution as nitrate (NO_3^-). Soil particles have no nitrogen-containing minerals. Nitrogen gas (N_2) makes up almost 80% of Earth's atmosphere, but plants cannot use this atmospheric nitrogen directly and depend on microorganisms to transform it into a usable form. Once nitrogen is in a form where it is combined with hy-

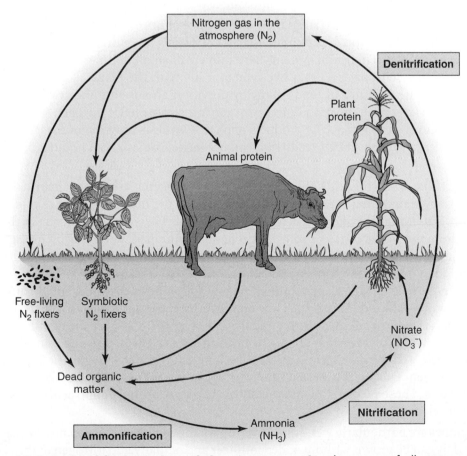

Figure 12.9 A simplified version of the nitrogen cycle. The source of all nitrogen in the biosphere is the N_2 in the atmosphere. This N_2 is fixed by free-living N-fixers or symbiotic N-fixers. Plants such as maize (shown on the right) depend on nitrate generated by soil bacteria; animals depend on plants.

drogen (as ammonia) or oxygen (as nitrate), plants can use it for synthesizing proteins and other macromolecules. Animals can eat the macromolecules, but eventually the proteins return to the soil as dead organic matter that is then degraded by decomposers. In addition, nitrate can be converted to N_2 by yet other microorganisms and released again into the atmosphere. Together, these various transformations of nitrogen make up the nitrogen cycle, as illustrated in **Figure 12.9**.

Nitrogen normally enters the living world through the process of biological nitrogen fixation, which converts nitrogen gas to ammonia. Only certain bacteria and fungi have the necessary enzymes to fix nitrogen, and they fix an estimated 100 million tons a year. Some of these nitrogen-fixing organisms live in lakes and rivers and others live in the soil. For agriculture, the most important nitrogen fixers are those that live symbiotically in the roots of certain plants, especially the legumes. Natural systems such as forests and grasslands rely heavily on the sustained activity of the nitrogen-fixing organisms in the soil to supply the ecosystem with usable nitrogen. Many soils are nitrogen poor, and it is precisely because legumes can fix nitrogen symbiotically that they compete well on such soils.

In the soil, most nitrogen is present in organic matter, from which it is slowly released as ammonia. Once released, ammonia is rapidly converted to nitrate by soil bacteria, and most plants normally take it up in this form. The enzymes for nitrate assimilation and its incorporation into amino acids are crucial for plant growth. The genes that encode these enzymes are highly regulated by light, hormones, and nitrate itself, so that nitrate assimilation can be integrated with other important aspects of cellular metabolism.

Most plants rely on soil nitrate as their main source of nitrogen, but soybeans and other legumes have, in addition to the ability to take up nitrate, an entirely different method of obtaining nitrogen. Their root systems have numerous small nodules (see next section), which are filled with bacteria that live symbiotically with the plants. The plants supply the bacteria with a source of food energy by providing them with photosynthetic products. The bacteria have the unique ability to convert atmospheric nitrogen (N_2) into ammonia by nitrogen fixation. The products of photosynthesis move from the leaves to the root nodules, and amino acids ready for plant use move from the root nodules back to the leaves.

In soils that are fertile and contain considerable nitrogen compounds that can be mineralized, soybeans and peanuts derive about a third of the nitrogen in their protein from the nitrogen-fixing activities of bacteria in their roots. They take up the rest of their nitrogen from the soil as nitrate. From an agronomic point of view, relying on nitrogen fixation may have drawbacks as well as advantages. The nodules that contain the bacteria take several weeks to develop, and the plants still depend on soil nitrate when they are young. In the spring, when temperatures are low and mineralization of soil organic matter is slow, available soil nitrogen is low and plants may be off to a slow start. However, the low levels of nitrogen cause better nodules to be formed, and once the nodules become active, the plants quickly catch up. As the plants get older, the nodules age, and when the seeds have set and the main period of seed development is about to begin, the bacteria stop fixing nitrogen. During this period, the plant must rely on proteins already synthesized in the leaves. These proteins are broken down and the resulting amino acids transported to the developing seeds.

Legumes are particularly well suited for crop rotations and intercropping. Precisely because they can be grown without added fertilizer and actually add nitrogen to the ecosystem, they are extremely valuable in all sustainable agricultural systems. Before the manufacture of relatively cheap chemical nitrogen fertilizers, planting legumes was widely recommended to add nitrogen to soil. Seed legumes such as soybeans or cowpeas, or perennial forages such as alfalfa or clover can be used for this purpose. Organic farmers still rely on legumes to provide much of the needed nitrogen for other crops.

12.8 Rhizobium bacteroids live as symbionts within the root nodules of legumes.

Especially important for nitrogen fixation both in natural ecosystems and for agriculture are the Rhizobium bacteria that live symbiotically inside the root nodules of leguminous plants. (The term *Rhizobium* is used here to describe the bacteria belonging to three different species in three different genera.) Scientific understanding of how this symbiosis is established has greatly advanced in recent years. Establishing an effective symbiosis requires an exchange of molecular signals that turn on specific genes in the two partners (**Figure 12.10**).

Rhizobia are free-living soil bacteria that attach themselves to the surface of root hairs in legume plants and cause the root hairs to curl around them and engulf them. By producing enzymes that partially dissolve the cell wall, they manage to penetrate the root hair. In response, the plant forms a tube or infection thread in which the bacteria multiply and are confined at the same time. The bacteria stimulate the root cortex cells to start dividing and form a new meristem. When this tube reaches the cortex cells, the bacteria are separated from it, surrounded by a membrane, and continue to live in these small vacuoles. In the meantime, the chemicals have also stimulated the cortex cells to divide and form a nodule. At the same time, the bacteria continue to divide and soon fill the cortex cells, living in small vacuoles within the cytoplasm. Most of the bacteria now cease dividing and begin making the enzymes and other proteins necessary for nitrogen fixation. Capable of fixing nitrogen, they are now called *bacteroids*. Some bacteria continue to divide and fill

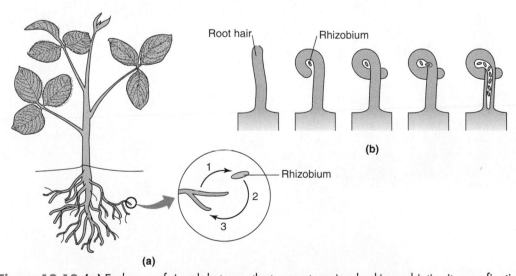

Figure 12.10 (a) Exchange of signals between the two partners involved in symbiotic nitrogen fixation. Step 1: legume roots release chemicals (flavonoids) that stimulate the expression of genes in Rhizobia and the synthesis of enzymes that make it possible to synthesize Nod factors. Step 2: Nod factors are secreted by the Rhizobia. Step 3: The Nod factors produced by the Rhizobia cause root hair curling and cell division to create a nodule. **(b)** Root hair curling permits the entry of Rhizobium bacteria into the root hairs. *Source:* After B. B. Buchanan, W. Gruissem, and R. L. Jones, eds. (2000), *Plant Biochemistry and Molecular Biology* (Rockville, MD: American Society of Plant Biology). Figures 16.16 and 16.17.

new nodule cells formed as the root nodule grows. Indeed, the root nodule is not just a swelling on the root but a modified lateral root with its own meristem and its own vascular tissue. When the nodule cells eventually die, the bacteroids die with them. They cannot survive in the soil and join their free-living cousins. However, chemicals secreted by the roots ensure that there will always be free-living Rhizobium bacteria in the soil.

The interaction between Rhizobium and its legume host is very specific. Usually one Rhizobium species can infect only one or a few closely related legume species. This specificity resides in the signaling molecules that both organisms use. The plant roots release flavonoids (complex organic molecules related to red and blue plant pigments) into the soil, and these can have many different structures, with different plant species producing different flavonoids. A particular Rhizobium species will respond best to the flavonoids of one or a few legume species, and not as well to the flavonoids of other plant species. Each Rhizobium species responds to these plant flavonoids by producing small lipid-carbohydrate molecules, called Nod (for nodulation) factors, with a unique structure; a particular legume species is triggered by only one type of Nod factor but not by any slightly different one. In this way, plants and bacteria signal their identity to each other in the underground. The synthesis of Nod factors requires several enzymes encoded by genes in the bacteria, and these genes are activated by flavonoids secreted by the legume.

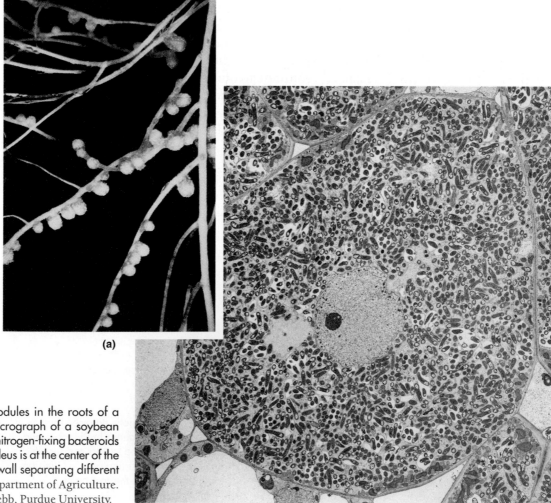

(a)

(b)

Figure 12.11 (a) Nodules in the roots of a legume. **(b)** Electron micrograph of a soybean nodule cell showing that nitrogen-fixing bacteroids fill the entire cell. The nucleus is at the center of the cell. Notice the thin cell wall separating different cells. *Sources:* (a) U.S. Department of Agriculture. (b) Courtesy of M. A. Webb, Purdue University.

The Nod factors in turn activate genes in the legume root that cause the formation of the root nodules (**Figure 12.11**).

The formation of the root nodule and the small vacuoles in which the bacteroids live requires plant cells to synthesize many new proteins and involves the expression of many plant genes not expressed in normal root cells. The most abundant of these proteins is the red protein leghemoglobin, a molecule closely related to muscle myoglobin and to red blood cell hemoglobin. Nodules contain so much leghemoglobin they actually look pink. The function of leghemoglobin is to diminish the concentration of oxygen in cells without making them totally anaerobic. The functioning of the bacterial enzymes that reduce atmospheric N_2 to ammonia requires that the oxygen level in the cells be very low. Other proteins made by the plant cells include special membrane transporters and enzymes that use the ammonia made by the bacteroids for synthesizing amino acids that can be transported through the vascular system to the rest of the plant.

There is an enormous flux of metabolites across the membranes of the bacteroid-containing vacuoles: products of photosynthesis in one direction, products of nitrogen fixation in the other. Transporters that allow these metabolites to be carried across the membranes are synthesized in response to the bacterial invasion. Many new genes are also activated in the bacteria as they form bacteroids capable of fixing N_2. Most important are the genes that encode enzymes for N_2-fixation, as well as proteins that regulate the activity of these genes.

12.9 Free-living bacteria and fungi also convert atmospheric N_2 into ammonia.

In recent years, scientists have described the many different free-living organisms that fix N_2 in soil and that live in the rhizosphere in close association with plant roots or that thrive on the soil surface. Some, such as the fungi that form root nodules on many trees, may contribute ammonia directly to the plant in the same way as the Rhizobia in legume root nodules. Bacteria that live in the sheaths surrounding the roots of many tropical grasses also contribute fixed nitrogen. Nitrogen-fixing bacteria that live in a crust at the soil surface are photosynthetic cyanobacteria that may form symbiotic associations with fungi called *lichens*. Measurements in the Sonoran Desert of Mexico showed that such cyanobacteria-lichen crusts annually contribute 7–18 kg per hectare of fixed nitrogen to the ecosystem. The input of fixed nitrogen from a legume–Rhizobium symbiosis is much greater, ranging from 50 to 200 kg/ha per year, depending on the legume species. Nitrogen-fixing bacteria growing in rice paddies have been shown to contribute anywhere from 5 to 80 kg/ha per year. This enormous range shows that there is much scope for increasing nitrogen fixation by soil organisms. The challenge for agriculturists is to encourage those nitrogen-fixing microbes without upsetting delicate food webs in the soil.

Not all Rhizobium–legume symbioses that are established have the same ability to fix nitrogen. A really effective strain forms healthy-looking nodules with a distinctive pink color. An ineffective strain forms colorless or greenish nodules. Once a plant has been colonized by one strain, others seem to be excluded, so it is important to start the plants off right. In many farming areas, the Rhizobia in the soil are not particularly effective for fixing nitrogen in symbiosis with the legume crops best suited to that area. However, effective strains are now available to farmers and can be applied at planting time as a seed coating.

Figure 12.12 Cover crop of clover, a forage legume on the floor of an almond orchard in California, is an integral component of sustainable agricultural practices. *Source:* Courtesy of R. Bugg, University of California, Davis.

Advocates of sustainable agriculture suggest that farmers should rely more on nitrogen fixation and less on chemical nitrogen fertilizers. The benefits of symbiotic nitrogen fixation can be realized in several different ways: by crop rotations (soybeans one year, maize the next year), intercropping (seeding vetch in a rice field that is close to harvest, or planting alternate rows of maize and beans) or green manuring (plowing under a fresh growth of alfalfa or clover [**Figure 12.12**]).

One often-publicized goal of plant genetic engineering is to transfer the ability to nodulate and to establish effective symbioses to nonlegume plants. This could save nitrogen fertilizer and prevent groundwater contamination by nitrate. However, because so many different proteins are needed to make functional nodules and because of the species-specific recognition between plants and Rhizobia, transferring the nitrogen-fixing capacity from legumes to nonlegumes will remain a distant goal. In addition, it is also important to remember that nitrogen fixation by bacterial symbionts is not "free" but depends indirectly on photosynthesis. Photosynthate must be diverted to the roots to provide energy for nitrogen fixation and is therefore not available for growth and seed production or other energy-requiring processes. Although reliable estimates are difficult to make, as much as 20% of the photosynthate in a nodulated legume plant may be used for nitrogen fixation.

12.10 Biological warfare goes on in disease-suppressive soils.

Take-all (isn't *that* an unusual name for a disease!) is a fungal disease of wheat roots caused by a pathogen with the unpronounceable Latin name *Gaeumannomyces graminis* var. *tritici*. It doesn't kill the plants outright, but the damaged root system makes the plants susceptible to drought stress. The plants mature early, and wheat yields can be severely reduced. The most characteristic symptom is the black, infected area at the base of the stem (**Figure**

12.13). In the northwestern United States, an important wheat-growing area, farmers lose 5–10% of the wheat crop to take-all disease. There are no commercially available chemical control agents, and there is no known resistance in related species that could be used for breeding resistant wheat strains. The fungal pathogen remains in the wheat stubble that stays on the field, and the disease gets steadily worse if wheat is grown year after year on the same land. The small time gap between wheat harvest in midsummer and sowing the next crop of (winter) wheat in the fall makes matters worse. The surviving fungus jumps from the decaying stubble to the new young plants, ready for another round of devastation. If a monoculture is continued for three to five years, infection severity greatly increases, but beyond that time—after more than five years—the disease begins to decline again. Researchers found that factors in the soil were responsible for the decline of the disease.

Figure 12.13 **Symptoms of take-all disease.** Take-all, a fungal disease of wheat, is most severe when wheat is grown several years in a row. The root system and the base of the plant rot and turn black, weakening the plants, which may fall over. Crop rotation is best defense against the disease. *Source:* Courtesy of Brian Hudelson, University of Wisconsin.

Soils that decrease disease severity are referred to as "disease-suppressive soils." Disease suppression is widely attributed to Pseudomonad bacteria that live and multiply in the root lesions caused by take-all fungus. The Pseudomonads probably produce antibiotics that inhibit the fungus growth.

In a disease-suppressive soil, a new equilibrium is established between the pathogen and one or more organisms that inhibit its growth. The pathogen has not disappeared, and the roots and plants (and farmers) still suffer from its presence. However, severe crop losses are curtailed, and an ecosystem brought into disequilibrium by continuous monoculture has found a new equilibrium. Crop rotation is the best way to avoid take-all disease, but the alternate crop may not be as financially rewarding as wheat or may require specialized planting and harvesting equipment that the farmer does not have.

One way to find out which microorganisms may be helpful in disease prevention is to examine the rhizosphere of healthy plants in otherwise heavily infested fields. If beneficial strains can be identified in laboratory and field studies, the microbes can be grown in fermentors and then applied to the infested fields as small pellets or as a coating on seeds. Seed companies already widely use seed treatments to enhance seed vigor or seedling performance by treating the seeds with certain chemicals (see Chapter 9).

Unfortunately, we still know almost nothing about the thousands of species of microorganisms or about their interactions. New molecular techniques may allow us to learn much more, even if we can't culture the microorganisms in the laboratory (see **Box 12.2**).

12.11 When they infect roots, nematodes cause serious crop losses.

Nematodes, also called *eelworms*, are tiny animals that live in the soil—as many as 500 in a spade-sized scoop. Many are free-living grazers that eat bacteria, fungi, protozoa, and microalgae, but others severely damage crops (**Figure 12.14**). Parasitic nematodes have a sharp mouthpiece with which they pierce the cell wall of the root epidermis.

Having gained entry (**Figure 12.15a**), nematodes have two different ways of parasitizing the plant. The root knot nematodes invade the roots and begin to multiply, causing tumorous growths on the roots (**Figure 12.15b**). The cyst nematodes remain on the

Box 12.2

Molecular Soil Ecology

Many important interactions between plants and their environment take place below ground, particularly if that environment is poor in nutrients, as in about a third of the world's soils. For example, whether plants can compete on nitrogen-poor soils depends on their ability to establish an effective symbiosis with nitrogen-fixing bacteria. Similarly, competition on phosphate-poor soils depends on the associated mycorrhizal fungi. Dynamic interactions among plant roots, microbes, and the microflora are crucial for plant growth. Although they have identified a few of the species involved in these interactions, ecologists are just now beginning to study underground biodiversity. They don't have a baseline, so they won't be able to figure out how rising CO_2 and global warming are affecting diversity underground. In the summer of 2000, the Ecological Society of America called for a major expansion of soil ecology research.

One reason people know so little about the diversity of soil organisms is that scientists don't know how to culture such organisms in the laboratory. Laboratory cultures have always been a prerequisite for identifying and studying new microorganisms. New molecular techniques can now help scientists identify species that they can't culture. For example, the polymerase chain reaction (PCR), which amplifies DNA, is being used to look at species diversity in bacterial populations. In such experiments, a species is identified only by the nucleotide sequence of a single gene (usually the gene encoding the large ribosomal RNA). Without knowing anything else about the species, scientists can determine how the abundance of a particular species changes. For example, they could determine the species complexity in groundwater and then find out how contamination of groundwater with manure runoff or pesticides changes that complexity.

To gain insights into metabolic processes of uncharacterized bacteria, scientists in Great Britain fed ^{13}C-methanol (^{13}C is a heavy stable isotope of carbon, which is normally ^{12}C) to a hydrated soil sample and allowed the bacteria to metabolize this methanol (certain species of bacteria use methanol as their principal carbon source). Some bacteria incorporated this heavy carbon into their DNA as they multiplied, and the scientists were able to physically separate this heavy DNA from standard bacterial DNA. Sequencing of the heavy DNA then allowed them to identify which bacterial species in the soil were able to metabolize methanol. Such molecular techniques allow valuable insights into underground species diversity.

Figure 12.14 Soybean field infected with soybean cyst nematodes. Whereas low-level infestation by nematodes may simply lower the final crop yield, large-scale invasions, as shown in the center of this field in Wisconsin, may result in yellowing and stunted plants. *Source:* Courtesy of Brian Hudelson, University of Wisconsin.

outside of the root and are attached to a single plant cell, which becomes very large. The plant feeds the growing nematode by continuously transferring photosynthate to the giant cell and then to the nematode via the mouthpiece. The body of the nematode swells up until it consists of a giant cyst containing thousands of eggs (**Figure 12.15c**). These eggs are released, and the cycle starts over again.

Most nematodes parasitize only one species of plant. As a result, their abundance in the soil, like that of pathogens, is greatly influenced by crop rotations. Continuous cultivation of the same crop leads to a buildup of nematodes in the soil that parasitize that particular crop.

On a worldwide scale, nematodes cause enormous crop losses, yet there are no chemicals that kill nematodes and are also safe to use. The nematocides produced by chemical companies are effective against nematodes only because they are extremely toxic to many organisms.

(a)	**(b)**	**(c)**

Figure 12.15 (a) Scanning electron micrograph of a nematode entering a root. **(b)** Carrots infected with root knot nematodes are stunted and deformed. **(c)** A female potato cyst nematode attached to a potato root. At this stage of its life cycle, the body of the female consists of a huge egg sac or cyst, still attached to the root by the head, which continues to feed off the plant. *Sources:* (a) U.S. Department of Agriculture. (b) University of Wisconsin. (c) Courtesy of Stephen Ohl.

They are, in essence, biocides, and their use is being phased out in the United States and other developed countries.

Effective control of nematodes can be achieved in several ways. Cultivation practices and crop rotations can do much to keep nematode levels down by encouraging their natural enemies and pathogens. A healthy soil may harbor as many as 60 species of predatory mites, microarthropods that feed mainly on nematodes. The different species of mites are extremely well adapted to different microhabitats in the soil. Conventional tillage (plowing the soil) upsets these microhabitats, disturbing the ecosystem, reducing the number of species as well as numbers of individuals. Much research is needed to understand the dynamics of these organisms in the soil. Perhaps if scientists learn to manipulate these mites in the laboratory and understand their feeding preferences, biotechnology can provide biological control methods. A return to farming methods that are ecologically sound could possibly achieve the same result. In addition, molecular geneticists are searching for genes that could kill a nematode when it invades the root system. Some *Bacillus thuringiensis* toxins (see Chapter 16) can act as nematocidal proteins.

12.12 The agronomic practices of sustainable agriculture promote a healthy soil ecosystem.

The soil is part of a complex ecosystem, and relations between the many different species may be harmed by modern agricultural practices, such as continuous monoculture, chemicals used for weed control and pest control, frequent tilling that reduces organic matter, and irrigation that raises soil salinity and pH. The complexity of the root ecosystem makes it impossible to predict how and when the balance will be tipped in

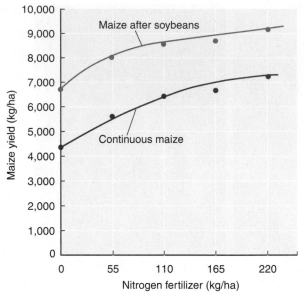

Figure 12.16 Maize yields in kilograms per hectare for "continuous maize" or "maize after soybeans" with different levels of nitrogen fertilizer. Crop rotation is clearly superior to continuous culture. The effect of soybeans on maize yield is equivalent to applying about 150 kg of nitrogen per hectare. However, this effect of crop rotation cannot be attributed solely to the capacity of soybeans to fix nitrogen. Crop rotations have many other beneficial effects, especially suppression of pests and pathogens.

favor of pathogens. A pathogen does not become a pathogen until its numbers get out of hand. Before that, it lives without doing any harm to the plant, helping to break down organic matter.

In the past 25 years, many studies have been done on the effect of agronomic practices on pathogens in the soil, and a few general principles are emerging. Clearly, the agronomic practices advocated by proponents of sustainable agriculture stabilize the soil ecosystem and ameliorate many of the problems farmers are now encountering.

Crop Rotation. Crop rotation generally increases crop productivity (**Figure 12.16**), and the most important benefit from crop rotation is a lower level of pathogens and pests in the soil. Crop rotation frequently reduces disease problems unless the other crop in the rotation is also a host. With take-all disease, for example, barley can also be a host, but potatoes, lentils, and peas can be used as "break" crops. However, a number of disease, insect, and weed problems are not diminished by rotation, because the microbial spores, insects, or weed seeds remain in the soil for many years no matter which crop is grown.

Tillage. Tilling the soil can have positive and negative consequences for crop productivity. Tilling mixes the crop residues into the top layer of soil, causing organic matter to decay faster. Soils with lower levels of organic matter tend to have less stable ecosystems because there is a lower overall level of microbial activity. Tilling destroys weeds, but it accelerates soil erosion by wind and water. Because of tilling's potential detrimental effects, some farmers practice no-till agriculture. *Direct drilling*, which involves dropping seeds in small holes by a mechanical planter, is an example of no-till agriculture. This method reduces erosion, especially if the stubble is left on the field but increases the severity of weeds, requiring the application of more herbicides.

Direct drilling has also been shown to increase the severity of three root rot diseases of wheat in the Pacific Northwest of the United States. The likely reasons are (1) it allows more pathogens to survive in the crop residues (stubble), and (2) the seeds are then placed in the soil layer richest in pathogens. The complexity of the situation is demonstrated by the fact that direct drilling of wheat in the same region greatly reduces damage from the cereal cyst nematode. A possible solution, and one way to achieve sustainable agriculture, is to abandon tilling in favor of no-till methods *and* to abandon continuous monoculture in favor of crop rotation.

Fertilizers. The type of fertilizer used, the time of its application, and where it is placed in the soil all can affect pathogens. If seedlings get off to a slow start because the soil is nutritionally deficient, they are more prone to infection by pathogens. However, the growth of the pathogens as saprophytic organisms—at first they all grow saprophytically—is often limited by the low nitrogen content of the stubble. If no nitrogen is available, they may be outcompeted by other saprophytes that grow well on low-nitrogen stubble. Some pathogens cannot survive on acid soils, so keeping the soil acidic controls them. However, legume–Rhizobium symbioses do not thrive on acid soils, and many crop rotations involve cereals and legumes. Liming acidic soils improves the growth of the legumes, but may increase problems with pathogens for the cereal in the crop rotation. Pathogens that are most damaging in acid soils are usually favored by the use of ammonium fertilizers, whereas pathogens that damage plants in neutral or basic soils are encouraged by nitrate applications.

Organic Matter. Adding organic matter (crop residue or manure) is always beneficial for the soil ecosystem. Farmers leave crop residues on the field during winter (**Figure 12.17**) or seed a special cover crop—often a legume—that may be harvested or plowed under (see Figure 12.12). The benefits of adding organic matter and not using inorganic fertilizers or synthetic pesticides are well known to organic farmers, who invariably experience an improvement in the condition of their soil when they switch from conventional to organic farming.

Figure 12.16 Crop residues are left on the field. In this maize field in central Michigan, crop residue captures the winter snow preventing it from being blown away and adding moisture to the soil. When the field is cultivated, the residue will add organic matter to the soil. *Source:* Courtesy of J. Hageman.

CHAPTER SUMMARY

Life underground is tremendously complex. Twenty years ago, people knew very little about the rhizosphere; today scientists know only a little of what is needed for devising intervention strategies to tip the balance in favor of beneficial organisms.

At least three general classes of microbes benefit plants. Certain bacteria and fungi establish symbioses with plant roots. They either live completely within root cells or grow in between cells, with their hyphae extending into the soil. Most important among these are the symbiotic Rhizobia and the mycorrhizal fungi. Other microorganisms keep pathogens in check and therefore contribute indirectly to plant growth. A third group, the saprophytes, is involved in mineralizing or breaking down organic matter, fixing lesser amounts of nitrogen compared with the symbiotic Rhizobia, and solubilizing inorganic minerals such as phosphate and iron. Many pathogens grow initially as benign saprophytes and become pathogenic when conditions favor them. The food for all these microorganisms and for the entire rhizosphere ecosystem comes from the plant: The products of photosynthesis are transported to the roots, which secrete sugars, organic acids, and complex carbohydrates. These chemicals modify the physical and the living environment in the immediate vicinity of the roots.

Agronomic practices greatly influence the rhizosphere ecosystem. Biotechnology can help by growing microbes in the laboratory that can be used to inoculate the soil. This is most easily achieved by coating the seeds just before they are planted. It is likely that in the future microorganisms will be specially selected or genetically engineered to more effectively promote plant growth and crop yield. Adopting agronomic practices that contribute to the stability of the soil ecosystem (crop rotation, intercropping to add more crop residues, no tillage, and so on) could eliminate some of the problems caused by present practices.

Discussion Questions

1. Consider global warming caused by rising CO_2 levels. How can global carbon storage in the soil be affected by (1) deforestation, (2) increasing CO_2, and (3) increasing temperature?

2. The soil contains many microbial species that cannot be cultured in the laboratory. What are some reasons why this is so? How then can scientists learn about them?

3. Draw a diagram of carbon flow in the soil. Who eats whom? Which elements contribute to humus, the more stable fraction of soil organic matter?

4. When using manure to fertilize vegetables, it should be cured, preferably at high temperatures. What happens during the curing process?

5. Discuss mineralization in temperate and tropical soils. Why are tropical soils quickly exhausted after they have been deforested. They seem to be nutrient poor although the vegetation is luxuriant (before it is removed).

6. Techniques of organic agriculture generally produce "healthier" soils. How might such techniques alter the balance of organisms in the soil.

7. Discuss rhizodeposition and its importance in feeding the underground system.

8. How does lack of oxygen affect soil chemistry and soil ecology? How can plants adapt to live in oxygen-poor water?

9. The use of nitrogen-fixing legumes is a major feature of nearly all agricultural systems. Trace the path of a nitrogen atom from N_2 in the atmosphere to protein in a maize seed.

10. Why do continuous monocultures change the soil ecosystem?

Further Reading

Allen, M. F. 1991. *The Ecology of Mycorrhizae.* Cambridge, U.K.: Cambridge University Press.

Baker, B., P. Zambryski, B. Staskawicz, and S. P. Dinesh-Kumar. 1997. Signaling in plant–microbe interactions. *Science* 276:726–733.

Bethlenfalvay, G. J., and R. G. Linderman, eds. 1992. *Mycorrhizae in Sustainable Agriculture.* ASA Special Publication No. 54. Madison, WI: American Society of Agronomy.

Campbell, R. 1989. *Biological Control of Microbial Plant Pathogens.* Cambridge, U.K.: Cambridge University Press.

Estabrook, E. M., and J. I. Yoder. 1991. Plant–plant communications: Rhizosphere signaling between parasitic angiosperms and their hosts. *Plant Physiology* 116:1–7.

Garbaye, J. 1991. Biological interactions in the mycorrhizosphere. *Experientia* 47:370–375.

Harman, G. E. 1992. Development and benefits of rhizosphere competent fungi for biological control of plant pathogens. *Journal of Plant Nutrition* 15:835–843.

Keister, D., and P. Cregan. 1991. *The Rhizosphere and Plant Growth.* Amsterdam: Kluwer.

Kuzyakov, Y., and G. Domanski. 2000. Carbon input by plants into the soil. Review. *Journal of Plant Nutrition and Soil Science* 163:421–431.

Legocki, A., H. Bothe, and A. Pühler, eds. 1997. *Biological Fixation of Nitrogen for Ecology and Sustainable Agriculture*, Vol. 39. NATO ASI Series G, Ecological Sciences. New York: Springer.

Lynch, J. M. 1990. *The Rhizosphere.* New York: Wiley.

Pankow, W., T. Boller, and A. Wiemken. 1991. Structure, function and ecology of mycorrhizal symbiosis. *Experientia* 47:311–322.

Parker, C., and C. R. Riches. 1993. *Parasitic Weeds of the World: Biology and Control.* Wallingford, U.K.: CAB International.

Schulten, H. R., and M. Schnitzer. 1998. The chemistry of soil organic nitrogen: A review. *Biology and Fertility of Soils* 26:1–15.

Smith, S. E., and D. J. Read. 1997. *Mycorrhizal Symbiosis.* San Diego: Academic Press.

Waisel, Y., A. Eshel, and U. Kafkafi, eds. 1996. *Plant Roots: The Hidden Half*, 2nd ed. New York: Dekker.

Ten Thousand Years of Crop Evolution

Paul Gepts
University of California, Davis

Agriculture is one of the major technological innovations of humankind. However, the appearance of agriculture was more than a mere technical advance. Since its inception some 10,000 years ago, agriculture has permitted civilizations to develop and converted wild landscapes into fields and pastures. The change from hunting-gathering to agriculture was so great that the term "Neolithic Revolution" is used to describe it. The Neolithic or New Stone Age was a phase in human evolution characterized by the development of more advanced stone tools. About the same time, humankind became settled down and started living in villages. Sedentarism, in turn, led to a reduction in birth spacing, which eventually caused higher population growth. Societies changed from an egalitarian to a hierarchical mode of organization. Most importantly, crops and farm animals were domesticated. Domestication is a change in the genetic makeup and, hence, morphological appearance of plants and behavior of animals, such that they fit the needs of the farmer and consumer. In this chapter, we review what is known about the origin of agriculture in different parts of the world and the genetic changes brought about by crop domestication. We also discuss some consequences of this knowledge for the conservation and use of crop biodiversity.

13.1 For 4 million years, people procured food by hunting and gathering.

Human origins can be traced back 4 million years when apelike ancestors of humans adopted an upright posture. Walking on its hind legs allowed *Australopithecus afarensis* to free its hands to manipulate objects. Nevertheless, subsequent evolution was still slow. It was not until 2.5 million years ago that *Homo habilis* appeared, with an enlarged body and brain. It is also around that time that the first stone tools appeared. The *Homo erec-*

tus descendants of *Homo habilis* lived from 2 million years ago to 400,000 years ago. The brain size of *Homo erectus* was larger than that of its predecessor, and body height was comparable to that of modern humans. Gathering and scavenging were the main food procurement activities. *Homo erectus* was the first hominid species to master fire, some 500,000 years ago, and to migrate out of Africa into Asia and across a land bridge into Indonesia.

Homo neanderthalensis was the immediate predecessor of *Homo sapiens* in Europe, the Near East, and northern Africa. They lived from approximately 250,000 to 30,000 years ago. Their body and brain were larger than those of modern humans. They lived in small bands practicing cooperative hunting of large game animals, which may have been made possible by their language capability. They may have been interested in music, as shown by the discovery of a flute in some of their remains. They cared for their sick and buried their dead.

Homo sapiens is not a direct descendant of *Homo neanderthalensis*, but rather a more gracile, nimble, and clever relative of the latter. This species originated in Africa some 50,000 years ago and also migrated to other continents. In Europe, they displaced or killed their Neanderthal predecessors. Their tools of stone, bone, and antlers were of standard size and shape, indicating a greater mastery of tool-making techniques. They also made multipiece tools such as harpoons, spear–throwers, and bows and arrows, and knew how to make ropes, which they used to construct nets, lines, and snares. They owned sewn clothing, adorned themselves with jewels, and buried their dead in elaborate rituals, indicating a belief in an afterlife. They are well known for their magnificent cave paintings and drawings in southern France and northern Spain, demonstrating both artistic inclinations and a keen sense of observation of their environment and particularly of their food sources (**Figure 13.1**). Approximately 15,000 years ago, hunter-gatherers started focusing their attention on a broader range of food items, including small game, waterfowl, fish, and wild

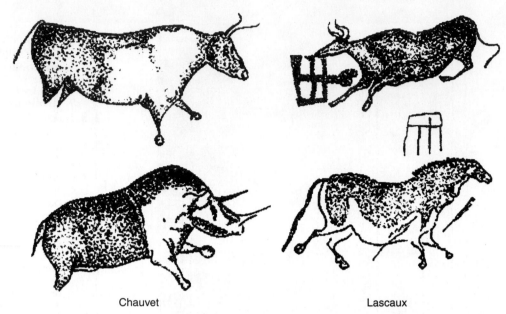

Chauvet   Lascaux

Figure 13.1 Cave art by hunter-gatherers (various caves in southern France). The extraordinary accuracy and artistic quality of these images show the hunters' awareness of their environment prior to domestication. *Source:* Courtesy of C. Zuechner. Available at http://www.uf.uni-erlangen.de/chauvet/chauvet.html. Accessed September 9, 2001.

grains. In addition, they developed special tools such as grinding tools and storage pits to deal with these foods. Whether this change took place by the hunter-gatherers' choice, or out of necessity because of dwindling numbers of large game animals, is still a matter of speculation.

Additional evidence on the food procurement practices of hunter-gatherers comes from the study of contemporary hunter-gatherers. They have an intimate knowledge of the animals they hunt and plants they gather. Their existence is far from precarious, as they can obtain a diverse and abundant diet, with relatively little work. For example, the !Kung San of the Kalahari Desert in southern Africa eat 23 of the 85 edible plant species and 17 of the 55 edible animal species they know. They spend two and a half days per week procuring food and the rest of the time as leisure. Similar observations have been made for other groups. For example, the Cahuilla native Americans of southern California used over 170 species, and their yearly harvest of oak acorns—their staple food—was copious enough to feed each family. In a classic experiment, U.S. geneticist J. Harlan sought to determine the amount of grain of a wild cereal that could be harvested with technology available to hunter-gatherers (**Figure 13.2**). He chose wild einkorn wheat, which grows in large, dense stands in the Near East (see Section 13.3). He constructed a primitive sickle and was able to harvest nutritious wild einkorn wheat in abundant quantities (2 kg/hr).

Why did the large majority of hunter-gatherers switch to agriculture as their principal food source? The answer remains uncertain, although it is one of the key issues archaeologists are investigating. The current consensus is that at some point during the warming phase after the last ice age, an imbalance appeared in some areas of the world between the supply of and demand for foods. One cause of this imbalance may be population increase above the local carrying capacity of the natural vegetation. A hectare of land can

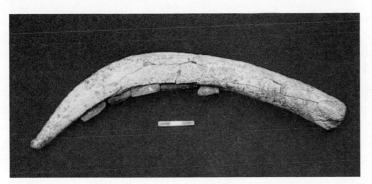

Figure 13.2 A replica of an ancient sickle made of a deer antler with embedded flint blades, excavated at Haçilar in Anatolia, Turkey, by J. Mellaart. Dated to the late 8th or 7th millennium B.C. *Source:* Courtesy of P. C. Anderson.

feed 10 to 100 times more herders and farmers than it can feed hunters and gatherers. Faced with a potential shortage of food from wild plants, gatherers may have turned to more intensive forms of plant management, such as seeding new stands of wild grains.

Another reason may be climate change. For example, in the Near East the warming trend after the last ice age was interrupted by a short-lived cold spell. During this cold spell, the gatherers may have intensified their management of the vegetation. Whatever the reasons may be, there was a gradual transition from hunting and gathering to herding and farming. Herders-farmers also continued to rely on natural vegetation to supplement their food.

13.2 Agriculture began in several places some 10,000 years ago and was a necessary condition for the development of civilizations.

A large part of human knowledge about the origin of agriculture comes from studies on the origin of individual crops. Two types of information are particularly relevant in this regard, as proposed in 1885 by Alphonse de Candolle, a Swiss botanist who initiated the studies on crop origins. The geographic distribution of wild relatives of a crop provides a general idea where a crop may have originated. Careful botanic explorations are necessary to determine the precise distribution of wild progenitors. It is within this area that a crop was probably first cultivated—provided, of course, that the contemporary distribution of wild crop relatives matches that of these relatives at the time of domestication. Additional genetic studies involving crosses between the crop and presumptive wild ancestors and a comparison of their DNA can then identify in more detail a specific region of domestication.

Archaeological studies provide a wealth of information on the transition from hunting-gathering to agriculture. For example, they can tell us more about the type of society in which agriculture was practiced: Were people living sedentarily, in settlements? What was the size of these settlements? Did people herd animals as well as raise crops? Were the plants raised still wild, and what was their age at the time of domestication? In wheat, as in many other cereals, a major difference between the wild progenitor and domesticated descendants lies with seed dispersal. Wild plants spontaneously shed their seeds at maturity, assuring their dispersal. In contrast, to minimize yield losses early farmers have generally selected domesticated plants to hold on to their seeds. For cereals in particular, archaeologists can distinguish the two dispersal modes by specific morphological traits of the seed (**Figure 13.3**).

Analyses of archaeological plant remains can also determine the approximate age of the materials. This is done through carbon-14 dating. As a result of these analyses, people have a much better idea of the age of the earliest remains of crop plants and in some cases their wild relatives (see **Table 13.1**). A striking observation from this table is that agriculture originated at similar times (approximately 10,000 years ago) in widely different regions

Figure 13.3 Comparison of wild and domesticated grains of wheat. *Source:* Courtesy of B. Smith, © 1995 Freeman. Figure from B. Smith (1995), *The Emergence of Agriculture.* New York: Scientific American Library, p. 73.

Table 13.1	Examples of the age of earliest crop plant remains	
Location	**Crop**	**Age (years before present)**
Mesoamerica	Squash	10,000
Mesoamerica	Maize	6,300
Central America	Cassava, *Dioscorea* yam, arrowroot, maize	7,000–5,000
Fertile Crescent	Einkorn wheat	9,400–9,000
Fertile Crescent	Lentil	9,500–9,000
Fertile Crescent	Flax	9,200–8,500
China	Rice	9,000–8,000

of the world, probably because of a climate change acting on a global scale. This climatic event was the end of the last ice age. The warming trend that followed the last ice age increased rainfall and led to changes in the geographic distribution of vegetation. In Mediterranean and monsoonal climates, where most crops originated, the warmer postglacial period led to wetter rainy seasons but also to longer, drier summers. These climatic changes favored annual plants that could complete their life cycle within the rainy season and could survive the dry season as seeds.

Where did our crops originate? As proposed in the 1920s and 1930s by Nicolai Vavilov, a Russian geneticist, there are a limited number of areas in the world where crops originated. These areas are generally located in tropical or subtropical regions, at middle elevation (approximately 1,000–2,000 m) in areas with varied topographies, including river valleys, hills and mountainsides, and plateaus. These regions often have a climate with distinct wet versus dry seasons, either Mediterranean climates (wet winter) or savanna or monsoon climates (wet summer). Six major independent agricultural centers of origin have been identified (**Figure 13.4**). These are (1) Mesoamerica (the southern half of Mexico and the northern half of Central America); (2) the Andes of South America, including its foothills region on the eastern slope; (3) the "Near East" (Southwest Asia); (4) the Sahel region and the Ethiopian highlands in Africa; (5) China; and (6) Southeast Asia. In addition, there are a few minor centers of domestication such as North America and Central Asia.

In each of these regions, people domesticated different sets of plants. However, in most regions they domesticated crops with similar uses. For example, they domesticated cereals in most of the regions, as well as crops of the legume family. Cereals and legumes complement each other nutritionally. Whereas cereals are deficient in lysine, an essential amino acid for the human body, legumes are deficient in methionine, cysteine, and tryptophan, three other essential amino acids. When eaten together, however, the two types of crops complement each other (see Chapter 7). Legumes also complement cereals agronomically, as they provide additional nitrogen obtained by symbiotic nitrogen fixation to cereal crops (**Table 13.2**).

Most plants were domesticated in a single region. However, people domesticated several crops in more than one place. Vicarious domestications are domestications in widely different places of similar plants, whether they belong to the same or different species. Examples are common bean (Mesoamerica and the Andes), cotton (Africa, India, the Andes, and Mesoamerica), and rice (China and Africa). When multiple domestications take place, people often select the same trait in different regions. Farmers have selected the similar seed color or color patterns in beans of the Andean and Mesoamerican areas.

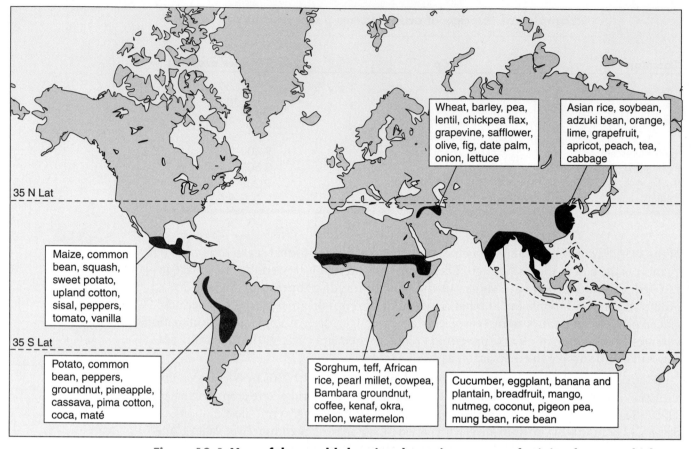

Figure 13.4 Map of the world showing the major centers of origin of crops, which are distributed mainly in tropical regions. *Source:* P. Gepts (2001), Origins of plant agriculture and major crop plants, in M. K. Tolba, ed., *Our Fragile World: Challenges and Opportunities for Sustainable Development* (Oxford, U.K.: EOLSS Publishers), pp. 629–637.

After domestication, the fate of each crop differed. Some became extinct or nearly so, because they were displaced by other crops introduced from other areas. This was the case for some crops domesticated in North America such as goosefoot or lambs-quarter (Chenopodium) and marsh elder (*Iva annua*). The cultivation of lambs-quarter, originally domesticated by Native Americans in eastern North America, was discontinued in the eighteenth century. It is now extinct. Other crops have survived to this day. However, the extent of their distribution differs very much from crop to crop. For some crops, the distribution remains limited to their original domestication center. Others have been dispersed to a limited degree outside their domestication center, and still others have acquired major production zones outside the domestication center. Finally, most major crops are now very widely cultivated. Most production for these crops takes place outside the respective domestication centers. Examples of the latter category are provided in the next three sections.

The hunting-gathering societies gradually receded and were replaced by farming and pastoral societies. Societies governed by immediate food rewards for their efforts were replaced by delayed-reward societies. Harvesting a crop (the delayed reward) is preceded by several operations including land clearing, plowing, and planting, all of which can be quite labor intensive. The need for additional labor as well as the reduction in birth spacing allowed by a sedentary lifestyle led to larger villages (**Figure 13.5**). To maintain co-

Table 13.2	Major centers of agricultural origins and their respective domesticated plants

	Major Centers of Agricultural Origin					
Crop Types	**Mesoamerica**	**South America**	**China**	**Southeast Asia (India, S. China, Indochina, Pacific Islands)**	**Near East**	**Africa**
Cereals	Maize		Asian rice, proso and foxtail millets	Asian rice	Wheat, barley, rye, oat	African rice, pearl millet, sorghum, tef, fonio
Pseudocereals	Amaranth, chenopodium, salvia (chia)	Amaranth, chenopodium				
Grain legumes	*Phaseolus* beans	*Phaseolus* beans, peanut, lupine, jack bean, *Inga* sp.	Soybean, adzuki bean	Pigeon pea, jack bean, winged bean, moth bean, rice bean	Pea, chickpea (garbanzo), lentil, lupine	Cowpea, Bambara groundnut, hyacinth bean, Kersting's groundnut
Roots and tubers	Sweet potato, jícama	Potato, arracacha, achira, cassava, jícama, oca, añu, yacón, ullucu, mashua, unchuc	Turnip, yams	Yams, arrowroot, taro	Turnip, carrot, radish	Yam
Oil crops	Cotton	Peanut, cotton	Rape seed	Coconut	Rape seed, safflower, flax, olive	Oil palm, castor bean
Fiber	Cotton, agave (sisal, henequén)	Cotton		Coconut, jute		Kenaf
Fruits	Papaya, avocado, guava, prickly pear	Cashew, pineapple, guanábana, cherimoya, Brazil nut, guava	Chinese hickory, peach, chestnut, apricot, quince, litchi, persimmon	Breadfruit, orange, lime, tangerine, grapefruit, mango, banana	Fig, walnut, date palm, almond, grape, apple, pear, plum	Baobab, watermelon, melon
Vegetables and spices	*Capsicum* pepper, squash, tomato, vanilla	*Capsicum* pepper, squash	Chinese cabbage, ginger	Cucumber, nutmeg, eggplant, plantain	Onion and relatives, lettuce, saffron, parsley	Okra, *Sesamum* sp. (leaves), *Solanum* sp.
Stimulants	Cacao	Coca, maté	Tea, ginseng, camphor		Poppy, digitalis, belladonna, licorice	Coffee

hesion in these larger groups, people performed religious rituals and festivals. Control of resources such as irrigation water created inequalities in these early agricultural societies. Eventually, more structured and hierarchical societies appeared, culminating in the creation of city-states led by a ruler. Examples of the earliest cities are Mari and Ebla in Mesopotamia, the area between the Tigris and Euphrates streams, which is surrounded by the Fertile Crescent. In this area, well-known early civilizations eventually came into being, including Sumer, Akkad, and Babylon. Agriculture played a major role in the development of these civilizations by generating a surplus of food, which allowed a part of society to pursue activities other than food procurement. These activities included spe-

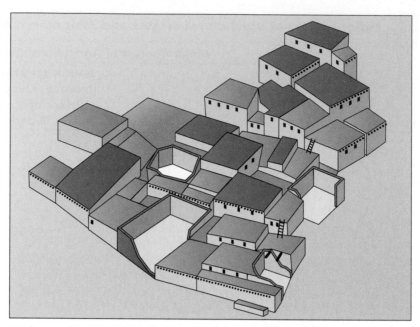

Figure 13.5 The early agricultural village of Çatal Hüyük in Anatolia, Turkey. This village already shows some level of urban planning with the alignment of the houses and the existence of courtyards. *Source:* Courtesy of Agropolis-Museum, Montpellier, France. Available at http://museum.agropolis.fr/english/pages/expos/fresque/la_fresque.htm. Accessed September 26, 2001.

cialized crafts, religion, the military, trade (including that of crop products), and administration. Thus the initiation of agriculture had a profound effect not only in biological terms through the domestication of plants and animals, but it also caused a profound change in human societies, a change that has lasted through today.

13.3 Wheat was domesticated in the Near East.

Wheat is one of the most important food plants in the world, principally as a source of calories. It is grown on an area of 215 million hectares on five continents. Total production in 2001 was 600 million metric tons. The word *wheat* is a generic term for a number of related grain crops belonging to the genus *Triticum*. Among the most important of these crops are bread wheat (*Triticum aestivum*) and pasta wheat (*Triticum durum*). An additional species—currently cultivated only as an animal feed in mountainous agricultural areas of Turkey, Italy, and Spain—is einkorn wheat (*Triticum monococcum*). This species was, however, the original domesticated wheat species. For millennia, it constituted the main staple crop in the Near East and surrounding areas. For example, the last meal of the man whose frozen 5,000-year-old remains were recently discovered in the Alps consisted mainly of einkorn.

Most wild relatives of wheat, which belong to the genera *Triticum* and *Aegilops*, are distributed in the Near East, a region encompassing modern-day Turkey, Lebanon, Israel, Jordan, Syria, Iraq, and western Iran. They are concentrated in the mountainous regions surrounding the alluvial plains of the Tigris and Euphrates, the region called Mesopotamia, on the west, north, and south. Because of its shape and role in the origin of agriculture, this region has been called the Fertile Crescent (**Figure 13.6**).

The genomes of these *Triticum* species underwent two major changes during their evolution. First, they went through a normal process of evolution and chromosome di-

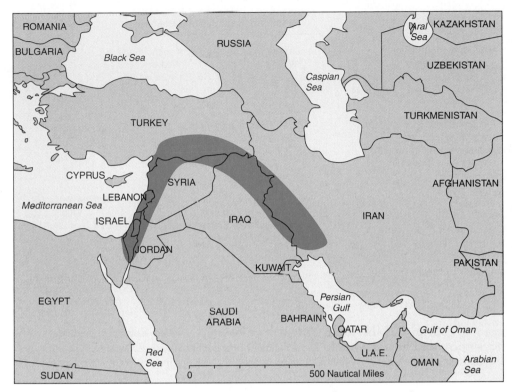

Figure 13.6 The Fertile Crescent in southwestern Asia.

vergence leading to different versions of the basic seven-chromosome set of wheat species. Geneticists have labeled these different sets with uppercase letters: *A, B, D,* and so on. Second, the *Triticum* species form a polyploid series that includes diploid species (two sets of seven chromosomes), tetraploid species (four sets of seven chromosomes), and a hexaploid species (six sets of seven chromosomes). These tetraploid and hexaploid species arose from two rounds of crosses between species with different basic chromosome sets, as shown in **Figure 13.7.** Normally, crosses between these species yield sterile hybrids. However, occasional doubling of the chromosomes, either in the gametes or in the progeny, led to a fertile polyploid progeny.

There are two evolutionary lineages in domesticated wheat. In the einkorn lineage, a wild, diploid species with an *AA* genome from the Fertile Crescent was the progenitor of the only domesticated wheat species, called *einkorn wheat*. It is a hardy species, although not high-yielding compared to its cousin, emmer wheat (see later discussion). It was therefore grown in more remote, often mountainous, areas. It was also an evolutionary dead-end because it did not lead to polyploid descendants as did the pasta and bread wheat lineage. In this lineage, two diploid wild species with an *AA* and a *BB* genome, respectively, crossed to give a wild tetraploid progeny, wild emmer wheat or *Triticum dicoccoides*, with an *AABB* genome. Domesticated emmer wheat (*Triticum dicoccum*) appeared some 9,000 years ago in the Fertile Crescent. The grains of this wheat were covered with tightly adhering leaflike structures called *glumes,* hence this primitive wheat is called "hulled-grain wheat." Because these glumes hinder processing the grain after harvest, farmers selected a variant with thin glumes that could easily be separated by threshing after harvest; hence the name "free-threshing" or "naked-grain wheat," also called "pasta wheat" (*Triticum durum*) because its grains are used to make pastas, such as spaghetti or macaroni.

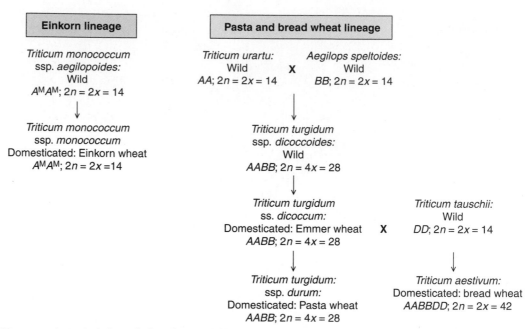

Figure 13.7 Origin of the three different cultivated wheat species: einkorn wheat, pasta wheat, and bread wheat. The evolution of bread wheat involved two independent hybridizations. Bread wheat is hexaploid, with three complete diploid genomes from three different species.

Emmer wheat became quite popular and spread within the Fertile Crescent and beyond. When it reached the region south of the Caspian Sea, it crossed with another wild diploid species with a DD genome (*Triticum tauschii*) to yield a hexaploid species with an $AABBDD$ genome, *Triticum aestivum* or "bread wheat." *Triticum tauschii* is adapted to the more continental climate of Central Asia, with both colder winters and hotter summers. Bread wheat therefore gained a broader adaptation beyond the Mediterranean climate favored by its forebear, emmer wheat. This accounts in part for the widespread cultivation of bread wheat around the world. Another reason may be its bread-making qualities. In contrast with tetraploid wheats, bread wheat grains contain sticky proteins that trap the CO_2 gas that originates from yeast fermentation, resulting in leavened bread. In contrast, bread made from tetraploid wheat, such as pita bread from the Middle East, is always unleavened or flat.

13.4 Rice was domesticated in eastern Asia and western Africa.

The rice most people know is actually of Asian origin and is called *Oryza sativa*. It plays an important role in human nutrition, because half of humankind relies on it for daily intake of calories. World production is around 600 million metric tons on an area of 150 million hectares. China and India are by far the largest producers. The highest per capita level of consumption takes place in Bangladesh, Cambodia, Indonesia, Laos, Myanmar (Burma), Thailand, and Vietnam.

In Asia, rice has two wild relatives, the annual *Oryza nivara* and the perennial *Oryza rufipogon*. These two wild species have widely overlapping distributions in subtropical Asia from the Indus valley across Southeast Asia and into China, the Philippines, New Guinea, and northern Australia. In these regions, wild rice grows in flooded sites owing to its unusual ability to transport oxygen from emerged leaves to flooded roots. Most cultivated rice is derived from *Oryza nivara*. Scientists generally think that the first domesticated varieties were paddy or lowland rices or varieties growing in inundated fields, a habitat similar to that of their wild ancestors. "Paddy" is a Malay word for threshed, unhulled rice and, by extension, for the shallowly flooded field in which rice is generally grown. From these evolved the rain-fed or upland rices that can be grown without irrigation in rainy climates. Deep-water or floating rices, which can have stems of up to 5 meters long, are generally grown in southern Asia and contain more genes from *Oryza rufipogon*.

The actual region where Asian rice was domesticated is still uncertain, partly because the two wild relatives are widely distributed in tropical and subtropical Asia, and partly because the archaeological record is sketchy in those areas: The hot and humid climate does not favor long-term conservation of plant remains. Archaeologists have identified the oldest remains of rice in the Yangtze and Yellow River valleys in China. These remains are some 8,000 years old. A possible clue to the actual domestication pattern in rice comes from the observation that there are two major types of rice cultivars. The *japonica* types, also called *sinica* or *keng* (sticky), have short grains and are generally adapted to more temperate climates. The *indica* types, also called *hsien* (not sticky), have long grains and are generally better suited for more tropical climates. These two groups may have actually resulted from separate domestications in different areas of the distribution of the wild relatives. Alternatively, the two groups could have resulted from a single domestication followed by divergence into indica and japonica types. Both groups contain paddy and upland varieties, but the deep-water rices are only found in the indica group.

From its center of origin in Southeast Asia, rice was dispersed over the entire world and eventually into the New World (the Americas) and Africa. In Africa, however, a local species, *Oryza glaberrima*, is grown mainly in western Africa. Farmers domesticated it from a local wild relative, *Oryza barthii*. West Africans grow it especially in remote areas, because it is more resistant to adverse conditions. Recently, because it is lower yielding and tends to drop its seeds, it has been losing ground to its Asian cousin. These multiple domestications of related rice species in different parts of the world are examples of vicarious domestication.

13.5 Maize and beans were domesticated in the Americas.

Maize (from the Arawak Indian name *maiz*, also called *corn* in the United States) is a food crop that is grown as such in many countries of the world. More recently, it has been grown for other uses such as animal feed and for industrial uses (such as alcohol production) as well. World production in 2001 was about 600 million metric tons. The major producer was, by far, the United States, followed by Brazil, France, Mexico, India, and Italy. Where did a crop with such a worldwide distribution arise? Several types of evidence, including botanical, archaeological, and folk oral history (**Box 13.1**), point to Mexico and Central America as the center of origin for this crop.

Maize was—and still is—the staff of life for people in Mexico and Latin America (as well as Africa and Asia). It is the main source of calories, but has other uses as well, for example, as forage and roofing material. There is an extensive vocabulary in native languages of Mexico designating maize or its products. The likely progenitor of maize, called *teosinte*, grows in dispersed populations on the western slope of Mexico and Central America, where their distribution parallels that of the ancient Mesoamerican civilizations. They grow in

Box 13.1

The Popol Vuh (Book of Dawn) of the Quiché Mayas and the Origin of Maize

A clue to the origin of maize is given by ancient texts that have survived the Spanish bonfires conducted in the Americas to destroy native cultures. In the Popol Vuh, the story of the Quiché Mayas of Guatemala about the dawn of life, the creator called Heart-of-Sky and other gods of the Mayas made initially unsuccessful attempts to create humans. On the first try they got creatures that growled, squawked, hissed, and did not pay their respects to the gods. Their descendants are the animals of today. They tried again with beings made of mud, but these could not move or walk and eventually dissolved themselves. On the third try, the wooden beings they created looked and talked like humans but failed to act in an orderly way and forgot to call on the gods in prayer. Eventually they endured the wrath of a hurricane and animals around them. Their only descendants are the monkeys who inhabit the forests today. The final attempt was a successful one when the gods used yellow and white corn to fashion human beings as we know them today. The Popol Vuh describes it in this way:

> . . . the making, the modeling of our first mother-father, with yellow corn, white corn alone for the flesh, food alone for the human legs and arms, for our first fathers, the four human works. . . .

This is the first person: Jaguar Quitze.
And now the second: Jaguar Night.
And now the third: Mahucutah.
And the fourth: True Jaguar.
And these are the names of our first mother-fathers.

They were simply made and modeled, it is said; they had no mother and no father. We have named the men by themselves. No woman gave birth to them, nor were they begotten by the builder, sculptor, Bearer, Begetter. By sacrifice alone, by genius alone they were made, they were modeled by the Maker, Modeler, Bearer, Begetter, Sovereign Plumed Serpent.

And when they came to fruition, they came out human:
They talked and they made words.
They looked and they listened.
They walked, they worked.

(From D. Tedlock's translation of the Popol Vuh © D. Tedlock 1996)

That according to this creation myth of native people the first humans were made out of maize illustrates the importance of this crop in the life of the Native Mexicans and attests to the antiquity of the crop. The existence of ancient documents, legends, and myths that mention a crop and are handed down over many centuries provides evidence that this crop has been cultivated for a long time and is not an introduction from another region of the world. Similar evidence is provided by the existence of words designating a crop in native languages.

regions receiving summer rains, often close to cultivated maize fields. Their flowering time tends to parallel that of maize (**Figure 13.8**), hence crosses between maize and teosinte are possible. Farmers cannot easily rid their maize fields of teosinte until the plants flower, when crossing may already have occurred. Teosinte and its hybrids with maize can therefore become weeds inside maize fields. A comparison of teosinte and maize DNA shows that teosinte is the closest relative of maize. Domestication of maize probably took place in the mountainous regions of western Mexico.

Maize is often cultivated with other crops such as beans and squash. Pole beans use maize as a support for growth, and squash plants cover the ground between maize and bean plants. This traditional way of growing these plants may have predated domestication. In several regions of Mexico, wild bean vines climb on teosinte plants. Therefore, the first farmers not only may have domesticated maize, beans, and squash, but may also have understood how to grow them from the way they grow in natural vegetation. Similar multiple-crop cultivation systems are still widely used in Africa (see Chapter 5).

Figure 13.8 Maize (left) cultivated in close proximity to teosinte, its wild progenitor (on the right). Close proximity of crops to their wild progenitors is a common occurrence in centers of domestication. Under these circumstances, genes flow between crops and their wild progenitors, and the formation of hybrids between the two is also frequent.

Although maize, beans, and squash have all been domesticated in Mesoamerica, there are differences as well among the domestication patterns of these crops. Maize was domesticated only once, in Mesoamerica. Beans, in contrast, were domesticated at least twice. In addition to the Mesoamerican domestication, which eventually gave rise to bean types such as the pinto, pink, and black beans, another domestication in the southern Andes led to kidney beans and green beans. Different species of squash were domesticated in different locations in Mexico and Central and South America. These are further examples of vicarious domestications.

Nowadays, maize, beans, and squash are grown worldwide, their distribution spreading since 1492 after the conquest of the Americas. The crop exchanges that followed the voyages led by Christopher Columbus (hence called the Columbian exchange) led to widespread dispersal of crops between the "Old World" (Eurasia and Africa) and the "New World" (the Americas). Old World crops such as wheat and rice became widely grown in the Western Hemisphere, and conversely crops from the Western Hemisphere such as maize, potato, and tomato became a significant part of Old World agriculture.

13.6 Domestication is accelerated evolution and involves relatively few genes.

The cultivated agricultural environment in which crop plants find themselves differs substantially from the natural environment in which the wild relatives of modern crops grew and still grow. The selection pressures the plants experience are very different in these two situations. Here we discuss just a few properties of plants.

Seed Shattering. In a natural environment, plants are responsible for their own propagation. They do so by shedding their seeds spontaneously at maturity, by various mechanisms depending on the plant species. In wild grasses, the rachis (the stem) of the inflorescence becomes brittle at maturity. This lets the seed drop to the soil around the plant. In wild legumes, fibers in the pod wall contract at maturity, causing the pod to open suddenly and like a slingshot propel the seeds around the plant (**Figure 13.9**). Imagine that a farmer arriving at a field ready for harvest discovers that half the harvest has already been scattered by the plant's seed dispersal mechanism! Thus a major change during domestication has been the disappearance of natural seed dispersal mechanisms. Domesticated plants keep their seeds on the plant.

Seed Dormancy. Wild plants are subject to a very variable environment compared to cultivated environments, particularly with regard to rainfall. To germinate, seeds need sufficient moisture, usually supplied by rainfall. Plants have developed a mechanism to deal with the erratic timing and amount of rainfall. Seed dormancy prevents premature germination of seeds in a moist environment. By delaying germination, dormancy gives seeds a chance to germinate in a moister, more favorable environment, whether in the year immediately after seed dispersal or in later years. Dormancy lets wild plants accumulate a seed bank in the soil (see Chapter 10) that allows such species to survive temporary, unfavorable conditions. A major goal of farming is to provide a more predictable and less competitive environment for crop plants (to the extent possible given the vagaries of year-to-year weather variability). Farmers plant their seeds in soils that have accumulated sufficient moisture following the first rains. To achieve a dense and regular stand of plants, most seeds must germinate more or less simultaneously and within a few days after planting. Thus domesticated plants have lost seed dormancy (**Figure 13.10**). Their seeds germinate readily when located in moist, sufficiently warm soils.

(a)   **(b)**

(c)

Figure 13.9 Seed dispersal mechanisms in wild progenitors of several crop plants. In the grasses **(a)** teosinte or wild corn (*Zea mays* var. *parviglumis*) and **(b)** *Aegilops tauschii* or previously *Aegilops squarrosa*, one of the wild progenitors of bread wheat, seed dispersal is achieved by a brittle stem, which releases seeds at maturity. In the legume family, dispersal is achieved by the explosive opening of the pods at maturity. **(c)** Three types of pod in common bean, *Phaseolus vulgaris*: at left are wild bean pods, unopened or opened (note the twisted, widely separated pod walls); in the middle are pods of dry beans (note the slightly separated, nontwisted pod walls, insufficient for seed dispersal); at right are pods of snap or green beans, which do not open, even at maturity. *Source:* P. Gepts (2001), Origins of plant agriculture and major crop plants, in M. K. Tolba, ed., *Our Fragile World: Challenges and Opportunities for Sustainable Development* (Oxford, U.K.: EOLSS Publishers), Figure 2.

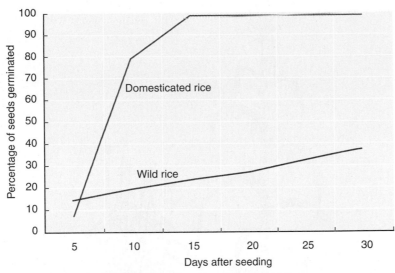

Figure 13.10 Germination of domesticated rice and wild rice. Domestication resulted in a loss of seed dormancy. The majority of the rice (domesticated variety) seeds germinated in 10 days, but the wild rice collected in Taiwan germinated gradually. Dormancy spreads germination over a long period providing an evolutionary advantage to the species.

Growth Habit and Harvest Index. Wild plants also confront strong competition for light, water, and nutrients from other plants, whether they belong to the same or a different species. To stave off this competition, at least in part, plants have developed special growth habits. The viney, climbing growth habit of some plants lets them grow on top of other plants and thus access light. Other plants have a very branched growth habit that allows them to shade out neighboring plants. Regardless of the growth habit, vegetative development (stems and leaves) is as important as actual seed production for survival of the species. Wild plants are therefore said to have a low harvest index, defined as the weight of the harvested part (usually seeds) divided by the sum of the total biomass, including the harvested and nonharvested aboveground parts of a plant. Typically, the harvest index for wild plants is around 0.2 to 0.3. In other words, a plant that weighs 1 kilogram (after drying) carries 200–300 grams of seed. Modern varieties have a harvest index around 0.4 to 0.5. To avoid excessive competition among domesticated plants, these usually have a more compact growth habit with fewer and shorter branches. Many crops with a wild relative growing as a vine have acquired an upright, non-climbing growth habit, more like a bush. For example, the ancestor of the common bean was a vine, and wild relatives of the common bean still have a twining growth habit. Most commercial production of common beans is from "bush bean." But pole beans are still widely grown in gardens all over the world. Highly branched plants have seen a reduction in the number and length of their branches, as illustrated by maize and its wild relative teosinte (**Figure 13.11**).

Adaptation to Photoperiod. The timing of flowering and seed production is also of great importance for wild plants. Many plants live in an environment that favors growth for only part of the year. This is the warm season in temperate regions or the humid season in subtropical areas. It is therefore important to schedule flowering so that seed development does not unduly extend into the unfavorable season. In addition, the dispersal mechanism of many seed plants relies on the drying phase associated with maturation to operate. This scheduling is achieved by timing flowering as a function of the length of the night. Some plants—called *short-day plants*—will flower when the night exceeds a certain number of hours. These are usually plants of tropical origin, where the length

Figure 13.11 Growth habit of maize and teosinte.

of day and night are more similar. Other plants—called *long-day plants*—will flower only in short nights of 8–10 hours, occurring in temperate, higher-latitude locations (see Chapter 8).

Gigantism and Diversity of Harvested Organs. The harvested organs of crop plants range from fruits such as apples or strawberries; seeds such as those of beans, soybean, rice, or maize; to leaves such as those of lettuce or spinach. Also, as Charles Darwin pointed out, the harvested parts of crop plants have acquired much broader diversity in shape and color, making them less camouflaged than their wild counterparts, as shown by the seeds or fruits of maize, beans, and squash (**Figure 13.12**). A part of these changes is driven by selection for novelty by farmers and consumers alike.

Resistance to Diseases and Pests and the Presence of Toxic Compounds. Wild plants are subject to the constant attacks of diseases and pests. Unlike animals, they cannot escape by moving away from their would-be attackers. Plants have, therefore, established a whole set of defense mechanisms that camouflage or make them unpalatable to attackers. These include spines, hairs, toxic subtances, and usually tan to dark seed colors. Examples of toxic compounds are morphine in poppies, digitalis from foxglove, cyanic-acid-producing compounds in cassava and lima bean, and glucosinolates in some plants of the cabbage family. The presence of these compounds has not prevented humans from domesticating some of these plants. People have adapted by selecting varieties with a reduced level or complete absence of the compound. Alternatively, they have developed processing techniques to remove the compounds altogether. Wild cassava contains high levels of chemicals that produce hydrogen cyanide. Domesticated varieties of this crop can be divided into two groups: sweet varieties that contain very low levels of these compounds and can be eaten after minimal processing such as boiling and drying, and bitter varieties that contain much higher levels of these compounds. The latter can only be consumed after more extensive detoxification.

Figure 13.12 Bright colors of flowers, fruits, and seeds among domesticated plants resulting from human selection for novelty items among crops: upper left, common bean; upper right, squash; lower left, maize; and lower right, 19th-century tulip varieties.

The traits distinguishing crops and their wild ancestors are quite similar in all crops; together, these traits are called the *domestication syndrome*. The change from the wild to the domesticated version of these traits—for example, from seed dispersal in the wild to seed retention in the domesticated state—is the result of a selection pressure resulting from human cultivation. This selection took place either unconsciously or consciously on spontaneous mutations in genes controlling the domestication syndrome in populations of wild relatives as they were being cultivated by the first farmers. Because of positive selection, the frequency of these mutations gradually increased until it reached 100%, at which point the crop could be considered fully domesticated. Domestication is therefore a selection process that results in genetic changes in plants and confers adaptation to a cultivated environment and acceptance by farmers and consumers.

Research in maize, bean, rice, tomato, and pearl millet has shown that the inheritance of these traits is usually simple and that only a few genes are involved despite the large phenotypic differences between crops and their wild relatives. These genes generally have a major effect on the phenotype, and their expression tends to be relatively independent of environmental influences. The genes appear to be concentrated in a few places on the chromosomes (loci) in the genome of domesticated plants. This relatively simple inheritance suggests that domestication could have happened fairly quickly over a time span of a few decades to a few centuries. The actual time it took to domesticate plants was probably longer, up to a millennium. It is likely that in the initial years of the transition to agriculture, plants were cultivated only occasionally. This decreased the selection pressure on plants and hence the speed of conversion to a domesticated type. In addition, a limiting fracture may well have been the combination of different domestication traits into a single plant through crosses within the cultivated fields. Nevertheless, the end result was a number of crops (and farm animals) that depended on humans for survival. Without planting and harvesting by humans, fully domesticated crop plants would most likely face extinction. Conversely, humans nowadays would be hard pressed to derive enough food by hunting and gathering. Only crops and farm animals can assure a sufficient production of food for humanity, currently and in the foreseeable future.

13.7 Crop evolution was marked by three major genetic bottlenecks for genetic diversity: domestication itself, dispersal from the center(s) of domestication, and plant breeding in the 20th century.

The preservation of biodiversity in the world's natural ecosystems has become of major concern to humankind, and this concern extends to the preservation of crop biodiversity. The genetic diversity of most modern crops is often limited compared to their wild ancestors. This reduction in diversity during crop evolution is by no means recent but began with the initial cultivation of plants. Selecting certain plants with desirable traits such as absence of seed dispersal, a more compact growth habit, and features increasing attractiveness to consumers reduced genetic diversity during the domestication process itself.

After domestication, crops were disseminated from their center of origin into other regions of the world. For many crops, these new cultivation areas had a very different environment from that of the domestication center. Most modern crops originated in tropical or subtropical areas but are now grown also in markedly temperate areas. This dispersal generally involved small samples of seeds or planting material, further reducing the genetic diversity of crops as they spread around the globe. In this new environment, farmers subjected crops to a second round of selection that adapted them to these new conditions.

Landraces. Crops grown in the original centers of domestication or elsewhere, especially in developing countries, did not result from scientific plant breeding until recently, hence they are called *landraces* or local varieties. Farmers recognize individual landraces based on their morphological traits, such as seed color or size or tuber shape. These landraces have usually been grown for many centuries by local communities and may therefore have been selected for adaptation to local conditions or tastes. They are generally believed to be better adapted to unfavorable or stress conditions such as infertile soils or drought conditions than cultivars resulting from breeding programs. In addition, farmers have selected landraces for specific uses. For example, farmers in the neighborhood of Mexico City grow a maize variety that produces husks (surrounding the ear of the maize plant), which can be sold as a food-wrapping material on the markets. Indians of the Bolivian highlands use special potatoes (*Solanum* x *juzepczukii*: triploid: $2n = 3x = 36$, *S.* x *curtilobum*: $2n = 5x = 60$) to prepare a freeze-dried food called *chuños*. The preparation takes special advantage of the particular environmental conditions prevailing where they live, namely the dry mountain air in that part of the Andes and the often freezing nights. Thus, landraces are characteristic of the marginal environments where they originated. Such environments are often environmentally, economically, and culturally heterogeneous. This heterogeneity is reflected in the genetic heterogeneity of the landraces.

The heterogeneity of the landraces may be one reason why farmers continue to grow them even after the initiation of breeding programs to develop new cultivars (see next paragraph and Chapter 14). Genetic heterogeneity buffers the effect of the spread of diseases and the year-to-year variability and even unpredictability of the weather. Their adaptation to local unfavorable conditions and to special consumer uses also explains their survival to this day. Nevertheless, as communications become easier, rapid changes in economic and social conditions of farmers and progressive reduction in cultural diversity threaten the continued existence of landraces. Proposals have recently been made to reward farmers for their conservation of landraces. However, the practical implementation of this proposal is a major stumbling block.

In addition to their use by local farmers, landraces are also used by plant breeders as a source of interesting traits they can introduce into modern varieties. Examples of such traits are the tolerance of Central American bean landraces to soils low in phosphorus, the high protein content of two Ethiopian barley landraces, and the drought tolerance (and thicker, deeper roots) in upland (rain-fed) rice from Africa, South America, Bangladesh, and Ethiopia.

Loss of Genetic Diversity. A third bottleneck developed during the 20th century. Scientific plant breeding, for reasons of practicality, focused on an ever-decreasing number of so-called elite varieties. These elite materials are ones shown in previous years to have superior characteristics such as higher yields, disease and pest resistance, and high-quality traits such as bread-making ability in wheat and long and strong fibers in cotton. This progressive reduction in genetic diversity is well illustrated by the situation in common bean (**Figure 13.13**). As indicated earlier, common bean was domesticated at least twice. All common bean cultivars therefore derive from one of two distinct lineages. One lineage originated in the wild beans growing in Mesoamerica, most likely Mexico, and the other from the wild bean populations in the southern Andes, somewhere between southern Peru and northwestern Argentina. The genetic diversity in both lineages was reduced, first by domestication and other events such as crop failures after domestication. In a second step, dissemination of beans to other regions of the world such as North America and Europe as well as the development of new bean cultivars further reduced genetic diversity.

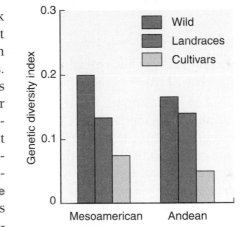

Figure 13.13 Stages in the reduction of genetic diversity during crop evolution as illustrated by the Andean and Mesoamerican evolutionary lineages of common bean. *Source:* Data from G. Sonnante, T. Stockton, R. O. Nodari, V. L. Becerra Velásquez, and P. Gepts (1994), Evolution of genetic diversity during the domestication of common-bean (*Phaseolus vulgaris* L.), *Theoretical and Applied Genetics* 89:629–635.

Because both the industry and the public require uniformity, a limited number of varieties in each crop now occupy a significant proportion of the land devoted to that crop (**Table 13.3**). Although high genetic diversity is not a precondition for high yields, reduced genetic diversity in crops carries some risks that have been dramatically exposed in recent history but that may have existed since domestication. The best known incident is the Irish potato famine, which started in 1845 and lasted for several years (see **Box 13.2**). Similar epidemics affecting genetically uniform crops have occurred in the past, such as the leaf rust epidemic of coffee on the island of Sri Lanka in the early 1870s and the corn leaf blight that struck the United States in 1970. The coffee plantations on Sri Lanka, destroyed by leaf rust, were replaced by tea plantations, which may be one reason why the English became enthusiastic tea drinkers. The 1970 epidemic of Southern corn leaf blight prompted plant breeders to increase the genetic diversity of their varieties.

Genetic Erosion. The loss of crop plant biodiversity, also called *genetic erosion*, affects all components of the genetic resources of crop plants: the wild progenitors, landraces, obsolete cultivars, advanced breeding lines, and elite cultivars. For example, wild progenitors suffer from a loss of habitat; landraces are lost, replaced by modern varieties and by market

Table 13.3 The diversity of our crops is limited because of cultivation of a few varieties over a large area

Crop	Total Number of Varieties	Major Varieties	
		Number	Acreage (%)
Beans, dry	25	2	60
Beans, snap	70	3	76
Cotton	50	3	76
Maize	197	6	71
Peanut	15	9	95
Soybean	62	6	56
Wheat	269	9	50

Source: National Research Council (1972).

demands for specific types that are only a subset of what a farmer grows, and modern varieties have a narrow genetic base because plant-breeding programs have limited resources.

Genetic erosion has existed since the origins of agriculture and throughout the different phases of crop evolution. Recently, however, the pace of loss has accelerated. To a large extent, the ultimate cause of this recent genetic erosion is the explosive human population growth in the last decades (see Chapter 1). This growth has fueled the need for ever-higher levels of food production in response to increased demand. As a consequence, deforestation has spread wider than ever before to make room for additional fields and pastures. Excessive grazing by domestic animals also destroys the natural habitats of wild relatives of crops. The introduction of higher-yielding varieties displaces existing farmers' varieties, and until recently, plant-breeding programs focused on an ever-shrinking genetic base to develop new cultivars.

Figure 13.14 Test tube conservation of cassava genetic resources. *Source:* Courtesy of W. Roca, CIAT, Cali-Colombia.

The international community's response to genetic erosion of crops has been the creation of both off-site (*ex situ*) and on-site (*in situ*) conservation programs. *Ex situ* programs typically include gene banks, which include cold storage facilities, where the seeds of hundreds of thousands of different lines are kept (see Chapter 9), and a kind of botanic garden where live specimens of trees (for example, fruit trees such as apple or avocado) are grown and propagated. In addition, thousands of varieties of certain plants such as cassava and potato that cannot be stored below freezing or in botanic gardens are kept in sterile culture flasks and propagated regularly (**Figure 13.14**). *In situ* programs seek to maintain landraces or their wild relatives where they grow originally, either on the farm or in native vegetation. An example of an *in situ* program is the Biosphere Reserve of the Sierra de Manantlán in western Mexico. In this reserve, maize and beans are grown in close proximity to their respective wild relatives.

Box 13.2

Genetic Uniformity and the Irish Potato Famine

Like many European countries, in the 19th century Ireland experienced a steady growth in population. Increasing attention to sanitation and other public health measures caused the death rate to drop without a concomitant drop in the birth rate (see Chapter 1). This was accompanied by improvements in agricultural technology such as the introduction of mechanical tilling and by increasing demand for agricultural products both from Great Britain and the United States. Because of Ireland's status as a British colony, a large proportion of the Irish people—most of them extremely poor farmers—relied on potatoes as their primary source of food: Adults ate 5 to 6 kilograms a day. Potatoes were abundant, cheap, and nutritious and a reasonably balanced food source, especially when combined with some vegetables and dairy products or fish. The potato crop in Ireland, as elsewhere, was genetically very uniform because potatoes are vegetatively propagated and farmers had come to emphasize one or two varieties that consistently produced high yields. To grow potatoes, farmers plant pieces of potato tubers saved from the previous harvest. After these sprout and grow, they produce genetically identical plants. Offered such a uniform crop, plant pathogens can spread rapidly though the field if weather conditions are right and if the plants are not genetically resistant to the pathogen. This is precisely what happened in Ireland (see figure).

The pathogen that attacked the Irish potato crop in 1845 was *Phytophtora infestans*, responsible for the disease called *potato late blight*. It originated in Mexico or in the Andes Mountains, where potato also originated. It made its way to Ireland in contaminated shipments of potato tubers for planting. The disease initially affects the older leaves but in later stages also affects the tubers, through which the pathogen can then infect next year's crop. The moist and mild weather in Ireland was very favorable for the development of the disease, which spread like wildfire, first in the latter part of the 1845 growing season and then in full force in 1846, when it destroyed nine tenths of the potato crop. In 1847, the crop remained free of disease. However, it was small because the quantity of healthy potato for planting from the previous year had been limited. Additional crop failures in following years further extended the devastation.

The late blight epidemic created havoc in the Irish population. Census data showed that in 1845, before the epidemic, the population stood at 8.5 million. By 1851, the population had dropped to 6.6 million. At least one million had died from starvation and disease; the rest had emigrated to England, continental Europe, and the United States. A great many Irish Americans trace their family's history to emigration caused by the potato famine, as the celebrations of St. Patrick's Day remind us every year.

The potato famine in Ireland, brought about by the potato late blight, caused widespread deaths on the island and led to massive emigration to Europe and North America. *Source:* Courtesy of S. Taylor. Available at http://vassun.vassar.edu/~sttaylor/FAMINE. Accessed September 26, 2001.

13.8 Hybridization has played a major role in the development of new crops, in the modification of existing ones, and in the evolution of some troublesome weeds.

Plants have a full range of reproductive systems from cleistogamy (self-fertilization within a closed flower bud) to obligate outcrossing, depending on the species. Even cleistogamous and other highly self-fertilized plants do cross at least occasionally with other plants, depending on environmental conditions and genetic factors. For example, spontaneous mutations can cause pollen sterility, which sharply increases outcrossing because all the progeny result from fertilization with pollen from other, male-fertile plants. Large populations of insect pollinators also increase outcrossing. On the whole, plants are much more promiscuous than animals. Like their wild counterparts, crops are also subject to outcrossing. This is especially true when crops and their wild relatives belong to the same species. A species is defined as the sum total of all individuals that can interbreed, whether actually, as in outcrossing species, or potentially, as in self-fertilized species.

Plant breeders have introduced the concept of a gene pool to describe the ease with which these crosses can take place (**Figure 13.15**). The first gene pool (GP1) contains the crop and its wild progenitor. Sexual crosses within this gene pool, whether among cultivars or between cultivars and their wild progenitors, are easy to make and yield viable and fertile progenies. The boundaries of this gene pool correspond to the concept of biological species. The second gene pool (GP2) contains species that can be crossed with GP1 but with a lower success rate. Progenies resulting from this cross have somewhat impaired viability and fertility. The third gene pool (GP3) contains species that can only be crossed with difficulty with GP1. Often special techniques such as tissue culture and embryo rescue must be used to secure progeny for the cross. The progeny have limited viability and fertility. The fourth gene pool (GP4) contains species that cannot be crossed with GP1. These contain plant species as well as species of other kingdoms. Although genes cannot be transferred by the traditional sexual crosses, they can be introduced by laboratory techniques such as Agrobacterium-mediated transformation (see Chapter 6). With these techniques, GP4 becomes an additional source of genetic diversity for improving crops. In addition, information from species in GP4, especially model species such as *Arabidopsis thaliana* and rice, can help breeders identify genes to improve crops.

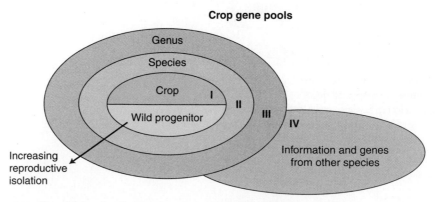

Crop gene pools

Genus

Species

Crop

I

II

III

IV

Wild progenitor

Increasing reproductive isolation

Information and genes from other species

Figure 13.15 The gene pool concept as used by plant breeders. *Source:* P. Gepts (1999), A phylogenetic and genomic analysis of crop germplasm: A necessary condition for its rational conservation and utilization, in J. Gustafson, ed., *Proceedings, Stadler Symposium* (New York: Plenum Press), pp. 163–181.

Hybridizations between and within species are important in shaping the genetic diversity of plants in general and of crops in particular. Hybridizations in crops take place between varieties, between varieties and their wild progenitors, and with other species altogether, whether they are wild or domesticated. Although crops generally tend to adopt a more pronounced selfing reproductive system during and after domestication, outcrossing remains occasionally important and can produce bursts of additional genetic diversity. In selfing crops, even low levels of hybridization can lead to genetic changes over several generations.

Hybridizations can also lead to new crops. We have already discussed the origins of wheat and the role that interspecific hybridization played in the evolution of hexaploid wheat. Another example is found in the bean family where the year bean (*Phaseolus polyanthus*) resulted from a cross between an ancestor of common bean (*Phaseolus vulgaris*) as a female parent and runner bean (*Phaseolus coccineus*) as a male parent followed by backcrossing to the runner bean parent. Other, more recent examples of hybridization leading to new crops include strawberry and triticale. The modern strawberry varieties sold in most stores and markets, with their large fruits, belong to the species *Fragaria ananassa*, which is an octoploid species with 56 chromosomes. This species is the result of a cross between two other octoploid species, *Fragaria chiloensis* (a species growing on the Pacific coast of North America) and *Fragaria virginiana* (a species growing in the eastern half of North America). Surprisingly, the cross between these two species did not take place in North America but in 18th-century France, where the two parental species had been introduced as agricultural curiosities. Antoine Duchesne, an 18th-century French botanist and agriculturist, was the first to surmise that a spontaneous cross had given rise to *Fragaria ananassa*. He repeated the cross and grew progeny, which matched *Fragaria ananassa* in appearance. His approach of making crosses to identify the potential parents of a crop is still practiced by modern scientists studying the origin of crop plants.

In contrast to the modern strawberry, which resulted from a spontaneous cross, triticale is the result of an interspecific cross between durum wheat (*Triticum durum*, AABB; see Section 13.3 of this chapter) and rye (*Secale cereale*, RR) made by plant geneticists starting in the second half of the 19th century (see also Chapter 14). The ABR hybrids from the initial crosses were sterile, because their gametes had an irregular number of chromosomes given the lack of pairing of chromosomes during meiosis. However, doubling of the chromosome number of the hybrids led to fertile progeny (AABBRR). Breeders capitalized on these fertile hybrids to make further crosses and selections that have yielded today's triticale varieties. These combine the higher yield of wheat with the vigor, winter hardiness, adaptation to acid soils, and higher seed protein content of rye. Most triticale production takes place in Eastern Europe. The crop is used primarily as animal feed (seeds) and forage (young plants).

When a crop and its wild progenitor can be easily crossed with each other, hybrid populations between the two can often be identified in the crop's center of origin. These hybrid swarms can be of varying sizes and longevities. Some are short-lived, whereas others persist for several years. Normally, they are not expected to last indefinitely, because their hybrid status renders them ill adapted to either the cultivated or the natural environment. However, backcrossing the hybrids to either parent can increase adaptation to either environment. For example, hybrids can acquire the same time to maturity and the same morphology as the domesticated parent yet still conserve the ability to disperse seeds. Before harvest, it is very difficult to distinguish the hybrid from the crop, especially before flowering, because there are few distinguishing characteristics in leaves and stems. During harvest, seeds of the hybrid mature but drop to the ground, where they remain until the next growing cycle. They then germinate and grow as a weed in whatever crops are grown in the next cycle. It is by this mechanism of hybridization and backcrossing that some of the most troublesome weeds have appeared (see also Chapter 17). Examples are shattercane, a weedy sorghum, and red rice, a weedy rice. Both are distributed not only in the crop center of origin (Africa and Asia, respectively) but through the seed trade have been distributed to other continents as well, including North America.

13.9 Natural and artificial polyploids lead to new crops and new traits.

Many wild plants are diploid—just like all animals—and contain two copies of each gene and of each chromosome in their cells. However, plants seem to be more flexible than animals in their genomic content: There are also many polyploid species. These contain more than two genomes in each cell: species containing four or six genomes are called *tetraploid* or *hexaploid*, respectively. Polyploids probably arise from spontaneous chromosome doubling in the diploid parent plant or from the union of "unreduced" gametes, which, through anomalies in the meiotic process, contain twice the usual number of chromosomes.

There are two basic polyploid types. In autopolyploids, the same basic set of chromosomes is repeated. As a consequence, chromosomes of one set can pair with comparable chromosomes of the other sets. Examples of autopolyploids are potato and alfalfa. Allopolyploids have sets of chromosomes that differ somewhat from each other and the chromosomes of one set usually do not pair with the chromosomes of the other sets. Wheat, cotton, tobacco, peanut, canola, coffee, and plantain (a starchy type of banana) are examples of allopolyploid species. Following the hybridization event that gives rise to the allopolyploid, the different chromosome sets gradually diverge. For ancient polyploidization events, the divergence is so pronounced that it becomes difficult to recognize duplicated chromosome sets. Only recently has evidence for their existence been obtained from DNA information. Examples of ancient allotetraploids are maize and soybean.

Polyploidy is a frequent phenomenon in plants, caused in part by the promiscuity of plants, already mentioned. Whereas hybridization may be a necessary condition for the formation of allopolyploids, it is by no means a sufficient condition for their survival, and other reasons have been proposed. For crops particularly, the larger size of polyploids compared to diploids is an advantage, because selection during domestication has generally favored larger harvested organs such as seeds. The combination of different genomes, each conferring adaptation to a specific environment, may have broadened the adaptation of polyploids and ensured their successful dispersal outside of their immediate centers of domestication.

An additional advantage of polyploids resides in the existence of multiple copies of the same gene. Each of these copies can then diverge and acquire slightly different functions. Also, new interactions can appear between genes belonging to the different genomes, which in turn lead to the appearance of novel, unprecedented traits or the enhanced expression of traits existing before the polyploidization event. An example of the former is the ability to make leavened bread from bread wheat (*Triticum aestivum*, an allohexaploid with $2n = 6x = 42$ chromosomes; *AABBDD* genome) but not from its progenitors, the domesticated allotetraploid wheat *Triticum durum* ($2n = 4x = 28$; *AABB* genome) and the wild diploid species *Triticum tauschii* ($2n = 2x = 14$; *DD* genome) (Figure 13.7). Somehow interactions between genes of the *A, B,* and *D* genomes led to biochemical changes in the seed proteins of bread wheat that are responsible for dough rising prior to baking. Diploid and tetraploid wheats are strictly used for making unleavened bread.

Cotton is an example of increased expression of traits. Most cotton production comes from two allotetraploid domesticated species, *Gossypium hirsutum* or upland cotton and *Gossypium barbadense* or Sea Island or pima cotton. Both species originated in the Western Hemisphere. Upland cotton may have been domesticated in the Yucatán peninsula of Mexico, and pima cotton was domesticated in South America. The two species have $2n = 2x = 52$ chromosomes and result from the cross between a maternal *A* genome and a paternal *D* genome ancestor some 1 to 2 million years ago, well before the advent of modern humans and the initiation of agriculture. Strangely enough, the *A* genome is actually native to the Old World and must somehow have been transported to the New World. Two related, diploid *A* genome species in the Old World, *Gossypium arboreum* in Asia and *Gossypium herbaceum* in Africa, produce fibers that are spinnable, that is, twine sufficiently so that individual fibers can be combined to make thread that can be woven into ropes or cloth. The diploid species have fiber yields that are 20–30% lower than their tetraploid relatives, which explains why they are no longer widely grown except in certain parts of the world such as India. The donor of the *D* genome has seeds covered with fuzzy fibers that cannot be spun. Yet genetic analyses show that the *D* genome in tetraploids provided genes that increase fiber production and quality. Again, scientists suggest that interactions between genes of the *A* and *D* genomes explain this observation.

Various species belonging to the genus *Brassica*, which includes cabbages, turnip, and rapeseed, form an interesting network of genetic relationships, which are illustrated by the Triangle of U, named after the Japanese agronomist N. U, who deciphered these relationships (**Figure 13.16**). The tips of the triangle contain three diploid species with differentiated genomes, *A*, *B*, and *C*. These include *Brassica campestris* (turnip, Chinese cabbage, and the oilseed crop turnip rape), *Brassica nigra* (black mustard), and *Brassica*

Figure 13.16 The Triangle of U, showing the relationships among diploid and tetraploid species of cabbage-related species. *Source: Brassica carinata,* P. H. Williams; *Brassica nigra,* M. Bleeker.

oleracea (different cabbages, kale, Brussels sprout, broccoli, and cauliflower). Along the three sides are the respective allopolyploids with the combination of genomes of the diploid species. Most important among these is *Brassica napus*, which includes rapeseed or canola, and is grown in cooler climates and is an important source of oil.

The various examples presented here clearly illustrate the important role polyploidization has played in the development of some of the world's most important crops. Polyploidization can lead to an increase in size of harvested organs, it can broaden the adaptation of crops to a wider set of environmental conditions, it can lead to the formation of new crops, and it can result in novel gene interactions that result in new traits.

13.10   The sequencing of crop genomes provides interesting insights into plant evolution and new tools to improve crops.

Until a few years ago, determining the nucleotide sequence of a piece of DNA was a cumbersome activity. Recent technical progress, however, has made it possible to sequence large stretches of DNA, including entire eucaryotic genomes, in a short period of time. The first complete genome sequences of two plant species, the small weed *Arabidopsis thaliana*, and two varieties (Indica and Japonica) of rice were obtained in 2001 and 2002, respectively. The DNA sequences of the genomes of other crops will undoubtedly be deciphered in the near future. These sequences constitute an additional genetic resource because they will provide information on how plant in general and crops in particular function. They will tell scientists how specific landraces or cultivars or their wild relatives resist diseases, tolerate insects, survive drought spells, or have certain nutritional characteristics. In turn, this information will help plant breeders develop better crop cultivars. A comparison of DNA sequences between wild progenitors and their domesticated descendants will also allow a more accurate determination of the domestication history of the crop. The next paragraphs present examples of this novel approach to illustrate its power.

One of the major differences between maize and its progenitor teosinte is the branching of the plant. Maize has a single stem, ending in a tassel on which the male flowers are located. A markedly shortened side branch ends in an ear where female flowers and therefore seeds are located. Teosinte, in contrast, has numerous side branches with tassels at their tips and ears as side branches. This major difference in appearance between maize and teosinte is controlled—surprisingly—by a single gene, *tb-1* or *teosinte branched-1*. A comparison of the sequences of *tb-1* and the DNA region immediately upstream of the gene has yielded interesting information: It confirms that domestication has induced a genetic bottleneck in the species. However, this bottleneck is only apparent in the upstream, regulatory region and not in the gene itself (**Figure 13.17**). This observation suggests that selection operating during domestication affected primarily the regulatory region. Whether this pattern of selection can be generalized to other domestication genes or genes in general during evolution, remains to be determined. The *tb-1* sequences of maize are very similar to each other and fell on the same branch of a genealogical tree. In addition, the *tb-1* sequences of teosinte that resemble most closely those of maize are those of populations from the Balsas River basin in western Mexico, confirming earlier evidence. It is remarkable that an area at a relatively short distance away in western Mexico has been proposed as domestication area for common bean. This suggests that early farmers may have domesticated the two crops (and other native crops) together.

Further information from DNA sequences of other maize genes suggests that domestication could have involved a small population of individuals and could have been completed

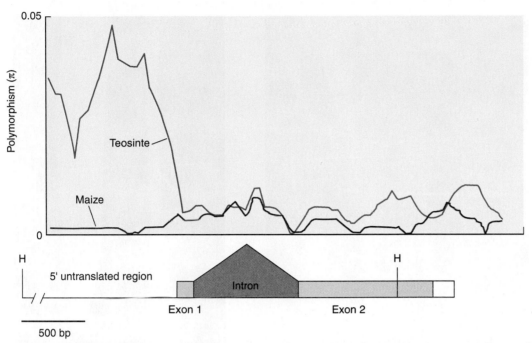

Figure 13.17 Diversity in the DNA sequence of the maize domestication gene *tb-1* and part of its upstream regulatory region. The height of the curve is an indication of the degree of genetic diversity. The diversity is the same for most of the gene except for the region immediately upstream of the gene, where a sharp drop in diversity is observed, further evidence of the genetic bottleneck occurring during domestication. *Source:* From the research of J. Doebley and colleagues: See R. L. Wang, A. Stec, J. Hey, L. Lukens, and J. Doebley (1999), The limits of selection during maize domestication, Nature 398:236–239; R. L. Wang, A. Stec, J. Hey, L. Lukens, and J. Doebley (2001), Correction: The limits of selection during maize domestication, *Nature* 410:718, Figure 1.

over a short period of time spanning anywhere between 300 and 1,000 years. However, no archaeological remains are yet available to independently verify this conclusion.

As explained earlier, crop plants typically have lost the ability to disperse their seeds. However, at plant maturity many crops still display a partial level of seed dispersal, which can obviously lead to losses at harvest time. A better understanding of the genes involved in seed dispersal will help breeders develop cultivars that lack this trait. Recently, researchers have identified two genes in *Arabidopsis thaliana* that cause explosive dehiscence (opening) of the fruit and, hence, dispersal of the seeds contained in it. The fruit of *Arabidopsis* resembles a pea pod and consists of two halves that are joined at the edges, referred to as valves. Two valves, one on each edge of the fruit half, are joined by a dehiscence zone. It is along this zone that the two halves separate after the fruit matures and dries. This explosive separation causes seed dispersal (**Figure 13.18**). The two genes identified in *Arabidopsis* are called *SHP1* and *SHP2* (*SHP* for "shatterproof").

In their active state, these genes are responsible for the development of fruit valve margins. The presence of a single active gene is sufficient to cause seed dispersal. Loss of function of the two genes blocks development of the dehiscence zone and deposition of lignin next to this area. As a consequence, the fruits remain closed and the seeds cannot be dispersed. *Arabidopsis* is a close relative of cabbage. In cabbage species that are grown for their oil-rich seeds, such as canola or rapeseed, premature opening of the fruits reduces crop yield. Learning the *SHP* gene sequences is a first step for canola breeders to isolate similar sequences in canola and develop improved cultivars without fruit shattering.

This last example shows how domestication of plants is an ongoing process. Although plants (and animals) were first domesticated some 10,000 years ago, plant breeders and

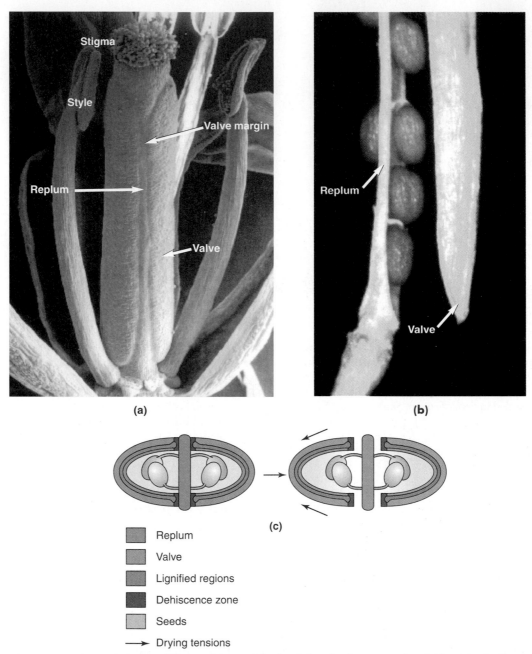

Figure 13.18 Anatomy of the ovary and fruit of the thale cress (*Arabidopsis thaliana*).
(a) The pistil (female reproductive organ) in the center of the picture has a stigma with papillae at the top, short style, and barrel-shaped ovary. The two parts of the ovary, the replum and the valve, are joined by the cells of the valve margin. **(b)** A mature seed pod in which the valves and the replum have separated, revealing the seeds inside. **(c)** Cross-sectional schematic showing the internal organization of a seed pod. *Source:* Courtesy of M. Yanofsky, © Macmillan Publishers Ltd. 2000. S. Liljegren, G. Ditta, H. Eshed, B. Savidge, J. Bowman, and M. Yanofsky (2000), SHATTERPROOF MADS-box genes control seed dispersal in *Arabidopsis, Nature* 404:766–770. Figure 1.

farmers continue selecting and developing new varieties that satisfy public demand for higher-yielding crops with stronger resistance to prevailing diseases and pests, and with improved nutrition, taste, and other useful traits. Our ability to find individual genes and use these for crop improvement by relying on plant breeding, including genetic engineering, has raised the question about the ownership of genes. Who owns the genes and genomes of landraces that native peoples have perpetuated for centuries (see **Box 13.3**)?

Box 13.3

Who Owns the World's Genetic Resources?

Over the last 20 years there has been, during various international treaty negotiations, a change in the attitude toward ownership of the world's agricultural genetic resources. The commonly accepted principle used to be that genetic resources were a common good of humanity and should be freely available to all without restrictions. Some people now hold an almost diametrically opposite opinion. Several factors have led to this change. In the United States at least, since a landmark ruling of the U.S. Supreme Court in 1980 awarding a patent for an oil-eating bacterium, it has become legally possible to patent life forms, including crop varieties, genes, and even DNA sequences. The European Union followed suit, but not Canada. Since then, patent offices have been overwhelmed by a tidal wave of patent applications on anything from genetically modified animals for use in medical research, transgenic plants, existing animal breeds and plant cultivars, chemicals naturally occurring in plants, and DNA sequences from a broad range of organisms, including humans, plants, and animals.

These patents provide legal protection to maintain ownership of new products, including new cultivars, diagnostic tests, and so on, developed wholly or in part using biotechnology. In addition, crop cultivars can also be protected by separate legislation, called Plant Variety Protection (PVP). The ability of biotechnology to manipulate individual genes with a well-defined sequence, when combined with intellectual property protection such as patents, provided a strong incentive to private companies to become involved in life science research and develop new products such as diagnostics and new varieties or breeds.

In this era of increasing globalization, private companies were not satisfied with the geographically limited protection provided by patents. The value of patents is limited to the country awarding the patent. If a company wants to export a product to another country, it must pursue a separate patent in that country. Private companies have therefore actively lobbied to include an intellectual property component in trade agreements, notably in the World Trade Organization agreement on Trade Related Aspects of Intellectual Property Rights (TRIPs), which came into effect in 1995. The TRIP agreement requires countries to adopt patent legislation or a *sui generis* (of its own kind) system to protect crop cultivars.

Concurrently with this move toward increased intellectual property protection pursued by powerful private companies from the developed world, assisted by their respective governments, a movement gained momentum among developing countries to seek compensation for their contributions to the maintenance and development of crop cultivars and animal breeds. As discussed earlier in this chapter, most crops were domesticated in tropical areas. These centers of domestication usually show high levels of crop genetic diversity, which is being maintained by farmers who plant these varieties year in and year out and exchange them with their extended family or neighbors. This genetic diversity is quite important, because it often represents a source of genetic diversity for improving existing cultivars. The world thus risks a classical unequal distribution of resources where developing countries, on one hand, contribute their genetic resources for free under the principle that genetic resources are a common good of humanity. On the other hand, these same countries would have to pay to purchase improved cultivars or other products developed in industrialized countries.

Several recent examples show that this asymmetry in the benefits derived from biological resources is real. Developed countries have awarded patents for biological resources that have been public knowledge for a long time in their countries of origin and therefore do not constitute an invention. These patents include patents for medical applications of the herb turmeric, potential new varieties of quinoa and nuña (popping) beans, native crops from Bolivia and Peru, a variety of yellow beans from Mexico, the insecticidal and fungicidal compound of the neem tree from India, as well as fragrant Basmati rice from the same country. Many of these patents have been challenged and overturned. However, the fact that they were awarded in the first place shows the current weakness of the patent system. It also takes substantial effort and funding to overturn patents once they have been awarded. In addition, patents or other forms of intellectual property protection, with their emphasis on rewarding the individual inventor, do not fit the community-based ownership and free exchange of genetic resources that often characterizes the domestication areas. Furthermore, patents exclude the farmers' exemption whereby farmers are allowed to save, exchange, and sell seeds produced on their land—again, a common practice in some countries.

Continued on the following page.

Box 13.3, continued

A number of international initiatives or treaties have been developed to deal with this asymmetry in intellectual property protection. The first one of these is the International Undertaking on Plant Genetic Resources for Agriculture, organized by the Food and Agricultural Organization (FAO) of the United Nations. The first version of this nonbinding agreement appeared in 1983. As of December 1997, 113 countries had adhered to the International Undertaking, with Brazil, Canada, China, Japan, Malaysia, and the United States as notable exceptions. Current negotiations aim to develop a system of common rules on conservation, free exchange, and benefit sharing with regard to crop genetic resources. They also expect to make this agreement legally binding. Difficult issues remain to be solved, including the conflict between free exchange and intellectual property rights and the development of a system of benefit sharing. Farmers in developing countries, in particular those in centers for domestication, would be awarded royalties for any commercial cultivar development that involved the genetic resources that they help conserve. Alternatively, royalties would be paid into an international fund whose goal it is to promote the conservation of crop biodiversity. A draft treaty was presented to the FAO council in 2001.

The second of these initiatives is the Convention of Biological Diversity, signed in 1992, but which the United States has not ratified. This convention has a triple goal: conservation of biological diversity, sustainable use of its components, and fair and equitable sharing of benefits from genetic resources, including appropriate access and transfer of the relevant technology. In contrast to the earlier paradigm proposing the free exchange of genetic resources, the convention confirms the principle of sovereign rights of states over their natural resources. It also considers the licensing of proprietary technology and sharing of research and development results.

It is fair to say that at this stage the issue of genetic resource ownership has not been settled. It is imperative to have an agreement in working order to stimulate the use of genetic resources and at the same time reward the players involved.

CHAPTER
SUMMARY

Hunting and gathering were the methods of food procurement when *Homo erectus* first appeared on the scene some 2 million years ago and also later when *Homo sapiens* started moving out of Africa 50,000 years ago. Early humans created tools for harvesting crops, and such harvesting by different groups on different continents led to the emergence of agriculture and to rapid genetic change in the plants collected. Wild plants became domesticated crops over the span of a few thousand years in different parts of the world: wheat and lentils in the Fertile Crescent of the Near East, rice in southwestern Asia, soybeans in China, maize and common bean in the Americas. Domestication involved both selection and, in the case of wheat, hybridization with related species. Domestication changed the properties of the plants (phenotype) and narrowed the genetic diversity present in the plant populations. Gradually landraces emerged, "varieties" that were adapted to local conditions of climate, diseases, and soil type. Similar changes occurred in all crops. Plant breeding in the 20th century caused further changes, narrowing the genetic base even more. Together with the crops, weeds emerged as a byproduct of domestication and agricultural practices. Some weeds were themselves the products of hybridizations. As scientists begin to unravel the genomes of crop plants, they will discover precisely how the plants evolved. Genome information will significantly aid crop improvement in the future.

Discussion Questions

1. Can there be domestication without cultivation? Can there be cultivation without domestication? Explain.
2. Describe the mutual dependence existing between humans and their crops and farm animals.
3. Why were so few plants domesticated (a few hundred species at most out of 250,000 plant species described)?
4. What are the major types of evidence scientists use to determine the center of a crop's origin?
5. Discuss the patenting of genes. How should the people who preserved genes be compensated by companies interested in those genes?
6. Groups opposed to GM technology claim that it is a major reason for the loss of biodiversity. Discuss the gradual loss of biodiversity resulting from agricultural practices.
7. Discuss harvest index in relation to domestication, plant breeding, and photosynthate allocation (Chapter 10).
8. Discuss *ex situ* and *in situ* conservation of germ plasm. What are the advantages and disadvantages of each method?
9. Explain the statement "Wheat is an allohexaploid."

Further Reading

Chapman, G. P. 1992. *Grass Evolution and Domestication.* Cambridge, U.K.: Cambridge University Press.

de Candolle, A. 1884. *Origin of Cultivated Plants.* New York: Hafner.

Damania, A., J. Valkoun, G. Willcox, and C. Qualset, eds. 1998. *The Origins of Agriculture and Crop Domestication.* Aleppo, Syria: International Center for Agricultural Research in the Dry Areas.

Diamond, J. 1997. *Guns, Germs, and Steel.* New York: Norton.

Frankel, O. H., A. H. D. Brown, and J. J. Burdon. 1995. *The Conservation of Plant Biodiversity.* Cambridge, U.K.: Cambridge University Press.

Harlan, J. R. 1992. *Crops and Man,* 2nd ed. Madison, WI: American Society of Agronomy.

Harris, D., ed. 1996. *The Origins and Spread of Agriculture and Pastoralism in Eurasia.* London: University College Press.

Ladizinsky, G. 1998. *Plant Evolution Under Domestication.* Dordrecht: Kluwer.

Piperno, D., and D. Pearsall. 1998. *The Origin of Agriculture in the Neotropics.* San Diego: Academic Press.

Sauer, J. 1993. *Historical Geography of Plants.* Boca Raton, FL: CRC Press.

Smartt, J. 1990. *Grain Legumes: Evolution and Genetic Resources.* Cambridge, U.K.: Cambridge University Press.

Smartt J., and N. Simmonds. 1995. *Evolution of Crop Plants.* New York: Wiley.

Smith, B. 1995. *The Emergence of Agriculture.* New York: Scientific American Library.

Vavilov, N. I. 1997. *Five Continents.* Rome: International Plant Genetic Resources Institute; St. Petersburg: N. I. Vavilov All-Russian Institute of Plant Industry.

Zimmerer, K. 1996. *Changing Fortunes: Biodiversity and Peasant Livelihood in the Peruvian Andes.* Berkeley: University of California Press.

Zohary, D., and M. Hopf. 1993. *Domestication of Plants in the Old World.* Oxford, U.K.: Clarendon.

From Classical Plant Breeding to Modern Crop Improvement

Todd W. Pfeiffer
University of Kentucky

In 1842, the Scot David Fife immigrated to Canada, bringing some wheat seeds along. He planted them on his homestead in what is now the province of Ontario, and the high yield and the good milling and baking qualities of the crop were remarkable. Soon Red Fife wheat had spread throughout the region. But it was susceptible to the early frosts that plague eastern Canada. In 1892, this wheat was backcrossed with wheat introduced from the Himalayas, to make it more frost resistant. A result of these crosses, available in 1907, was Marquis wheat, which had all the good traits of Red Fife with the addition of frost resistance. Marquis was sent to the Canadian and U.S. prairies, and was an instant success. By 1917, a decade after the first seeds were sown, its North American crop was 7 million tons per year.

David Fife was building on the achievements of his predecessors, who for 10,000 years selected wheat plants and in the process contributed to its slow genetic modification (see Chapter 13). Others continued this work after him. Indeed, plant breeding is a never-finished business. Changes in environment, production technology, pest pressure, consumer preferences, as well as human population growth all require new, altered, and improved varieties. Varieties come and go; the scene has been described as a variety relay race, one variety passing the production baton to the next worthy variety. For example, in four decades, the plant breeders at IRRI have released 318 new rice varieties to the farmers of Asia. This process, which exemplifies human ingenuity so well, will continue as long as there are humans on Earth, because, as N. W. Simmonds and J. Smartt wrote in 1999 in *Principles of Crop Improvement*, "The perfect variety has never been bred. Successful varieties are merely less imperfect than their predecessors."

Remarkable increases in crop productivity have occurred in the past 50 years (see Chapter 3). Half of these increases stem from deliberate selection of more productive varieties; the remainder, from adopting crop production technologies that improve the environment in which the plants grow.

14.1 Plant breeding involves introduction, selection, and hybridization.

Plant introduction is generally considered the first phase of plant breeding. The voyages of discovery by the European explorers (A.D. 1400–1600) spurred an irreversible change in much of the world's agriculture. The intercontinental exchange of crops greatly altered the global distribution of many crop species. Crops domesticated in one region often found an equally or even more suitable growing area on an entirely different continent.

Calculations for the food crops of the world, made by J. Kloppenburg, Jr., and D. L. Kleinman and shown in **Table 14.1**, document the enormous interdependence of the world's regions with respect to genetic resources. These researchers divided the world into 10 production regions and calculated the percentage of the 20 most important food crops grown in each region that were also domesticated there. They found that 70% of the food crops grown in West Central Asia were also domesticated there (principally wheat and grain legumes), whereas only 12% of the crops grown in Africa were domesticated in Africa (principally sorghum and cowpeas). Crops that originated in one region are now grown all over the world; the most important contributing region to each production region is shown in the last column of Table 14.1. Thus 82% of the crops grown in Australia originated in West Central Asia (principally wheat and other small grains), whereas 52% of the crops grown in Africa came from Latin America (principally maize, beans, and cassava). None of the world's 20 most important food crops are indigenous to North America or Australia.

If these data are considered from the standpoint of the relationship between developed and developing countries, it is clear who is indebted to whom: The developing coun-

Table 14.1	Food crops and their regions of origin.			
Region of Production	**Percentage of Crops Grown in This Region of Production That Also Originated Here**	**Origin and Percentage of Crop Area Devoted to Crops from That Area in This Region of Production**		
China-Japan	37	Latin America	41
Indochina	67	Latin America	32
Australia	0	West Central Asia	82
Hindustan	51	West Central Asia	19
West Central Asia	70	Latin America	17
Mediterranean basin	2	West Central Asia	46
Africa	12	Latin America	56
Euro-Siberia	9	West Central Asia	52
Latin America	44	China-Japan	19
United States and Canada	0	West Central Asia	36

J. Kloppenburg, Jr., and D. L. Kleinman divided the world in 10 regions of production and calculated the percentage of the 20 most important food crops grown in each region that were also domesticated there.

tries have contributed 95% of the genetic resources that humanity needs for food, and the advanced industrial countries have contributed only 5%. Thus financially poor nations rich in genetic crop resources have given those resources to economically rich nations poor in genetic food crop resources. Selection within landraces is the second phase of plant breeding. As described in the previous chapter, hundreds of landraces were created for each crop species. As people emigrated and as botanists and plant explorers exchanged materials, landraces were tested in new environments. Depending on the time of introduction, farmers, seedsmen, and scientists helped select crops for adaptation to different growing environment and production methods. These selections were made not only among different landraces, but also for superior individual plants within particular landraces.

An example of selection is the history of alfalfa in the United States as recounted by M. D. Rumbaugh. This is an important forage crop, that is, a crop eaten by animals. Once a landrace was found that could survive the cold winter, alfalfa could thrive in the U.S. Midwest. Such a landrace grew in the cold, dry steppes of Russia, and in 1906, N. E. Hansen in South Dakota obtained seeds from the Imperial Agricultural College in Moscow. The dry seasons of 1911–1913 in South Dakota demonstrated the value of this germ plasm, named Cossack, as Cossack survived and thrived in the climate similar to its native region. In 1916 over 1,000 bushels of seed were available, and Cossack alfalfa benefited farming in South Dakota.

Selection leads to genetic stability. Although this can be desirable in the short run, in the long run genetic homogeneity makes a crop less adaptable to a changing environment. Certain environments gradually expose the specific weaknesses in a variety. To overcome these weaknesses, genes from other varieties must be added. Hybridization is the addition of these genes by cross-pollinating between two varieties that complement each other's weakness. Because it involves the deliberate crossing of plants with specific characteristics, hybridization has led to spectacular and rapid improvements.

14.2 Genetic variation manipulated by selection is the key to plant breeding.

A plant breeder looks at a plant or looks at the data obtained by measuring a characteristic such as seed production and thinks, "I like what I have found so far; maybe I should evaluate this again. If true, this would be great." But even with the sense of excitement comes a note of caution, because what you see isn't always what you get. What you see or what you measure is the *phenotype* of the plant. The phenotype is controlled by two factors, the underlying *genotype* and how the environment affects the expression of that genotype. A simple equation is used to express this relationship:

$$P = G + E \text{ (or phenotype = genotype + environment)}$$

When selections are made, plant breeders measure large populations of plants grown together. Then they base selection on the phenotype of each plant compared to the others. These phenotypes often vary and can be plotted on a graph as a distribution (**Figure 14.1**). The plants with the desired phenotype along this distribution are then selected for further growth.

But what the farmer really wants to select is not just the phenotype, which can vary with the environment according to the preceding equation. What the farmer wants is to select the underlying desirable genotype. And, just as you can use $P = G + E$ as the expression for the value of a single plant, you can use

$$\text{Variation of } P = \text{variation of } G + \text{variation of } E$$

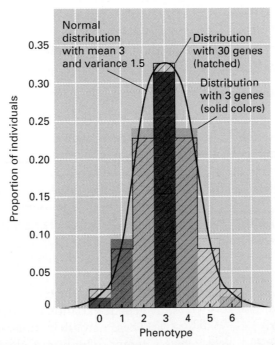

Figure 14.1 Continuous distribution of a phenotype. The bar graphs show the theoretical distribution of phenotypes for a characteristic determined equally by 3 genes (shaded bars) or 30 genes (open bars). Both distributions are continuous (line). *Source:* D. L. Hartl (1996), *Essential Genetics* (Boston, MA: Jones and Bartlett), Figure 14.5, p. 407.

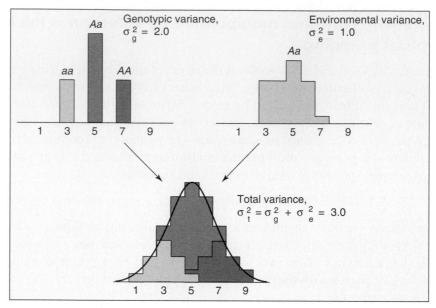

Figure 14.2 The combined effects of genotype and environment. Upper left: Population affected only by genotype. Upper right: one genotype affected only by the environment. Bottom: Population affected by both genotype and environment. The total variation is the sum of the genetic and environmental variation. *Source:* D. L. Hartl (1996), *Essential Genetics* (Boston, MA: Jones and Bartlett), Figure 14.7, p. 408.

as the expression of the *differences among the individuals* in the population (**Figure 14.2**). The change that occurs because of selection depends on how much genetic variation is present. This is what gets passed on to the next generation of plants. Phenotypic variation does not guarantee that genetic variation is present (see **Box 14.1**). The proportion of variation that is genetic is called the *heritability* of the trait:

$$\text{Heritability} = \text{genetic variation/phenotypic variation}$$

Heritability can range from 0 (no genetic component of variation) to 1 (variation caused only by genetics). Obviously, the higher the heritability for a characteristic, the more likely that a selection protocol will succeed in reliably altering the phenotype. Unfortunately, most heritabilities are less than 1 (**Table 14.2**). How can a crop be improved if the heritability is low? At first glance, the solution is obvious: Either increase genetic differences or decrease environmental effects. But in practice these two changes may be difficult or expensive. For example, bringing in new genes to increase heritability involves finding plants for crossing in which the alleles at the genes controlling the character are different, so that genetic variation in the progeny will increase. These new alleles must be better, not just different, which isn't always easy or possible. Indeed, when genetic variation is low, breeding progress is slow. If the variation caused by the environment is high, this can be reduced by growing the plant populations in many

Table 14.2	**Heritabilities of crop characteristics**	
Characteristic	**Crop**	**Heritability**
Yield	Soybean	0.03–0.58
Yield	Maize	0.14–0.28
Seed protein	Barley	0.53
Seed weight	Sunflower	0.33–0.60
Seed oil	Sunflower	0.40–0.75
Lodging	Soybean	0.43–0.75
Maturity	Soybean	0.75–0.94
Heading date	Barley	0.75

Box 14.1

Johannsen and the Princess Bean: Defining Variation for Plant Breeders

In 1903 W. L. Johannsen, the scientist who later coined the term *gene*, conducted a series of experiments that defined the sources of variation in a self-pollinated crop. Johannsen studied the inheritance of seed size in Princess, a commercial bean variety of his day. This variety had beans of many sizes. Johannsen chose large and small seeds and self-pollinated the plants that grew from those seeds. Large seeds produced plants that produced large seeds (the largest averaged 640 mg/seed), whereas small seeds produced plants that produced small seeds (the smallest averaged 350 mg/seed). However, within these averages was a lot of variability, and in both cases there were some large and small seeds. These were now planted and more seeds produced by self-pollination.

In a genetic model, one would expect that seeds from plants from small seeds would still be small, and large seeds would be large, and so on. But this is not what happened. For example, the seeds from a single plant weighed 200, 300, 400, and 500 mg. The offspring of these seed classes averaged 475, 450, 451, and 458 mg. Seed size differences seen in the parental generation did not repeat in the progeny generation.

Both a genetic component and an environmental component controlled seed size. The seed size mixture of the original variety consisted of many different genetic types such that large seeds gave rise to large seeds and small seeds gave rise to small seeds. However, because Princess bean had been produced for many generations and bean is naturally self-pollinated, the individual plants were homozygous. Differences in seed size among seeds on one plant were entirely due to environmental effects. No genetic differences existed within a pure line. Because there were no genetic differences within a pure line, selection for large and small seed sizes was futile. Johannsen's conclusion was that genetic differences can be passed to progeny, but environmental differences cannot.

different environments and then selecting those plants with the desired phenotype in all these places. Again, this is time consuming and expensive. Plant breeders are always trying to balance acceptable costs for parent selection, testing schemes, and expenses to maximize heritability and selection progress. An example of the rigorous application of the principles of plant breeding to crop production is the steady increase in soybean yields in the United States (**Figure 14.3**).

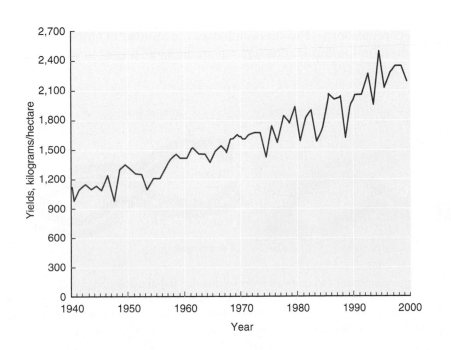

Figure 14.3 Increase in soybean yield in the United States. Soybean has a narrow genetic base because only a few genotypes were initially imported from China. While yield increases for soybean have not been as great as for other crops such as wheat, maize, or rice, breeding progress has still been consistent.
Source: Data from USDA.

14.3 Hybridization creates variability by producing different sets of chromosome segments via sex.

Plant breeders rely primarily on meiosis, which occurs during sexual reproduction, to produce the genetic variation that is necessary if selection is to be effective. Parents are chosen, and no matter whether the plants are naturally self-pollinating or cross-pollinating, in deliberate breeding the individual parents are cross-pollinated (or crossed, or hybridized) to eventually produce progenies that have genetic differences. Because of this, the early plant breeders were called the "plant hybridizers." If plants are normally self-pollinating, then pollination must be prevented usually by removing the anthers or making the plant male-sterile, and then allowing pollen from a different plant to be deposited on the stigma (see Chapter 6).

Meiosis, the movement of chromosomes during the production of sexual gametes, naturally creates different chromosome and genetic combinations from genetically variable individuals. Within the somatic (nonreproductive) cells of an individual, the chromosomes exist in pairs: two homologous chromosomes that are similar in length, in position of the centromere, and in order of the genes located on them. Half of the chromosomes—one chromosome in each pair—originally came from the individual's maternal parent, whereas the other half came from the individual's paternal parent. When these chromosomes are distributed to the gametes, they are not distributed in a pattern such that one gamete gets all paternal or all maternal chromosomes. Instead, chromosome distribution to gametes is random with regard to maternal and paternal chromosomes. The outcome of meiosis is that gametes have half the chromosome number as the somatic cells with, specifically, one chromosome from each pair.

Meiosis occurs in two sequential divisions, meiosis I and meiosis II (see Chapter 6, Figure 6.10). Initially the chromosomes duplicate themselves and become double stranded. Meiosis continues as the two homologous chromosomes come in contact with each other; they become aligned and interconnected side by side throughout their length in a process called *chromosome synapsis* (**Figure 14.4**). The homologous pairs align in the center of the cell, with one member of each pair connected by protein fibers to opposite poles. One member of each pair is pulled to the opposite poles. At each pole there can be from 0 to n maternal chromosomes (where n is the number of chromosome pairs), with the remainder being paternal chromosomes. This division halves the number of chromosomes per cell. In the second division the chromatids separate in the same way as in mitosis, producing four gametes (Figure 6.10).

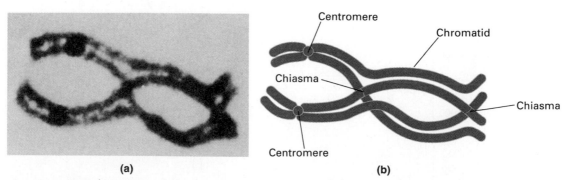

(a)   (b)

Figure 14.4 Light micrograph **(a)** and interpretative drawing **(b)** of a pair of homologous chromosomes showing crossing over where the two chromosomes are exchanging pieces. *Source:* F. W. Stahl (1964), *The Mechanics of Inheritance* (Upper Saddle River, NJ: Prentice Hall), courtesy of James Kezer). As modified into Figure 4.9 in D. L. Hartl (1996), *Essential Genetics* (Boston, MA: Jones and Bartlett), p. 146.

The number of maternal and paternal chromosomes in each gamete depends on the orientation of each synapsed chromosome pair when they align in the center of the cell in the first meiotic division. In the second division the duplicated chromosomes divide, and the maternal/paternal chromosome makeup remains the same as at the end of the first meiotic division. This maternal/paternal chromosome makeup resulting from the independent assortment of chromosomes provides a large source of genetic variation (**Figure 14.5**). The amount of variation depends, in part, on the number of chromosomes in the plant.

Each plant species has a specific number of chromosomes in the somatic cells. For example, the three major cereals, although evolutionarily quite closely related, differ in chromosome number: Wheat has 42 chromosomes, rice has 24, and maize has 20. This number ($2n$) consists of two times the number of each chromosome, and meiosis reduces this number to the gametic chromosome number (n). Because in meiosis the different chromosome pairs align at random with respect to the maternal and paternal chromosomes, the number of different possible chromosome combinations in the gametes is 2^n. So, for maize ($n = 10$) the number of possible chromosome combinations is $2^n = 2^{10} = 1,024$, whereas wheat has $2^n = 2^{21} = 2,097,152$ possible maternal/paternal chromosome combinations. At fertilization, one random female gamete combines with one random male gamete. Therefore, after sexual reproduction the number of possible chromosome combinations in progeny from genetically different parents is the number of possible gametes from parent A (2^n) multiplied by the number of possible gametes from parent B (2^n). This can be simplified as 4^n: for maize $4^n = 4^{10} = 1,084,576$ and for wheat $4^n = 4^{21} = 4,398,046,511,104$! Thus homologous chromosome pairing and separation in meiosis followed by sexual reproduction can create an incredible amount of variation. But meiosis doesn't stop here. Another mechanism for creating variation occurs during meiosis: chromosome crossing over. Although homologous chromosomes are alike in the genes comprising them, they do not both have the same alleles of those genes. If genetic variation resulted only from the independent assortment of chromosomes, new combinations of alleles *within one chromosome* would not be produced. However, during the time of synapsis of homologous chromosomes in meiosis I, the aligned interconnected chromosomes can exchange segments and new allelic combinations can be produced within chromosomes (Figure 14.4). This genetic recombination

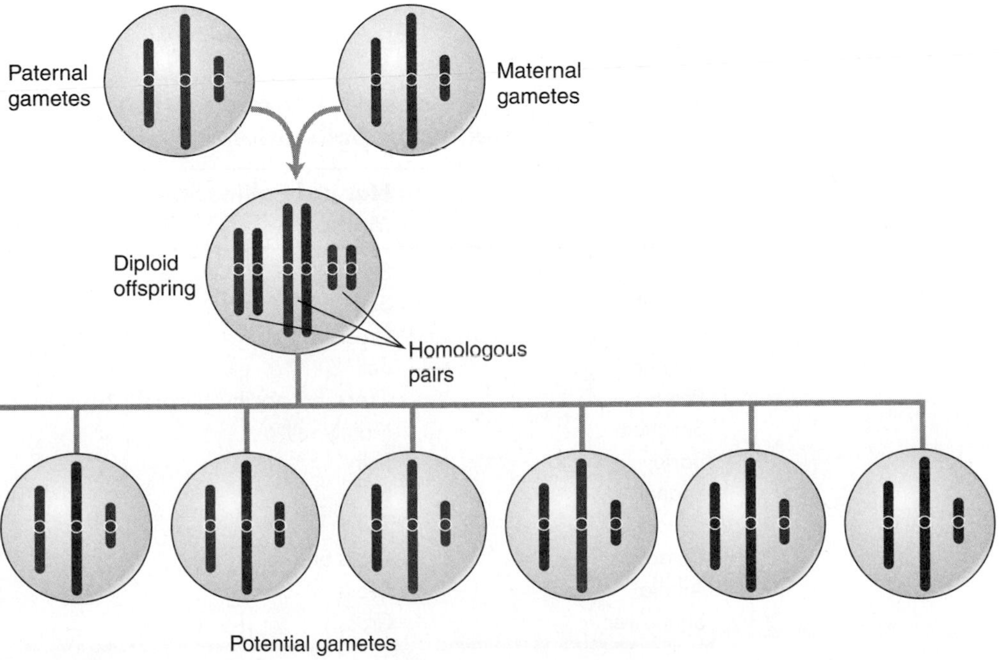

Figure 14.5 Meiosis leads to genetic variability. In this example, there are three pairs of chromosomes, one member of each from each parent (P and M). Each gamete formed from the offspring has one chromosome of each pair. So 8 different gametes are possible with respect to P and M.

Paternal gametes

Maternal gametes

Diploid offspring

Homologous pairs

Potential gametes

between homologous chromosomes greatly increases potential genetic variation from sexual reproduction.

14.4 The plant-breeding method chosen depends on the pollination system of the crop.

Since the early 1900s plant breeders have improved crops by following a well-defined process of four steps:

1. Choose parents that have individual traits of interest.

2. Hybridize these parents.

3. Follow an appropriate breeding scheme that includes selecting among the progeny of those parents to recover the favorable characteristics from both parents.

4. Release the best progeny as a variety (see Chapter 9).

For steps 3 and 4, the appropriate breeding scheme and the type of variety released depend on the method of pollination and reproduction specific to the crop, either cross-pollinated or self-pollinated (**Table 14.3**).

Self-pollinating crops have perfect flowers, which means that male and female reproductive organs are in the same flower. Pollen from the anthers in one flower fertilizes the ovules in that same flower. Self-pollination or selfing is the most severe form of inbreeding, which also occurs when different but related plants pollinate one another. Cross-pollinated crops have mechanisms to limit self-pollination. Some, such as maize have imperfect flowers, with male and female reproductive organs in different flowers, whereas others have genetic systems that prevent self-pollination, called *self-incompatibility systems*. For example, the female reproductive structure (the stigma) may reject the pollen that lands on it from the same flower but allow pollen from a different flower to germinate after it lands on the stigma and to fertilize the ovule.

According to their method of pollination, plants differ in their genetic states. Individuals of a self-pollinating plant are primarily homozygous (the two alleles of each gene being identical), whereas the individuals of a cross-pollinating plant are heterozygous (the

Table 14.3 **Types of varieties released by plant breeders depend on method of pollination**

Crop	Natural Pollination Method	Principal Cultivar Type
Rice	Self	Pureline and hybrid
Wheat	Self	Pureline
Maize	Cross	Hybrid
Soybean	Self	Pureline
Potato	Cross (but not in cultivars)	Clonal
Sorghum	Mainly self	Hybrid and pureline
Barley	Self	Pureline
Peanut	Self	Pureline
Bean	Self	Pureline
Cassava	Cross and self	Clonal
Alfalfa	Cross	Population
Sunflower	Cross	Hybrid and population

two alleles of a gene being different). In some types of varieties, all plants have the same genotype: The variety is genetically homogeneous. In other types of varieties, the individuals differ from each other in their genotype: The variety is genetically heterogeneous.

In breeding self-pollinated crops, the initial hybridization among parents may be difficult. The flowers often must be emasculated (pollen-producing anthers are removed) early in floral development, to prevent self-pollination. Then pollen is collected from a specific second parent and mechanically transferred to the stigma, the female reproductive structure, in the emasculated flower. This produces the genetically variable progeny that are propagated in subsequent generations by self-pollination. Of course, once self-pollination occurs for several generations, inbreeding sets in. In fact, the heterozygosity formed by the initial hybridization event is reduced by half in each succeeding generation. Eventually, all the individual plants become nearly homozygous.

Nonhybrid varieties of self-pollinated crops are marketed as pure lines; they are pure because they consist of one genotype. A pure line is a homogeneous population of homozygous plants. There is no genetic variation within such a variety, and the seeds the farmer harvests are produced by self-pollination, which prevents the introduction of new genetic variation. Farmers can save seed from their crop and maintain the variety's genetic identity and performance in subsequent years. Because of this, special laws such as the Plant Variety Protection Act in the United States have been enacted to ensure that plant breeders can receive an economic return from developing pure line varieties.

In breeding naturally cross-pollinated crops, the difficulty during hybridization often lies in preventing unwanted pollen from fertilizing the parent. Breeders must keep plants apart by planting them in separated areas or by isolating flowers in close proximity to each other by setting up mechanical barriers to pollen movement. Varieties of cross-pollinated crops do not trace to one individual plant. Most cross-pollinated crops have evolved and adjusted to being genetically heterozygous; they are sensitive to the increased homozygosity associated with inbreeding and exhibit inbreeding depression. Increasingly reduced vigor is a result of increasing homozygosity (**Figure 14.6**). To create a variety, multiple plants are selected and advanced to the next generation by cross-pollination. Because multiple plants are involved in pollination, varieties of cross-pollinated crops can be marketed as populations that are heterogeneous mixtures of heterozygous plants. Seed production must take place in fields that provide adequate isolation from other varieties of the same crop, to prevent pollen intermingling and gene transfer.

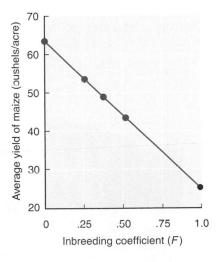

Figure 14.6 Inbreeding depression for yield in maize. As the offspring of a crop are bred to each other, yield decreases. *Source:* N. Neal (1935), The decrease in yielding capacity in advanced generations of hybrid corn, *Journal of the American Society of Agronomy* 27:666. As modified in D. L. Hartl (1996), *Essential Genetics* (Boston, MA: Jones and Bartlett).

14.5 F₁ hybrids yield bumper crops.

F₁ hybrid varieties can be produced from either cross- or self-pollinated crops. The creation of hybrids involves first the production of inbred (pure) lines by self-pollination, followed by controlled cross-pollination between two selected inbred lines. This produces a homogeneous F₁ hybrid variety in which individual plants are heterozygous. F₁ hybrids have many genetic advantages, and the story of the development and adoption of hybrid maize is one of the great success stories of plant breeding.

In 1908 the U.S. plant breeder G. H. Shull reported the results of crossing maize plants from two different inbred lines. The cross between two highly homozygous lines produced heterozygous offspring, because the lines were homozygous for different sets of genes. The result was astonishing. The two inbred lines had each produced about 20 bushels of maize per acre in the last crop. Their outbred offspring quadrupled this yield, to 80 bushels per acre! This unanticipated strength in the heterozygous outcross was called *hybrid vigor,* and such hybrids have played an important role in increasing maize yield.

A major problem with applying Shull's experiment (**Figure 14.7**) on a massive scale was that the hybrid had to be recreated each year. In contrast to pure lines, the farmer could not use the seeds harvested from the hybrid plants for next year's planting. The reason for this follows the principles of genetics. Suppose the gene combination *AaBb* is a desirable combination for a hybrid. This can be achieved by crossing two inbred lines: *AAbb* and *aaBB*. In such a cross, all the progeny are *AaBb*. However, crossing *AaBb* plants among themselves gives only one quarter of the resulting plants as *AaBb*. Many of the progeny plants have some recessive homozygosity, with its resulting yield depression. This heterogeneity is bad news for the farmer, but good news for the company that supplies seeds to the farmer each year.

After its introduction in the 1920s, hybrid maize spread rapidly. By the 1940s, virtually all maize plants grown in the United States were F_1 hybrids. Yields increased fourfold between 1920 and 1990, with most of that increase occurring after 1940, when the spread of the hybrids had already been completed. Some analysts have argued that up to 75% of the increase in maize productivity can be ascribed to introducing hybrid maize, and that the remainder can be accounted for by other technologies (fertilizers, pesticides, and mechanization). It is more accurate to say that after hybrids were introduced, all breeding efforts focused on producing hybrids compatible with the new technologies that became available after 1940. For example, after agronomists realized that maize responds to nitrogen fertilizers with increased growth and crop production, hybrids were bred

Figure 14.7 G. H. Shull's hybrid maize growing in 1906 at the Cold Spring Harbor Laboratory, New York. Note the bags covering the tassels (male organs) of some plants to prevent the spread of pollen. This is the same laboratory where Barbara McClintock did her Nobel Prize-winning research on maize genetics. *Source:* Cold Spring Harbor Laboratory Archives.

specifically to respond to high levels of nitrogen fertilizers. Very likely plant breeders could also have selected open-pollinated strains of maize that respond to fertilizers. However, there was no financial incentive to do so.

Starting in the 1950s, maize breeding shifted gradually from public institutions to private companies, and the fact that farmers cannot replant the maize they harvest from their F_1 hybrids, because it is genetically heterogeneous, provided the financial incentive for those companies to breed hybrids. Because farmers can plant the seeds they harvest from open-pollinated varieties, there was no financial incentive to improve those varieties. Wheat, which is sold to farmers as pure lines, can be replanted after harvest, so wheat improvement has remained the responsibility of public institutions.

14.6 Backcrossing comes as close as possible to manipulating single genes via sexual reproduction.

Hybridization of pure lines adds many genes from each line to the offspring. But sometimes a plant breeder wants to add to a crop just one characteristic, such as resistance to a particular disease. When the plant breeder finds a parent that can contribute the desired character, and one or two genes control that character, backcross breeding is used. In this method, a single desirable characteristic of one parent (the *donor parent*) can be added to the genetic makeup of a superior variety (the *recurrent parent*). This gets around the problem of trying to select simultaneously for many traits at the same time among extremely variable progeny. Backcrossing is as close as breeders can come to manipulating single genes through sexual reproduction, although in reality they manipulate many genes at once, often in large chromosome segments.

The introduction of genes for short stems into traditionally long-stemmed wheat provides a good example of backcrossing a morphological character. When tall wheat plants are heavily fertilized with nitrogen, they fall over at maturity because the slender stalk cannot hold up the heavy load of grain. As a result, much of the harvest can be lost, especially if the wheat is harvested by a combine machine. For this reason, plant breeders have tried to breed shorter strains of cereals.

Suppose a breeder wants to introduce by backcrossing a gene for "shortness" into a normal, high-yielding, tall wheat variety. This is done in several steps (**Figure 14.8**):

1. The breeder finds a suitable wheat parent that is short. It may not have any other desirable characteristic, such as high yield, but that does not really matter.

2. The short parent is crossed with the high-yielding tall variety, and the resulting seeds are planted to produce new plants. These plants have 50% of their genes from the tall variety and 50% of their genes from the short donor parent. These plants are allowed to self-pollinate so that the gene for shortness will be homozygous. The seeds produced by these plants are planted again, and now the breeder selects those plants that are short but otherwise most closely resemble the high-yielding parent.

3. These selected plants are then crossed with the high-yielding tall recurrent parent. Progeny plants are expected to have 75% of their genes from the high-yield variety. Selfing is followed by selection for shortness and resemblance to the high-yield variety parent.

4. This backcrossing procedure is repeated until a new variety of wheat emerges that has all the desirable characteristics of the original high-yield variety plus the gene for shortness.

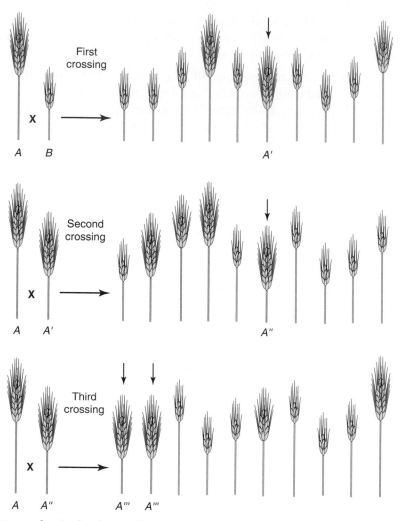

Figure 14.8 Hypothetical scheme for backcrossing. The objective is to introduce the gene for shortness from strain *B* into the high-yielding wheat strain *A*. From the first cross of *A* × *B*, plants (*A'*) are selected that resemble *A* but are shorter. This selection is done after the progeny of the first *A* × *B* cross have been allowed to self-fertilize to make sure that the gene for shortness is homozygous. These are then crossed to the initial strain, *A*. Once again, plants (*A"*) are selected that resemble *A* but are shorter. These in turn are crossed back to strain *A*. After repeated crossing and selections, the gene for shortness of strain *B* becomes part of strain *A*.

Genetically, the purpose of backcrossing is to retain as many genes as possible from the recurrent parent and to introduce the fewest possible number of genes, other than the gene of interest, from the donor parent. The more the backcrosses, the fewer the genes contributed by the donor parent. After the second, third, fourth, and fifth backcrosses, the donor parent contributes on the average 12.5%, 6.25%, 3.12%, and 1.56% of the genes. A breeder releases a new variety to the farmers when enough backcrosses have been made so that the new variety has both the new desirable trait and good agronomic characteristics.

Ideally, the breeder would like to get the number of genes from the donor parent down to the absolute minimum, perhaps one gene. In practice that is not possible because the breeder is really moving a chromosome segment broken by recombination during meiosis, not a single gene. Even so, this method is most useful for characteristics determined by a single or a few genes. If many genes on different chromosomes are involved in determining a desirable phenotype, it is next to impossible to backcross successfully.

14.7 Quantitative trait loci (QTLs) are more complex to manipulate than qualitative genes.

Unlike traits controlled by single genes that can be improved by backcrossing, many characteristics, or *characters*, that plant breeders would like to improve are controlled in a complex manner. These characteristics are called *quantitative traits,* because a biologist can't look at a plant and easily classify it as having or lacking the favorable allele(s) for the characteristic. The trait must be measured (quantified). Quantitative traits are expressed in progeny populations in continuous distributions. These distributions may be determined by heredity due to multiple genes (Figure 14.1) or by environment (Figure 14.2). The locations on chromosomes of each of the genes controlling a quantitative trait are called quantitative trait loci. Because many of the characteristics, such as yield, that plant breeders attempt to improve are quantitative traits, manipulating them is central to plant breeding.

If a characteristic is controlled by many genes, each of them with many alleles, the goal of plant breeding is to get the best alleles of each gene into a single plant variety. With traits controlled by multiple genes, the ultimate goal of the plant breeder is to compile the most favorable allele of each and every gene into one variety. This is a daunting task (**Table 14.4**). Just identifying the desirable alleles is difficult, and to begin to find them plant breeders make a large number of crosses among varieties that have similar phenotypes. It is common for a breeder to make more than 100 crosses each year

Table 14.4	Conventional breeding program for a new cereal	
Generations		**Evaluation/Selection/Testing**
P	Selected parents	
F_1	800 crosses	
F_2	2 million plants	Disease resistance
F_3	400,000 plants	Disease resistance and
F_4	12,000 lines	Field characteristics
F_5	1,200 lines	
F_6	300 lines	Disease resistance
F_7	50 lines	Yield
F_8	5 lines	Field characteristics Industrial uses
F_9	3 lines	Uniformity Quality testing
F_{10}	2 lines	
F_{11}	1 line!	

A conventional breeding program for a new cereal starts with selected parents and careful hand pollination to produce 800 crosses. This F_1 progeny is uniform. Then, 2 million plants are grown (2500 from each parent cross) and evaluation begins. After 10 generations the breeder ends up with one line.

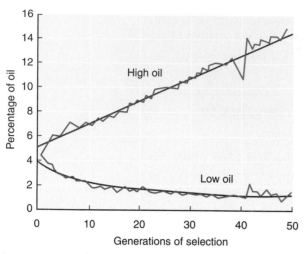
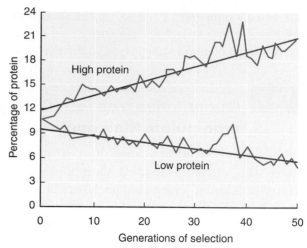

Figure 14.9 Fifty generations of selection for oil content and protein content in maize. The jagged red lines are actual data; the blue, straight lines are trends. In a comparatively short time people can use genetic variability to select plants with altered desired phenotypes. *Source:* Data from U.S. Department of Agriculture.

from among a group of 50+ selected individual varieties. Why so many? Why not just hybridize the two best varieties? Plant breeders don't know which favorable alleles of which genes are present, producing a variety's desirable phenotype. If two parents that are good because they have the *same* favorable alleles are crossed, transgressive progeny (progeny better than both parents, because they have accumulated more favorable alleles than either parent) will be rare. However, if the two parents are good because of *different* alleles, transgressive progeny may result from combining these different favorable alleles in one offspring.

Just as breeders do not know which favorable alleles of what genes are present in the parents, they also do not know which favorable alleles of what genes are in the offspring. The plant breeder must once again select the best performing varieties based on phenotype and cross those among each other. This long-term ongoing process has been the heart of plant breeding. Selecting for quantitative traits has produced continuous but incremental improvement (**Figure 14.9**). As long as enough genetic diversity remains available so that some subsets of parents differ in the favorable alleles they contain, improvements from selection are possible.

14.8 The Green Revolution used classical plant breeding methods to dramatically increase the yields of wheat and rice.

Perhaps the crowning achievement of breeding crops by crossing has been the Green Revolution. A leader in this effort, Norman Borlaug, was awarded the Nobel Peace Prize (there is no Nobel Prize for agriculture) for his work on developing high-yielding wheat in Mexico. A parallel effort in the Philippines resulted in the breeding of new high-yielding varieties of rice. The increased cereal production that these varieties made possible in the 1960–1980 period accounts for much of the steady rise in food production that was discussed in Chapters 1 and 2.

For both wheat and rice, a decade of repeated crossing introduced into single strains alleles that produced the following characteristics:

- *High yield.* This is clearly determined by many genes (a quantitative trait). For example, the new strains assimilate soil nutrients better than the previous ones. They also have a higher harvest index and a greater biomass accumulation (see Section 14.13).

- *Fast maturation.* In wheat, this means that the new strains are the fast-growing spring wheats, rather than winter wheats, which require a period of cold weather. In rice, this means that the growth from planting to harvest occurs in 125 days instead of the usual 210 days. In both cases, if the climate is right (as in Asia), rapid growth allows more than one crop cycle per year. This alone doubles the amount of food a given piece of land can produce.

- *Semidwarf growth habit.* The more grain a plant produces, the heavier it is, and a plant's spindly stems are not strong enough to carry the extra weight of grain. Although this is advantageous to the plant (the seeds fall to the ground to grow the next season), it is disastrous for the farmer. When the head of grain is on the ground, it is extremely difficult to harvest; moreover, moisture on the ground encourages the growth of fungal spores on the grain. Semidwarf varieties (in wheat they are 90 cm tall at maturity instead of the typical 120 cm) have strong stems and do not fall over.

- *Disease resistance.* Fungal diseases (for example, those that cause wheat rust and rice blast) wreak havoc on growth and yields. Alleles conferring resistance have been identified in certain wheat and rice strains and crossed into the high-yielding strains.

- *Adaptability to local conditions.* Once the four characteristics just mentioned have been introduced into single strains, these strains can be crossed to local varieties adapted to the growth conditions and consumer desires of a given region. The first "miracle rice" was designated IR8 (**Figure 14.10**). As breeding and the release of new strains continue, breeders at the International Rice Research Institute have now reached well above IR100 in their numbering of new lines.

The adoption of high-yielding wheat and rice varieties has followed similar patterns in many countries. Initially, the farmers primarily grew traditional varieties, most selected from landraces (stage I). These plants were often well adapted to their specific environment but had problems, such as lack of resistance to new diseases or low yield. To improve these varieties, local plant-breeding stations crossed in genes from other strains and landraces (stage II). But these improved varieties were still not yielding enough to feed the expanding population or to meet market demands. At this stage, international research centers entered the sequence by breeding the semidwarf varieties (stage III). Their high yields and fast maturity made them attractive to many countries. Since then, local breeders have been extensively crossing these plants with local varieties to improve their adaptation (**Figure 14.11**).

Experience shows that the adoption process for a new agricultural technology, be it seeds, fertilizers, or tractors, often follows an S-shaped curve. After a new advance (such as a variety of seed) is introduced, the first farmers to adopt it take the greatest economic risk. If the new method fails on their farms, they could lose their entire crops, or at least produce less. However, if the technology is successful, they also reap the greatest benefits.

The first U.S. farmers to adopt hybrid maize thus received the highest increase in income, which more than offset the additional price they had to pay for the seed maize. New technologies are adopted first on the biggest and most prosperous farms. These are managed by farmers who keep abreast of the new developments in agriculture and can obtain

(a)

Figure 14.10 Green Revolution rice varieties. (a) The first rice of the Green Revolution (IR-8) and its parents, Peta and Dee-gee-woo-gen. **(b)** The effect of nitrogen fertilizer on yield by the tall-stemmed Peta and the semidwarf IR-8. The semidwarf does not fall over when the heads are heavy (high yield) and can be harvested. The harvest from Peta is lost when the plants fall over. *Source:* Courtesy of Eugene Hettel, IRRI, and data from the International Rice Research Institute.

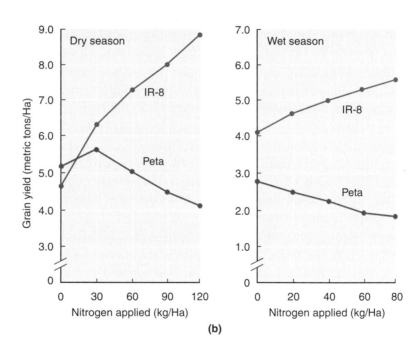

(b)

or have the capital needed to profit from such developments. The later a farmer adopts a new technology, the less will be the economic benefit.

In the case of the high-yielding varieties of wheat and rice, the farmers had to adopt not only the new seeds, but an entire technology package that included fertilizers, insecticides, herbicides, equipment for irrigation, and tractors to till the land. Indeed, the new varieties made it possible to grow two or even three crops per year, thereby potentially increasing the demand for labor. Thus labor-saving technologies (tractors and herbicides) had to go hand in hand with the new strains. Furthermore, nitrogen fertilizer was needed so that the new varieties would yield up to their potential (Figure 14.10). A maximum yield response to fertilizer was the central improvement in those varieties.

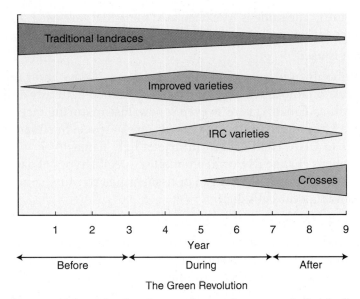

Figure 14.11 The sequence of adoption and use of crop varieties before, during, and after the Green Revolution. The figure shows how each set of varieties came into use and then declined again as better varieties became available. The vertical width of each bar reflects the extent of their use. Improved varieties = locally bred, but still tall. IRC = International Research Center short-stemmed varieties. Crosses = crosses between IRC varieties and local varieties. *Source:* Data collected by D. Dalrymple, U.S. Department of Agriculture.

Cycles of adoption quickly followed each other in different countries, starting with the adoption of improved wheat in Mexico (see Chapter 3). In March 1968, the director or the U.S. Agency for International Development, William Gaud, used the term "Green Revolution" to describe the great gains in production that the semidwarf varieties plus the associated agricultural technologies were achieving. In Asia, adoption of new rice varieties started in the late 1960s and was completed 12–15 years later. The gains in overall yield were quite impressive (**Table 14.5**) and are largely responsible for food production outstripping population growth in those areas, as discussed in Chapter 1.

Table 14.5	The result of the Green Revolution: yields (metric tons per hectare) of wheat and rice in India and China		
Country	**Crop**	**1963**	**1983**
India	Wheat	0.9	1.7
India	Rice	0.9	2.2
China	Wheat	1.0	2.5
China	Rice	2.0	4.7

Source: Data from Food and Agriculture Organization.

14.9 Induced mutations can produce new crop varieties.

Conventional plant breeding uses natural genetic variations to improve crops. Although there is much variation in crops and their wild relatives, this may not be what farmers want for human purposes. Mutagens—agents that cause heritable changes in genes—can be used to generate further variations, from which the ones humans want can be selected.

Behind this mutate-and-select strategy are many decisions:

■ What plant organ or tissue is used?

Seeds are most common, as they are easy to treat in large numbers, store, and transport. However, they are multicellular, so there will be many cells with different mutations and some cells with none at all. The plant that grows from a treated seed is a genetic chimera (mixture). The same is true if isolated meristems are used. In contrast, treatment of gametes such as pollen, or somatic cells in culture leads to nonchimeric plants.

- What mutagen is used?

The most commonly used mutagens involve radiation. X-rays and gamma rays, products of the decay of radioactive elements, can be obtained naturally or from a radioactive source such as cobalt. There is even a new, fast-maturing rice variety that the Chinese selected from seeds exposed to cosmic rays in space. In either case, there is extensive damage to DNA. High-energy ultraviolet light damages certain bases in DNA. Chemical mutagens change DNA and often cause specific types of damage. Of course, the challenge in mutating a plant organ or tissue is just to cause some mutations, not to cause so many that the cells die.

- How are the desired varieties selected?

If seeds or complex organs are treated, the plants that grow up will be heterogeneous (different plants carry different mutations), heterozygous, and chimeric. When these plants are either self-pollinated or intercrossed with varieties already in use, the single mutations become more apparent. Selection may begin at this point or at later generations.

- What new variations are desired?

The International Atomic Energy Agency lists 2,252 new crop varieties now in use that have come from mutation breeding (**Figure 14.12**). In almost all cases, these varieties now have characteristics that did not exist in the crop in nature. The new characteristics fall into three general classes:

Agronomic traits: Plant architecture, earlier or later flowering time, disease resistance.
Chemical composition: Oil content, fatty acid, protein and amino acid composition, starch quality
Reproductive factors: Male sterility, self-incompatibility

Figure 14.12
Radiation breeding. The popular Texas red grapefruit variety Rio Red, released in 1988, was developed by radiation of seedling budwood of the Ruby Red variety, which arose as a spontaneous mutant in 1926. In these grapefruit, the gene encoding an enzyme that converts the red pigment into a colorless pigment has been knocked out. Worldwide, there are now more than 2,000 crop varieties that have been produced by radiation. *Source:* Courtesy of R. Morillon.

14.10 Tissue and cell culture facilitate plant breeding.

Cell and tissue culture are laboratory-based methods for manipulating plant embryos, organs (roots, shoots), and tissues. These methods began in 1934 with the discovery that tomato root tips could grow indefinitely in the laboratory. Later, scientists defined the chemical signals that allow plant tissues in culture to dedifferentiate (to form a callus of parenchyma cells) and then redifferentiate to form an embryo and plantlet. Cloning for plants is much more a reality than it is for vertebrate animals.

Three major technologies of plant cell culture apply to breeding:

- *Embryo rescue* is the laboratory culture of embryos. It is especially useful in interspecific crosses, where an embryo may form but often cannot grow in the maternal plant. So the scientist dissects out the embryos, "rescuing" them, and grows them in a culture medium until they can be transplanted. These new plants can then be used for further breeding. An important example is the development of the new crop Triticale (**Figure 14.13**). Wheat (Triticum), with its nutritional superiority, was crossed with rye (Secale), with its environmental hardiness. The resulting embryo had a set of chromosomes from each plant and so was haploid. It would not grow in either mother plant, so the breeders rescued it and grew the embryo in the lab. After chromosome

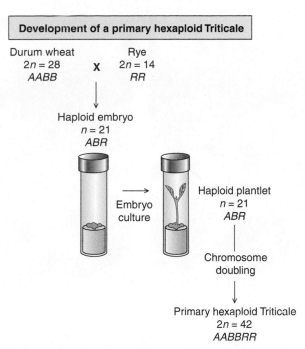

Figure 14.13 **Development of a new crop, Triticale.** Embryo rescue and micropropagation were used to make a new crop that combines the nutrition of wheat with the adaptability of rye.

doubling (see later discussion), the plantlets were transplanted to the field and a new, successful crop was born.

- *Anther or pollen culture* produces haploid cell cultures, containing only the male chromosome set. This culture shows many mutations from lab-based mutagenesis, desirable ones are easily selected. For example, in a recent experiment in Japan, Dr. Toshiro Kinoshita and colleagues exposed rice anthers and pollen to gamma irradiation. The resulting plantlets showed several new mutations for early maturation and dwarfness. When treated with colchicine, a drug that blocks cell division, the haploid cells become diploid. Thus a homozygous cell line results and the plants regenerated from these cells are the same as pure, inbred lines that take six to eight generations of selection in the field.

- *Micropropagation* is the growth of entire plants from plant parts or cells (**Figure 14.14**). This is especially useful for producing disease-free plants in crops that are vegetatively produced, such as potatoes, yams, and cassava. Thousands of identical plantlets from a genetic cross can be grown in the lab and then transplanted. This is most valuable in tree crops, where a breeding experiment might take years to evaluate. Micropropagation is widely used in 150 plants, most of them forest and fruit trees, and in plantation crops such as date palm and coffee.

Although it is easier for a farmer with fields to breed crops there rather than set up a laboratory, the technologies involved in tissue culture are inexpensive and simple, and well within reach of developing countries or small companies that want to participate in crop improvement.

(a)   (b)

(c)

Figure 14.14 Regeneration of agave. Many crops are vegetatively propagated once a superior variety (clone) has been identified. The Centro de Investigación Científica de Yucatán (CICY) in Mexico has undertaken a major project to vegetatively propagate *Agave fourcroydes* (common name, henequen) that is used for fiber production in arid regions of Mexico (a different agave is used for tequila production). **(a)** Regeneration of shoots from pieces of the agave meristem; **(b)** growth of young plants in sterile boxes; **(c)** An experimental field of vegetatively propagated agave clones. *Source:* Courtesy of Manuel Robert from CICY.

14.11 The technologies of gene cloning and plant transformation are powerful tools to create genetically modified (GM) crops.

Genetic engineering depends on the fact that the same molecule, DNA, makes up the genes of all organisms. This means that a gene from a fungus (made of DNA) can be inserted in a rice plant's chromosome (also made of DNA) and, with luck and experimental skill, can be induced to function. The ability to transform crops with genes from any other organism opens up the entire community of life as the source for new genes for a crop (see Chapter 6). Although the process of genetic modification started long ago (see Chapter 13), only crops that are genetically engineered are referred to as genetically modified (GM) crops. Some writers prefer the term *genetically enhanced* or *improved*. The process of introducing a new gene into an organism is also known as *transformation*.

Transformation involves several steps:

- A useful gene must be isolated, usually from an organism different from the crop species.

- The gene must be transferred into plant cells. These systems differ from plant to plant. In most cases a crop specific method has been discovered and optimized.

- The new gene must be integrated into the crop plant genome.

- Where a new gene settles in the host plant's DNA greatly affects its expression. So the breeder's old tool of selection still must be applied in the creation of GM crops.

- Fertile plants must be regenerated from the transformed cells. This is done by micropropagation.

- The transgene must be expressed in the crop plant in the correct tissue at the right time. This requires knowledge about the details of gene expression in the host plant. For example, if a seed protein is the transgene a promoter that activates an adjacent gene only during seed formation might be used.

- The new gene must be transmitted to the next generation when the crop plant reproduces. This essentially means that the plant has been genetically transformed.

There are several advantages of GM technology over other methods. First and foremost, it is specific: Unlike crosses where many genes are transferred during sexual reproduction, here the scientist transfers only a single gene (or a few genes) to the crop plant. Second, the method is fast; as in any micropropagation technique, many plants can be generated quickly. Finally, the genes themselves are an advantage. Unlike any other breeding methods, plant biotechnology allows a scientist to put virtually any gene from any organism into a crop plant. In practice, scientists create a large number of transformed crop lines for a specific gene and then carry all these lines through to many rounds of conventional plant breeding, to ensure that the new line has exactly the same characteristics as the parent line, with the exception of the introduced characteristic. Thus GM technology does not eliminate plant breeding but supplements it just like other laboratory-based methods. Examples of successful GM technology are given throughout this book:

- Vitamin A-rich "Golden Rice" (Chapter 7)
- Improved amino acid balance of animal feed (Chapter 7)
- Plants with modified water deficit tolerance (Chapter 10)
- Male sterility for more convenient hybridization (Chapter 8)
- Slowing down of fruit ripening (Chapter 8)

Figure 14.15 Papaya trees resistant to papaya ringspot virus, produced by GM technology. A healthy papaya tree loaded with fruit is in the foreground. A virus-infected tree is to the left of it. *Source:* Courtesy of Stephen Ferreira, University of Hawaii.

In the 1980s, papaya ringspot virus disease had gained a foothold in the Hawaiian papaya production region, and by 1994 nearly half the state's papaya area was infected. A resistant variety was needed, but conventional breeding had been unsuccessful in producing one. Researchers at Cornell University and the University of Hawaii decided to try a form of pathogen-induced resistance. They used GM technology and inserted the papaya ringspot virus's coat protein gene, isolated from the virus itself, into the papaya DNA. The presence of the virus coat protein in the plant did not cause the disease, because the additional viral genes necessary to cause infection were not present, but it did prevent additional virus infection, creating a type of plant immunity. After the necessary safety reviews were approved, the papaya growers switched to the GM papaya variety to save their industry and keep the fruits available for consumers. (**Figure 14.15**).

As described later, there are many applications of transgenic technology to pest resistance in crop plants. These include inserting genes that cause the plant to make proteins that confer resistance to pests, as well as genes that confer resistance to herbicides. As discussed later, herbicide resistance allows farmers to spray a field once with herbicides to destroy weeds, without affecting the transgenic crop plants (see Chapter 17).

14.12 Marker-assisted breeding helps transfer QTLs and major genes.

In the 1980s and 1990s another technical revolution occurred in genetics. Different types of molecular markers were developed that greatly improved genetic maps, making these maps useful for plant breeders.

Molecular markers are stable changes (often affecting only a single base) in short DNA sequences that occur at specific locations on chromosomes. Geneticists have developed different methods to identify these changes (**Figure 14.16**). When individuals have different sequences in the DNA at a particular chromosome region such that the scientist can distinguish among the individuals, the molecular marker is said to be *polymorphic*. Molecular markers usually follow the same rules of inheritance as genes. This allows them to be placed on genetic maps, pictorial representations of the order of genetic locations on a chromosome. Molecular markers are essentially signposts at many locations along the chromosome, far more than there are genes. Maps can be generated that have molecular markers covering all the chromosomes, at reasonably close proximity to each other.

Two features of genetic maps make them useful in marker-assisted breeding. If an interesting gene that is difficult to detect phenotypically is located very close to a molecular marker, geneticists say the two loci are tightly linked. This means that almost always, the molecular marker is inherited along with the allele the breeder is interested in. Instead of analyzing the transmission of the interesting gene, the plant breeder can substitute the easier analysis of the transmission of the molecular marker. A second use of molecular markers is in determining if two alleles influencing a single characteristic are from the same

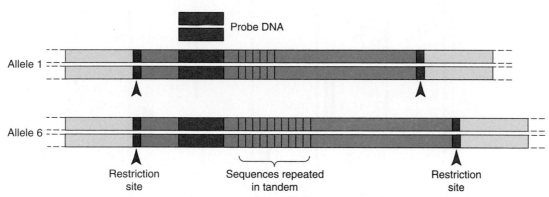

Figure 14.16 An example of a molecular marker. Allelic variation may result from a variable number of repeating DNA units; the length of the DNA fragment depends on the number of repeats. This serves as a molecular marker when the length variation can be visualized by cutting the DNA between the two arrows and separating the fragments by gel electrophoresis according to size. Often a marker such as this is near an agronomically important gene and can be used to "flag" the presence of that gene.

or different genes. Recall from our earlier discussion that one aim of plant breeders is to get the good alleles from different genes into the same plant. Molecular markers can be used to map alleles, and if they map at different locations, they can't be the same gene. So if two potential parents have the same phenotype but the genes controlling those phenotypes are located in different places on the genetic map, it is possible to combine those two different genes in one progeny with the potential for an improved phenotype.

Plant breeders combine phenotypic testing and molecular marker mapping to dissect the inheritance of quantitative traits. Once the genetic control of a quantitative trait is broken into its component quantitative trait loci (QTL) it will be possible to reassemble all these QTL into one progeny, producing a superior progeny. An example of genetic mapping of QTL for three quantitative traits is shown in **Figure 14.17**. The goal of that research is to find and transfer a favorable allele from a wild tomato species into cultivated tomato. Researchers cataloged progeny with a set of molecular markers, and measured the phenotypes for fruit weight, acidity and soluble solids. They carried out statistical analysis of the data to detect marker alleles associated with phenotypic differences in these traits. They identified six QTLs affecting fruit weight, four affecting soluble solids, and five controlling acidity level. For example, the QTL at marker locus CD34A identified a favorable allele from the wild species *L. chmielewskii* that increased acidity. Without QTL analysis it would have been nearly impossible to identify and transfer this one favorable allele that can improve acidity, an allele that was found in a less than desirable overall genetic source.

14.13 Plant breeders have a long wish list for crop improvement.

The goal of plant breeding is to increase crop productivity, the amount of useful crop that can be grown in a particular location. Although the potential yield is the major characteristic selected, many individual characteristics that are each genetically determined interact to produce yield. Furthermore, the selected characteristics of a crop interact with the economics of crop production: It may be more expensive to produce a high-yielding variety, so the overall effect of the high-yielding genes on actual crop production may not be so impressive. And the extensive ecological management needed to coax high yield may

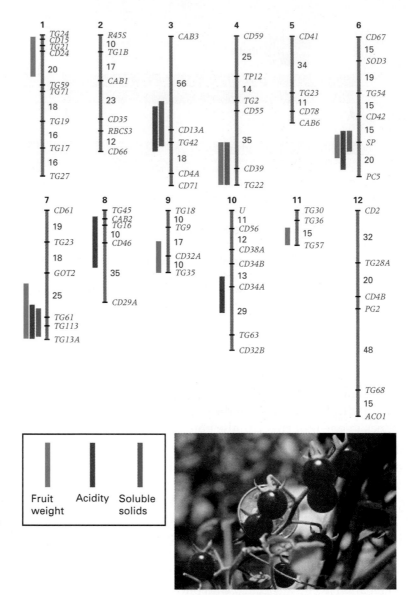

Figure 14.17 Location of QTLs for several quantitative traits in the tomato genome. The genetic markers are shown for each of the 12 chromosomes. The numbers in red are distances in map units between adjacent markers, but only map distances of 10 or greater are indicated. The regions in which the QTLs are located are indicated by the bars—green bars, QTLs for fruit weight; blue bars, QTLs for content of soluble solids; and dark red bars, QTLs for acidity (pH). The data are from crosses between the domestic tomato (*Lycopersicon esculentum*) and a wild South American relative with small (~1 cm in diameter) fruit (*Lycopersicon chmielewskii*), shown in the photograph, which remain green at maturity. The F₁ generation was backcrossed with the domestic tomato, and fruits from the progeny were assayed for the genetic markers and each of the quantitative traits. *Source:* D. L. Hartl and E. W. Jones (1998), *Genetics*, 4th ed. (Sudbury, MA: Jones and Bartlett), Figure 16.16, p. 694. Data from A. H. Paterson, E. S. Lander, J. D. Hewitt, S. Peterson, S. E. Lincoln, and S. D. Tanksley (1988), Resolution of quantitative traits into Mendelian factors by using a complete linkage map of restriction fragment length polymorphisms. *Nature* 335:721–726. Photograph (inset) courtesy of Steven D. Tanksley.

ultimately degrade the environment, so that the yield potential may be harder to reach. Some of these relationships are shown in **Figure 14.18**.

The physiological bases for food production by plants, as well as the environmental requirements for the optimal functioning of plants are described elsewhere in this book. Now, using the tools of plant breeding and current agronomic knowledge, agricultural scientists are conceiving, designing, and developing new crop varieties to increase agricultural

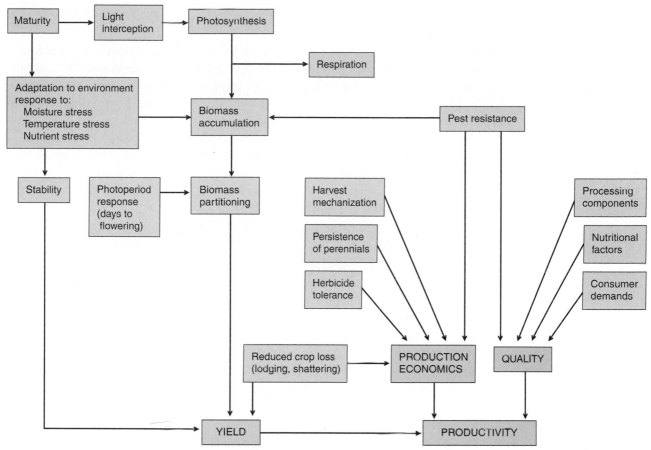

Figure 14.18 **A concept map integrating different selection criteria as they have the potential to improve crop yield, quality and production economics and as they relate to the overall plant-breeding goal of increasing agricultural productivity.**

productivity. In *Principles of Crop Improvement*, N. W. Simmonds and J. Smartt succinctly stated the overall goal of plant breeding as "Harvesting the maximum yield of useable products with minimum inputs and the least hazard to the environment."

1. *Increased biomass accumulation:* Increased biomass accumulation is a long-term necessity for increasing yield, and the key component is increased photosynthesis by the crop. Many of the factors described later in this list impact biomass accumulation by increasing photosynthesis. For example, pest resistance can maintain leaf area, maturity alterations match crop growth to the environmental cycle, and changes in leaf architecture increase light interception. In the long term, biotechnological manipulations that reduce photorespiration may eventually achieve greater efficiency in photosynthesis.

2. *Increased harvestable yield:* One way to increase harvestable yield is to reduce crop loss. As you've read, a key feature of domestication was selection against shattering, and a key success of the Green Revolution wheat and rice varieties was reduced falling over, or lodging. Diseases such as stalk rots and stem borers increase lodging, so increased pest resistance can also impact this aspect. A second way to increase harvestable yield is to increase the flow of biomass into the plant parts that people eat, while maintaining plants that are strong enough to bear the crop. That is, it does not do much good for a maize plant to produce more leaves; people and animals consume the grain. The harvest index (HI)—the percentage of biomass harvested—has been successfully increased in rice and wheat from 30% to nearly 50%. Once the HI

is greater than 50%, however, a renewed emphasis must be placed on biomass accumulation (Figure 14.19).

3. *High nutritional quality:* A crop is more beneficial to people if the nutritional qualities of the crop are improved. Proportions of essential amino acids and the total protein in cereal grains could be increased to improve their nutritional quality, as in CIMMYT's Quality Protein Maize project. The same nutritional factors should be improved in root crops such as potatoes, sweet potatoes, and cassava. Plant breeding can also eliminate toxic compounds. For example, cassava produces linamarin, a cyanogenic glucoside. The removal of linamarin would reduce the extensive processing that is now required before cooking. However, plants without toxic compounds would be more susceptible to insect attack, so researchers must weigh the advantages and disadvantages of all projects.

Iron deficiency is the world's most widespread nutritional disorder and affects about 5 billion people. Worldwide, 39% of preschool children and 52% of pregnant women are anemic; more than 90% of these women and children live in developing countries. Iron deficiency anemia causes 20% of maternal deaths during childbirth. Scientists at IRRI have selected natural variants of rice that accumulate more iron (21 ppm, two times normal) and zinc; the zinc enhances the body's ability to absorb iron. Supplying the experimental rice variety IR68144 in diets has increased the level of ferritin (an iron-containing compound) in the blood two to three times. Other attempts are under way to manipulate the iron level by adding three transgenes to rice: (a) an enzyme to break down phytate, which ties up 95% of the iron; (b) an iron storage protein that doubles the iron level; and (c) a sulfur-rich protein to help humans absorb the iron.

4. *Resistance to pests:* If crop plants had the genes to be resistant to all pests, food production would rise, crop quality would increase, and production costs and envi-

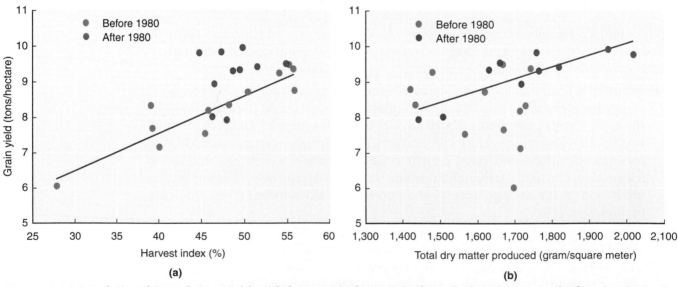

(a) (b)

Figure 14.19 Relationships of rice yield with harvest index (a) and total plant biomass (b) for rice varieties released before 1980 and rice varieties released after 1980. In part **(a)** there is a high positive correlation between yield and harvest index (HI) for varieties released before 1980 as breeders concentrated on short-stemmed rice varieties; short stems increase the HI. No such correlation exists for varieties released after 1980. The reverse is seen in part **(b)**, where yield does not depend on increasing plant biomass in varieties released before 1980 but where yield increases in varieties released after 1980 are closely related to increases in total biomass accumulation. *Source:* S. Peng, R. C. Laza, R. M. Visperas, A. L. Sanico, K. G. Cassman, and G. S. Khush (2000), Grain yield of rice cultivars and lines developed in the Philippines since 1966. *Crop Science* 40:307–314.

ronmental concerns associated with pesticide use would decrease. Along with direct selection for yield, breeding for pest resistance is the most widespread plant-breeding activity. Breeding for pest resistance is an ongoing concern, because new strains of pests, especially fungi, always arise and attack the varieties that were resistant to the old strains. Durable resistance based on multigene instead of single-gene pest resistance is a major objective of plant breeding. Combining a crop's evolved resistance genes with transgenic resistance could help.

5. *Response to inputs and production economics:* High-yield results from growing plants with high-yield potential in a "high yield" environment. To maximize the production economics, a variety must be able to respond to the level of inputs a farmer provides, that is, what the farmer can reasonably afford. As we noted earlier, the ability to respond to additional inputs was a hallmark of the Green Revolution grain varieties. The initial GM crops have been focused on improving production economics. Crops made tolerant to herbicides by transgenic technology reduce costs of weed control; cotton, potato, and other crops transformed with the Bt genes to make them resistant to insect pests reduce insect control costs. The search for novel genes that can reduce production costs and constraints will continue because both farmers and consumers will benefit. For example, since 1960 rice production costs per unit have declined 30% and the price of rice adjusted for inflation has declined 40%.

6. *Performance in high-stress environments:* Not all agricultural environments are "high yield" environments. Many areas of the world have insufficient or poorly distributed rainfall, and water availability is often the limiting environmental factor. Crop plants that use water more efficiently and that are more drought tolerant will greatly improve crop productivity in many regions of the world. Matching a crop's maturity to the annual environmental cycle improves productivity by avoiding moisture and temperature stresses. Tolerance to soil nutrient stress (aluminum tolerance, salt tolerance) is also beneficial and interacts with drought tolerance through improved root growth and water uptake. Plant breeders are trying to match crops and varieties to specific environments, although this entails an expansion of the breeding effort. CIMMYT now points its breeding efforts to nine defined target environments with four different types of maize.

7. *Plant architecture:* The number and positioning of the leaves, the branching of the stem, the height of the plant, and the positioning of the organs to be harvested are all important to crop production. They often determine how well plants will intercept light, how closely they can be planted, and how easily the crop can be harvested mechanically.

8. *Photoperiod response:* Many crops flower and set seed in response to day length. This limits the crop to a certain climatic zone and limits harvest to once a year. Selecting varieties that are insensitive to photoperiod or are adapted to a variety of photoperiods would allow multiple cropping during a growing season. Altering a crop's photoperiod response such that plants flower earlier and have a longer reproductive period can increase biomass partitioning to seeds.

9. *Processing characteristics:* The value of some crops comes not from their direct harvested state, but from their processed state. For example, oilseed rape is not consumed directly; instead, its valuable products are from its seed: vegetable oil and the residual protein meal. Rapeseed (Brassica rapa or B. napus) has become more valuable as it has developed into two crops: canola and industrial rapeseed. Oilseed rape contains glucosinolates in the meal and erucic acid in the oil. Both are negative nutritional factors, but erucic acid is a valuable industrial component. Canola was bred as a food crop by eliminating these two components, which greatly increased

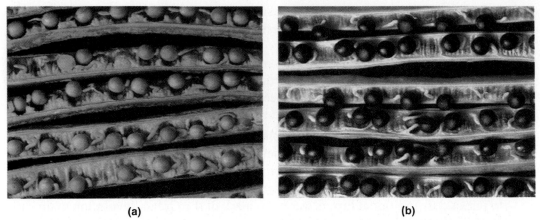

(a)　　　　　　　　　　　　　　　　(b)

Figure 14.20 GM technology will allow the production of pro-vitamin A-enriched canola oil. (a) Control seeds and **(b)** seeds rich in pro-vitamin A. *Source:* Courtesy of Toni Voelker, Monsanto.

the edible value of its oil. Canola is being further improved by adding genes to produce an oil that is high in vitamin A (Figure 14.20), similar to the development of Golden Rice. Industrial rapeseed was bred by increasing erucic acid to 55%, thus making processing of its oil for industrial uses much more efficient. At the same time, the glucosinolates were removed so that the residual meal could still be used as a protein food source. The two separate crops are more valuable than the original crop with its dual uses. In a move toward a bio-based economy, many crops are being targeted for producing industrial and pharmacological components (Chapter 19).

CHAPTER SUMMARY

Increased crop productivity through genetic improvement of plants has significantly impacted world agriculture and the world's populations. Plant breeding has followed the general pattern of introduction, selection, and hybridization. Crop introduction has been crucial for world agriculture, because many of the world's crops are produced outside their region of domestication. For centuries, selection has been based on the genetic variation created naturally via meiosis. Planned hybridizations have provided a degree of control over the variation produced. Furthermore, plant breeders have improved selection progress by using quantitative genetics to better understand the phenotype/genotype/environment relationship. Now they are adapting techniques from molecular biology to further enhance selection of quantitative traits.

One of the great success stories in agriculture has been the tremendous yield increases achieved by coupling high-yield varieties and improved agricultural production technologies, creating the Green Revolution. Plant breeders are working to extend the Green Revolution by intensifying selection, developing hybrid varieties in more crops, and increasing the range of plant functions through mutation and transgenic breeding. Crop plants can still be improved in many ways, including, in part, (1) increased yield through improved plant growth processes, (2) improved crop quality, (3) increased pest and stress tolerance, and (4) enhanced processing characteristics. Plant breeding will continue to play a crucial role in crop improvement because the needs are many, the techniques are expanding, new genetic combinations are limitless, and the successes of the past illuminate the potential of the future.

Discussion Questions

1. How are mutation and transgenic breeding used in similar ways to improve crops?

2. Discuss how crop movement has played a crucial role in world agriculture. Do you think crop movement had a diversifying or homogenizing effect on world agriculture? Explain.

3. Visit http://www.futureharvest.org/news/maizepressrelease.shtml. What is Quality Protein Maize? How will this improve the value of maize in developing countries?

4. Discuss the advantages and disadvantages of hybrid varieties for farmers and plant breeders.

5. Visit http://www.cimmyt.cgiar.org/research/abc/10-FAQaboutGMOs/htm/10-FAQaboutGMOs.htm. Using the information presented in this article, along with your own knowledge and information from other readings, comment on what you think are long-term prospects are for GM crops.

6. Visit http://www.cgiar.org/irri/AR2000/Casehistories.pdf. How does achieving success through plant breeding require more than knowledge of genetics and plant growth?

7. Plant breeders can select for improved plant response to predictable environmental factors. Describe three environmental constraints that you believe breeders should change plants to respond better to.

8. Describe three physiological factors (discussed in other chapters) affecting plant growth, which you think would be beneficial for plant breeders to improve.

Further Reading

Boerma, H. R. 2000. *Integrating DNA Markers in a Breeding Program.* Available at http://mars.cropsoil.uga.edu/~hrb/CSSA2000/HRB.html. Accessed September 28, 2001.

Fehr, W. R., and H. H. Hadley, eds. 1980. *Hybridization of Crop Plants.* Madison, WI: American Society of Agronomy.

Luther Burbank, Plant Pioneer. Available at http://www.lutherburbank.org/pioneer.html. Accessed September 28, 2001.

Mann, C. 1997. Reseeding the Green Revolution. *Science* 277:1038–1043.

Morris, M. L., and M. A. López-Pereira. *Impacts of Maize Breeding Research in Latin America 1966–1997.* Available at http://www.cimmyt.cgiar.org/Research/maize/map/impacts/index.htm. Accessed September 28, 2001.

Peng, S., K. G. Cassman, S. S. Virmani, J. Sheehy, and G. S. Khush. 1999. Yield potential trends of tropical rice since the release of IR8 and the challenge of increasing rice yield potential. *Crop Science* 39:1552–1559.

Peng, S., R. C. Laza, R. M. Visperas, A. L. Sanico, K. G. Cassman, and G. S. Khush. 2000. Grain yield of rice cultivars and lines developed in the Philippines since 1966. *Crop Science* 40:307–314.

Plant Breeding for Enhanced Micronutrients. Available at http://www.sph.emory.edu/PAMM/IH552/cnyhus2000/runpage.html. Accessed September 28, 2001.

The Rewards of Rice Research: IRRI Annual Report 1999–2000. Available at http://www.cgiar.org/irri/AR2000/Rewards.htm. Accessed September 28, 2001.

Simmonds, N. W., and J. Smartt. 1999. *Principles of Crop Improvement.* Oxford, U.K.: Blackwell Science.

Stix, G. *Resistance Fighting: Will Natural Selection Outwit the King of Biopesticides?* Available at http://www.sciam.com/1998/0598issue/0598techbus2.html. Accessed September 28, 2001.

Tanksley, S. D., and S. R. McCouch. 1997. Seed banks and molecular maps: Unlocking genetic potential from the wild. *Science* 277:1063–1066.

Crop Diseases and Strategies for Their Control

Andrew F. Bent
University of Wisconsin–Madison

Plants are marvelously productive under ideal growing conditions, but ideal growing conditions rarely occur. More typically, plants encounter abiotic stresses such as water shortage, lack of nutrients, or excess soil salinity, or biotic stresses in the form of plant pathogens, insect pests, or weeds. In this chapter we discuss (1) plant diseases caused by microbial pathogens, (2) the intriguing biology of plant–pathogen interactions, and (3) new biotechnology approaches that create plants with enhanced disease resistance. Insects and weeds are considered in the subsequent two chapters.

15.1 Viral, bacterial, and fungal infections diminish crop yields.

There are hundreds of thousands of different virus, bacteria, and fungus species in the world, and thousands of these are pathogens that infect plants. Any one pathogen can severely depress the yield of a given crop. Pathogens may reduce yield by causing tissue lesions; by reducing leaf, root, or seed growth; or by clogging vascular tissues and causing wilt. Young seedlings can be overwhelmed by a pathogen and die soon after germination. Even in the absence of obvious symptoms, pathogens can cause a general metabolic drain that reduces plant productivity. Pathogens may also cause pre- or postharvest damage to the harvested product that can range from cosmetic blemishing to total decay.

Given this arsenal of threatening microbes, you may think it a miracle that a crop can be produced at all. But the interaction of plants and pathogens is often very specific. Although some pathogens can infect many species of plants, many are highly specific as to the crop species and the part of the plant that they infect (root, fruit, leaves, stem, and so on) (**Figure 15.1**). Furthermore, plants have specific defense mechanisms that keep infections under control. Like animals, after they have experienced an infection plants can

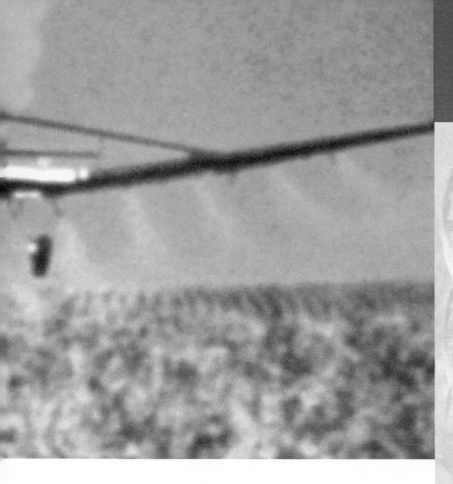

even develop some immunity against some pathogens. Plant breeders devote significant efforts to producing plant varieties with strong resistance to disease. Still, for any given crop the number of economically relevant pathogens is typically in the range of a few dozen species. See, for example, the tally for rice pathogens in **Table 15.1**. As you read through this chapter, you will see that the different plant diseases illustrate many different biological phenomena.

The microbes that do cause disease to a given crop can be devastating. From earliest recorded history to the present, every civilization carries stories of crop disease outbreaks that have caused famine, economic upheaval, mass migration, and death. For example, the Irish potato famine of 1845 and 1846 killed hundreds of thousands of poor Irish people. They could not afford to buy the more expensive wheat, so they died of malnutrition and starvation. This famine caused the first major wave of Irish emigration to the United States. The Bengal famine of 1943 was substantially attributable to an epidemic of brown spot disease on rice. Plant diseases can also alter the landscape dramatically, as has happened with the eradication of American Chestnut and American Elm trees across North America by single, destructive pathogen species. Less severe crop disease epidemics occur in most farming regions every few years, challenging the financial stability of farm families and farm communities.

Table 15.1	Rice diseases	
Agent	**Plant Organ Attacked**	**Number of Diseases**
Virus	Leaf	12
Bacteria	Leaf	4
Bacteria	Grain	3
Fungus	Leaf	11
Fungus	Stem, root	10
Fungus	Seedling	5
Fungus	Grain	10
Nematodes	Root	11

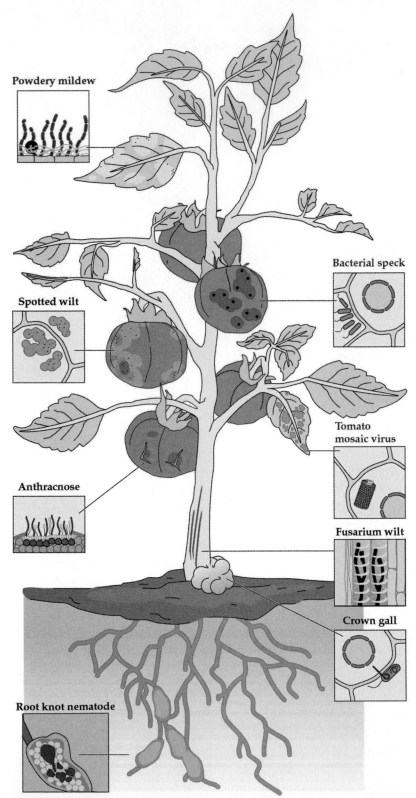

Figure 15.1 Most microbes attack only a specific part of the plant and produce characteristic disease symptoms. Tomato, shown here, can be attacked by more than 100 different pathogenic microorganisms. *Source:* B. B. Buchanan, W. Gruissem, and R. L. Jones, eds. (2000), *Plant Biochemistry and Molecular Biology* (Rockville, MD: American Society of Plant Physiologists), p. 1104.

15.2 Disease epidemics are caused by the convergence of multiple factors, many of which are commonly present.

What causes a **disease epidemic**? First, a virulent strain of pathogen must be present in sufficient numbers at the right place and time to start off the epidemic. Second, susceptible plant varieties must be widely present. Third and equally important, because pathogens are often sensitive to factors such as temperature, humidity, and wind, weather conditions must be just right (or "just wrong") for an epidemic to occur. These three factors make up the **disease triangle** that is crucial to the occurrence of crop diseases (**Figure 15.2**).

There is a subtle and malevolent corollary to this disease triangle. Because severe epidemics do not occur unless all three factors are favorable, unwise farming practices can become established and proceed without penalty until the setting is ripe for an epidemic. For example, a major epidemic of southern corn leaf blight ripped through the maize crop of the United States in 1970, reducing the harvest by an estimated 15% ($1 billion) and shaking the confidence of people who take for granted plentiful food production in the United States. Luckily, the cause of this epidemic was pinpointed fairly rapidly. Most maize varieties in use at the time shared an identical gene, and *Cochliobolus heterostrophus*, a formerly minor pathogen of maize, could readily infect maize plants carrying that gene. When maize breeders and seed suppliers shifted to varieties with a more diverse genetic makeup, the disease subsided immediately.

The maize blight example reveals two major and partially preventable causes of disease epidemics: monoculture and genetic uniformity. **Monoculture** is the growth of a single crop species on a large piece of land, but also implies a strong regional emphasis on a given crop. This type of crop specialization is often economically advantageous, but it does allow pathogen populations to build by fostering plant-to-plant and field-to-field movement. Regular rotation among different crops can prevent excessive buildup of pathogens. However, some pathogens can remain in the soil for many years, and crop rotation does not always work. In addition, some pathogens may infect several crops (**Figure 15.3**).

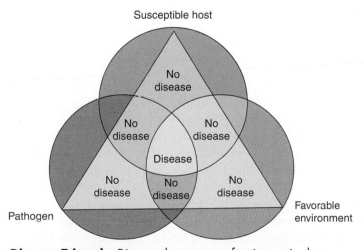

Figure 15.2 The Disease Triangle. Disease, the outcome of an interaction between a plant and a potential pathogen, depends on three key factors: the plant (genotype and planting configuration), the pathogen (genotype and prevalence or mode of introduction), and the environment. *Source:* Courtesy of Paul Williams, University of Wisconsin. Available at http://www.plantpath.wisc.edu/pddceducation/MasterGardener/General/Slide2.htm. Accessed May 4, 2001.

Figure 15.3 Monocultures promote disease. Scab, caused by infection of cereals by the scab fungus *Fusarium graminearum*, results in the presence of many small and wrinkled kernels. Such infections cause concern because the fungus produces toxins that can cause a wide range of disorders in humans and animals. Rotation with a crop that is not a cereal is the best control strategy. *Source:* Courtesy of Brian Hudelson, University of Wisconsin.

Permitting **genetic uniformity** is perhaps the more dangerous practice. As plant breeders and farmers gravitate toward the most successful varieties of a given crop, there is a tendency to use fewer and fewer plant genotypes. This can promote disease epidemics because plants and pathogens are often co-adapted to the extent that the pathogen, or specific strains of the pathogen, often can cause disease only on specific genotypes of a given plant species. If most varieties of a crop species grown in a given region share a very similar genetic makeup, then the pathogen that "solves" this genetic puzzle will have the key to causing disease on a vast area, with potentially devastating consequences. The solution is to maintain genetic diversity among the different popular varieties of a given crop species.

In some cases disease-favoring weather, virulent pathogens, and susceptible plants are common in a given region, and disease prevention tools such as fungicides or disease-resistant plant varieties are either too expensive or not available. Growers must then tolerate substantial annual losses, grow a different crop, or adopt cultural practices that greatly minimize the disease problem. This is true with rice blast disease across large regions in Asia. In the late 1990s, the Chinese conducted a grand experiment in the Yunnan Province, in which thousands of farmers participated. To counteract the negative effects of genetic uniformity, each farm grew mixtures of different rice varieties with different levels of resistance to rice blast (**Figure 15.4**), the most serious rice disease. The level of rice blast infection was dramatically decreased, and in many areas the farmers were able to stop using fungicides. It is likely—although this was not examined—that other diseases were also reduced. This type of benefit has been observed in wheat mixtures as well. So why is this simple method not used more widely? First, the rice (or other crop) that is harvested is not of uniform quality, because many different lines are used; and second, mechanical harvesting is not possible, because the different varieties do not all mature at the same time.

Figure 15.4 Rice blast disease, caused by *Pyricularia oryzae* (the sexual stage is known as *Magnaporthe grisea*). One of the most destructive diseases of rice worldwide, blast is favored by high nitrogen fertilization and rainfall or irrigation—common conditions in modern rice cultivation. The pathogen can girdle and/or cause breakage of the panicle neck node, reducing grain filling or killing developing panicles and grain. In this photo the diseased plants are yellow (early senescence/maturation, with poor grain filling) and the healthy plants are green and growing. Rice blast is controlled by using resistant rice varieties and, if disease pressure is very high, fungicides and other cultivation practices. *Source:* Courtesy of Eugene Hettel, International Rice Research Institute.

15.3 Viruses and viroids are parasites that have only a few genes.

Viruses and viroids are the smallest infectious agents that cause diseases. They are so small that people can only see them with an electron microscope. Viruses are not even cells, and plant virus particles often consist only of a nucleic acid (RNA or DNA) molecule wrapped up in a coat of protein. For example, the tobacco mosaic virus is a rod-shaped particle (**Figure 15.5a**) consisting of a coiled RNA molecule coated with spirally arranged coat protein molecules (**Figure 15.5b**). The nucleic acid of most viruses has only enough genetic information for a few proteins. Viroids are even smaller, consisting of only a relatively small RNA molecule. Both viruses and viroids require the cellular machinery of their host to reproduce themselves. The majority of plant viruses have a single-stranded RNA genome, and the life cycle of a typical (+)-strand RNA virus is shown in **Figure 15.5c**. Normally, viruses and viroids never quite kill their hosts, but because they divert the cellular functions of RNA and protein synthesis for their own ends (replication), they weaken the plant and diminish crop yields.

Plant tissues can often be infected by one or more viruses and yet show no outward symptoms of this infection. However, many viruses cannot propagate in meristems (see

Figure 15.5 (a) Electron micrograph of tobacco mosaic virus particles. **(b)** Model of a virus particle showing the nucleic acid core surrounded by a protein coat. **(c)** Simplified life cycle of an RNA virus.

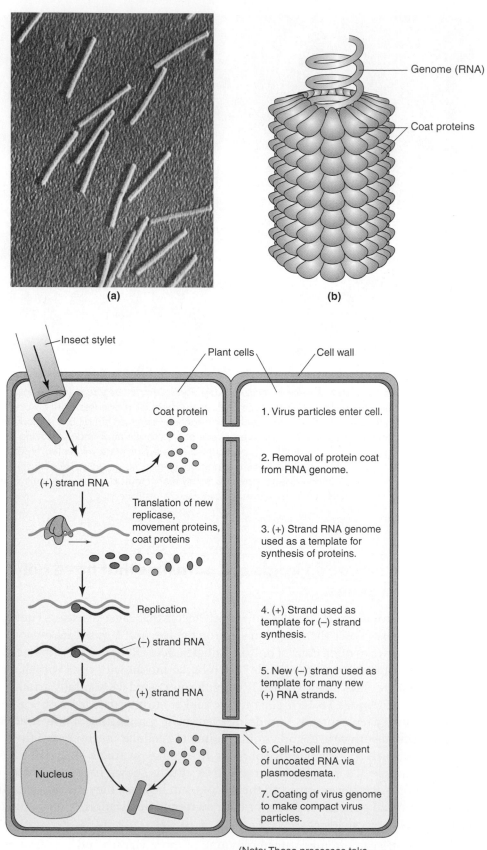

(a)

(b)

Genome (RNA)

Coat proteins

Insect stylet

Plant cells

Cell wall

Coat protein

(+) strand RNA

Translation of new replicase, movement proteins, coat proteins

Replication

(−) strand RNA

(+) strand RNA

Nucleus

1. Virus particles enter cell.

2. Removal of protein coat from RNA genome.

3. (+) Strand RNA genome used as a template for synthesis of proteins.

4. (+) Strand used as template for (−) strand synthesis.

5. New (−) strand used as template for many new (+) RNA strands.

6. Cell-to-cell movement of uncoated RNA via plasmodesmata.

7. Coating of virus genome to make compact virus particles.

(Note: These processes take place in cytoplasm, not in nucleus.)

(c)

Figure 15.6 Virus diseases are often propagated by insect vectors. Aphids and other sucking insects transfer viruses from one plant to another and are responsible for the spread of virus diseases. Shown here are aphids on the stem of a barley plant. These aphids spread the barley yellow dwarf virus, a common disease of cereals. *Source:* Courtesy of Brian Hudelson, University of Wisconsin.

Chapter 8). Plants regenerated from meristems are virus free, so horticultural industries use this type of clonal plant propagation to make virus-free plants. That viruses diminish the vigor of the plant is shown by the fact that the virus-free plants produced via meristem culture usually grow more vigorously than normal (virus-infected) plants. Plant viruses don't harm people, and because plant viruses are everywhere, most people eat plant viruses on a regular basis.

The nucleic acid molecule of a virus carries only a few genes. Its DNA or RNA typically contains genes for the viral coat protein, for an enzyme that allows the nucleic acid molecule to replicate itself, and for a "movement" protein that lets the virus move from cell to cell. Viruses may enter the plant through a small wound, and they are often transmitted by insects such as aphids (**Figure 15.6**) or leafhoppers (see Chapter 16). They then multiply within the infected plant cell. The plasmodesmata that connect plant cells exclude viruses, but the movement proteins encoded by some viral genomes modify these channels so that the viral nucleic acids can move from one cell to another. Many viruses use the phloem to move rapidly throughout the plant.

By disrupting the cell's metabolism or directing it toward their own growth and multiplication, viruses diminish the plant's capacity for photosynthesis and balanced growth. When they occur in large numbers, viruses can cause yellow spots or sectors on leaves, and virus are often named after these and other characteristic coloring patterns. Tobacco mosaic virus, cucumber mosaic virus, barley yellow dwarf virus, and papaya ringspot virus all are important plant pathogens that cause disease not only in the plants after which the virus is named, but in many other plant species as well (**Figure 15.7**).

Citrus tristeza virus illustrates the devastation that a single virus disease can cause. Citrus tristeza virus lives in the stem xylem, and its spread in the tree rapidly erodes growth and

Figure 15.7 Symptoms of virus infection. Symptoms caused by virus diseases can include a mosaic-patterned yellowing of the leaves. In addition, the leaves may be distorted and curled. The squash fruit shown here is mottled and has raised bumps, all typical symptoms of viral infections. *Source:* Courtesy of Brian Hudelson, University of Wisconsin.

productivity. Infected plants must be removed and burned. To spread, many plant viruses require insect vectors, and the tristeza virus is spread by aphids, especially the tropical citrus aphid. In the state of São Paulo in Brazil, 75% of the orange trees died within 12 years of the appearance of the tristeza virus. The same virus also decimated the California Citrus industry in the late 1950s and early 1960s. In 1995 the brown citrus aphid, an insect that very efficiently transmits Citrus tristeza virus, became established throughout Florida. Virus control programs often focus on insecticide use to control the insects that spread the virus. Active efforts to control this disease have been in place for decades, but the threat continues because insect control is only partially effective.

Citrus tristeza virus also spreads when citrus trees are grafted. Citrus trees are normally propagated in nurseries by grafting twigs from elite trees that produce good fruit onto a rootstock (a young seedling cut off near the ground) of a citrus variety that will produce a strong root system. Citrus trees must be replaced at regular intervals, so millions of new grafted trees are needed every year. It is important, therefore, to certify that the trees for grafting are free of virus and that the rootstocks used are virus resistant.

15.4 Plant-pathogenic bacteria cause many economically important diseases.

Some bacteria play vital roles in crop production, such as those that fix nitrogen, transform nitrogen and sulfur compounds in the soil, or contribute to the decay of organic matter and the process of mineralization. However, many species of bacteria cause important diseases and contribute substantially to the annual 10–15% crop loss caused by diseases worldwide. Because they are microscopic in size, bacteria can enter the plant through natural openings or wounds. In some cases, they secrete enzymes that help dissolve the protective barrier of the epidermal cell wall and penetrate the plant in this way.

Many bacterial diseases show up as spots on fruits, leaves, and stems. These spots, local infestations of the bacteria, diminish the photosynthetic capacity of leaves or disfigure fruits, thereby decreasing their commercial value. Other bacteria cause soft rots by secreting enzymes that digest the cell walls of the plants, causing the plant tissues to become very soft or even liquid. Anyone who has left vegetables in the bottom drawer of the refrigerator for a month has witnessed the destructive action of *Erwinia* bacteria. Several other members of this genus cause postharvest losses of fruits and vegetables.

Bacteria that invade the vascular conductive tissues of plants can clog such tissues so that water and minerals cannot get from the roots to the leaves. Such bacteria cause plants to wilt, and the diseases are accordingly referred to as *wilts* (**Figure 15.8**). These bacteria also secrete enzymes that dissolve cell walls of conductive tissues and the disease then spreads to other cells, quickly killing the plants. The most economically important wilt disease is caused by *Ralstonia solanacearum*, a bacterial species that can infect plants from 44 different genera. The disease occurs especially in warmer regions and affects many crop plants, including bananas, tomatoes, and potatoes.

Figure 15.8 Bacterial wilt disease of cucumber. When bacteria or fungi invade the vascular system (xylem), blockage of water transport causes the plants to wilt. Bacterial wilt of cucumber is caused by *Erwinia tracheiphila*, which are spread by the cucumber beetle when it feeds on the plants. *Source:* Courtesy of Brian Hudelson, University of Wisconsin.

Pierce's disease of grapes is caused by *Xylella fastidiosa*, a bacterium that clogs up the xylem of grapevines. The same bacterium causes a serious disease of citrus called citrus variegated chlorosis and infects apples, pears, and almonds. These bacteria are spread by a leafhopper (insect), the blue-green sharpshooter, that sticks its proboscis into the conductive tissue of the plant (see Chapter 16). Another insect vector with a new geographic range, the glassy winged sharpshooter, recently appeared in California. In the absence of effective control measures for this insect, the disease is threatening the California grape and wine industries. Because of the economic importance of the citrus industry to Brazil, scientists there chose *Xylella* as the first plant pathogen to have its entire genome sequenced (**Box 15.1**). By knowing the genes that *Xylella* needs to complete its life cycle and pathogenic activities, we should be able to generate better strategies to combat the disease.

Bacterial pathogens can also cause blights and cankers. Blights are characterized by rapidly spreading infections that kill infected cells and tissues. Using fruit trees again as an example, a species of *Erwinia* causes fire blight of apples and pears. The disease, which can kill young trees in one season, devastated U.S. pear orchards in the 1930s and still kills numerous trees annually. Citrus canker, a disease caused by a *Xanthomonas* bacterium, devastated the citrus industry in the 1920s. In 1984 it allegedly reappeared in some of Florida's largest nurseries, and millions of citrus seedlings and young trees were immediately destroyed because the agricultural authorities were afraid there might be a repeat of the earlier devastation. A new outbreak of citrus canker is again spreading through Florida as this book goes to press. There are no chemical sprays to effectively contain such outbreaks, and the only way to keep the disease from spreading at present is to destroy (burn) all trees suspected of being infected.

The genus *Agrobacterium* includes soil-dwelling bacteria that can cause plant tumors (crown gall disease) or other outgrowths (hairy root disease). *Agrobacterium* transfers a segment of its DNA from a plasmid to the genome of the infected plant cell, causing the plant to express genes encoding enzymes that create unique food compounds usable by *Agrobacterium* but not by other microbes or organisms. Other transferred genes specify

Box 15.1

The Genome of *Xylella fastidiosa*

Xylella fastidiosa, a bacterium that lives in the xylem of plants, causes a range of economically important plant diseases, including citrus variegated chlorosis in commercial sweet orange varieties and Pierce's disease in grapes. A group of Brazilian molecular biologists selected this pathogen for genome sequencing because of the importance of citrus to the Brazilian economy. The fruits from affected trees are small and have no commercial value. The genome of more than 2.5 million base pairs (DNA is double-stranded, and bases are paired) encodes 2,904 predicted genes. Putative functions were assigned to half of these genes based on comparisons with genes from other organisms.

The following types of genes involved in pathogenicity of *Xylella fastidiosa* were identified:

- Proteins involved in synthesis and secretion of toxins
- Enzymes to break down the plant cell wall
- Enzymes to detoxify plant defense chemicals
- Efficient sugar transporters to support existence in the nutrient-poor xylem sap
- Regulatory proteins that help adjust gene expression for growth in different environments
- Synthesis and secretion of extracellular polysaccharides (slime)
- Efflux of antibiotics that might be produced by the plant to kill the bacteria
- Uptake and sequestration of iron and other metals

---

the synthesis of the hormones auxin and cytokinin that cause these "food factory" plant cells to proliferate, forming tumors. Plant biotechnologists are now using this fascinating capability of *Agrobacterium* to transfer DNA to the host plant (discovered in the mid-1970s) to introduce desirable genes into plants (see Chapter 6).

15.5 Pathogenic fungi and oomycetes collectively cause the greatest crop losses.

Fungi and oomycetes are a group of microscopic organisms that include molds, mushrooms, mildews, and yeasts. Most of the approximatey 100,000 known species of fungi are strictly *saprophytic*, living on dead organic matter that they help decompose. A small minority—about 100 different species—cause diseases of humans and animals, but more than 8,000 species cause plant diseases, some of which are extremely damaging. There are also fungi that benefit plants. Mycorrhizal fungi live in close association with plant roots and help in the acquisition of phosphate and other key minerals (see Chapter 12). Certain endophyte fungi live within the plant and produce substances that protect the plant against grazing animals or bacteria.

Fungal pathogens are of concern not only because they cause yield reductions. Some fungi produce chemicals that are toxic or carcinogenic to people. When these fungi infect the part of the plant that is consumed, the health of humans and animals can be adversely affected. Some fungi that infect ears of maize produce such mycotoxins (**Figure 15.9**). People generally will not eat such infected ears when they purchase sweet corn, but infected ears may contaminate batches of maize used in processed foods.

The vegetative body of a fungus, called a *mycelium*, consists of a mass of long filaments or hyphae. Fungi propagate by means of spores that may be formed asexually (vegetatively) or as a result of a sexual mating process between two individuals. The sexual/asexual distinction is important for plant disease control because asexual reproduction generates additional copies of essentially the same individual, whereas sexual reproduction generates

Figure 15.9 Maize ear rot is caused by fungi that produce mycotoxins. Several fungi can infect ears of maize and cause decay of the kernels. Two ears infected with *Gibberella zeae* are on the left, and two ears infected with *Fusarium moniliforme* are on the right. Both fungi produce mycotoxins that are harmful to the animals and people who eat the maize. *Source:* Courtesy of Brian Hudelson, University of Wisconsin.

offspring with new gene combinations that may have different disease-causing abilities. Spores are microscopic and easily carried away by water, wind, people, and animals. Pathogens such as the rust fungi are particularly problematic because they reproduce rapidly by sexual means and produce wind-borne spores that can spread over wide geographic regions.

Other properties of fungal pathogens are important determinants of how a disease persists and is propagated. Almost all plant–pathogenic fungi spend part of their lives on their host plants and part in the soil or on plant debris on the soil. But some need to spend part of their lives on dead host tissues in order to complete their life cycles. Others live continuously on their hosts and only their spores may land on the soil, where they remain inactive until a new host appears. Still others are "obligate parasites" that cannot live in the absence of plant host tissue. Other fungi have a saprophytic stage and can live on any organic matter in the soil. These latter pathogens, which often have a wide host range, can survive in the soil for years, even in the absence of their hosts.

How do fungi enter the plant and cause disease? Like bacteria, they can enter plants through natural openings or wounds, or by "forcing" their way into the plant. Fungi secrete enzymes that break down the large macromolecules of the cell wall, and if the plant's defense is not quick or strong enough, the fungal hyphae quickly grow and spread from cell to cell. Once inside, they can affect the plant in many ways. Some produce toxins that alter the permeability of the cell membrane, destroying its ability to regulate what goes in and out of the cell. Many fungi secrete slime that accumulates in the vascular tissues, preventing transport of water and nutrients from roots to the shoot, causing the plant to wilt and die. Like similar diseases caused by bacteria, these diseases are called *wilts*. Other fungi produce plant hormones, and as a result the plant loses control over its own developmental processes. Some soil-borne fungi attack seedlings as soon as they germinate, killing the root system. Seedlings that look healthy one day are gone the next, and many seedlings die even before they emerge from the soil. This type of disease is known as "damping off." Many fungi simply cause necrotic (dead) spots on leaves, stems, fruits, and seeds of the plants they attack. As with viruses and bacteria, they decrease the vigor of the plant, render the seeds and fruits less fit for human consumption, or decrease the market value.

The rust diseases caused by different Basidiomycetes are among the most destructive diseases known to humanity. They have caused numerous famines and economic depressions and still reduce the world grain harvest by an estimated 10% per year. Various subspecies of *Puccinia graminis* infect different cereals such as wheat, oats, corn (maize), and barley,

Figure 15.10 Wheat stem rust (left) and leaf rust (right) diseases. These are two of the most destructive plant diseases known. These rusts have an interesting life cycle: They spend part of their life on wheat and part on barberry. At one time scientists thought eradicating barberry would control this pathogenic fungus, but this has only worked to a limited extent. Spores of the rust fungus travel hundreds of miles through the atmosphere. Resistant wheat varieties are the best control option, but new races (variants) of the rusts are constantly evolving. *Source:* Courtesy of Brian Hudelson, University of Wisconsin.

and produce long, narrow rust-colored blisters on the leaves and stems—hence the name *rust*. Heavy infections greatly diminish plant growth and seed size (**Figure 15.10**). The only practical control of wheat stem rust (*Puccinia graminis* var. *tritici*) is through breeding varieties resistant to infection by the pathogen. Researchers have identified resistance genes in the wild relatives of wheat and then introduced these genes into domesticated wheat by crossing and backcrossing. The genotypes of this rust fungus continually change because the fungus mates and produces new progeny, so new pathogen "races" regularly appear that are not daunted by the resistance genes being used. Some wheat breeders continue to work with race-specific resistance genes, seeking more complex and durable resistance gene combinations. Other wheat breeders focus on forms of rust resistance that are less effective, but which are durable because they act against any race of the pathogen. For plant breeders who battle cereal rust diseases, the job is never finished.

Rice blast, a disease that was discussed previously (see Figure 15.4), is another example of a very serious fungal disease. The causal fungus produces a toxin that kills plant cells wherever it grows. Rice blast was recorded as a disease as far back as the Chinese Ming dynasty in 1637 and is widespread in Asia. The blast fungus thrives on plants that have been fertilized with nitrogen, and the high-yielding varieties introduced as part of the Green Revolution are therefore especially vulnerable. These rice varieties require nitrogen fertilization to reach their high yields, but rice blast generally reduces by 20% the yield increase that nitrogen fertilizers induce in rice.

15.6   Chemical strategies for disease control can be effective but are sometimes problematic.

An important advance in agriculture was achieved in the 1880s with the discovery of "Bordeaux mix." This blend of copper sulfate and lime was originally applied to maturing grapes to deter thieves, but was subsequently found to suppress the very problematic grape downy mildew disease. This was one of the first **fungicides**. Over the last hundred years, numerous other compounds with antifungal or antibacterial activity have been discovered. These compounds have gone a long way toward allowing a cheap and predictable supply of food and other plant products. Unfortunately, the expense and toxicity of these compounds often complicates their use.

Fungicides may be applied as sprays (see Figure 20.7) or as dusts on plants in the field, or as a seed coating prior to sowing. Many of the older broad-spectrum fungicides must be present before infection, because contact with the growing pathogen is important and the fungicide does not move beyond the plant surface. These compounds often must be applied many times during a growing season. Some newer fungicides can move systemically within the host plant, allowing "curative" action against existing infections. Many newer fungicides are also less broadly toxic, affecting a narrow set of target organisms. The development of fungicides parallels the development of pesticides, from broadly toxic to toxic against many fungi to specifically toxic to a few target species.

Fungicides can be expensive to use, so they are often impractical for grain crops. However, they are widely used on vegetable, fruit, and flower crops. For example, 95% of all grapes and potatoes grown in the United States are treated with fungicides, but fungicides are not used on most maize, rice, or wheat. In these grain crops, disease-resistant varieties keep fungal diseases at bay.

Antibacterial compounds include copper or sulfur sprays, and also well-known antibiotics such as streptomycin or tetracycline. Because of their expense, these compounds also are used almost exclusively on fruit and vegetable crops.

An additional problem arises with repeated use of some antibacterial compounds and fungicides: **pathogen resistance**. For example, some bacteria carry genes that allow the bacteria to degrade, export, or otherwise resist specific antibiotic compounds. When an antibiotic is used for many years continuously, strong selection pressure is applied, permitting the initially small population of resistant bacteria to become dominant in the pathogen population of that area. The antibiotic is then no longer effective and other control strategies are needed. Bacterial movement on contaminated seeds, seedlings for transplant, or even shoes or farm equipment can then spread these antibiotic-resistant strains to other farms or regions.

The human or broader environmental toxicity of fungicides and antibacterial treatments remains a significant concern. A number of compounds that were formerly used are no longer acceptable. Fungicides are normally "registered for use" by government agencies for certain crops, and then only under certain use restrictions (for example, no application within a month of harvest). Agricultural chemical companies devote very substantial resources to the synthesis and testing of new compounds. They search for chemicals that will have efficacy (the ability to kill a pathogen or slow its growth not only in the laboratory but also under realistic field conditions). But these compounds, designed to be toxic to pathogens, must also have acceptably low toxicity to humans and other nontarget organisms such as fish or birds. Also, they may have low persistence so that they do not remain present in their toxic form. Even after a compound with efficacy has been identified, it can cost millions of dollars to carry out toxicity testing and certification. Companies take on these costs, however, because an acceptable compound can return many millions in profit if the marketplace accepts it.

Increasingly, and often with good reason, society at large is rejecting the use of many broadly toxic fungicides and other pesticides. As with so many other issues in society, it is necessary to balance the benefits of pesticide use with the associated costs. There is widespread agreement that the most noxious pesticides should be banned, but where do people draw the line? As methods to detect toxicity become more refined, should we ban highly useful compounds that pose a very small but detectable health risk? If researchers can determine that larger doses of a compound are toxic, can we tolerate the presence of extremely small amounts of the compound on food? Is the use of toxic compounds more acceptable in developing nations that are more pressed to produce food? Experience shows that when toxic pesticides are banned, the manufacturers protest vociferously, but alternative ways of dealing with the pests or diseases soon emerge.

Clearly, chemical strategies for disease control are extremely important, but also problematic. Fungicides and antibiotics help control destructive plant pathogens and have been extremely valuable in allowing reliable, large-scale production of food crops. But these pesticides can be expensive to use. In addition, they do not always work particularly well, because pathogens may become resistant to a given compound, and the compounds may be toxic to nontarget organisms. Much attention has therefore returned to an older method of disease control, also imperfect, but easier for the grower to use. That method is the planting of crop varieties with inherent, genetically determined disease resistance.

15.7 Plants defend themselves by using preformed defenses and by turning on defense genes.

Plants have coexisted with pathogens for millions of years, and as a result they have evolved many different defense mechanisms. Defenses that are always in place include thick cell walls that are difficult to penetrate and a waxy cuticle (leaf/stem surface) that tends to dry out rapidly, providing less support for the growth of fungi and bacteria (**Figure 15.11**). In some cases farmers may be able to avoid a pathogen, for example, by planting after the cool, moist soil conditions that favor certain pathogenic fungi have largely passed. In addition to physical barriers that block infection, plants contain a diverse array of antimicrobial compounds (Figure 15.11). The coevolution of plants with pathogens

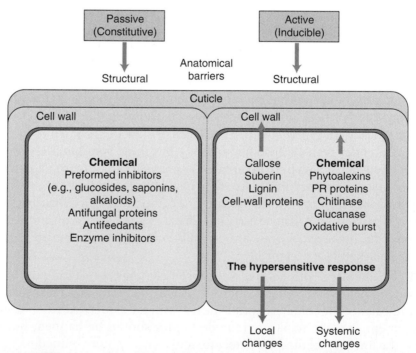

Figure 15.11 Different types of plant defense, based on existing anatomical or biochemical features of plant cells (left) or active changes induced after a challenge by pathogens (right). *Source:* Lucas, J. A. 1998. *Plant Pathology and Plant Pathogens*, 3rd ed. Malden, MA: Blackwell Science.

and insects has resulted in the production by plants of an incredible array of secondary metabolites that include phenolics, tannins, glycosides, flavonoids, and many other chemical families (see also Chapters 16 and 19). Some of these compounds attract pollinators or promote seed dispersal, but others function mainly as defense compounds. Antimicrobial peptides or enzymes are also present in plants.

It is important to note that some antimicrobial compounds are always present, but the synthesis of many others is induced only after the plant has been infected with the pathogen (Figure 15.11). Infection-induced antimicrobial compounds are also called *phytoalexins*. The response of a plant to microbial infection is multifaceted and involves

- Increased expression of a large number of genes in cells at the site of infection
- Activation of pre-existing enzymes
- Strengthening and cross-linking of the cell walls
- Secretion of phenolics into the cell walls
- Generation of signaling molecules that will move locally or systemically to activate defenses in other plant cells
- In some cases, the hypersensitive response, a beneficial plant cell death response

The genes that are expressed more actively when a plant is infected are sometimes termed **defense genes**. A better name, also in use, is **pathogenesis-related genes** (PR genes). This name is better because genes expressed upon infection by a pathogen do not necessarily contribute to defense against this pathogen. But many of the most prominent PR genes do in fact encode proteins with known antimicrobial properties that help build the defenses of the plant. These include chitinases and glucanases that can degrade the cell walls of invading pathogens, antimicrobial peptides that are not enzymes but rather are directly toxic to microbes, and enzymes that control the pathogen-induced biosynthesis of antimicrobial compounds. For other pathogenesis-related genes, scientists do not understand how if they contribute to plant defense.

Why are antimicrobial responses activated only after infection? Why don't plants just make these compounds all the time? It turns out that many antimicrobial responses are costly to the plant: They consume valuable energy and mineral resources; they may also be mildly or severely toxic to the plant. Scientists have generated plants in the laboratory that express antimicrobial responses all the time, but these plants are often stunted and produce less seed. Clearly, it is advantageous for the plant to be able to quickly sense a pathogen invasion and to respond with a massive defense effort rather than to express this response continuously.

15.8 Early plant recognition of a pathogen can allow effective defense gene activation, but successful pathogens elude the plant's defenses.

How does the plant know a pathogen is present? Recognition often involves the binding of a molecule coming from the pathogen to a molecule of the plant. This is similar to human use of antibodies, which allow specific cells of the human immune system recognize a foreign substance. The plant genes that encode these recognition proteins are called **resistance genes**. The details of the process are still under study, but scientists believe that each resistance gene encodes a protein that recognizes a specific pathogen compound and then activates host defense responses (**Figure 15.12a**). The protein encoded by a resis-

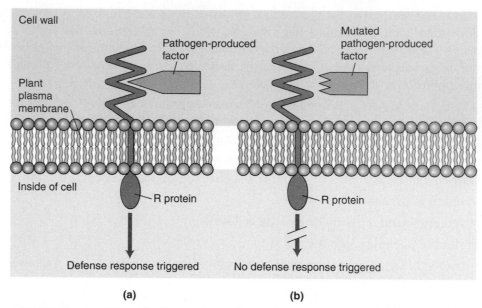

Figure 15.12 Recognition of a factor from the pathogen by the resistance gene (*R*-gene) protein. (a) The pathogen factor binds to the *R* protein and activates the defense response inside the cell. **(b)** The pathogen factor has mutated and is not recognized by the *R* protein. No defense response is elicited. The *R*-gene product shown here has part of the protein outside the cell and part inside the cell. Other resistance proteins are completely inside the cell and the factor from the pathogen needs to enter the plant cell.

tance gene may recognize a virus coat protein, or a bacterial virulence factor that is secreted into the host, or perhaps a fungal protein that is present on the pathogen surface. Early detection of the pathogen is followed by a strong and rapid induction of defense responses in the cells immediately surrounding the infection.

Plants that have undergone a strong, resistance gene-mediated defense response often form numerous necrotic spots that are evidence that the plant has "committed suicide" on a microscopic scale. The plant actively sacrifices infected cells as part of a larger defense response that contains the pathogen in a small walled-off region of dead cells (**Figure 15.13**). This rapid **hypersensitive response** programmed cell death process may be adaptive not only by walling off the pathogen but also because it releases antimicrobial compounds, releases signaling molecules that elicit defenses in other host cells, and kills off a host cell that might otherwise support the growth of a virus or a biotrophic fungus. The infected cell dies, but the plant is saved if the pathogen is contained.

Different resistance gene products control defense activation by detecting extremely different pathogens: viruses, bacteria, fungi, or even nematodes or insects. When the first plant disease resistance genes were characterized in 1993–1995, an interesting finding emerged. The proteins encoded by these genes share similar protein structures. The mechanisms for pathogen recognition are highly conserved across many different plant species and diseases. What is so interesting about that? It implies that the different specifics of pathogen recognition probably evolved from a small number of progenitor resistance genes. New resistance genes with new pathogen recognition capabilities arise over time. This evolution has been crucial for the ongoing battle of plants to keep pathogens at bay.

It may seem odd that a pathogen would produce compounds that allow the plant to sense the pathogen's presence and activate barriers against it. We noted earlier when discussing fungicides and antibacterial chemicals that pathogen populations can become dominated over time by individuals that are resistant to such chemicals. These chemicals exert strong

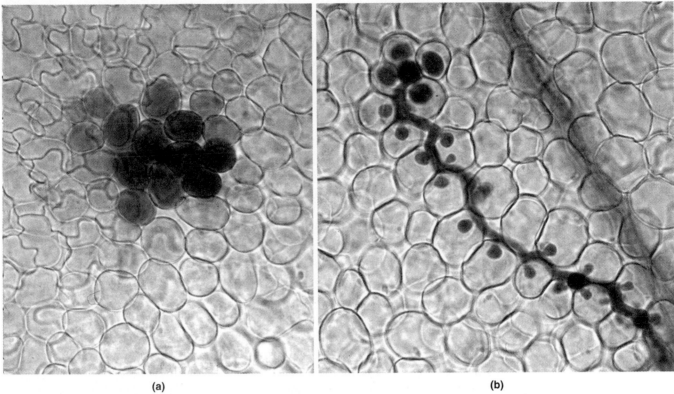

(a)   (b)

Figure 15.13 The hypersensitive response associated with disease resistance. Panels **a** and **b** both show *Arabidopsis thaliana* mesophyll tissue that is infected by *Peronospora parasitica*, the causal agent of downy mildew disease. **(a)** Resistant plant showing hypersensitive response. A resistance gene allowed the plant to recognize and block the infection of the pathogen. A small number of plant cells at the site of initial infection have taken up blue stain because they have died as part of the hypersensitive response, while surrounding plant cells remain healthy. The pathogen, which is not visible, has stopped growing. **(b)** Susceptible plant with continuing infection. This plant did not have a resistance gene with specificity for this pathogen. There is no hypersensitive response to infection. Instead, the pathogen (stained dark blue) has started growing through the leaf, establishing contact with each cell as it passes, and disease will soon follow. *Source:* Courtesy of Jane Parker, Max Planck Institute, Cologne, Germany.

selection pressure, encouraging a resistant population to emerge. The same phenomenon occurs when pathogens face resistance genes. In this case, the pathogen does not become resistant to the plant's many different antimicrobial responses, but rather evolves to elude the plant's recognition system. Pathogen individuals emerge with slight changes (mutations) in the compound that allowed it to be detected (**Figure 15.12b**). If the plant-recognized compound is dispensable, the pathogen may even lose an entire gene responsible for synthesizing that compound. But some compounds are important for pathogen growth and virulence, and their structure is constrained by particular functional demands (for instance, if the recognized pathogen compound is an enzyme that must catalyze a specific biochemical reaction). Plant resistance based on recognition of a conserved and essential pathogen structure is much more likely to be long-lived ("durable") over many years in the field.

If pathogens can evolve to elude the plant's defenses, they can still respond. Recent studies have shown convincingly that they do. A given plant carries a few hundred different resistance genes—as many as 0.5% to 1% of all the genes in the plant. And although most genes in an organism are highly conserved across many generations, resistance genes change over surprisingly short time periods.

15.9 Classical strategies of crop protection against pathogens have relied on identifying resistance genes in wild plant accessions and older crop varieties.

Plant breeders must keep ahead of pathogen evolution if they are to keep farmers supplied with disease-resistant plants. In some cases, pathogens seem to evolve slowly and resistance genes in the crop "last" for decades. It is only a small challenge for plant breeders to maintain those genes in the best varieties of a particular crop. But other pathogens evolve rapidly to escape plant resistance. If appropriately effective resistance genes are not available in the breeding lines that are in use, plant breeders must extend their search and access other plant sources. In traditional plant breeding, those plants must be interfertile (sexually compatible) with the crop variety—they must be of the same or very closely related species. Wild relatives, also called wild accessions, serve as an important source of new disease resistance traits. Introducing resistance genes into elite cultivars involves extensive backcrossing to eliminate all the undesirable genes from the wild plant (see Chapter 14).

An improvement on the use of wild accessions comes when breeders can use older varieties or landraces of the crop in question. These landraces, which often come from local farming cultures where the varieties were saved from generation to generation, serve as a valuable reservoir for disease resistance traits. As world commerce brings increasing uniformity across wide areas, many local farmers have switched to the latest new varieties and the older varieties have been lost. But academic breeding programs, companies, and governments now recognize that the older varieties are a precious genetic resource, and germ plasm collections (seed banks) that maintain at least some of these older varieties exist for virtually all crop species.

Some pathogens, such as the rust fungi of wheat mentioned earlier in the chapter, evolve at alarming speeds. In these cases, single resistance genes effective against the current pathogen population have often lost effectiveness in the first year or two after a new crop variety is released, or even before it is released. In light of such failures, many plant breeders have shunned these strong resistance genes in favor of less effective but more broadly and durably active resistance traits. These forms of resistance are often controlled by multiple genes that each make a small contribution to resistance. Breeding for this type of resistance can be very difficult, but it may be preferable if strong resistance genes prove ineffective.

15.10 New strategies may use genes that encode specific antimicrobial compounds or genes for defense-activating "master switch" proteins.

With the availability of recombinant DNA genetic engineering methods, a number of new approaches to plant disease control have become very promising. In one of the simplest approaches, a gene for a single antimicrobial compound or protein is expressed in the plant. The gene may encode a chitinase, for example, that degrades the chitin-containing cell walls of many fungi. This antimicrobial compound approach has been tried, but success has been limited because no single protein that is highly effective at stopping pathogen growth without damaging the plant has been identified. There are indications that resistance

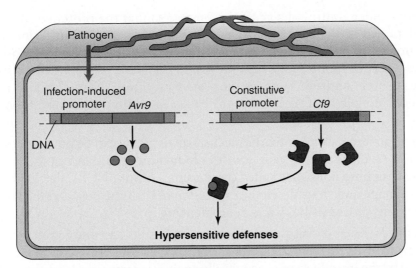

Figure 15.14 New approach to protecting plants against infection using gene technology. The plant expresses a native resistance gene (*Cf-9*). The plant has been transformed to carry a new gene construct: a plant gene expression promoter fused to a gene for *Avr9*, the pathogen peptide recognized by plants that express the *Cf-9* gene. The *Avr9* gene is usually not expressed, but the promoter can be activated by any of a variety of pathogens, causing *Avr9/Cf-9* protein interaction that activates strong plant defense responses against a wide variety of pathogens.

can be improved by expressing two or more such antimicrobial proteins in the same plant. The approach is very appealing in principle, and successes are likely to emerge as more compounds are tested in the future.

A number of other creative approaches are being used to engineer plants with improved disease resistance. For example, the pool of resistance genes available to plant breeders is widened substantially if genetic engineers can move resistance genes between plant species that are not sexually compatible. This type of transfer has succeeded, for instance, between tomato and pepper and between tobacco and tomato. When a tobacco gene that confers resistance to tobacco mosaic virus was incorporated into tomato, the tomato plants became resistant to the same virus. But this may be just the beginning. It is becoming feasible to design or select new resistance genes in the laboratory in response to specific disease organisms, and to then place these genes in crop plants. The advantage of working with resistance genes or other defense-activating proteins is that they act as the "master switch" that turns on the plant's own multifaceted defense responses.

The feasibility of a related approach was recently demonstrated, in which an unmodified plant resistance gene is used to provide widely targeted disease resistance. This strategy allows the crop to recognize infection by an array of previously virulent pathogens, as described in **Figure 15.14**.

15.11   Genetic engineers can make plants resistant to viruses by using genes from the virus itself.

Strategies for genetically engineering resistance against virus diseases have been successful enough to be released in commercially available crop species. If researchers introduce genes of a virus, such as the gene that encodes the virus coat protein, into the plant genome under control of a general promoter so that the protein is always made, virus development is often disrupted. There is a very low level of infection, but virus multiplication is severely inhibited. Scientists do not yet understand how or why this works. The presence of too much coat protein in the plant cells may make it difficult for the virus to ever "undress" (remove the protein coat from the nucleic acid molecule) and start replicating in the cell, because the balance is shifted toward an excess of coat protein.

As an alternative, a natural antivirus mechanism of plants can be put to work. If an introduced gene (such as a virus gene) causes production of mRNA or other RNA intermediates that can form double-stranded (base-paired) forms of RNA, the plant interprets this gene as foreign and shuts down the expression of both the host and the virus copies of that gene. This process is known as **gene silencing**.

Virus-derived resistance has given very promising results and one commercial success. Papaya ringspot virus threatened the Hawaiian papaya crop with extinction; the industry was saved by the introduction of genetically engineered papaya plants that express the coat protein of this virus in their cells. In the past the industry controlled this disease by using pesticides to kill the insects that transmit the virus. In Hawaii, the insects had become resistant to the pesticides, making this type of control impossible. The GM (genetically modified) papayas are now marketed as "pesticide free." These results are particularly important because, although fungicides and antibacterial pesticides are available, no effective chemical control strategies that are directly antiviral have been invented for plants. Genetically determined disease resistance, plant cultivation methods, and insect control remain the only viable option for controlling virus diseases.

15.12 The plant immune system can be activated so that it meets subsequent infections with a stronger defense response.

As noted, plants, just like humans, defend themselves by activating defense responses when infected, and resistance genes in the plant allow early recognition of the pathogen and rapid induction of a strong defense response. Infected plants also undergo a systemic induction of elevated defensiveness that lets them meet the next infection with a stronger defense response (**Figure 15.15**). A plant can be infected on one leaf or region and then, over the next day or two, can activate defenses all over the plant. This process, known as **systemic acquired resistance**, does not require the pathogen-specific resistance genes discussed earlier. It can be effective against pathogens to which the plant is not initially resistant. One leaf may be infected and badly damaged due to a poor plant defense response, but later infections by the same pathogen strain on other leaves cause much less disease damage because the plant meets them with a stronger defense response.

Salicylic acid (a chemical closely related to aspirin, which is acetyl salicylic acid) is a key mediator of the cellular responses that activate systemic acquired resistance. For example, salicylic acid causes the plant protein NPR1, which normally dwells in the cytoplasm, to move into the nucleus and interact with transcription factors that activate defense gene expression. Intriguingly, salicylic acid does not turn on high expression of all antimicrobial defense genes throughout the plant. Instead, it potentiates the plant, plac-

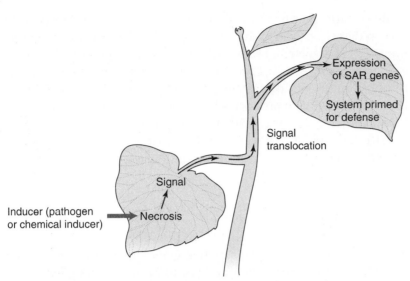

Figure 15.15 Model for the transmission of the signal that brings about systemic acquired resistance (SAR). Infection by a pathogen that generates cellular necrosis will cause release of a signal that moves through the plant and activates defenses in distant tissues. These activated tissues then exhibit disease resistance that is stronger than the resistance present at the original site of infection. Synthetic chemical inducers have been identified that will activate a similar SAR response when sprayed onto plants.

ing it in a state of alert or watchfulness so that the plant turns on defenses quickly when the next infection occurs.

Although systemic acquired resistance is normally activated by pathogen infections, simple spraying of salicylic acid onto plants activates a similar response. This is hard to do in a practically beneficial way because too much salicylic acid is toxic to the plant, but scientists have identified compounds that are structurally similar to salicylic acid and that last longer and have less potential for damaging the plant. These represent a new concept in chemical disease control: The compounds are not directly toxic to microbial pathogens or other organisms, but instead activate plant defense responses that constrain the pathogen. An alternative approach is being tested that does not use sprays, but instead involves transgenic plants that express the NPR1 protein at high levels. As with the sprays, this elevated NPR1 expression potentiates the plant, promoting a more effective defense response against pathogens that might otherwise cause serious disease.

CHAPTER SUMMARY

Plant diseases caused by viruses, bacteria, or fungi greatly diminish crop yields throughout the world every year and attract substantial attention from farmers, breeders, chemical companies, and scientists. The severity of their impact has been increased by certain agricultural practices such as monocultures and failure to rotate crops. Genetic diversity of the crop can minimize the threat of severe disease epidemics. Some disease problems can be reduced by practices such as rotating crops, using fungicides and other chemicals, and especially by planting disease-resistant plant varieties.

Plants respond to pathogens by inducing a variety of defense responses. Plants have a sensitive pathogen detection system that allows early and strong activation of defenses. However, pathogens can evolve to escape detection, hence ongoing efforts are required to maintain disease resistance in crop varieties. Genetic engineers are developing methods that allow improved disease control. These include simple expression of antimicrobial compounds, or expression of "master switch" proteins that turn on the multifaceted plant defense responses. Other approaches, such as engineered expression of a virus gene within the plant, can reduce virus multiplication. Farms often operate on narrow profit margins, and farmers who most successfully escape or minimize disease losses by exploiting these tools are at a distinct advantage.

Discussion Questions

1. Compare the types of microorganisms that cause human and plant diseases and discuss the implications. How likely are they to jump the "kingdom barrier"?

2. Discuss the advantages and disadvantages of monocultures.

3. Compare the human immune system with the way plants defend themselves. Are there more similarities or more differences?

4. Despite the wide distribution of pathogens, epidemics are not common. Why is this so? What constrains epidemics?

5. Discuss the relationships among generation time, evolutionary pressure, and the emergence of resistance, as these relationships pertain to microbes, insects, plants, and animals.

6. What are some of the features of a pathogen, and of the pathogen molecule recognized by the plant, that might make the resistance conferred by any particular plant resistance gene more or less durable?

7. Discuss the dangers that may come from pesticide residues and those that come from mycotoxins, which are more abundant when there is more insect damage.

8. The use of a broadly popular, higher-yielding new crop variety in a given area may cause growers to move away from use of indigenous varieties that yield less well, but which yield more reliably when conditions are favorable for a locally prominent disease. Discuss the pros and cons of each variety, and any possible solutions.

9. In what ways or instances might it be good or bad for growers to shift away from saving their own seed and instead paying money to seed companies for newer crop varieties?

Further Reading

Agrios, G. N. 1997. *Plant Patholog*, 4th ed. San Diego, CA: Academic Press.

Bent, A. F., and I-C. Yu. 1999. Applications of molecular biology to plant disease and insect resistance. *Advances in Agronomy* 66:251–298.

Ellis, J., P. Dodds, and T. Pryor. 2000. Structure, function and evolution of plant disease resistance genes. *Current Opinion in Plant Biology* 3:278–284.

Hammond-Kosack, K., and J. D. G. Jones. 2000. Responses to plant pathogens. In B. B. Buchanan, W. Gruissem, and R.L. Jones, eds., *Biochemistry & Molecular Biology of Plants*. Rockville, MD: American Society of Plant Physiologists, pp. 1102–1156.

Hartman, G. L., J. B. Sinclair, and J. C. Rupe. 1999. *Compendium of Soybean Diseases*. St. Paul, MN: American Phytopathological Society.

Lucas, J. A. 1998. *Plant Pathology and Plant Pathogens*, 3rd ed. Malden, MA: Blackwell Science.

Salmeron, J. M., and B. Vernooij. 1998. Transgenic approaches to microbial disease resistance in crop plants. *Current Opinion in Plant Biology* 1:347–352.

Schumann, G. L. 1991. *Plant Diseases: Their Biology and Social Impact*. St. Paul, MN: APS Press.

Strategies for Controlling Insect, Mite, and Nematode Pests

John H. Benedict

Texas A&M University System and Texas Agricultural Experiment Station

Despite today's best management efforts, the insects, mites, and nematodes that feed on plants reduce the amount of food and other crops that farmers produce. Therefore, people generally consider these invertebrates "pests." Crop losses occur regardless of whether crops are grown in organic, conventional high-input, or subsistence farming systems. Obviously, farmers want to minimize these losses, to get a higher return (food or money) for their capital and labor. Much controversy and confusion, however, surrounds the various approaches used today to control agricultural pests.

A number of strategies have been developed to reduce crop losses from invertebrate pests. Multiple strategies can be woven together in *integrated pest management* (IPM). A specific IPM program may be developed for each crop species and geographic production region. The primary goal of integrated pest management is to keep pest levels below the *economic injury level*, the initial point at which crop losses from the pest can justify the cost and use of a control strategy. Many pest control strategies exist; one of the most widely used is extermination with chemical pesticides (**Figure 16.1**). Because of the adverse health effects, environmental impact, and lack of specificity of many such chemicals, chemical control is less desirable than biological control. Biological control

Figure 16.1 Aerial pesticide spraying. Synthetic pesticides remain an important weapon in the fight against pests that decrease crop yields. However, despite spraying many different pesticide chemistries for over 50 years, pests continue to seriously damage crops. Is this the best that we can do? *Source:* USDA. Photo by Tim McCabe.

seeks to suppress outbreaks of pests in agricultural crops by using living organisms that kill the pests, such as predators, parasites, and disease. Another effective and ecologically sound strategy is to breed pest-resistant plants. This approach requires that genes for desired resistance traits be transferred as genes into the susceptible crop variety from other plants or organisms. Combining classical plant breeding with genetic engineering can improve pest resistance in crops. Genetic engineering is also being used to improve the effectiveness of biological control agents, biopesticides, and chemical pesticides.

Two major differences between agroecosystems and natural untouched ecosystems are that (1) agroecosystems are usually simpler, with fewer species of plants and animals, and (2) agroecosystems are managed by humans (see Chapter 3). Considerable controversy exists over how to best manage agroecosystems for long-term sustainability, which strategies of pest control should be used, and how and when they should be applied. Thus many biological, societal, and economic factors require that pest control strategies and integrated pest management programs be dynamic, changing from one geographic crop production region to another, and in some cases from one year to the next. Pest control approaches used in high-input systems in technologically advanced countries differ from those of subsistence farmers in less developed countries.

16.1   Worldwide, crop yields are greatly diminished by a wide variety of pests.

Insects, mites, and nematodes reduce crop yields worldwide each year by more than 20% on average. However, individual fields may sustain losses of 50 to 100% from one or more pests. Insect pests are the most damaging, responsible for 50 to 60% of these losses; next come nematodes, which cause 40 to 50% of the losses; least damaging are mites, responsible for 10 to 20% of the losses. The USDA estimates that 3% of the annual U.S. wheat crop

Table 16.1	Crop losses in farming from insect and mite pests worldwide		
	% Crop Losses		
Crop	**1965**	**1988–1990**	**Change in Loss**[a]
Barley	3.9	8.8	+4.9
Maize	13.0	14.5	+1.5
Cotton	16.0	15.4	−0.6
Potatoes	5.9	16.1	+10.2
Rice	27.5	20.7	−6.8
Soybeans	4.4	10.4	+6.0
Wheat	5.1	9.3	+4.2
Average	10.8	13.6	+2.8

[a] Change in percentage losses (1988–1990 minus 1965). Includes losses due to viruses transmitted by insect vectors.

Source: Modified from N. Duck and S. Evola (1997), Use of transgenes to increase host plant resistance to insects: Opportunities and challenges, in N. Carozzi and M. Koziel, eds., *Advances in Insect Control: The Role of Transgenic Plants* (Bristol, PA: Taylor & Francis), p. 8.

is lost to insects alone: enough bread to feed two million people for more than 20 years. Pest damage to crops is greatest in the tropics, with losses averaging 20% from insect and mite pests, and an additional 20% from nematodes. Less well known is the fact that insects and mites also damage stored agricultural products, such as the primary food grains (rice, maize, and wheat) and root crops (potatoes and sweet potatoes) and legumes (beans). In developing countries, especially in the tropics, crop storage losses from insect pests alone average 30%, whereas the crop losses during storage are much lower, averaging less than 10% in developed nations located in temperate climates. Thus developing countries in the tropics, with the greatest need for food, suffer the greatest losses from pests—losing on average about 50% of their crops to combined pre- and postharvest pests.

Surprisingly, since the initial widespread use of synthetic insecticides in the 1950s the annual percentage of crops lost to insects and mites has remained relatively constant, at about 12%. However, if you look at individual crops worldwide, you see that the percentage lost to these pests was greater in 1990 than in 1965 (**Table 16.1**). On average, the percentage lost to all invertebrate pests is the same today as in 1990.

The amount of pest-caused yield loss or decrease in yield quality farmers tolerate depends on their economic circumstances. For subsistence farmers yield loss means less food, whereas losses for commercial farmers in developed countries mean less profit, or a revenue loss that puts the farmer out of business. Invertebrate pest outbreaks, and the resulting yield losses and high control costs are a major cause for the low profit and riskiness of farming. Worldwide, analysts valued total preharvest losses for the 1988 through 1990 period at US$90 billion for eight principal food and cash crops (barley, coffee, maize, cotton, potato, rice, soybean, and wheat). You can see that more effective methods of invertebrate pest control are needed to reduce pest-induced losses. Using genetically engineered pest-resistant crops may be the best strategy for today's farmer to achieve sustainable improvement in pest control and yields.

How Do Pests Damage Crops? Invertebrate pests can damage crops in a number of ways. Some insects such as certain beetles, grasshoppers, and caterpillars are chewers that consume plant tissues (**Figure 16.2a**); especially desired are young leaves and immature flower buds and fruit because they are better nutritionally and contain fewer or lower levels of defense chemicals. Chewers may directly reduce yield by damaging or consuming

(a)

(b)

Figure 16.2 Insect pests damage different plant parts. (a) Larvae of the cotton bollworm, shown here on a cotton boll (developing fruit), eat immature plant parts, especially flowers and developing seeds and fruits. **(b)** Whiteflies have a very broad host range, and the larvae or nymphs damage crops by pushing their sharp, knifelike mouthparts directly into the phloem and sucking out the phloem sap. Some species are very difficult to control with insecticides. **(c)** Larvae of bruchid beetles damage dry seeds during storage or while developing on the plant. *Sources:* (a) and (b) courtesy of USDA-Agricultural Research Service; (c) courtesy of M. F. Grossi de Sa, Cenergen, Brazil.

(c)

plant parts to be harvested, or indirectly, by reducing photosynthesis and growth. Other plant-feeding insects such as plant bugs and thrips, and all plant-feeding mite and nematode species, use sharp mouth parts called *stylets* to penetrate specific tissues; they inject salivary secretions and microorganisms, then suck up the partially digested plant cell components (**Figure 16.2b**). These stylet users are specialists at these sipping and sucking techniques that damage leaves, fruit, and roots, reducing the plant's ability to grow and produce fruit. Aphids use another feeding strategy, inserting their stylets into the phloem and sipping the photosynthate directly. As noted earlier, some pests damage seeds when they are stored (**Figure 16.2c**).

Some pests are crop specialists, feeding only on one or two crop species, such as the famous boll weevil, which feeds and reproduces almost exclusively on cotton. Others, such as the two-spotted spider mite and many nematode species, are generalists feeding on many crop species. Many insects, such as the corn rootworm, and nematodes live and feed below ground on plant roots. Some insects and mites burrow in leaves, stems, or fruit, feeding and living inside plant tissues, making these pests difficult to control with conventional synthetic pesticides.

Plant-feeding nematodes are the most abundant invertebrates in many agroecosystems, reaching densities of 30 million per square meter of soil. They are smooth, tiny roundworms, usually less than 2 mm long, and so thin they are almost invisible to the naked eye (see Chapter 12). Many live unseen in the soil, as much as 10 cm or more below the surface, where they feed on roots and underground tubers. Their feeding can result in wounding, scaring, galling, and death of roots, reducing the ability of roots to contribute to plant growth, thus reducing plant yield.

Pests Transmit Plant Pathogens. Do pests damage crops in ways other than by feeding on them? A few leafhoppers feeding on a sugar beet plant will cause little yield loss. However, if these insects are carrying the curly top virus, a serious pathogen of sugar beets, feeding by the leafhoppers can infect the entire plant with the virus resulting in plant stunting and yield loss. Worse yet, the disease will be spread throughout the crop, causing considerable yield and economic losses. Thus in addition to crop damage caused by feeding, insect, mite, and nematode pests can cause additional yield losses by carrying and infecting crops with disease-causing pathogens. Worldwide, over 200 plant diseases are known to be transmitted by insects, mites, and nematodes, three quarters of which are virus diseases (**Table 16.2**). Many of these diseases dramatically reduce crop yields and yield quality (see Chapter 15). Pest-transmitted diseases have been especially destructive in tropical countries. For example, African cassava mosaic virus, transmitted by whitefly insects (Figure 16.2b), causes crop losses of 30 to 40% in all African cassava-producing countries.

Another way that pests injure crops is through the saliva they inject into the plant during feeding. Saliva commonly contains various chemicals, including enzymes that can injure plant cells or stimulate abnormal plant growth and/or abscission (loss) of flowers or fruit, resulting in greater reductions in photosynthesis and yields than from the feeding damage alone.

How many pest species attack a crop? The number of invertebrate pest species found in a field on a particular day may be as high as 50; however, in temperate climates only 5 to 10 will be doing damage, and of these only 1 to 5 would be doing sufficient damage to be considered for a control strategy. This complex of damaging pests usually consists of one nematode, one mite, and four insect species. As the crop grows and matures, different pests appear and disappear in response to changes in nutritional quality of the crop, weather, natural enemies, and diseases (**Figures 16.3a** and **16.3b**). In the tropics, a particular field may be infested with twice the number of pest species found in temperate climates.

| Table 16.2 | Examples of crop diseases carried by pests |

Disease	Disease Agent	Pest Vector	Crop Infected
Alfalfa mosaic	Virus	Insect: aphids	Alfalfa, beans, tobacco
Aster yellows	Mycoplasma	Insect: leafhoppers	Wheat, barley, celery, squash
Barley yellow dwarf	Virus	Insect: aphids	Barley, oats, wheat, rye
Cucurbit wilt	Bacteria	Insect: beetles	Squash, cucumber
Grapevine fanleaf	Viroid	Nematode: dagger	Grape
Pierce's disease of grape	Bacteria	Insect: leafhoppers	Grape
Rice tungro	Virus	Insect: leafhoppers	Rice
Sugar beet yellows	Virus	Insect: aphids	Sugar beet, lettuce, spinach
Sugar cane mosaic	Virus	Insect: aphids	Maize, sugar cane, sorghum
Tobacco ringspot mosaic	Virus	Nematode: dagger	Soybean, tobacco, blueberry
Wheat streak mosaic	Virus	Mite: eriophyids	Wheat, maize

Sources: Modified from D. J. Borrer et al. (1992), *An Introduction to the Study of Insects* (Philadelphia: Saunders, p. 13; and G. N. Agrios (1988), *Plant Pathology* (New York: Academic Press), p. 713.

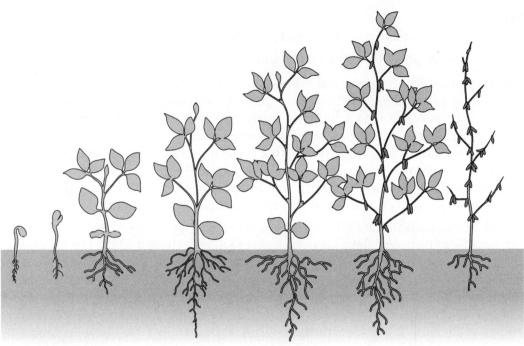

Bean leaf beetle	Bean leaf beetle	Bean leaf beetle	Bean leaf beetle	Bean leaf beetle
Cutworms	Blister beetles	*Diabrotica* spp.	Corn earworm	Grasshoppers
Leafhoppers	Green cloverworm	Grasshoppers	Grasshoppers	Cyst nematode
Mexican bean beetle	Mexican bean beetle	Green cloverworm	Stink bugs	
Seedcorn maggot	Spider mites	Mexican bean beetle	Yellow woollybear	
Soybean thrips	Cyst nematode	Spider mites	Cyst nematode	
Cyst nematode		Tarnished plant bug		
		Thrips		
		Yellow woollybear		
		Cyst nematode		

(a)

(b)

Figure 16.3 **(a)** The pests that attack soybean plants change throughout the season as the plants develop. At each growth stage, the crop is attacked by a changing complex of insect, mite, and nematode pests. **(b)** The larva of a Mexican bean beetle, a devastating pest of the common bean, is about to be eaten by the spined soldier bug. Changes in predator populations can account for changes in pest populations throughout the season. *Sources:* (a) Modified from R. L. Metcalf and W. H. Luckmann, eds. (1994), *Introduction to Insect Pest Management*, 3rd ed. (New York: Wiley), p. 21; and (b) USDA-Agricultural Research Service.

16.2 Pest outbreaks are natural but have been increased by our agricultural practices.

Why do pests suddenly appear and cause economic crop loss? Populations of a particular herbivorous insect, nematode, or mite species in both natural plant communities and agricultural crops can suddenly increase to tremendous numbers. These outbreaks are a result of the changes in biological or environmental conditions that would normally hold these pest populations in check (**Table 16.3**). The most important natural population-regulating factors are weather and abundance of food and natural enemies, such as predators, parasites, and disease. The human population is a good example of a population surge caused by increased food availability and reduced disease (see Chapter 1).

Under conditions that are ideal for pest reproduction, pest populations build up rapidly, because of their very short life cycle from egg to reproducing adult. The life cycle of a pest may be as short as eight days for two-spotted spider mite and many aphid species, or several years for some beetle and cicada species. In the tropics, where the weather is warmer and the growing season is longer, there can be continuous pest reproduction year-round, dramatically increasing pest problems compared to those experienced in temperate climates. When the weather is warm, a good average for the generation time of most agricultural pests is thirty days. During a typical crop-growing season of four or five months, pests may complete three or more generations. Pests can produce 30 to 3,000 offspring per female per generation, thus leading to tremendous population explosions as the growing season progresses. For example, only 1,000 greenbug aphids per hectare of newly emerged wheat can lead to millions of greenbugs per hectare in only four generations—truly staggering numbers of aphids.

The higher the pest density and the longer pests feed on the crop, the higher the losses in yield and quality. The duration of time over which this injury occurs is just as important as the level of injury—short-duration injury usually has less effect on yield loss or quality than injury of longer duration.

Agricultural activities over the centuries have increased the opportunities for pest outbreaks in crops. Some of the most important activities favoring such outbreaks are

- Growing crops as a monoculture that provides large areas of the pest's favorite food crop
- Improving crop suitability for the pest with irrigation, fertilizer, and herbicide
- Growing the crop in areas where no natural enemies of the pest exist
- Growing the crop in a new area causing native species of insects, mites, or nematodes to feed on the crop and become pests

Table 16.3 Pest outbreaks on crops occur with changes in the environment

- Physical environment becomes more favorable for pest reproduction, survival, and growth.
- Food plants become more abundant and/or nutritious for pests.
- Natural enemies of the pests become less common.
- Diseases of the pest become less common.
- Competing species become less common.
- Environmentally induced physiological or reproductive or genetic changes in the pest make it better adapted to the environment and/or crop.
- A combination of the preceding factors, which is most common.

Source: Modified from A. A. Berryman (1987), The theory and classification of outbreaks, in P. Barbosa and J. C. Schultz, eds., *Insect Outbreaks* (New York: Academic Press), p. 4.

(a)   (b)

Figure 16.4 (a) The alfalfa plant bug is a typical nonnative plant pest. Insects that may not be pests in their country of origin can become pests when they move without their predators. **(b)** The brown planthopper, which causes major problems in rice, is a typical example of a secondary pest. This insect became a serious pest in Indonesia after the government subsidized pesticide use by rice farmers. *Sources:* (a) USDA-Agricultural Research Service; (b) courtesy of Eugene Hettel, International Rice Research Institute.

- Unknowingly introducing pest species of insects, mites, or nematodes from one geographic area of the world to other areas where the crop is grown, and here they adapt to the new cropping environment (**Figure 16.4a**)

- Accidentally removing the natural enemies of pests through farming practices such as cultivation, weeding, irrigation, monoculture, and planting of nursery crops, and the use of herbicides, plant growth regulators, broad-spectrum pesticides (which are toxic to a broad range of invertebrates) (**Figure 16.4b**)

- Continued use of the same chemical pesticide, resulting in pests adapting to the pesticide so that it is no longer toxic; this is known as *pesticide resistance*

Farming Practices May Contribute to Pest Damage. As you can see, many of the farming practices that enhance yields also contribute to increased pest problems. The widespread use of broad-spectrum pesticides is one of the greatest factors aggravating pest outbreaks. Farmers have found that repeated application of pesticides over 20 or more generations of the pest all too frequently results in the development of pest populations that are resistant to the pesticide. In other words, resistant pests can feed, grow, and reproduce normally in the presence of the pesticide.

Pesticides kill not only the crop pests but also the insects, mites, and nematodes that function as natural enemies of the pest. When the natural enemies are removed, the crop pest targeted with the pesticide may rapidly return to even more damaging numbers than before it was treated; this is known as *pest resurgence*. Further, loss of natural enemies may induce the outbreak of *secondary pests*; these are pests that were held in check by natural enemies and caused no economic damage until broad-spectrum pesticides were applied. A good example of this problem was the epidemic of brown planthopper (Figure 16.4b) in Indonesia after the government started subsidizing pesticide use up to 75% of the cost of the pesticides. Before the use of nonspecific chemical pesticides, brown planthopper populations were held in check by their natural enemies. Pest resurgence and/or outbreaks of secondary pests result in the need for more pesticidal sprays throughout the growing season. This pesticide-induced paradigm has been called the *pesticide syndrome* or *pesticide treadmill*, and once a farmer is caught on the treadmill it is difficult to get off. The best way to avoid the treadmill is to minimize use of broad-spectrum pesticides.

The high frequency of pest outbreaks in crops is most commonly caused by (1) the accidental introduction of foreign pest species, without their natural enemies, to regions of the world where they were never present before, such as the pink bollworm introduced along with cotton seed to Australia, and (2) aerial migration of large numbers of pest insects and mites that move rapidly from one field or crop to another or from native vegetation to cropland, over distances less than a kilometer to greater than 100 kilometers. Aerial migration is a common survival strategy for many pests like plant bugs, armyworms, aphids and spider mites, but not nematodes. Neither introduced nor migratory pests are controlled by natural enemies because these enemies were left behind on the previous crop or native vegetation or in another country. To remedy the lack of natural enemies in many crop ecosystems, programs have been conducted to locate effective natural enemies, and release them in the pest-infested crop. These natural enemies are also called *biological control agents* and the pest management strategy that uses them to control crop pests is known as *biological control* (see Section 16.11).

16.3 Pesticides are valuable and effective pest management tools but must be managed wisely.

Pesticides have been and will probably continue to be the most effective method to quickly stop pest outbreaks that threaten to destroy crops. Comparisons of pesticide effectiveness for different pests invariably show the dramatic effect of pesticide use: Without pesticides, crop damage ranges from 35 to 100%, whereas with pesticides, crop losses drop to 0 to 20%, depending on the pest species, pesticide, and crop. These differences assume cultural practices, such as monocultures, that do not necessarily minimize pest damage. The detrimental effects of some pesticides on wildlife, the environment and human health have been well documented and brought to the public's attention, first in the classic book *Silent Spring*, written in 1962 by the biologist Rachel Carson. Later, special interest groups such as Greenpeace, the Union of Concerned Scientists, and the Audubon Society focused the public's attention on these pesticide safety issues. In response to public concern over pesticide risks, a new governmental agency, the U.S. Environmental Protection Agency (EPA), came into being in the 1970s as the watchdog over the environment. The EPA is also responsible for evaluating the use, risks, benefits, and registration of pesticides and other potential environmental hazards. Because of these events, the chemical nature of pesticides has changed dramatically since the 1940s, when DDT was the insecticide of choice for every insect and mite problem. Many such high-risk pesticides have been banned from use in agriculture in the United States, Europe, and a number of other countries. They were replaced gradually, beginning in the 1950s and 1960s, with safer organophosphate- and carbamate-based pesticides. Many of these in turn were replaced in the 1970s and 1980s with yet safer pyrethroid-based pesticides. Synthetic pyrethroids are widely used today because of their safety for people, wildlife, and the environment; however, they are broad-spectrum, killing not only the target pest but most of its natural enemies as well (see **Box 16.1**). Synthetic pyrethroids are being replaced today with even safer, more se-

Box 16.1

Are Natural Pesticides Safer Than Synthetic?

This question cannot be answered by a simple yes or no. In the United States, safety of pesticides is determined primarily by the Environmental Protection Agency (EPA) through evaluation of a wide range of tests, usually conducted by an independent research firm. The EPA is required by law to show that all registered pesticides (only registered pesticides can be sold) do not present unreasonable risks to human health and the environment. EPA evaluates toxicity of each pesticide to humans and representative wildlife species, such as mammals, birds, fish, amphibians, invertebrates, and microorganisms, that inhabit water and land. Residues of the pesticide remaining on the crop and soil after application are also determined. Pesticide toxicity to humans is evaluated using a single very-high-dose exposure (acute exposure) that could cause death, and a long-term low dose (sublethal) exposure (chronic exposure) to detect risks from long-term exposure. Because they cannot test pesticide toxicity on humans, researchers use rats and mice to estimate how humans would respond to acute and chronic pesticide exposure— on average the test animals used are many times more sensitive to pesticides than humans. The long-term exposure in chronic exposure tests typically shows whether a pesticide causes cancer.

Analysis of many chronic exposure feeding tests shows that both natural and synthetic chemicals caused cancer; 48% of natural chemicals tested (mostly secondary compounds from plants) and 61% of synthetic chemicals (from pesticides) caused cancer. This is not surprising, if you consider that many synthetic pesticides are modified versions (by chemists) of natural plant, animal, and micro-bial toxins. However, many naturally occurring toxins in foods have never been tested, because it is not required by law; some of these toxins are natural components of the food we eat every day, and thus exposure is considered unavoidable. If one compares the acute toxicity of natural and synthetic pesticides one finds a wide range of toxicities; some toxins in both groups are almost nontoxic, whereas others are very toxic to people at low doses.

Many natural pesticides can disrupt natural biological control in agriculture, just as synthetic pesticides can. Moreover, many natural pesticides are just as toxic to vertebrates as they are to invertebrate pests; that is, they are not selectively toxic just to pests. Some of the safest of all pesticides are the transgenic Cry proteins produced by current *Bt* crops. They are selectively toxic to only a few pest species.

So in answer to the question, are natural pesticides safer than synthetic, scientists can answer maybe. It depends on which natural pesticide and synthetic pesticide are being compared. When evaluating the risk and safety of any pesticide, remember that (1) toxicity is a function of its chemical structure, not of its origin (natural versus synthetic makes no difference); (2) safety of a pesticide also depends on how frequently and how long someone is exposed, and how high the exposure dose is; and (3) perceived risks are not always consistent with the actual risks of a pesticide once the facts are known.

lective microbial toxins, new natural and synthetic pesticides and transgenic pesticide-producing crops. This shift in pesticide use from one chemical class to another is caused by the newer pesticides being safer for humans and the environment, as effective or superior at controlling the targeted pest, and safer for natural enemies of the pests. Also, safer application methods have evolved, from dusts and sprays to granular, underground, and plant-produced transgenic pesticides.

According to EPA statistics, the amount of insecticides and miticides used in U.S. agriculture decreased by 50% from 1979 to 1997, from 85,000 tons down to 37,000 tons, in part because many older organophosphate and carbamate insecticides have been replaced by pyrethroid insecticides. An additional 5 to 10% reduction in conventional pesticide usage is estimated globally, for 1997 to 2000, because of genetically engineered insect resistant crops partially replacing synthetic pesticide sprays.

The newest, safest, and most specific pesticides include the following:

1. Crystal proteins (Cry proteins), also called *endotoxins,* or *Bt proteins*, of the bacterium *Bacillus thuringiensis*. These toxins attack the insect's digestive tract, resulting in death.

2. Synthetic hormones that mimic the effects of molting hormones. These chemicals cause immature insects to prematurely molt and die. Synthetic hormones are very specific and affect only a few related insect species.

3. Azadirachtin is gaining acceptance globally, especially on vegetable crops. Made from extracts of seeds from the neem tree of India, the EPA registered it in 1993 for commercial use against insect pests in agriculture in the United States.

When insecticides are very specific, as are *Bt* proteins, the farmer must mix two or more pesticides to control the two to five pest species that occur simultaneously in a field. This approach is more expensive than using a single broad-spectrum pesticide.

Although pesticides can be very effective, primarily because they are toxic and kill more than 90% of the targeted pest, crop losses still occur. Economic estimates for the benefits of insecticide used in the United States suggest a return of $3 to $5 for every dollar invested in these chemicals. These returns go to farmers but also have helped make food less expensive for consumers. However, the farmer does not pay the societal costs of pesticide usage such as loss of fish and other wildlife, damage to water supplies, health care costs for the people suffering from pesticide poisoning, and the cost of regulating and monitoring pesticides. These costs are currently borne by society at large, and are estimated at between $4 billion and $10 billion per year in the United States.

New technologies and society's desire for a safer environment have inspired new and safer pesticides, application methods, and pesticidal crops for use in agriculture. Novel pesticidal technologies commonly take ten years to develop, evaluate, and commercialize. Today's transgenic pesticidal plants and new low-risk natural and synthetic pesticides will be important tools to stop pest outbreaks and increase crop yields for the coming decade.

16.4 The goal of integrated pest management is to avoid pest outbreaks, using multiple strategies.

The recognition of widespread pest resistance to pesticides and the unacceptable consequences of excessive pesticide use led to the development of IPM approach. It is an ecologically based philosophy and methodology that emphasizes the need for pest management strategies to minimize pest outbreaks and minimize pesticide use. It is an approach that relies primarily on natural control factors such as pathogens, predators, parasites, and weather, and integrates these with other pest-suppressing tactics such as cultural practices and pest-resistant crops (**Table 16.4**). Broad-spectrum chemical pesticides are used, but only as a last resort to manage pest outbreaks. Moreover, pesticides are used only when absolutely justified by expected crop losses and when all other means of control (sustainable biological control strategies, including cultural control) have been exhausted.

Although IPM programs have as their primary objective the maximization of economic profit on the farm, proponents of IPM see many other goals. These include protecting the environment by minimizing pesticide use, increasing regional cooperation among farmers, substituting local skill and inputs for imported inputs (such as pesticides), especially in developing countries, creating new products, and prolonging pesticide life span. For more than 50 years, pest control specialists have been developing IPM programs that carefully describe strategies and knowledge for controlling agricultural crop pests. Development of sophisticated programs, and their adoption by farmers and their crop consultants, has been greatest in the industrialized nations and least in the developing nations, where farmers have less access to practical IPM information tailored for their cropping systems.

How and Why Do Pest Populations Fluctuate? The key to successful IPM is understanding what regulates the population fluctuations of a pest in a particular region and environment and implementing strategies to suppress outbreaks. No general recipe for pest control works for every pest. In a typical IPM program, pest and natural enemy densities are monitored at specific intervals during the growing season (usually weekly). Additional data gathered in the field (such as age of plants, temperature, humidity, pest densities, densities of certain pathogens that attack the pest, and so on) and other data concerning cost of the pest control intervention, price of the crop, and expected crop yield are all evaluated (this may require analysis using computerized programs). Then, if necessary, pest management action is taken based on these data and past experiences. The real question is whether the pest, at this level of population density and in this environment, is likely to cause economic injury. The level of pest density at which action is taken is called the *action threshold* or *economic threshold* and is below the economic injury level. Examples of fluctuations in pest density around an average density (AD) and its position relative to the action threshold (AT) and the economic injury level (EIL) are seen in **Figure 16.5**. The EIL is the lowest level of pest density that if unchecked would result in economic loss sufficient to justify the expense of pest control intervention. Until farmers can predict that pest injury will exceed this level, they have no economic justification for using a control strategy, such as applying a pesticide. The EIL is the central guidepost for decision making in pest control today.

Understanding biological factors affecting fluctuation of pest populations is crucial for pest control, as shown by the work of scientists from Iowa State University who studied outbreaks of the green cloverworm in Iowa soybean fields. This pest, like many others, does not overwinter in Iowa but arrives each spring as a migrant from the southern United States. If green cloverworm arrives late in spring, its population never builds up and the soybean crop suffers no damage. If it arrives too early in the spring, a local fungus attacks it and causes the cloverworm population to collapse before it can cause harm. However, in the years when it arrives at just the right time, green cloverworm densities become high enough to cause economic damage. This example makes clear that pest control programs should be fine-tuned to take into account each particular situation and that there is no general prescription even for controlling a single pest species.

What Is the Role of the Farmer in IPM? Ideally, the farmer should determine the population density of a pest species by counting the number of individuals on crop plants from representative samples taken in the field. With flying insects, this can be done using traps baited with volatile substances normally emitted by females to attract males (known as *pheromones*). For some pest species, counting their numbers may not be feasible. For example, corn rootworm does considerable damage to maize grown in continuous culture in the upper Midwest United States. Counting rootworms in the same year as the maize is growing is too expensive, and using only the previous year's damage data is unreliable. Thus farmers must base their rootworm control strategies on the assumption that dam-

Table 16.4	Strategies used in integrated pest management

Biological Control
  Introduction and permanent establishment of natural enemies
  Repeated mass release of natural enemies (predators and parasites)
  Conservation or enhancement of natural enemies
  Application of mass-cultured parasitic nematodes and bacterial, fungal, or viral diseases of pests (biopesticides)

Cultural Control
  Planting, harvesting, and irrigation timing
  Cultivation and tillage practices
  Crop rotations and disposal of crop residues
  Trap crops
  Pest-resistant varieties
  Intercropping and strip harvesting

Mechanical and Physical Control
  Screens, traps, barriers, heating, cooling, light, and energy types to attract, repel, or kill pests

Reproductive and Genetic Control
  Chemicals to alter pest reproduction
  Mass release of sterile individuals
  Introduction of harmful genes into pest population

Chemical Control
  Pesticides
  Behavior-disrupting pheromones that attract or repel
  Pest hormones that disrupt growth, survival, or reproduction
  Pesticide resistance management

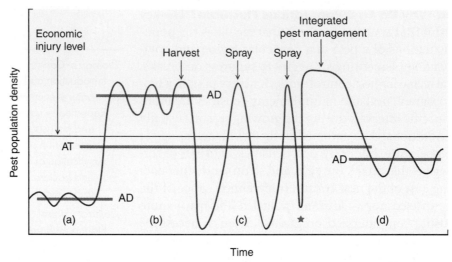

Figure 16.5 Changing pest population density in an agricultural ecosystem. (a) Before agriculture, the pest population density fluctuates around an equilibrium level, AD = average density. **(b)** After agriculture begins, the additional food for the pest that is provided by the crop allows the pest to establish a higher equilibrium level. The pest population rises above the economic injury level, causing damage to the crop. When the crop is harvested, the pest population falls. **(c)** When pesticides are introduced and sprayed, they kill the pests but also many of their natural enemies; the population of the pest thus rises again until the pesticide is reapplied. Pesticide-resistant pests occasionally appear (★), causing the pest population to rise more rapidly than before. **(d)** When integrated pest management is introduced, including biological control and pest-resistant varieties, a new equilibrium density well below the economic injury level is established. AT = action threshold. *Source:* Modified from W. Kilgore and R. Doutt, eds. (1967), *Pest Control* (New York: Academic Press), p. 365.

age will occur if the field has a long history of economically damaging infestations; this is less than ideal IPM.

How does a farmer predict whether a specific pest infestation will occur and cause a crop loss that justifies treating the crop with a biological or chemical control agent? A key goal of IPM is to provide the farmer and his or her consultant with sufficient infor-

Figure 16.6 Soybean crop response curves for defoliation by the green cloverworm. In years of normal rainfall (wet) defoliation by green cloverworm, *Plathypena scabra*, has much less effect on yield loss than in low rainfall years (dry). Note the differences in the number of insects required to cause economic damage (yield loss) between wet years and dry years. *Source:* Modified from L. Pedigo (1989), *Entomology and Pest Management* (New York: Macmillan), p. 260.

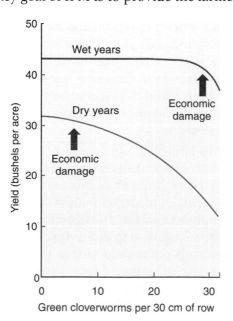

mation to make this decision. Many factors affect the outcome, including pest species, weather, crop management activities, crop variety, age of crop, when attacked and abundance of natural enemies. The decision to treat is easily made when the crop is very valuable, cost for the control method is relatively low, and a pest outbreak is occurring, threatening great economic loss if uncontrolled. Crop losses are more difficult to predict when pest densities and injury are low. Surprisingly, crops respond to pest injury in a variety of ways. For example, in wet years, less yield is lost for the same density of green cloverworm in soybean than in dry years (**Figure 16.6**). Moreover, very light plant injury in many crops can result in no yield loss. In fact, in some crops, such as cotton, light injury early in the growing season may result in a slight increase in yield, a response known as *plant compensation to injury*. The downside of light pest injury and compensation is that the injury may cause delayed crop maturity and harvest. The curves in Figure 16.6, called *crop response curves*, are useful to farmers in deciding when to apply a control tactic in the practice of IPM. For example, a soybean farmer can determine from these crop response curves that in a wet growing season the crop can tolerate up to 30 green cloverworms per 30 cm of row without serious economic loss, whereas in a dry year only 5 green cloverworms per 30 cm can be tolerated before control is economically justified.

Many farmers may not have access to the knowledge necessary to monitor and understand the interactions of their pest–crop agroecosystem and implement a successful IPM program. This can result in using pesticides as "insurance" against crop loss, so farmers make applications of pesticides that may not be necessary. Global Integrated Pest Management Facility, Extension Services, World Bank, FAO, USAID, and other national and international organizations sponsor pest control educational programs to help improve and tailor IPM practices, safety, and crop productivity for all farmers worldwide. Because of the complexity of IPM programs, swiftly changing pest status in the field, new pest control technologies, and regulatory requirements surrounding the application of high-risk pesticides, the majority of farmers in developed countries hire licensed and highly trained IPM consultants to monitor and manage crop pests. Farmers in developed countries who choose to manage crop pests without an IPM consultant usually seek IPM training and leadership from local and national agricultural extension agents and university IPM training programs.

16.5   Modifying agronomic practices can keep pest populations down.

Pest control strategies can be classified as *curative* or *preventive*. The crisis use of chemical pesticides to knock a pest population outbreak down to tolerable levels is a curative strategy. The most ecologically sound methods of pest control are preventive and are used first to keep pest populations from exploding so emergency curative strategies are not needed. Many of the cheapest and most effective control strategies are preventive cultural controls and require nothing more than smart and timely modifications of everyday agronomic practices, such as which variety and crop to grow; when to plant, irrigate, or cultivate; and whether to destroy crop residues. The farmer can select, mix, and match these agronomic practices (preventive IPM tools) to achieve the most inexpensive, effective, and sustainable pest control for each situation.

Crop rotation is probably one of the most widely used cultural control methods in the IPM toolbox. As noted earlier, many kinds of pests thrive only on specific crop species or groups of crop species. Populations of such pests tend to build up if the same crop is planted in the same field year after year. Rotation works well for many soil-dwelling nematode and insect pests that are present in the soil before the crop is planted. For example, growing a field of maize one year, then following it with soybean the next year is an excellent method for controlling corn rootworm in the U.S. Midwest. However, for some tropical root-knot nematodes that feed on a wide range of crop species, rotation is not a practical option. Surprisingly, almost all agronomic practices can have a positive or negative effect on which pest species and natural enemies are present, and on their population buildup. For example, use of high irrigation, high fertilization, and high plant density of cotton create an attractive and optimum food source for boll weevil, tobacco budworm, and bollworm, thus increasing potential for outbreaks and economic damage. Many pests emerge in the spring at a specific time, so farmers can avoid these pests by planting the crop at an

(a)   (b)

Figure 16.7 Together pest-resistant varieties and parasites greatly reduce pest populations. (a) A parasitic wasp (*Lysiphlebus testaceipes*) getting ready to lay eggs in greenbug aphids. **(b)** Population size of the greenbug aphid, *Schizaphis graminum*, after 4 weeks on a susceptible and a resistant variety of barley, either without (green bars) or with (orange bars) the greenbug parasite, *Lysiphlebus testaceipes*. *Sources:* (a) Courtesy of Peter Bryant; (b) Data from C. R. Carroll, J. H. Vandermeer, and P. M. Rosset (1990), *Agroecology* (New York: McGraw-Hill), p. 307.

appropriate interval before or after pest emergence to avoid pest buildup. For example, delayed planting of winter wheat reduces damage from Hessian fly, and delayed planting of cotton reduces boll weevil damage.

In the tropics, many subsistence farmers practice *intercropping* to reduce pest damage; they grow two or more crops together in a field (see Chapter 5). Today this practice is almost nonexistent in temperate industrialized farming systems of developed nations. A review of pest abundance in intercrops found 49 of 69 pest species had reduced abundance in intercrops compared to associated monocultures. From an ecological view, intercropping is thought to reduce pest abundance by increasing habitat diversity, which encourages greater abundance of natural enemies and reduces pest population growth rates and crop injury.

One of the most effective strategies for preventing pest outbreaks is growing *pest-resistant varieties* that sustain less pest injury. Such varieties are the foundation of an IPM program. Pest-resistant varieties can also improve effectiveness of natural enemies of pests (**Figure 16.7**). This becomes a winning combination for inexpensive biorational and sustainable pest control. Molecular techniques can be used to identify pest resistance genes that can be introduced into the best commercial varieties to improve pest resistance (see Sections 16.6, 16.7, and 16.9).

Optimum pest suppression is usually achieved in IPM programs through a combination of strategies. For example, the potato cyst nematode was accidentally introduced from the South American Andes into the United Kingdom, along with its natural plant host, potato, in the 1700s. Today the potato cyst nematode is a major pest in the United Kingdom, infesting 40% of all potato fields and reducing the value of United Kingdom crop yields by an estimated US$60 million per year. Farmers use nematode-resistant varieties, crop rotation, destruction of crop residues, and nematicides in IPM programs to control this pest. However, crop rotation is difficult, because eggs of the potato cyst nematode can live for two to five years in the field and only hatch when potatoes are grown. In addition, some nematode populations are becoming resistant to the pest-resistant cultivars; and the most effective nematicides are banned because of their high human toxicity. New nematode-resistance genes to improve plant resistance and new safer nematicides, or other effective control strategies, are needed globally for nematode control.

16.6 Plants defend themselves against pests with chemical warfare.

If plants are such good food sources, and if pests are so abundant and voracious, why don't you see pests devouring every green thing? Over millions of years, plants have evolved effective physical barriers and a mind-boggling array of chemical defenses to protect themselves against herbivorous animals (**Table 16.5**). Plants produce more than 50,000 specialty chemical compounds, called *secondary metabolites* or *secondary plant chemicals*, that repel, kill, or reduce pest growth and reproductive potential. However, volatile plant chemicals can also function as attractants, and insects have evolved olfactory mechanisms to identify these chemicals. This is how insects find and are attracted to their food source, the crop plant. Some plant chemicals even attract insects that are predators or parasites of plant pests. When an insect predator "smells" a particular plant chemical produced during pest feeding, the predator expects to find on that plant its own food—the pest insect.

Herbivores have developed mechanisms for coping with these chemical defenses, resulting in a never-ending game of plants producing new defensive compounds and herbivores succeeding in overcoming the plant's defenses. This is known as *coevolution*. Pest

Table 16.5	**Plant defenses against herbivores**
Defenses	**Effects on Pests**
Chemical (secondary metabolites) Nitrogen compounds Phenolics Polyacetylenes Proteins Terpenes Others	Poison and/or repel pests; or reduce pest growth rates and survival by various toxic effects or impairing digestion; also may increase pest susceptibility to natural enemies
Morphological Bark and cell wall thickening Spines and hairs Waxy and silica covered surfaces	Physical barriers that prevent injury, and reduce pest feeding, survival, and reproduction; also may contain toxic chemicals
Physiological Compensate for yield loss Rapid repair of injured tissue	Plant is able to grow and reproduce in spite of pest injury

Source: Modified from F. G. Maxwell and P. R. Jennings, eds. (1980), *Breeding Plants Resistant to Insects* (New York: Wiley).

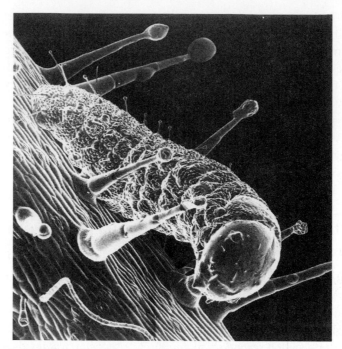

Figure 16.8 Scanning electron micrograph of an alfalfa weevil larva trapped by the defensive glandular hairs on the surface of a leaf. The secondary plant chemicals produced by leaf hairs help trap and kill the larva. Plants produce many specialty chemicals and morphological structures to defend themselves from pests. *Source:* U.S. Department of Agriculture.

adaptation to plant-produced toxins is similar to the adaptation of pests to synthetic pesticides. A major difference between the plant's use of natural pesticides and human's use of synthetic pesticides is that people commonly use a single toxin to which the pest can quickly adapt after a few years of exposure, whereas plants produce many toxic chemicals that together make pest adaptation much slower, taking thousands of years.

Plants and herbivores have become specialists. Any one plant species synthesizes a limited number of defensive chemicals (maybe 50 to 100 compounds), and each pest species has adapted over time to feed on this narrow range of plant toxins. Thus, each herbivore species can feed on only a limited number of plant species. Some pests have become so specialized that they can only survive and reproduce on the few plant species that produce the toxins to which they are adapted.

Plant defensive compounds may be present in special glands, located on the epidermis of the plant (**Figure 16.8**), on waxy surface layers, and deep within leaves, roots, or stems. In most plants, chemical defenses are present in all tissues but they are especially concentrated on the surface to deter pest damage as early as possible. Plants use their chemical defenses in two ways: (1) like an ever-present minefield, called *constitutive defense*; and (2) *inducible defense*, where the defense chemicals are produced only when the plant is under attack and being wounded by the pest—why waste precious energy building all plant defenses when they may not be needed? Many secondary plant metabolites are volatile, and pests use them as cues to help locate and identify the crop species they are adapted to feed on. Understanding pests and their interaction with secondary plant metabolites offers great potential to improve future pest control for most crops. Altering secondary plant chemistry of a specific crop not only can improve pest resistance but can cause a crop to become repellent or unattractive to the pest, or cause it to be more attractive to natural enemies of pests.

Box 16.2

The War Against the Hessian Fly

The Hessian fly is the most destructive insect pest of wheat (*Triticum*) in the world, and wheat breeders have waged war against it for 40 years. Its Latin name *Myetiola destructor* is well chosen, because it has destroyed millions of bushels of wheat all over the world. This black, two-winged fly, about the size of a small mosquito, is thought to have arrived in North America during the American Revolution. Its larvae or eggs may have been present in wheat straw brought by Hessian mercenaries (from Germany). The adult female lays 200 to 300 eggs on the leaves of a young wheat plant, and the larvae eat their way into the leaves and stems, completely stunting the growth of the plant.

Winter wheat, which is planted late in the summer and grows about 20 cm tall before winter arrives, is often killed outright, because the injured plants do not survive the winter. If 20% of the plants are affected, the farmer will suffer economic loss, and before the 1950s up to 70% of the wheat was infested in bad years. At that time, cultural practices were the farmer's only defense against this pest.

In the late 1940s and early 1950s, plant breeders from Purdue University and Kansas State University began releasing the first Hessian fly–resistant wheat varieties. These were obtained by crossing domestic wheats with wild relatives of wheat that resist the Hessian fly. The percentage of wheat plants affected by Hessian flies soon dropped below 20%. However, after a few years the new varieties proved less and less resistant, so new resistance genes had to be found. Researchers screened the large wild wheat collection (gene bank) of the University of California at Riverside, which contains more than 2,000 strains of *Triticum* and *Aegilops* species that were originally collected in the Middle East. Many of the resistant strains apparently came from a region around the Caspian Sea in northern Iran and Azerbaijan, the area believed to be the original home of the Hessian fly. Genetic engineers are continuously transferring new resistance genes from the wild strains to common wheat, to keep up with the continuing evolution of the Hessian fly to pest-resistant varieties. Resistance management strategies, discussed in Section 16.10, should be used to delay Hessian fly adaptation and conserve these valuable genetic sources of natural pest resistance.

Hessian fly. *Source:* Courtesy of USDA.

16.7 Breeding pest-resistant crop varieties can keep some pests at bay.

Plant breeders can improve pest-resistant crops by finding individual plants in wild populations that show the desired resistance trait and by transferring the genes for desired expression of the trait to conventional crop varieties (see Chapters 13 and 14 for further discussion). Archeologists think humans have been improving crops for thousands of years by selecting plant types that resist pest injury. Only very recently has it become possible to identify the specific genes responsible for a pest-resistant trait. Two early success stories in the development of pest resistance are the Hessian fly on wheat (**Box 16.2**), and the grape phylloxera on European grapes. The grape phylloxera insect, which feeds on the roots of grape vines, was accidentally introduced into European vineyards. In the United States, its native country, the phylloxera had co-evolved with native grapes as an unimportant insect herbivore. However, European grapes had no defenses against the insect and were extremely vulnerable. By 1884, most of France's wine industry was failing because of phylloxera outbreaks attacking and killing the roots of the vines. A program to develop pest-resistant European grapes was begun. The most phylloxera-resistant root stocks from grapes native to the United States were selected and sent to Europe where European grape varieties were grafted on to them. Within a few years, this practice saved the French wine industry by eliminating outbreaks of grape phylloxera.

Today, well over 100 insect-, mite- and nematode-resistant crop varieties have been bred in the United States. Possibly twice that many pest-resistant varieties, in all major crops, are grown worldwide. Over half of these resistant crop varieties are for pests of the major grain food crops: rice, maize, sorghum, and wheat. Pest-resistant varieties are developed by teams of scientists that include entomologists (insect and mite specialists), nematologists, plant pathologists, geneticists, agronomists, plant breeders, and recently biotechnologists. In developing countries, international agricultural research centers have been the primary developers of pest-resistant crops (see Chapters 2 and 14).

Using pest-resistant crops in IPM is advantageous because they (1) provide season-long pest control; (2) are usually less expensive than pesticides; (3) require no expensive, labor-intensive applications and equipment; (4) produce higher dollar returns to the growers than pesticides; (5) are compatible with biological control and pesticides; (6) are environmentally safe and usually harmless to people; and (7) provide cumulative benefits over time in that the longer and more widely a pest-resistant variety is grown, the more effective the pest suppression.

16.8 Pests can evolve to overcome plant defenses.

In many cases, pests evolve to overcome the defense mechanisms introduced by plant breeding. When only a single trait for pest resistance is present in the crop variety, the pest may quickly adapt to the pest-resistant trait, such as a higher level of a secondary plant metabolite, or a genetically engineered protein toxin. The source of the chemical resistance trait is not the cause of pest adaptation. Pests are challenged to adapt to any toxin they are frequently exposed to; the toxin acts as a selective agent driving the pest population to become more resistant so that the pest population survives. A limited number of pest species (fewer than 100) have developed resistance to natural plant toxins in pest-resistant crops, far fewer than have developed resistance to conventional pesticides (over 500 species). For example, after a period of 5 to 10 years of growing greenbug-resistant wheat varieties, strains of greenbug began to adapt and flourish on the resistant wheat varieties. Resistant pest populations, such as resistant greenbugs, are called *biotypes* and commonly look physically different from susceptible individuals of the same species. Resistant biotypes have developed in 14 insect species attacking a dozen crops, from alfalfa and maize, to rice and vegetables. In some cases, such as Hessian fly, as many as eight separate biotypes have developed consecutively over time that overcame the consecutive releases of eight new resistant wheat varieties. As new biotypes of a pest develop to overcome a pest-resistant crop variety, plant-resistance programs have to find new genetic sources of pest-resistance and breed them into current crop varieties to produce a new crop variety that is resistant to the new pest biotype. The development of resistant biotypes again makes the point that pests can adapt to any toxin, regardless of its source.

This stepwise human-managed "coevolution" of crop protection chemicals, whether natural or synthetic, whether applied as topical sprays or genetically produced in the resistant plant, will continue to challenge IPM. Many scientists believe that certain approaches can be used to slow the time it takes for a pest to adapt to the chemicals in pesticides and pest resistant varieties (see Section 16.10). On average, pest adaptation may occur 10 to 20 years after a new pest-resistant variety is grown commercially. However, many pest-resistant crop varieties have been grown for more than 20 years without resistant

Figure 16.9 Cotton plants resistant to root knot nematode also have reduced damage from fungal disease. Cotton plants growing in soil infested with root knot nematodes, *Meloidogyne incognita*, and Fusarium wilt, show stunting and defoliation (left) compared with nematode-resistant plants (right). Nematode resistance works synergistically to reduce the effects of Fusarium infections associated with nematode feeding injury to the roots. *Source:* Modified from R. D. Riggs and T. L. N. Block (1993), Nematode pests of oilseed crops and grain crops, in K. Evans, D. Trudgill, and J. Webster, eds., *Plant Parasitic Nematodes in Temperate Agriculture* (Oxon, U.K.: CAB International), p. 223.

pest biotypes developing. Scientists cannot accurately predict if or when resistant pest biotypes will develop in a specific pest–crop agroecosystem. The diversity in crop genes needed to repeatedly combat development of new pest biotypes is another reason for maintaining extensive crop gene banks and natural biodiversity; they are essential to maintain sustainable crop productivity with pest-resistant varieties.

Pest-resistant varieties work by reducing target pest densities and damage compared to susceptible varieties. Even pest-resistant crops are not immune to damage. They do not kill or repel all pest individuals from the crop; their goal is to keep pests below economically damaging numbers. Studies on insect pest reductions in 10 major crops using pest-resistant varieties showed pest densities averaged 12 times lower on resistant compared to susceptible crop varieties. Growing pest-resistant varieties also may aid in controlling the spread of plant diseases transmitted by pests. For example, a nematode infestation of a crop encourages the incidence and severity of fungal diseases such as Fusarium, and verticillium wilt (**Figure 16.9**). Pest-resistant varieties also slow down pest population development, giving farmers more time to implement other IPM tactics and giving natural enemies and diseases of the pests more time to overtake pest populations and further suppress their density and damage.

16.9 Genetic engineering can be used to improve plant resistance to pests.

If growing crops that are resistant to pests is so widespread and effective, why do farmers still use pesticides? The answer is that pest-resistant crops do not reduce all pests to levels that farmers can tolerate, whereas properly applied pesticides do. Also, well-adapted resistant varieties are not available for all pests on all crops. Conventional pest-resistant varieties have been estimated to reduce pesticide use by more than 30% a year for all pesticides used in the United States on maize, barley, grain sorghum, and alfalfa; however, there are still unacceptable crop losses if conventional pesticides are not used. Beginning in the late 1980s, molecular biologists found new pest-resistance genes that could be used to achieve a higher level of target pest control, greater than 80%, than achieved with conventional pest-resistant varieties. This is equal to or greater than the level of control achieved with conventional pesticides. The first genes available for genetic engineering of crops for pest resistance were the Cry genes of *Bacillus thuringiensis*, also called *Bt genes*. When this bacterium forms spores, each spore contains a large protein crystal that has a mixture of proteins encoded by different genes. Each type of protein has a slightly different toxicity for insects and other invertebrates.

The first crops with Cry protein pest-resistant varieties—cotton, maize, and potato—were released in the mid-1990s. Growers found these transformed insect-resistant varieties effective in controlling some of their worst pests (**Table 16.6**), and possessing all the advantages of traditional pest-resistant crops that were discussed earlier (Section 16.7). Farmers worldwide have rapidly adopted these transgenic varieties. In 2001, commercial varieties of insect pest-resistant *Bt* maize (**Figure 16.10a**), *Bt* cotton (**Figure 16.10b**) and *Bt* potato were planted on 12 million acres in the United States, Argentina, Canada, China, Australia, South Africa, Mexico, Spain, France, Portugal, Romania, and the Ukraine. Farmers worldwide have used these insect-resistant varieties to control insect pests that have become resistant to conventional insecticides, such as budworms and bollworms (many species worldwide), or insects such as stalk-boring worms in maize that are difficult to kill with conventional pesticides because they feed inside plant tissues where they cannot be reached by insecticides. *Bt* crop technology also has value for farmers in developing countries. In 1999, 1.5 million small farmers in China adopted pest-resistant *Bt* cotton varieties to grow on their farms. The compelling reasons were the effective control of the insecticide-resistant bollworm (*Heliothis armigera*), the insecticide savings and the dramatically increased yields, income, and personal safety. In 1992 bollworm outbreaks, insecticide-resistance, insecticide applications and crop losses cost China approximately US$1.2 billion. The cotton industry was reduced 40% and China began to import cotton. For China and its subsistence farmers, *Bt* cotton is an incredibly beneficial and valuable

Table 16.6	**Level of caterpillar pest control provided by the CryIAc protein in transgenic *Bt* cotton, Bollgard® varieties NuCOTN33B and NuCOTN35B, in the United States**

Pest	Percentage Kill[a]
Bollworm, preblooming	90
Bollworm, blooming	70
Cabbage looper	95
Cotton leaf perforator	85
European corn borer	85
Pink bollworm	99
Saltmarsh caterpillar	85
Tobacco budworm	95

[a]Measured as percent mortality of newly hatched larvae on *Bt* cotton.

Source: Modified from G. C. Moore, J. H. Benedict, T. W. Fuchs, and R. D. Frieson (1999), *Bt* cotton technology in Texas: A practical view. Texas Agricultural Extension Service L-5169 (Revised) (College Station, TX: Texas Agricultural Extension Service), p. 6.

IPM technology. The Chinese farmers who grew *Bt* cotton in 1999 saw yield increases averaging 25% over conventional cotton, with less than one insecticide spray on the *Bt* cotton, compared to 14 on conventional cotton. In fact, about 70% of Chinese *Bt* cotton farmers applied no insecticide sprays on their *Bt* cotton in 1999. They also observed a 25% increase in natural enemies of cotton pests on *Bt* cotton. In the United States, *Bt* cotton also has been shown to reduce insecticide use by about 2.2 sprays while increasing yields about 10% and raising farm profits about $100–$200 per hectare, compared to conventional cotton.

(a)

(b)

Figure 16.10 Transgenic pest-resistant crops are controlling insects and increasing yields. *Bt* maize and cotton are controlling pests by producing a bacterial protein that is toxic to certain caterpillar pests. **(a)** In *Bt* maize, tunneling in stalks and ears by stalk borers is greatly reduced (see split stalks on right side of inset photograph) compared to conventional maize (left). Stalks with tunneling break and fall over (stalks on left side in main photograph). **(b)** In *Bt* cotton, caterpillar injury to flower buds (see right side of inset with small dead caterpillar) and fruit is reduced compared to conventional cotton (left). *Bt* cotton produces more cotton fibers (white spots seen on left side of main photograph) than conventional cotton (right). Cotton fiber is the hair produced by the seed, and every 200 flower buds or fruit lost means about 1 lb of fiber lost by the farmer. *Source:* Courtesy of Monsanto Company.

Benefits of *Bt* Crops. For any pest-resistant crop variety, the pest control and economic benefits to farmers vary from field to field, year to year, and region to region. These benefits can result from (1) reduced conventional insecticide costs, (2) increased yields and crop value (an undamaged crop has greater value), or (3) both. Other, less economically definable benefits to the farmer are (1) better natural biological control of target and nontarget pests; (2) reduced environmental pollution; (3) safer farm environment to live and work in; (4) less risk of crop loss from all pests; (5) less risk of lawsuits from accidental poisonings; (6) complementation of other IPM strategies by *Bt* crops for a more sustainable farming operation; (7) better relationships with neighbors, especially home/city/school owners and administrators; and (8) greater peace of mind. One of the most effective ways to reduce conventional pesticide use on crops and allow natural enemies to become more abundant and effective at controlling pesticide-induced pests is to develop pest-resistant varieties targeting the key pests and those resistant to conventional pesticides. Current transgenic *Bt* crops do just that.

Many species of caterpillar pests damage maize, some by boring into the stalks and ears, where they cannot be satisfactorily controlled with current pesticides or pest-resistant varieties (Figure 16.10a). Yield losses to such pests reduce preharvest yields by more than 10% worldwide. Caterpillar tunneling and feeding also increase incidence of bacterial and fungal infection, thus increasing the incidence of some microbial toxins (mycotoxins) in stalks and grain. This of course reduces grain quality and value. One of these mycotoxins, aflatoxin, produced by the common fungus *Aspergillus flavus* in maize kernels pre- and postharvest, is extremely toxic to humans and livestock. Reducing crop pest injury to crops such as maize and rice by using pest-resistant *Bt* varieties or other types of pest resistance has the potential to significantly reduce mycotoxins in food.

The first Cry genes used were particularly effective against certain lepidopteran insects (that is, caterpillar stage of some moth species). Since then, Cry proteins that are toxic to beetles, mosquitoes, and nematodes have been discovered. This opens up the possibility that genetic engineering with Cry genes will target many other plant pests in the future.

New Genes for Insect-Resistant Transgenic Crops. Possibly the next major breakthrough in commercially available engineered pest resistance will be maize varieties that are resistant to damage from beetle larvae known as *corn rootworms*. Farmers control soil insects and nematodes with the insecticide/nematicide terbufos, and in the United States alone some 6,000 tons are applied annually to the maize crop. Replacing this insecticide/nematicide with a safer, more effective transgenic pest-resistant crop would be desirable. The source of new pest-resistant genes is likely to be the bacterium *Bacillus cereus* that produces insecticidal proteins, known as *Vip proteins*, during its vegetative growth stage. These proteins make up another new class of pesticidal chemicals, the Vip class, for "vegetative insecticidal protein." The initial field trials with transgenic Vip maize for control of corn rootworms look promising.

Another approach to slowing down insect growth is to use genes that encode protease and amylase inhibitors, defensive compounds that are abundantly found in seeds. Insects use amylases and proteases to digest the starch and protein in their food. Plants have evolved amylase and protease inhibitors that inhibit the digestive amylases and proteases of insects. Often such inhibitors have considerable specificity, inhibiting the digestive enzymes of some insect species but not of others. A specific amylase inhibitor protein found in the common kidney bean completely inhibits growth of the pea weevil and the cowpea weevil. When researchers introduced this bean gene into peas and the protein was expressed in peas at the same level as is normally found in beans, the development of pea weevil larvae was completely inhibited (**Figure 16.11**). Similarly, genetic engineers are inserting genes for protease inhibitors to make plants resistant to specific insect pests.

Is the level of pest control always the same with pesticidal crops? Something unexpected occurred during the first years when *Bt* cotton was farmed on a commercial scale—the level of target pest control was not as consistent as expected. The level of target pest control appeared to change (1) under different environmental conditions, (2) during different stages of plant growth, and (3) with different *Bt* varieties. For example, control of bollworm is greater for preblooming *Bt* cotton than for blooming (Table 16.6). Exactly how these factors influenced the level of target pest control is unclear; however, they are thought to apply to all natural and transgenic pest-resistant crops produced now and in

Figure 16.11 Weevil injury to peas is stopped by the amylase inhibitor gene from common bean. Seed-feeding weevils such as the pea weevil, *Bruchus pisorum*, and cowpea weevil, *Callosobruchus maculatus*, are serious pests of stored seeds, such as dried beans and peas. A gene encoding an amylase inhibitor was transferred from common kidney beans to peas using techniques of genetic engineering. In the photograph, the peas on the right possess the amylase inhibitor gene and are undamaged by weevils, in contrast to the control peas on the left, which do not have this gene. Notice the darkly colored weevils and the holes in the control peas. *Source:* Courtesy of T. J. V. Higgins, CSIRO, Australia.

the future. Stated simply, the level of pest control produced by a plant trait can be influenced by any genetic or environmental factor that affects primary or secondary metabolism of the crop. Thus genetic background of a variety, weather, soil fertility, moisture stress or insect injury, to name a few, can all influence effectiveness of plant-produced pesticides, or any other gene product of the crop plant.

16.10 How can we stop the development of pests that are resistant to pesticides or transgenic pesticidal crops?

We cannot stop some pests from developing resistance to conventional pesticides, biopesticides, or new transgenic pesticidal crops. Evolution of pests that are resistant to a particular toxin is a gradual process, requiring on average six or more years. Scientists already know that genes for resistance to Cry proteins exist in wild populations of some target pests of the new *Bt* crops, and researchers can therefore predict with some confidence that resistance will develop. In reality, all pesticides, whether natural or produced by humans, are valuable limited resources, and their loss results in serious social and economic costs. Methods to delay or avoid development of pests that are resistant to control tactics are necessary if agriculture is to become more sustainable. The likelihood of developing resistant populations of a pest is very difficult to predict for a specific pest and toxin; however, based on more than 500 real-world cases, scientists can make some generalizations. Resistance is likely to develop rapidly if all the following occur at the same time:

1. The pesticide is very effective.

2. The pesticide is widely used.

3. The pest does not travel over long distances (many miles) and specializes on a few plant species.

4. The pesticide is applied frequently for a long time.

5. The pest population harbors many resistance genes, even if these occur in rare individuals.

When the Environmental Protection Agency first registered *Bt* cotton, *Bt* maize, and *Bt* potatoes in the United States for commercial production in the mid-1990s, the developer of these products, Monsanto Company (St. Louis, Missouri), filed plans that it hoped would delay the development of pest resistance to the specific Cry protein insecticides in Monsanto transgenic *Bt* crop varieties (**Table 16.7**). The Environmental Protection Agency has made such resistance management plans a requirement for registering transgenic insecticidal crops. The question of how to manage transgenic *Bt* pesticidal crops to minimize selection pressure on target pests is a perplexing and controversial question for which there is not a single best answer.

How does insecticide resistance develop in insect pests? When a specific population of insects is exposed to an insecticide, all insects die except those that are partially or fully resistant. Contrary to popular belief, pests don't suddenly mutate when exposed to a toxin, and develop an entirely new toxin-resisting trait that was never present before. Rather, the trait for resistance to a specific insecticide *is already present* in at least one individual in the treated population. Repeated treatments of the same population of a pest with the same pesticide, and increased matings of partially resistant individuals with other partially resistant individuals, produce a larger and larger number of highly resistant individuals in a field population. The methods outlined in resistance management plans for transgenic plants try to avoid matings of partially resistant insects with other partially resistant pests because part of their offspring would be much more resistant to the specific toxin than would the parents. These strategies are most successful if pests that are genetically heterozygous for pesticide resistance are susceptible or only partially resistant, so they are easily killed by the dose of the pesticide used.

Table 16.7 Current Monsanto deployment strategy to minimize development of pests with insecticide resistance to CrylAc in Bollgard® cotton

Short Term (at commercialization 1996–1997)
High-dose expression of crystal protein to control pest insects heterozygous for resistance alleles (gene)
Refuge of non-*Bt* cotton to produce susceptible insects (4 non-*Bt* acres per 100 *Bt* acres, or 25 non-*Bt* acres per 100 *Bt* acres)
Agronomic practices that minimize insect exposure to crystal protein in *Bt* cotton
Integrated pest management to increase beneficials, and reduce conventional insecticide use
Monitoring target insect populations for susceptibility to crystal protein
Report on *Bt* cotton performance, especially any "failures," and investigate cause

Medium Term (2–5 years after commercialization)
Continue all short-term strategies above, plus:
Combine 2 insecticidal genes within the plant with different target sites/modes of action for the target lepidopterans

Long Term (>5 years after commercialization)
Continue all short- and medium-term strategies above, plus
Incorporate additional natural pest-resistance traits into *Bt* cotton
Incorporate other novel insecticidal genes into cotton for polygenic pest-resistance

Source: Modified from W. Mullins (1999), Monsanto Company, St. Louis, Missouri, personal communication.

Scientists have suggested several approaches to reduce the selective pressure from a pesticide that causes the pest population to adapt to it over time. *Reducing selective pressure delays resistance.* The approaches are:

1. In mixtures, two or more toxins (pesticides) are combined in the same field by mixing varieties that produce different toxins or combined (referred to as "pyramided") in the same plant.

2. In *rotations,* different toxins are used in rotation, one year one toxin-producing variety and the next year another toxin-producing variety.

3. *Mosaics* are special patchworks of toxin-producing varieties where varieties with different toxins are planted in adjacent fields.

4. *Refuges* are variety plantings with no pesticidal plants that are expected to be colonized by and produce susceptible pests that mate with any rare partially or fully resistant pest individuals produced on a pesticidal crop. The susceptible pests are expected to maintain overall population susceptibility by these matings with partially or fully resistant pests.

5. *High-dosage* is the use of pesticidal crops that produce a dose of the pesticide in their tissues that is several-fold higher than needed to kill all susceptible and partially resistant pests in the population.

These approaches can be combined. For example, refuges can be used with mixtures and planted in mosaics. Mixtures and rotations of conventional broad-spectrum spray pesticides have been used with limited success to delay development of pesticide resistance in some insect and mite pests.

Plants in nature do not rely on a single genetic trait (*monogenic resistance*) to resist a pest but rather on many resistance traits (*polygenic resistance*), so scientists would be wise to take a tip from what works in nature and develop polygenic resistance in crop varieties with at least two or more pest-resistant traits pyramided together in each variety, and planted in mosaics, mixtures, or rotations with a refuge. Pyramided traits are expected to be most effective when combined with a fully implemented integrated pest management program that uses all available pest suppression methods in a truly holistic crop management approach. In the United States, the current resistance management programs for *Bt* crops use many of these strategies (Table 16.7).

16.11 Biological control relies on the use of predators, parasites, and pathogens to stop pest outbreaks.

Why is biological control so valuable to successful long-term pest management? The answer is that natural biological control of pests is the most ecological, sustainable, effective, inexpensive, and long-term pest control method available to farmers. In many cases, the farmer need do nothing to make biological control work on the farm except minimize use of conventional broad-spectrum pesticides. However, biological control commonly does not provide a high enough level of control for every pest in a farmer's field to avoid economic damage. Frequently, the pests most poorly controlled by natural biological control are introduced foreign pests and migratory pests, as discussed earlier. Biological control agents move into a farmer's field from three sources: (1) Native parasites, predators, and diseases naturally occur in the geographic region where the farmer grows the crop, known as *natural biological control* or *natural enemies*. (2) Introduced predators, parasites, or diseases are collected from foreign lands, commonly the native home of an introduced pest, and then released and established in a new geographic crop production region where these biological control agents never existed before, known as *classical biological control*. After release they survive by feeding on the introduced pest. Eventually, populations of the introduced pest and introduced natural enemy reach a new population balance at a lower density and hopefully the introduced pest no longer causes outbreaks and serious economic losses (**Figure 16.12a**). The aim is to permanently lower the population density of the introduced foreign pest and thus eliminate or reduce the need for conventional broad-spectrum pesticides. Such a lowering can often be achieved with perennial crops, such as orchard fruits, that are part of a stable agroecosystem, or in other stable natural ecosystems, such as forests and rangelands. However, decreasing pest density is very difficult to achieve with high-value annual row crops because they do not constitute a stable ecological system. (3) Additional biological control tactics used are *augmentation* of naturally oc-

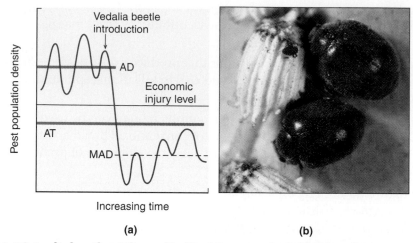

(a)   (b)

Figure 16.12 Lady-beetles (also called ladybugs or ladybird beetles) were introduced to control an introduced pest of oranges in California. (a) Pest population density before and after adding Vedalia beetles to the orange tree ecosystem. AT = action threshold; AD = average density; MAD = modified average density. **(b)** The adult Vedalia beetle, *Rodolia cardinalis,* is shown feeding on cottony-cushion scale. This lady-beetle was introduced into California orange orchards to control cottony-cushion scale pest in 1889. Both the pest and its natural enemy the Vedalia beetle came from Australia, their native home. *Source:* (b) Courtesy of J. G. Morse, University of California.

curring biological control by the periodic release of natural enemies that have been reared in the laboratory, or the *conservation* and encouragement of natural enemies already in the field. These tactics may be more appropriate for annual row crops.

The 1997 guide to beneficial insects, mites, and nematodes of North America (C. D. Hunter, 1997, Suppliers of Beneficial Organisms in North America. California Environmental Protection Agency [Sacramento, CA]) lists approximately 106 insect, 17 mite, and 6 nematode species that are cultured and available commercially for augmentative biological control of crop pests. However, their use in commercial agriculture has been limited because of their lower effectiveness and high cost relative to conventional broad-spectrum pesticides.

Invertebrate predators, such as lady-beetles (also known as ladybugs or ladybird beetles) (**Figure 16.12b**), capture and quickly kill and eat the pests. On the other hand, parasites are insects and nematodes that act like a "fatal disease" in that they deposit a number of their eggs or immature offspring, in or on a single pest, and the immature parasites slowly feed and grow over a long period, days to weeks, on the living pest, eventually killing it and becoming a free-living adult (**Figure 16.13**). Parasitic insects are commonly small wasps (Hymenoptera) that feed on nectar as adults and are some of the most abundant and effective of the natural enemies of crop pests.

A dramatic recent example of classical biological control comes from sub-Saharan Africa where a "new" pest, the cassava mealy bug, began attacking cassava, the staple for 200 million Africans. In the early 1980s, the mealy bug caused enormous crop losses. Researchers at the IITA research center in Nigeria, searching for enemies of the mealy bug in its South American homeland, found a small parasitic wasp that kills cassava mealy bugs by laying its eggs in the mealy bug's body. Periodic distribution of the wasp in the 28 African countries of the "cassava belt" is now the responsibility of the center. The dollar value of

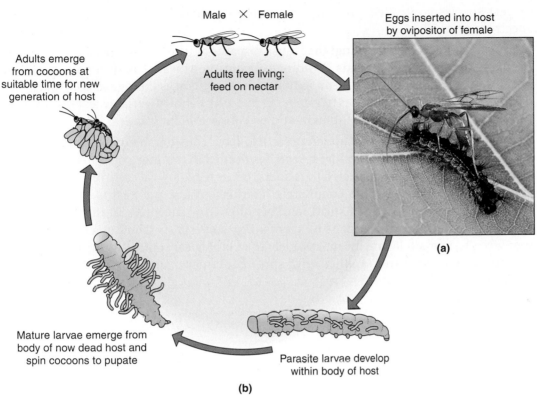

Male ✕ Female

Adults free living: feed on nectar

Eggs inserted into host by ovipositor of female

Adults emerge from cocoons at suitable time for new generation of host

(a)

Mature larvae emerge from body of now dead host and spin cocoons to pupate

Parasite larvae develop within body of host

(b)

Figure 16.13 Parasitic wasps are important biocontrol agents. (a) *Aleiodes* wasp laying eggs on a gypsy moth caterpillar. **(b)** Typical life cycle of a parasitic wasp. Female parasitic wasps (Hymenoptera) lay eggs in or on agricultural pests where they hatch into minute grubs that slowly feed on the living pest, eventually growing to adulthood and killing the pest. The entire wasp life cycle usually takes about 15 days. *Source:* USDA photo by Scott Bauer.

the program has been calculated at $150 returned to the farmers and society for each $1 invested, making it a very successful pest control program.

Microorganisms that cause diseases of invertebrate pests can also be used as biopesticides. These microbial control agents include pathogenic bacteria, fungi, and viruses, and are applied much like sprayable synthetic pesticides. These disease agents are commonly cultured in the laboratory of pharmaceutical companies using the same fermentation equipment that is used for culturing the bacteria and fungi used to produce medical antibiotics. Sprayable *Bt* and *Saccharopolyspora spinosa* are produced in this way, and are the most widely used biopesticides. The oldest microbiological agent in use are the spores of *Bacillus thuringiensis* (*Bt*). It has been sprayed commercially since the 1950s in the United States for control of certain species of lepidopteran caterpillars (butterflies and moths). Different formulations of *Bt*, with different strains producing different Cry proteins, are widely used in vegetable crops in Hawaii, Florida, and California, and especially by home gardeners and organic farmers everywhere. Overuse of *Bt* sprays has led to emergence of pests such as the diamondbacked moth, which can resist certain Cry toxins. Sales and usage of traditional biopesticides such as *Bt* and baculovirus account for less than 2% of the total global insecticide/miticide/nematicide market. The reason for their low usage is mainly their relatively low effectiveness at controlling the target pest, narrow spectrum of pest species controlled, and higher expense compared with conventional broad-spectrum pesticides.

Entomopathogenic fungi (fungi that cause diseases of insects) illustrate the potential for using other microorganisms in biological control. At least 750 different species of such fungi have been identified, and attempts to use them as biological control agents of insects, mites, and nematodes began in the late 19th century. Variable test results have plagued the commercial development and use of such biopesticides.

Nematodes and *viruses* that attack pests also have considerable potential as biological control agents, but much more research is needed to produce cost-effective, usable formulations that can be marketed to farmers. Because biological control is species specific rather than general, the commercial market for any one pest is much smaller than for multiple pests. Such small markets offer little incentive to industry for developing highly specialized control agents. For some pests, such as cutworms and root weevils, nematodes are the most widely used biological control agents because they can be easily mass-reared, applied with spray equipment, and provide rapid and satisfactory control.

CHAPTER SUMMARY

Despite today's best pest management efforts, insects, mites, and nematodes that feed on plants are a serious constraint on crop yields, reducing yields on average 20% globally. Invertebrate pests also feed on the stored portion of crops, reducing it by an additional 20% or more worldwide. Integrated pest management aims to control populations of insect, mite, and nematode pests and to avoid economic damage to crops. The primary strategies used are pest-resistant crop varieties, natural and synthetic pesticides, cultural practices, and biological control. Worldwide, thousands of species of invertebrates can be found in crops and soils under crops, but for any geographic production region only 10 to 20 species per crop are sufficiently damaging to require continuous monitoring and use of control strategies. In fact, many invertebrates and microorganisms found in crops are beneficial in defending the crop against invertebrate pests and diseases. Some of the important crop pests are generalist herbivores and can feed and reproduce on a number of crop species, whereas others are specialists that can feed and reproduce on only one or a few crop species.

Outbreaks of pests occur in all agroecosystems whether farmed as subsistence, organic, or high-input commercial agriculture. Crop loss from pests is greatest in the tropics both in the field and in storage after harvest. Often pest outbreaks occur because damaging pests were introduced from another country without their natural enemies to constrain them. Moreover, damaging pest outbreaks in agroecosystems can be induced or aggravated by many of the same agronomic activities that have produced the high yields seen in today's agriculture, especially frequent use of broad-spectrum pesticides. These pesticides cause outbreaks because they coincidentally eliminate the natural or introduced biological control agents along with the pest. Without these natural enemies to check pest populations, the key pests quickly rebuild to damaging densities and new pest outbreaks occur. Excessive use of broad-spectrum pesticides has also caused pests to develop resistance to the toxins in these products. Pests have the potential to develop resistance to most control strategies if given enough time and exposure. Hundreds of pest species have become resistant to natural and synthetic chemicals, and even to pest-resistant crops and toxins from bacteria in biopesticides. The most effective, long-term or sustainable strategies are biological control and pest-resistant crop varieties. Both have been in use for more than 100 years. However, managing crop pests is most effective and sustainable when integrated into pest management programs that are designed to manage pest-resistance, agroecosystem sustainability, safety, and profitability.

Recent advances in biotechnology and genetic engineering provide methods to improve pest management strategies, making them more ecological, safe, effective, and sustainable, as well as increasing yields. Genetic engineering is greatly expanding the sources from which pest-resistance genes can be obtained to improve crop resistance to pests. For some pests, transgenic Bt crops are quickly replacing conventional insecticidal sprays. New pesticides are clearly safer, more environmentally benign, and more sustainable than many of the older broad-spectrum organophosphate and carbamate pesticide chemistries. Use of the most disruptive older pesticides is declining in developed countries. Because organisms adapt to chemicals used in pest control, whether they are produced by pest-resistant varieties or applied as sprays of synthetic or natural chemicals, plant breeders, geneticists, genetic engineers, the agrochemical industry, and IPM practitioners will always be challenged to keep pest management strategies one step ahead of pests.

Integrated pest management of crop pests involves a complex integration of biological and economic data, requiring a thorough understanding of pests, beneficial organisms, crop phenology, cropping systems ecology, and pest control strategies. During the cropping season, successful IPM requires repeated and thorough field monitoring of pests to accurately judge the need for pest control action. This requires considerable sophistication on the farmer's part. Most farmers in developed nations use trained crop consultants. The greatest challenge today is for farmers in developing countries, especially subsistence farmers in the tropics, to replace conventional pesticide spray programs with knowledge and sustainable practices of integrated pest management tailored for their unique environment.

Discussion Questions

1. What is integrated pest management, and what is its goal? In what sense is it a philosophy as well as a set of techniques?

2. Why are *Bt* crops not "the" solution to pest control? What are the advantages and disadvantages both for the farmer and the environment?

3. What is biological control, and how does it operate? Can't people just quit trying to control pests in crops? Why or why not?

4. Synthetic pesticides are toxic chemicals. Are they more or less toxic than naturally occurring chemicals and pesticides? What does this toxicity mean for the agroecology of the farm?

5. In 1960 Rachel Carson did people a great service by focusing attention on the side effects of pesticides. Discuss the evolution of pest control since that time. What are pest-resistant plants?

6. Can the strategies used in integrated pest management make agriculture more sustainable? Explain.

7. What is biological control of pests, and how does it work?

8. Why do pesticide-resistant pest outbreaks occur?

9. Can pests become resistant to transgenic pesticidal crops such as *Bt* maize? Explain.

10. Are there ways to stop or delay the development of pests that are resistant to current pesticidal *Bt* crops? Explain.

11. How do native plants growing in natural habitats defend themselves from plant-feeding invertebrates?

12. Discuss the multiple roles of secondary plant metabolites in the complex interplay between plants, their pests, and the predators of those pests. Discuss coevolution between plant genes responsible for the synthesis of secondary metabolites and insect genes that determine pest success.

Further Reading

Altman, A., ed. 1998. *Agricultural Biotechnology*. New York: Dekker.

Barker, K. R., and S. R. Koenning. 1998. Developing sustainable systems for nematode management. *Annual Review of Phytopathology* 36:165–205.

Carozzi, N., and M. Koziel, eds. 1997. *Advances in Insect Control: The Role of Transgenic Plants*. Bristol, PA: Taylor & Francis.

Carroll, C. R., J. H. Vandermeer, and P. Rosset, eds. 1990. *Agroecology*. New York: McGraw-Hill.

Casida, J. E., and G. G. Quistad. 1998. Golden age of insecticide research: past, present, or future? *Annual Review of Entomology* 43:1–16.

Clement, S. L., and S. S. Quisenberry, eds. 1999. *Global Plant Genetic Resources for Insect-Resistant Crops*. Boca Raton, FL: CRC Press.

Coots, J. R. 1994. Risks from natural versus synthetic insecticides. *Annual Review of Entomology* 39:489–515.

Croft, B. A. 1990. *Arthropod Biological Control Agents and Pesticides*. New York: Wiley.

Denholm, I., J. A. Pickett, and A. L. Devonshire, eds. 1999. *Insecticide Resistance: From Mechanisms to Management*. Oxon, U.K.: CAB International; and London, U.K.: Royal Society.

Evans, K., D. L. Trudgill, and J. M. Webster, eds. 1993. *Plant Parasitic Nematodes in Temperate Agriculture*. Oxon, U.K.: CAB International.

Fenoll, C., F. M. W. Grundler, and S. A. Ohl, eds. 1997. *Cellular and Molecular Aspects of Plant–Nematode Interactions*. Norwell, MA: Kluwer Academic.

Gould, F. 1998. Sustainability of transgenic insecticidal cultivars: Integrating pest genetics and ecology. *Annual Review of Entomology* 43:701–726.

Hall, F. R., and J. J. Menn, ed. 1999. *Biopesticides*. Totowa, NJ: Humana Press.

Kogan, M. 1998. Integrated pest management: Historical perspectives and contemporary developments. *Annual Review of Entomology* 43:243–270.

Mengech, A. N., K. N. Saxena, and H. N. B. Gopatan, eds. 1995. *Integrated Pest Management in the Tropics: Current Status and Future Prospects*. New York: United Nations Environment Programme (UNEP); Wiley.

Metcalf, R. L., and W. H. Luckmann, eds. 1994. *Introduction to Insect Pest Management*, 3rd ed. New York: Wiley.

National Research Council Report. 2000. *Genetically Modified Pest-Protected Plants: Science and Regulation*. Washington, DC: National Academy Press.

Persley, G. J. 1996. *Biotechnology and Integrated Pest Management* (Oxon, U.K.: CAB International).

Poinar, Jr., G., and R. Poinar. 1998. Parasites and pathogens of mites. *Annual Review of Entomology* 43:449–469.

Walter, D. E., and H. C. Proctor. 1999. *Mites: Ecology, Evolution and Behavior* (Oxon, U.K.: CAB International).

Weeds and Weed Control Strategies

Patrick J. Tranel
University of Illinois

Weeds reduce crop yields primarily by competing with crop plants for light, water, and nutrients. In regions where farmers cannot afford herbicides and farm machinery, weeds are possibly the most important factor limiting increased food production. Hand weeding in such places consumes countless hours and is usually done by women and children. In the United States, farmers spend an estimated $10 billion annually to control weeds. The economic importance of weeds is emphasized further by the fact that herbicides comprise as large a share of the worldwide agrochemical market as all other pesticides combined.

As with insect and disease management, development of effective weed management strategies starts with understanding the pest. Thus we begin this chapter by defining what a weed is, discussing the attributes of plants that make them weedy, and describing how weeds interfere with crop plants. We then focus on the general strategies for managing weeds, and discuss how weeds have fought back against human attempts to control them. We conclude with a discussion of new opportunities for weed management that are being offered by biotechnology.

17.1 Weeds can be defined from anthropocentric, biological, and ecological perspectives.

Everyone has some notion of what a weed is, yet defining weeds with a short statement that is both complete and accurate is surprisingly difficult. A widely used definition is "a plant growing where it is not wanted." The appeal of this definition is its inclusiveness. Common lamb's-quarters growing in a soybean field reduces soybean yield, and therefore a farmer considers it a weed. Similarly, volunteer maize plants are not desirable in a soybean field because they too reduce soybean yield (**Figure 17.1**). Volunteer plants are, therefore, rightfully defined as weeds even though one does not normally think of a major crop species

as a weed. To take this a step farther, common lamb's-quarters is considered a weed of worldwide importance; young leaves of this plant, however, can make a tasty salad. Downy brome is a troublesome weed to wheat producers, but valuable forage for livestock producers. Morning glory growing up a trellis is an attractive flowering plant, but twining around maize plants it is a damaging weed. What emerges from these examples is that whether a plant is a weed depends on your perspective of that plant and where it is growing. Thus the anthropocentric (human-centered) definition of a weed, a plant growing where it is not wanted, is appropriate. Any plant can be a weed if a human decides it is a weed.

Figure 17.1 Any plant can be a weed if it is growing where it is not wanted. Crop seed left behind during harvest may produce plants the following year. These "volunteer" plants are considered weeds. Volunteer maize is a common weed of soybean in cropping systems that employ a maize-soybean rotation. *Source:* Weed Science Society of America.

Simply defining weeds as plants growing where they are not wanted, however, fails to reveal anything about the biological characteristics common to most weeds. Furthermore, although by this definition any plant *can* be a weed, few plant species actually are *troublesome* weeds. Of over 250,000 plant species in the world, only a few hundred are troublesome weeds. What biological characteristics of these relatively few plant species make them weeds?

Although no single factor determines how weedy a plant is, most weeds share two common characteristics: competitiveness and persistence. On a given area of land, a finite amount of resources (light, water, nutrients, CO_2) is available to support plant growth. Modern agricultural practices try to capture all available resources for crop growth, but frequently weeds can secure some resources for their growth. How extensively weed species draw these resources away from the crop is a measure of their competitiveness. Weed persistence is their ability to survive year after year on a given area of land, despite farmers' attempts to control them.

Rapid seedling establishment, high growth rates, prolific root systems, and large leaf areas are traits contributing to the competitiveness of weeds (**Figure 17.2**). Many weed species are not especially competitive on an individual plant basis; instead, they exert their competitiveness through sheer numbers (**Figure 17.3**). Such species produce several thousand seeds per plant, and some may produce a million or more per plant!

High reproductive output contributes not only to weed competitiveness by numbers, but also contributes to weed persistence. Even if nearly all weeds in a field are killed, the few survivors will produce ample seed to ensure infestations in following years. Seed dormancy, however, is probably the single most important factor contributing to the persistence of weeds. Seeds lie dormant in the soil, where they form a soil seedbank (see Chapter 9). Seed dormancy is a mechanism that has evolved to ensure the survival of the species. Cultivated soils typically contain thousands of seeds per square meter of surface area, waiting for the right opportunity to germinate.

Figure 17.2 A competitive weed. Common cocklebur's large leaf area and extensive use of soil moisture combine to make it one of the world's most competitive weeds.

Figure 17.3 Many weeds exert their competitiveness through sheer numbers. This field is so heavily infested with waterhemp that the rows of maize are barely visible. If the waterhemp plants are not controlled, they will cause near complete yield loss. *Source:* Courtesy of Aaron Hager, University of Illinois.

Many perennial weed species reproduce vegetatively from their roots or from specialized structures such as stolons, tubers, and rhizomes. Canada thistle, for example, has a deep, spreading root system capable of vegetative reproduction (**Figure 17.4**). This plant is extremely persistent; even after repeated destruction of the aboveground portion of the plant, the root system will send up new shoots. In fact, patches of Canada thistle and other perennial weeds are often spread by tillage, which breaks up and disperses the vegetative propagules.

Another key trait contributing to the persistence of weeds is developmental plasticity. This is the ability of a plant to alter its growth and development in response to the environment, to optimize reproductive output. For example, many weeds respond to moisture stress and other stresses by initiating flowering even at a very young age. Furthermore, these weeds can then go on to produce mature seeds just a few days after flowering.

Figure 17.4 Vegetative reproduction contributes to the severity of many perennial weeds. Canada thistle sends up numerous shoots from its extensive root system, forming dense patches. *Source:* Courtesy of Phil Westra, Colorado State University.

How did weeds come to possess the traits just described? In other words, where did weeds come from? You can better understand what it is about plants that makes them weeds when you consider the ecological and evolutionary histories of weeds. Before doing so, however, take a look at the taxonomic classification of weeds. Taxonomists have grouped all plants into about 450 different families. About two thirds of the world's most troublesome weeds, however, appear in only 12 of these families. Three example weeds from each of these 12 families are listed in **Table 17.1**. Nearly all major crops,

Table 17.1 **Example weeds and an example crop from each of 12 plant families that comprise the majority of the world's worst weeds**

Plant Family	Example Weeds	Example Crop
Grass (Poaceae)	Bermudagrass (*Cynodon dactylon*) Barnyardgrass (*Echinochloa crus-galli*) Johnsongrass (*Sorghum halepense*)	Wheat
Sedge (Cyperaceae)	Purple nutsedge (*Cyperus rotundus*) Yellow nutsedge (*Cyperus esculentus*) Smallflower umbrella sedge (*Cyperus difformis*)	No major crops
Sunflower (Asteraceae)	Canada thistle (*Cirsium arvense*) Common cocklebur (*Xanthium strumarium*) Common dandelion (*Taraxacum officinale*)	Sunflower
Buckwheat (Polygonaceae)	Wild buckwheat (*Polygonum convolvulus*) Curly dock (*Rumex crispus*) Red sorrel (*Rumex acetosella*)	Buckwheat
Pigweed (Amaranthaceae)	Smooth pigweed (*Amaranthus hybridus*) Spiny amaranth (*Amaranthus spinosus*) Redroot pigweed (*Amaranthus retroflexus*)	Grain Amaranth
Mustard (Brassicaceae)	Shepherd's-purse (*Capsella bursa-pastoris*) Hoary cress (*Cardaria draba*) Wild mustard (*Brassica kaber*)	Canola
Legume (Fabaceae)	Black medic (*Medicago lupulina*) Sensitive plant (*Mimosa pudica*) Kudzu (*Pueraria lobata*)	Soybean
Morning glory (Convolvulaceae)	Field bindweed (*Convolvulus arvensis*) Field dodder (*Cuscuta campestris*) Swamp morning glory (*Ipomoea aquatica*)	Sweet potato
Spurge (Euphorbiaceae)	Leafy spurge (*Euphorbia esula*) Garden spurge (*Euphorbia hirta*) Spotted spurge (*Euphorbia maculata*)	Cassava
Goosefoot (Chenopodiaceae)	Common lamb's-quarters (*Chenopodium album*) Russian thistle (*Salsola iberica*) Nettleleaf goosefoot (*Chenopodium murale*)	Sugarbeet
Mallow (Malvaceae)	Venice mallow (*Hibiscus trionum*) Velvetleaf (*Abutilon theophrasti*) Arrowleaf sida (*Sida rhombifolia*)	Cotton
Nightshade (Solanaceae)	Black nightshade (*Solanum nigrum*) Jimsonweed (*Datura stramonium*) Cutleaf groundcherry (*Physalis angulata*)	Potato

Source: Adapted in part from L. G. Holm and others (1977), *The World's Worst Weeds, Distribution and Biology* (Honolulu: University Press of Hawaii).

including the grass crops such as wheat, rice, and maize; legumes such as soybean, common bean, and peanut; and other crops listed in Table 17.1, also come from these 12 plant families. Weeds and crop plants are not as different as you might have thought! That weeds and crops are not entirely different makes sense from an ecological perspective, because to be a weed, a plant must adapt to the same environment to which the crop is adapted.

Ecologically, weeds are often classified as "ruderals." Ruderals are early successors of disturbed environments. Consider a riverbank that is often disturbed by floodwaters. One strategy for occupying such an environment is to have a rapid life cycle, so that the plant can produce seed before the next flood occurs. In addition, it is important to produce numerous, dormant seeds: numerous because most seeds will be swept away and never land on a favorable environment, and dormant because all the seed should not germinate at the same time, as the environment is so unpredictable. These are the same traits that adapt a plant to human cropping systems, which also undergo frequent disturbance. Plants that are both ruderal and competitive are the worst weeds.

Another ecological concept, the niche theory, is also helpful in understanding why weeds succeed. A niche is simply a species' place within a community. An element of the niche theory is that species are specialized to occupy specific niches, and therefore no one species can use all available niches. This is why most natural ecosystems are characterized by numerous species. Cropping systems based on a single crop species, however, provide many unoccupied niches (for example, the space between crop rows, and the area above the crop canopy). In nature, niches seldom go unoccupied. Consequently, monoculture cropping systems are destined to have weeds!

17.2 Weeds reduce crop yield by interfering with the growth of crop plants, and weeds may also reduce yield quality.

The primary way in which weeds exert their negative effects on crops is by reducing crop yield. They do so through plant–plant interactions that include competition, allelopathy, and parasitism. Competition is the most important type of plant–plant interaction for the vast majority of weeds and therefore is a key topic in weed science. The severity of yield loss caused by weed competition depends on numerous factors:

- The crop grown
- What weed species are present
- Density and spatial distribution of the crop
- Densities and spatial distributions of each weed species
- The duration of competition (when the weeds emerge relative to the crop)
- Availability of resources (light, water, nutrients, and CO_2) for which plants are competing
- Various climatic factors (such as temperature)
- Various edaphic (soil-related) factors (such as soil texture)
- Presence of insects and diseases (which may be general or specific to the crop or certain weeds)

Box 17.1

Weeds That Fight Unfairly

Most weeds fight against crop plants by competing for limited resources. Some weeds gain an additional edge by producing compounds—allelochemicals—that inhibit the crop. Parasitic weeds, however, are in a class all their own. These plants, of which the witchweeds, broomrapes, and dodders are the most agronomically significant, form direct attachments to crop plants to steal water and minerals, and sometimes photosynthate.

Witchweeds, which comprise several species in the genus *Striga,* are commonly encountered throughout much of Africa and Asia (see Figure 5.8). Seeds of witchweed germinate in response to strigol, which is produced by the host plants. The germinated seedling attaches to roots of its host through specialized connections called *haustoria.* During the first month of development, witchweed seedlings remain underground, all the while extracting water and nutrients from the host. Consequently, substantial crop yield reduction occurs before the farmer even knows the weed is present (hence the name "witchweed"). Its underground existence also makes it a very difficult weed to control. After emerging from the soil, witchweed becomes photosynthetically active but still relies on its host for water.

Broomrapes, in the genus *Orobanche,* also are root parasites; unlike witchweeds, however, which primarily attack grass crops, broomrapes attack a wide variety of legume and vegetable crops. Broomrapes are especially problematic in Mediterranean countries. Dodders, in the genus *Cuscuta,* are stem parasites that attack a wide variety of plants but are most problematic on legumes. After emerging from the soil, dodder twines around the host plant and forms haustoria connections. The portion of the dodder stem leading out of the soil eventually withers and dies. Also, because dodders are nonphotosynthetic, they are entirely dependent on their host for resources.

Because parasitic weeds are difficult to control, infestations often increase in severity, and high yield losses result. Under drought conditions, parasitic weeds can cause complete yield loss, and even under good growing conditions 50% yield loss can be expected. The potential severity of these weeds is illustrated by the fact that the United States initiated an effort to eradicate witchweed when it was first detected in the Carolinas in the 1950s. At that time, it occupied an area of about 150,000 hectares in 27 counties. By strictly enforcing quarantine of the area and through an extensive education and eradication effort (carried out for nearly 50 years!), the infestation has now been reduced to a couple thousand hectares.

Because of this myriad of interacting factors, it is difficult to predict yield loss simply based on weed density. This is one reason why economic treatment thresholds are not widely used in weed management.

In most cropping systems, preplanting operations such as tillage or herbicide applications remove all existing vegetation. During the first stages of crop growth, competition does not occur because resources are sufficient for both crop and weed seedlings, and because the seedlings are not spatially overlapping. As both crop and weed plants grow larger, competition arises. If weeds are not killed now, they will reduce crop yield. If, however, weeds are killed during the first few weeks the crop is establishing, subsequently emerging weeds usually will not be strong competitors with the much larger crop plants. The window of time beginning when weed competition starts and ending when the crop can outcompete any newly emerging weeds is the critical weed-free period. The significance of the critical weed-free period is that it defines a relatively narrow window of time during which weeds must be controlled to prevent yield loss. In particular, for competitive crops such as maize and soybean, the critical weed-free period is usually no more than a 4-week period that begins a week or two after planting. Less competitive crops, however, such as sugarbeet and onion, have critical weed-free periods of 10 weeks or more.

Allelopathy, a type of plant–plant interaction distinct from competition, is the production of a chemical (allelochemical) by one plant that hinders the growth of another. Many weeds are suspected of producing allelochemicals; because of the difficulty in studying allelopathy, however, its role in agricultural systems is poorly understood. In practice, the effects of competition and allelopathy often are not differentiated, and in both cases the net result is decreased crop yield. Parasitic weeds are physically attached to crop plants, siphoning off resources rather than competing for them (**Box 17.1**).

In addition to competition, allelopathy, and parasitism, weeds can interfere with crop plants through indirect interactions. Certain weeds may increase crop disease severity by acting as a host for one or more stages of a pathogen's life cycle. For example, wheat producers consider barberry a weed because it is an alternate host for the pathogen that causes wheat stem rust.

Besides reducing crop yield, weeds may also interfere with crop harvesting and thereby reduce the *effective* yield (**Figure 17.5**). And weeds present in the crop at harvest can reduce crop quality:

- Weeds in cotton fields at harvest can stain cotton fibers.
- Juice released from nightshade berries during harvest stain soybeans.
- Aerial bulblets of wild garlic harvested with wheat impart an off-flavor to the processed flour.

Negative economic effects of weeds also occur through livestock poisoning, by affecting human health and well-being and through negative aesthetic effects. Homeowners seem to prefer weed-free lawns! Similarly, golfers demand weed-free greens, and golf course owners invest heavily in weed-control chemicals. The aesthetic impact of weeds also plays a role in crop production. In highly managed and productive cropping systems, for example, farmers may apply excessive herbicide treatments because they don't want to see *any*

Figure 17.5 Weeds cause harvest losses. Heavy infestations of weeds make harvesting the crop more difficult and less efficient. In particular, viny weeds, such as the bur cucumber shown, can clog combines and cause inefficient grain separation. *Source:* Courtesy of Aaron Hager, University of Illinois.

weeds in their fields. In some cases, the land is owned by one person and farmed by another, and the landlord insists that the fields be weed free.

17.3 Three approaches to weed management are prevention, control, and eradication.

In the preceding sections, we considered weeds from anthropocentric, biological, and ecological perspectives, and discussed their negative impacts. Now, how should weeds be managed? There are three general approaches to managing weeds: prevention, control, and eradication. Because weeds are persistent, once they become established in a field it is nearly impossible to eradicate them. For this reason, the use of preventive strategies is an important—although often overlooked—component of weed management.

The goal of preventive weed management strategies is to keep weed seeds and vegetative propagules from being introduced into a field. Weed seeds are often harvested along with the crop (**Figure 17.6**). Consequently, farmers who plant their crops using seed

Figure 17.6 Weed seed are often harvested along with crop seed. Seeds of jointed goatgrass are similar in size and weight to wheat seeds. Consequently, jointed goatgrass is commonly disseminated by farmers planting contaminated wheat seed. *Source:* Courtesy of Phil Westra, Colorado State University.

from a previous year's harvest often inadvertently sow numerous weeds. Thus adequately cleaning crop seed or buying weed-free seed is an example of a weed prevention strategy, and the first component of an overall weed management approach. By considering how weed seed might be introduced into a field, the farmer can implement other effective weed prevention strategies. As examples, new weed species may be brought into a field via farm machinery, dispersal from weeds growing in field borders, and irrigation water. Thus, cleaning implements between fields, mowing weeds in field borders to prevent seed production, and installing screens in irrigation systems are weed prevention strategies.

The enormous cost and effort necessary to eradicate a particular weed species is justified only in rare circumstances. Such circumstances include an especially noxious weed that has been introduced only recently into a new area. Once a new species becomes more firmly established (has increased in numbers, increased its range of infestation, and has contributed numerous seeds to the soil seedbank), eradication is impractical if not impossible. One of the few examples of taking an eradication approach to weed management is the attempted eradication of witchweed from the United States (see Box 17.1). Successful eradication of a weed species requires a long-term, concerted effort consisting of several preventive and control strategies.

17.4 Weed control is achieved by cultural, mechanical, biological, and chemical means.

Despite preventive strategies, some weeds are always present, and as just stated, eradication is rarely a practical goal. Thus the primary weed management approach is to use weed control steps to reduce the negative impact of established weed populations. Weeds are controlled through cultural, mechanical, biological, and chemical means.

Cultural weed control includes all practices that promote the growth of the crop over the growth of weeds. Planting high-quality seed at the proper depth and when soil moisture and temperature are adequate will give the crop a head start over the weeds. Over the past few years, many soybean producers in the midwestern United States and elsewhere have adopted narrow-row planting. In addition to improving soybean yield, narrow rows benefit weed management. The narrower the rows, the earlier in the growing season leaves of plants in adjacent rows will overlap, shading the soil and hindering growth of weeds. Rotating crops from year to year is also a very effective cultural weed management strategy. Annual weeds compete well with and persist in annual crops but are less persistent in perennial crops. Consequently, rotating a perennial forage crop with annual crops such as maize and soybean reduces the severity of annual weed infestations.

Cultural weed management strategies also include growing two crops together, or "intercropping." For example, oats are often seeded with perennial forages such as alfalfa. The rapidly establishing oat crop competes with and suppresses weeds while the perennial crop is becoming established. Intercropping is a common strategy in developing countries for maximizing food production per unit area and decreasing the amount of hand weeding required. The benefit of intercropping, in terms of weed management, is that two crops are better able than one at occupying niches that would otherwise be available to weeds. Organic farmers often include in their crop rotations the planting of a highly competitive plant species that is grown but not harvested. Such "mulch crops" are used not only to improve soil quality and add nitrogen via nitrogen fixation but also to manage weeds. The weed management benefits of this strategy are, in some cases, partly caused by allelopathy. As noted earlier, allelopathy is a way in which some weeds hinder crop growth. Allelopathy can also be used to hinder weed growth. Rye, for example, is a commonly used mulch crop that is allelopathic to weeds.

Mechanical weed control, the second of the four basic weed control strategies, includes the oldest forms of weed control: hand pulling and hand hoeing weeds. In countries where herbicides and farm machinery are not affordable, hand weeding is still the predominant form of weed control (**Figure 17.7**). Even in more developed countries, hand

Figure 17.7 Hand hoeing of weeds. Hand hoeing is backbreaking and time consuming but is still the primary means of weed control in developing countries. This couple in the Luang Prabang province of Laos is weeding upland rice. Note the numerous weeds among the young rice plants. If not removed at this stage the yield will be lost. *Source:* Courtesy of Eugene Hettel, International Rice Research Institute.

Figure 17.8 Mechanized tillage. Large machinery enables farmers to mechanically control weeds over numerous hectares. *Source:* Available at www.farmphoto.com. Accessed October 8, 2001 (used with permission).

weeding is still commonly used in high-value, small-acreage crops. The disadvantage of hand weeding, of course, is that it is very labor intensive: it takes one person over 200 hours to hand-weed a one hectare field! In contrast, farmers using large machinery can till 10 or more hectares per hour (**Figure 17.8**). Primary tillage operations, which break apart and loosen the soil, as well as secondary tillage operations, which are used for seedbed preparation, are effective at destroying weeds. In addition, after planting the crop farmers use selective cultivation, such as tillage between crop rows, to destroy weeds (also see **Figure 17.9**). An important disadvantage of soil tillage is that it greatly increases the soil's susceptibility to wind and water erosion. No-till cropping systems reduce soil erosion, but also eliminate one of the most effective weed management options.

Figure 17.9 Selective cultivation. A rotary hoe can be used to selectively destroy weed seedlings before or shortly after the crop has emerged. Because the rotary hoe disturbs only a shallow layer of soil, the soybean plants shown in the photo will not be significantly damaged. *Source:* Courtesy of Mike Owen, Iowa State University.

Mowing, another example of mechanical weed control, is an effective way of preventing seed production, but it does not kill most weeds. Perennial weeds, in particular, are rarely killed by mowing. By its broadest definition, mechanical weed control also includes flooding (commonly practiced to control weeds in rice), nonliving mulches (such as plastic sheeting or dead organic material, commonly used in high-value crops), and the use of fire (occasionally used to control weeds in noncrop areas and for initial clearing of land).

Biological weed control is the third of the four basic weed control strategies. Although this can be a successful strategy (**Figure 17.10**), to date biocontrol of weeds has been used primarily in pasture and noncrop systems. One reason for the limited use of weed biocontrol

(a)   (b)   (c)

Figure 17.10 Biological control of weeds. Most successes of biological weed control have occurred with weeds in noncrop systems. Purple loosestrife, although possessing attractive purple flowers, is an invasive weed of wetlands and displaces native vegetation. A beetle, *Galerucella calmariensis* **(a)**, was released to control the purple loosestrife shown in **(b)**. Six years after release, purple loosestrife was controlled, and cattail and other native vegetation have returned **(c)**. *Source:* Courtesy of Illinois Natural History Survey.

in cropping systems is that weed suppression by biocontrol agents usually occurs slowly and incompletely, and therefore the agents do not suddenly halt weed competition. A second reason is that most biocontrol agents have a very narrow host range, and therefore multiple biocontrol agents are needed to control the multiple weed species in any one field. Despite these limitations, efforts to develop new weed biocontrol strategies continue, and a few biocontrol strategies are currently used in cropping systems.

Mycoherbicides, which are microbial organisms formulated to be applied in a similar manner as are herbicides, have been developed to control specific weeds. For example, Collego® is a formulation of a fungus used to control northern jointvetch in rice and soybean. And biocontrol agents are not limited to microbes and insects: For example, certain breeds of geese selectively eat grasses and nutsedge in broadleaf crops, so a few farmers use them as weed biocontrol agents.

The fourth and final weed control strategy is *chemical weed control,* or the use of herbicides. Herbicides offer many advantages over other forms of weed control:

- They control multiple weed species.
- Some can move into roots and thereby kill perennial species.
- Some possess soil residual activity and thus continue to protect the crop by controlling newly emerging weeds long after herbicide application.
- They usually cause little or no injury to the crop plant, and thus can kill weeds growing right next to crop plants (whereas mechanical control of such weeds would injure the crop plants).
- They can be applied to large areas of land in a short period of time.
- They control weeds quickly, bringing competition with the crop to an abrupt halt.
- They allow farmers to control weeds without tillage, thereby reducing soil erosion.

Of course, anything that sounds too good to be true probably is, and herbicides are not without their disadvantages:

- They may be acutely and chronically toxic to humans and other nonplant species.

- They may move with surface waters or leach through the soil and thereby contaminate water supplies.

- Some herbicides persist in the soil long enough to damage a susceptible crop grown in the same field the following season.

- Improper herbicide application or unusual weather conditions may result in crop injury or poor weed control.

- Drifting of herbicides during application may result in off-site, nontarget plant injury (such as the neighbor's prized roses).

- Herbicides may negatively impact the soil ecosystem.

- Repeated use of a herbicide may result in a population of weeds that is resistant to that herbicide.

A further disadvantage of herbicides is that because they are so effective, farmers tend to overrely on them and to abandon other good farming practices. For example, since the advent of herbicides many farmers have abandoned the practice of rotating annual grain crops with perennial forage crops. A decrease in diversity of cropping systems leads to increases in disease and insect problems and to a decrease in soil quality. Despite the disadvantages of herbicides, however, they have been used extensively since the middle of the twentieth century, and will continue to be used well into the foreseeable future. Worldwide, farmers spend some $15 billion annually on herbicides. In the United States, more than 95% of all maize and soybean hectares receive at least one herbicide application annually, resulting in the use of 100 thousand tons of herbicide active ingredient! Because of the importance of herbicides to weed control, we discuss them in greater detail in the next section.

17.5 Herbicides kill plants by disrupting vital, and often plant-specific, processes.

Herbicides are chemical molecules that can move into plants and disrupt a vital process. They exhibit phytotoxicity at low dosages: The use rate of many herbicides is in the range of 1–2 kilograms of active ingredient per hectare, and some of the more potent herbicides are used at rates as low as a few grams per hectare. The first steps in herbicide phytotoxicity include uptake and translocation of the herbicide by the plant. Soil-applied herbicides may be taken up by plant shoots as they grow through the soil, or by roots. Other herbicides are applied to plant foliage and can penetrate cuticles to gain access into plants. Once inside a plant, most herbicides move with the xylem stream and are distributed to transpiring tissue. Some herbicides are translocated through both the xylem and phloem. These are called *systemic herbicides,* because they are distributed throughout the plant system. Systemics are effective at controlling perennial species because these herbicides are translocated to, and cause death of, the root system. A few herbicides, called *contact herbicides,* are not translocated in plants. Effectiveness of contact herbicides depends on thorough coverage of the foliar tissue. Regardless of whether a herbicide is xylem-, phloem-, or non-mobile, it must, at some point, move into individual cells and disrupt a vital process.

What vital processes do herbicides disrupt? Most herbicides bind to, and thereby block the activity of, a specific enzyme. The protein with which a herbicide interferes is referred

Table 17.2	Classification of commonly used herbicides			
Mode of Action	**Site of Action**	**Chemical Family**	**Example Herbicides**	**Activity**
Inhibition of photosynthesis	D1 protein in photosystem II	Triazines Triazinones Ureas (and others)	Atrazine Metribuzin Linuron	Foliar and soil, control dicot and some grass weeds
Light-dependent membrane destruction	Photosystem I (herbicide acts as electron acceptor)	Bipyridyliums	Paraquat Diquat	Foliar, nonselective
	Protoporphyrinogen oxidase	Diphenylethers (and others)	Acifluorfen Lactofen	Primarily foliar, control primarily dicot weeds
Induction of abnormal growth by mimicking auxin	Probably multiple auxin receptors	Phenoxys Benzoic acids Carboxylic acids	2,4-D Dicamba Picloram	Foliar and soil, control dicot weeds
Inhibition of branched-chain amino acid synthesis (valine, leucine, and isoleucine)	Acetolactate synthase	Sulfonylureas Imidazolinones (and others)	Chlorsulfuron Metsulfuron Imazethapyr Imazaquin	Foliar and soil, control primarily dicot weeds
Inhibition of aromatic amino acid synthesis (phenylalanine, tryptophan, and tyrosine)	5-enolpyruvyl-shikimate-3-phosphate synthase	—	Glyphosate	Foliar, nonselective; however, can be used selectively in glyphosate-resistant crops
Inhibition of lipid synthesis	Acetyl-CoA carboxylase	Aryloxyphenoxy-propionates Cyclohexanediones	Diclofop Sethoxydim	Foliar, control grass weeds
	Unknown	Thiocarbamates	EPTC Triallate	Soil, control grass and some dicot weeds
Inhibition of cell division	Tubulin	Dinitroanilines	Trifluralin Pendimethalin	Soil, control grass and some dicot weeds
	Unknown	Chloroacetamides	Alachlor Metolachlor	Soil, control grass and some dicot weeds
Inhibition of pigment synthesis	4-hydroxy-phenyl-pyruvate dioxygenase	Isoxazole (and others)	Isoxaflutole	Soil, control several dicot and grass weeds

to as the herbicide's *site of action*. By binding to their sites of action, herbicides inhibit or disrupt specific physiological processes and thereby cause plant death. A herbicide's *mode of action*, which is a more general term than site of action, refers to the process by which herbicides kill plants. The modes of action of many commonly used herbicides are listed in **Table 17.2**.

The modes of action of many herbicides involve plant-specific processes—in particular, processes that take place in chloroplasts. Consequently, many herbicides have low toxicity to humans. For example, it stands to reason that a herbicide that targets photosynthesis would not likely be highly toxic to humans. Similarly, herbicides that block synthesis of essential amino acids (those not synthesized by humans) should, a priori, be nontoxic to humans. The fact that many herbicides target processes not present in humans explains why, in general, herbicides are less toxic to humans than are insecticides, which usually *do* target processes shared by humans (see **Table 17.3**). This does not mean, however, that

Table 17.3	Representative acute oral toxicities of common herbicides and insecticides	
Compound	**LD$_{50}$^{a} (mg/kg)**	**Human LD$_{50}$^{b} (g)**
Herbicides		
Paraquat	150	10.5
2,4-D	666	46.6
Acifluorfen	1,300	91
Atrazine	1,780	125
Glyphosate	4,300	301
Chlorsulfuron	5,545	388
Insecticides		
Terbufos	4.5	0.32
Carbofuran	11	0.77
Chlorpyrifos	96	6.7
Carbaryl	246	17.2
Permethrin	430	30.1
Malathion	1,375	96.3

^{a}The LD$_{50}$ is the dose of the product (in mg per kg of body weight) required to kill half of the test population (rats). The smaller the LD$_{50}$, the more toxic the compound.
^{b}The human LD$_{50}$ is calculated by multiplying the rat LD$_{50}$ by 70 kg.

Source: Rat LD$_{50}$s from B. L. Bohmont (2000), *The Standard Pesticide User's Guide*, 5th ed. (Upper Saddle River, NJ: Prentice Hall), pp. 212–214.

herbicides are nontoxic to humans; herbicides may have acute effects in humans unrelated to their herbicidal effects. Also, long-term exposure to some herbicides may result in carcinogenic or other toxic effects.

Herbicides are often grouped based on similarities in chemical structures. Herbicides belonging to the same chemical family have the same site of action and often share many other herbicidal properties (see Table 17.2). Different herbicides within the same chemical family can vary in crop selectivity, species of weeds controlled, and soil persistence and mobility. Small modifications to a chemical molecule can significantly alter its herbicidal properties. Compare the structures of 2,4-D and 2,4-DB, both shown in **Figure 17.11**. 2,4-D is an active herbicide molecule, whereas 2,4-DB, per se, is not. However, most dicot plants metabolize 2,4-DB to 2,4-D, so when 2,4-DB is applied to these plants, the net result is that it *is* herbicidal. Legume plants do not rapidly metabolize 2,4-DB to 2,4-D; consequently, 2,4-DB can be used as a selective herbicide on alfalfa and other legumes. 2,4-D, in contrast, kills most legumes. This example illustrates how tinkering with the structure of a herbicide molecule can result in slightly different herbicides with new uses. When a company involved in herbicide discovery identifies a promising new herbicide chemistry, it evaluates thousands of molecules that are variations of that chemistry, looking for those with the best herbicidal properties (broad-spectrum weed control, crop safety, and low use rates) and the best toxicological and environmental profiles.

The 2,4-DB example just given also illustrates the basis for the crop selectivity of many herbicides. That is, differential metabolism of herbicides between crop and weed species usually accounts for crop selectivity. Most herbicides do not remain unaltered in plants. Rather, many biochemical reactions metabolize herbicide molecules, usually resulting in the degradation of the herbicide into nontoxic molecules. The ability of crop species to metabolize certain herbicides into nontoxic metabolites faster than weed species metabolize them allows the selective use of these herbicides. The 2,4-DB example is one of a few exceptions to the norm, because a herbicide metabolite (2,4-D in this case) is more phytotoxic than the parent herbicide. Usually herbicide metabolites are *less* phytotoxic than the parent herbicide. Plant metabolism of herbicides may also result in the chemical combining of the herbicide molecules with sugars and other organic molecules, rendering them nontoxic. For example, the enzyme glutathione-S-transferase conjugates atrazine and many other herbicide molecules with glutathione, a tripeptide molecule.

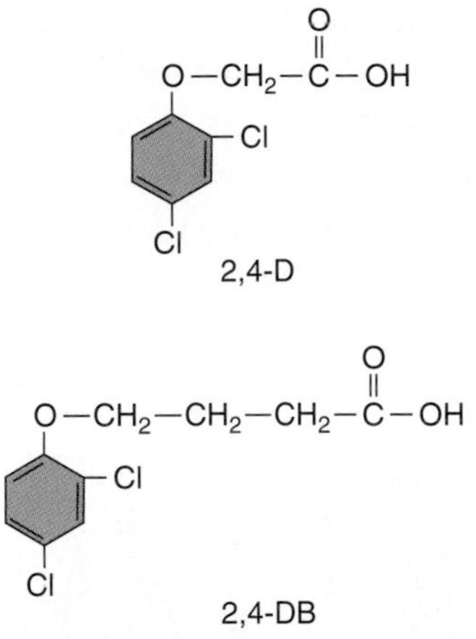

Figure 17.11 Chemical structures of 2,4-D and 2,4-DB. Although these herbicides share a common base structure, the different side chains result in different herbicidal properties. Most plants metabolize 2,4-DB, which is not phytotoxic, to 2,4-D, which is phytotoxic.

Some herbicides are nontoxic to the crop because the crop has a different site of action from that of the targeted weed species. This is true for the aryloxyphenoxy-propionate and cyclohexanedione herbicides on dicot crops. The site of action of these herbicides, acetyl-CoA carboxylase (ACCase), catalyzes the first step in fatty acid biosynthesis, the formation of malonyl-CoA. Dicots have two forms of ACCase—one is plastid-encoded, the other is nuclear-encoded—whereas grasses have only the nuclear-encoded ACCase. The aryloxyphenoxy-propionate and cyclohexanedione herbicides inhibit only the nuclear-encoded ACCase. Consequently, farmers can use these herbicides on many dicot crops to control grass weeds.

17.6 Weeds continue to adapt to attempts to control them.

During human attempts over the years to control them, weed populations have not remained static. Strategies that farmers use to kill weeds have resulted in numerous examples of Darwin's theory of natural selection in action. Because weed species evolve in response to weed management practices, some weeds are becoming harder to control.

A common consequence of repeated use of a particular weed control strategy or a change in cropping systems is a change in the predominant weed species that are present. These so-called *weed species shifts* occur because certain weed species are more tolerant than others of particular control strategies. Shifting from conventional tillage to no-till cropping systems usually increases perennial and biennial weed infestations. Also, small-seeded weed species, such as pigweeds and lamb's-quarters, become more common under no-till because the small seeds depend less on burial by tillage for germination. Heavy use of a particular herbicide also causes weed species shifts. For example, repeated use of 2,4-D, which controls only dicot weeds, may result in an increased prevalence of grass weeds. This shift occurred in cereal crops in the 1950s and 1960s, when 2,4-D was about the only herbicide option available.

Weed species shifts are changes in the types of species; evolution in response to weed control may also occur *within* weed species. Most weed species show substantial genetic variation. Variants within a species that possess genetic attributes allowing them to survive a particular weed control strategy increase in frequency on repeated selection by that control strategy. Weed scientists often use the term "biotype" to distinguish such variants from the species as a whole.

One way in which weed species have evolved to avoid control is to look more like the crop. Because these crop mimics are more difficult to distinguish from the crop, they are likely to be overlooked by farmers during hand weeding. The common practice of hand-weeding rice fields has resulted in the selection of a biotype of barnyardgrass that looks more like rice than does normal barnyardgrass (**Figure 17.12**). Sometimes crop mimicry results because a weed hybridizes with the crop species. For example, cultivated rice occasionally hybridizes with wild rice, resulting in wild rice strains that mimic cultivated rice. A purple-leafed strain of cultivated rice was developed so farmers could easily identify the crop during hand weeding. This solution was short-lived, however, as hybridization between this strain and wild rice soon resulted in purple-leafed biotypes of wild rice! Another example of crop mimicry is the evolution of weed seeds that are more similar in size and shape to crop seed. This occurs because the result is that the weed seeds are less likely to be removed during harvesting and cleaning of the crop seed. Consequently, the weeds are planted with the crop the following year. Examples include the evolution in lentil fields of a biotype of common vetch that produces flat rather than round seeds; and the evolution in flax fields of a biotype of flaseflax that flowers at the same time as the crop, produces fruits that do not shatter (and hence the seeds are harvested with the crop), and produces seed nearly identical in size and shape to the crop seed.

Figure 17.12 Example of a crop mimic. Hand-weeding in rice has selected for a biotype of barnyardgrass (center) that looks more similar to rice (left) than to normal barnyardgrass (right). *Source:* Reprinted with permission from Spencer C. H. Barrett, Crop mimicry in weeds, *Economic Botany* 37:255–282. Copyright © 1983, The New York Botanical Garden.

Because herbicides are so effective at controlling weeds, they exert a strong selection pressure. Consequently, weed populations have evolved that are resistant to herbicides (**Figure 17.13**). There is no evidence that herbicides have ever actually created a mutation that conferred herbicide resistance. Rather, the evolution of herbicide-resistant weed biotypes is analogous to the evolution of crop mimics. Herbicide-resistant individuals preexist in weed populations because of natural variation, and herbicides simply select for those individuals. After repeated herbicide selection, alleles of genes conferring herbicide resistance increase in frequency within weed populations, eventually resulting in resistant populations.

The first example of a problematic herbicide-resistant weed population was reported in 1970 with the identification in Washington State of common groundsel resistant to triazine herbicides. Continued use of these herbicides has led to several additional triazine-resistant weed populations (**Figure 17.14**). Widespread use beginning in the 1980s of herbicides that target acetolactate synthase (ALS) and of herbicides that target ACCase quickly led to many more herbicide-resistant weeds. By 2001, over 150 weed species had evolved resistance to at least one herbicide, and for nearly every major herbicide there is at least one weed that has evolved resistance to it. In some cases, biotypes of weeds have evolved resistance to multiple herbicides with distinct modes of action (see **Box 17.2**).

Figure 17.13 Example of a herbicide-resistant weed. Repeated selection by a herbicide often results in a population that is no longer controlled by that herbicide. The photo shows a sensitive plant (left) and a resistant plant (right) two weeks after treatment with imazethapyr, a herbicide that inhibits the ALS enzyme. The resistant plant was not injured by the herbicide because it contains a single nucleotide substitution in its ALS gene, rendering the enzyme insensitive to the herbicide. In the presence of imazethapyr selection, the resistant plant has an obvious fitness advantage; thus, its genotype would increase in frequency over time.

The mechanism of herbicide resistance is usually a change in the herbicide target site that reduces its affinity for the herbicide. In other words, the herbicide is no longer able to bind and inhibit the target site. Often the change is very subtle—a single amino acid—and thus requires no more than a single base mutation within the gene that encodes the target site. Although subtle, the change is specific: The change must be sufficient to abolish herbicide binding, yet it cannot severely alter the normal function of the target site. The basis for triazine resistance in weed populations typically is a substitution of glycine for serine at amino acid position 264 in the D1 protein of photosystem II in the chloroplast, which is the herbicide target site. This change renders the D1 protein insensitive to triazine herbicides, yet the protein is still able to carry out

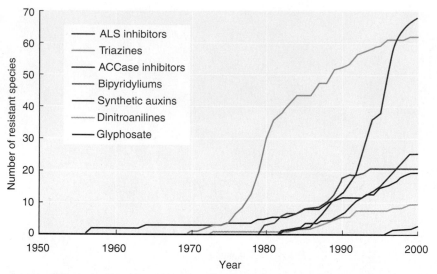

Figure 17.14 Evolution of herbicide-resistant weeds. Widespread use of a particular herbicide is often soon followed by the identification of weed biotypes resistant to that herbicide. In particular, note the rapid occurrence of resistance to ALS-inhibiting herbicides, which have been used only since 1982. *Source:* I. Heap, *The International Survey of Herbicide Resistant Weeds.* Available at www.weedscience.com. Accessed March 27, 2001.

its normal function of transferring electrons through photosystem II. A cost is associated with this change, however, in that the triazine-resistant D1 protein is not as efficient as the wild type protein in electron transfer, which in turn reduces the photosynthetic efficiency of the plant. Consequently, triazine-resistant plants are slightly less fit than wildtype plants in the absence of triazine herbicides. This explains why triazine-resistant individuals are initially at a very low frequency in natural weed populations.

17.7 The speed with which herbicide resistance evolves depends on the herbicide's site of action.

The development of herbicide-resistant weed populations has been most problematic for herbicides that target the enzyme acetolactate synthase (ALS). This is partly because several, single amino acid substitutions can occur in ALS that render the enzyme insensitive to ALS-inhibiting herbicides but have little to no effect on the enzyme's normal catalytic activity. Populations of weeds resistant to ALS-inhibiting herbicides have arisen in as few as four years after repeated use of these herbicides. Other herbicides are much less likely to select for resistant weeds. Glyphosate, for example, is the world's most widely used herbicide, yet only a few glyphosate-resistant populations have evolved. The likelihood for resistance-development to a particular herbicide strongly depends on that herbicide's site of action, and on whether point mutations in that site of action will abolish herbicide binding but not normal activity of the site of action. For this reason, sustainable use of herbicides (management of herbicide resistance) requires knowledge about the physiology and biochemistry of how herbicides work and about the genetics of herbicide resistance.

Box 17.2

When a weed population develops resistance to a particular herbicide, the farmer is forced to use an alternate control strategy. Usually this consists of switching to another herbicide that has a different site of action. Subsequent selection by the second herbicide may lead to the weed population acquiring multiple herbicide resistance. If this process is reiterated, the farmer will eventually run out of suitable herbicides. In nearly all cases so far, however, herbicide-resistant weeds are resistant to herbicides from only one or two different chemical families and thus other herbicide options still exist. An exception, however, is rigid ryegrass. This weed, which is particularly troublesome in southern Australia, is winning nearly every herbicide battle!

The combination of very large farms and the potential for wind erosion of the dry soils has encouraged the adoption of minimum-tillage, herbicide-intensive weed management in the wheat fields of southern Australia. These fields are heavily infested with rigid ryegrass—due in part to the intentional planting of this species throughout Australia because of its value as a forage crop. Also, rigid ryegrass is an outcrossing species and therefore possesses substantial genetic variation. The combination of these factors—intensive herbicide use, heavy infestations, and genetic variability—have led to widespread occurrence of herbicide resistance in rigid ryegrass.

Several biotypes of rigid ryegrass have evolved resistance to herbicides from more than one chemical family. In the most severe case, a single biotype is resistant to herbicides from nine chemical families that comprise five different sites of action! Specifically, this biotype is resistant to ACCase-inhibiting and ALS-inhibiting herbicides, herbicides that bind D1 protein, chloroacetamides, and thiocarbamates. Multiple resistance in this biotype is conferred by multiple mechanisms: It has both altered sites of action and the ability to metabolize particular herbicides.

The Invincible Weed from Down Under

Metabolism-based resistance is common among herbicide-resistant rigid ryegrass biotypes. In some cases, metabolism-based resistance selected by one herbicide also confers resistance to other herbicides with different sites of action. This occurs because the herbicide-metabolizing enzyme selected (probably a cytochrome P450 monooxygenase) has low substrate specificity and thus can metabolize diverse herbicide molecules. Thus when a herbicide application fails because of resistance in a rigid ryegrass population, it is difficult to predict to what other herbicides that population might also be resistant.

A few rigid ryegrass populations have developed resistance even to glyphosate; some experts had previously claimed that glyphosate was a herbicide to which weeds would never evolve resistance. In fact, two glyphosate-resistant biotypes have been identified in Australia and a third glyphosate-resistant ryegrass biotype was identified in California. Thus rigid ryegrass really has won nearly every herbicide battle. Paraquat is one of the few herbicides to which rigid ryegrass has not yet evolved resistance. Paraquat is a nonselective herbicide, however, and can only be used to control ryegrass before the wheat crop has emerged.

So is there a "silver lining" to this story? Many farmers faced with herbicide-resistant rigid ryegrass are more frequently rotating crops. Specifically, they are rotating wheat with hay or mulch crops. And, if nothing else, this invincible weed from Down Under has taught people what to expect if they rely exclusively on chemical weed control.

Site of action-mediated herbicide resistance is caused by a single gene change and, if it involves a nuclear gene, may be dominant or recessive. Resistant ALS genes, for example, are dominant (or semidominant), and this further explains why resistance to ALS-inhibiting herbicides evolves so rapidly. For a recessive resistance gene, the resistant allele is only selected if it exists in a weed in a homozygous condition. Geneticists estimate that alleles of genes that confer herbicide resistance exist in natural populations typically at a frequency of 10^{-6}; the likelihood that a weed will be homozygous for such a low-frequency allele is exceedingly small, especially for outcrossing weed species. Site of action-mediated resistance to dinitroaniline herbicides is recessive. Not surprisingly, resistance to dinitroanilines has occurred only rarely.

Once herbicide resistance develops in a weed population, it may spread to other populations. For dominant, nuclear-encoded resistance, such as resistance to ALS-inhibiting herbicides, pollen movement readily spreads the trait. As noted earlier, resistance to triazine herbicides usually arises via a change in the D1 protein. The gene encoding this protein is in the plastid genome. Because plastids of most plant species are inherited maternally, triazine resistance normally spreads only by seed movement, not via pollen.

The previous discussion of herbicide resistance has focused on site of action-mediated resistance. Although this is the most common mechanism of resistance in weed populations, other mechanisms of herbicide resistance have evolved. The second most common mechanism is metabolism of the herbicide. As noted, herbicide metabolism is often the basis for crop selectivity of herbicides. Sometimes weeds evolve the ability to rapidly metabolize herbicide molecules, rendering them less sensitive to those herbicides. Although the magnitude of resistance is often much less than is seen with site of action-mediated resistance, metabolism-based resistance may result in resistance to multiple herbicides with different sites of action (see Box 17.2).

What can farmers do to prevent the evolution of herbicide resistance and preserve the effectiveness of currently available herbicides? As previously mentioned, herbicide-resistance management requires that farmers understand how and why herbicide resistance develops. Herbicide resistance evolves in response to herbicide selection pressure. Therefore, the best way to prevent resistance development is to avoid repeatedly applying the same selection agent. Farmers are encouraged to use herbicides with different sites of action from year to year. In Australia, the government sought to help farmers do so, by requiring that herbicide containers be labeled with an alphabetic code indicating that herbicide's site of action. Other countries are now considering similar requirements. Besides rotating herbicides, farmers are also encouraged to use nonherbicidal weed control strategies to prevent development of herbicide resistance. In practice, however, farmers are often driven by the philosophy of "It worked last year, so I am going to use it again this year" and feel driven by economics to use the same one or two herbicides year after year. Consequently, herbicide resistance will continue to be a major problem confronting farmers in their ongoing efforts to control weeds.

17.8 Biotechnology offers new strategies for managing weeds.

Because weeds will continue to find ways to adapt to weed control efforts, farmers will always need new strategies. Biotechnologists have developed new weed management tools, and many more are in progress. Before we describe these tools, however, it is important to note that they are just tools. As people involved in crop production continue to learn over and over, for pest management there is no easy solution. Weeds, like any other pest, must be managed with multiple strategies. Strategies offered by biotechnology simply add to a farmer's collection of weed management tools.

By far, the predominant new tools for weed management arising from biotechnology have been herbicide-resistant crops (HRCs). In fact, HRCs are currently the biotechnology products that dominate all crop production. There are at least two reasons for HRC dominance over other crops created from biotechnology. First, compared to the manipulation of other traits, such as crop yield or crop composition, HRCs can be created with a single, easily selected gene and thus is a very straightforward process. Second, there is a large economic incentive for a company to develop HRCs: By developing a crop resistant to a herbicide that the company markets, the company can capture return on its investment by selling both the crop seed and the herbicide.

What exactly are HRCs, and how are they developed? HRCs are simply crop varieties that are resistant to a herbicide that is normally lethal (or highly injurious) to that crop. Researchers develop them through either a selection or gene insertion process. Selection can be accomplished simply by screening large numbers of plants with the herbicide of interest, looking for a naturally occurring variant within the crop species that is resistant. In practice, however, it is more efficient to create variation by subjecting the crop to a mutagenic agent (chemical or radiation) before performing the selection. Crops that are resistant to sulfonylurea, imidazolinone, and cyclohexanedione herbicides have been obtained by selection and are now commercially available. HRCs obtained by selection are analogous to herbicide-resistant weeds and have the same mechanisms of resistance. In fact, the occurrence of herbicide-resistant weeds is what inspired scientists to develop HRCs by selection!

Scientists use gene insertion (transgenic) approaches when they cannot get an acceptable level of resistance to a particular herbicide from selection. The inserted gene may encode a form of the herbicide target site that is insensitive to the herbicide. Genetic engineers have obtained resistance to glyphosate by inserting into crop plants a modified gene from an *Agrobacterium* species that encodes a resistant form of the herbicide target site, 5-enolpyruvyl-shikimate-3-phosphate synthase (EPSP synthase). Researchers isolated this bacterium as resistant to glyphosate, and subsequent research revealed that the unusual form of EPSP synthase caused the resistance. The gene was isolated and modified (to ensure adequate expression in plant cells and to ensure targeting of the encoded EPSP synthase to chloroplasts, where it functions) and then transferred into plants. Resulting crops are called Roundup Ready® because they are resistant to Roundup® herbicide, which contains glyphosate as the active ingredient.

Transgenic HRCs that metabolize particular herbicides have also been developed and commercialized. LibertyLink® crops are resistant to glufosinate (sold as Liberty® herbicide) because these crops have been engineered to metabolize this herbicide. Glufosinate, also called *phosphinothricin*, is essentially a synthetic, truncated version of a toxin produced by *Streptomyces hygroscopicus* (see **Figure 17.15**). To prevent autotoxicity, this organism

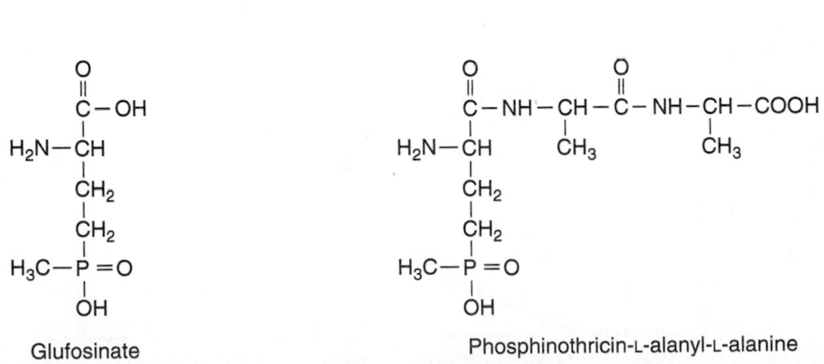

Glufosinate

Phosphinothricin-L-alanyl-L-alanine

Figure 17.15 Chemical structures of phosphinothricin-L-alanyl-L-alanine (also called *bialaphos***), a natural compound produced by** *Streptomyces hygroscopicus,* **and the herbicide glufosinate (also called** *phosphinothricin***).** Both chemicals are used as herbicides, although bialaphos is probably converted by plants to phosphinothricin. Phosphinothricin is toxic because it inhibits a key enzyme in amino acid metabolism. The microorganism has a gene that encodes an enzyme that detoxifies phosphinothricin. When this gene is introduced into crops, the crops are resistant to the herbicide.

also produces an enzyme, phosphinothricin-N-acetyl transferase, which acetylates phosphinothricin into a nontoxic metabolite. Genetic engineers used a gene encoding this enzyme to make transgenic, LibertyLink® crops.

Herbicides already exist that can be used selectively for growing major crops, so what new advantages do HRCs offer? Essentially, HRCs enable "better" herbicides to be used on crops. Glyphosate, for example, is a good herbicide in that it effectively controls most major weeds (including troublesome perennial species), is essentially nontoxic to humans, has low environmental persistence, has a low likelihood to select for resistant weeds and has foliar activity (and therefore farmers can use it based on need, after determining the severity of the weed infestation). Furthermore, because glyphosate is so effective, it reduces the need for tillage and thus can be used to reduce soil erosion. Given all these advantages, it is not surprising that within five years after their introduction, Roundup Ready® varieties comprised over 50% of the U.S. soybean market (**Figure 17.16**). A survey of 452 farmers in 19 U.S. states released by the American Soybean Association in the autumn of 2001 showed that about half the U.S. soybean growers are indeed using conservation tillage practices in combination with Roundup Ready® soybeans and making fewer tillage passes over their fields compared to 1996. These practices not only limit soil erosion, but they reduce production costs as well. In some areas of South America, farmers are planting these soybeans without authorization from their government or from Monsanto, the producer of these genetically modified (GM) soybeans, because of the cost savings they provide.

What are the disadvantages of HRCs? Much of the opposition to HRCs centers on fears and concerns common to people's reactions toward all transgenic crops or "genetically modified organisms" (GMOs). We discuss these fears and concerns in detail in Chapter 20. Remember that many HRCs have been obtained by selection and not by gene transfer, and thus are not classified as GMOs. Nevertheless, potential disadvantages are associated with HRCs, especially if they are not used correctly.

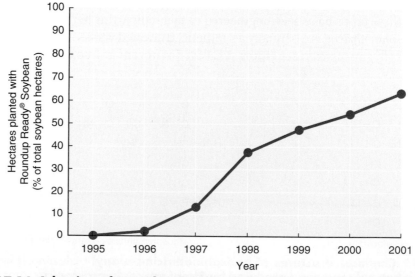

Figure 17.16 Adoption of Roundup Ready® soybean in the United States. Since its introduction in 1996, U.S. farmers have rapidly adopted soybean genetically modified to tolerate glyphosate. *Sources:* USDA and National Center for Food and Agricultural Policy.

One of the major potential disadvantages of HRCs is that they may be overused. Overuse may result in farmers abandoning other weed management strategies, further reducing diversity of crop systems and creating new problems. The availability of HRCs tempts farmers to rely on a single herbicide, which, as discussed before, can lead to the selection of herbicide-resistant weeds and can bring about shifts to more tolerant weed species. Herbicide-resistant weeds may also arise by gene flow from HRCs. Although transmission of herbicide resistance between crops to weeds has been possible before the advent of HRCs, the single-gene nature of resistance in HRCs increases the likelihood for resistance transmission. Another potential problem is that HRCs themselves may become weeds that are difficult to control. For example, a farmer growing Roundup Ready® maize one year and Roundup Ready® soybean the next would have volunteer maize growing in the soybean crop that glyphosate would not control. Resistance in volunteer crops can be compounded when different herbicide-resistant varieties of a particular crop are grown in proximity. In Alberta, Canada, crossing among three types of herbicide-resistant canola has resulted in triple-resistant volunteer canola. These plants resist glyphosate, glufosinate, and imidazolinones!

Biotechnology offers weed management more than just the development of HRCs. As described in Box 17.1, parasitic weeds can be especially difficult to control, and some use chemical signals from the host plant for inducing both seed germination and attachment of haustoria. Biotechnology may enable geneticists to manipulate crop plants so that they do not produce these signals. In fact, traditional breeding efforts have already had some success in developing sorghum varieties that do not produce strigol and thus resist witchweed. Biotechnology also offers exciting new opportunities to modify allelopathic properties of crop plants and thereby develop novel weed management strategies: In essence, this could amount to using crop plants that produce their own herbicide! Although this might seem farfetched, the production of both a herbicide and an enzyme that degrades that herbicide, by *S. hygroscopicus* (described previously), illustrates that genes for herbicide production and then to safeguard the plant are available in nature. Biotechnology also offers the potential to increase the effectiveness of weed biocontrol agents.

Scientists have already learned, using herbicide-resistant weeds as models, how to develop herbicide-resistant crops. And weeds have more to teach people. For example, in the future scientists may use competitive weed species as models to develop more competitive crop species. Because weeds are, in many respects, some of the world's most successful plants, they are potentially a rich source of genes for crop improvement.

CHAPTER SUMMARY

Weeds can be defined simply as plants that are somehow deleterious to human activities. The most common way weeds adversely impact human activity is by reducing crop yields by competing with the crop for resources necessary for growth and development. In addition to being competitive, most weeds are also persistent, which enables them to survive repeated attempts by humans to control them.

Farmers use multiple weed control strategies—including cultural, mechanical, biological, and chemical methods—to reduce the negative impacts of established weed populations. Worldwide, mechanical weed control, including hand weeding, is still a predominant method of controlling weeds. In developed countries, however, herbicides have largely replaced other forms of weed control. The effectiveness of herbicides and the ability of many to be used selectively in crops have enabled farmers to efficiently control weeds and increase crop yields. Overreliance and misuse of herbicides, however, have contaminated food and the environment, reduced diversification of farming systems, and provoked development of herbicide-resistant weed populations.

Within the last decade, commercialization of herbicide-resistant crops has provided farmers new tools for managing weeds. Used wisely, this technology can benefit both farmers and consumers by further increasing crop yields, decreasing soil erosion and fostering use of herbicides with the fewest adverse effects. No *single* tool, however, is the right tool to control weeds. Sustainable cropping systems require an integrated approach to managing weeds, in which multiple weed control strategies are used judiciously.

Discussion Questions

1. What is a weed? Give a definition from different perspectives—even from the plant's perspective!

2. What would you expect were the native habitats of plant species we now call *weeds*?

3. If you were to create the world's worst weed, what traits would it need to have?

4. Besides reducing crop yields, how do weeds negatively impact humans and human activities?

5. If you were leading a governmental effort to eradicate a new weed species recently introduced into the country, what steps would you take? Why would you take those steps?

6. How and why do weeds adapt to weed control strategies?

7. What characteristics of a weed species would increase the probability that it would evolve herbicide resistance?

8. How does a herbicide's target site influence that herbicide's likelihood to select resistant weed populations?

9. Should herbicides continue to be used? Explain.

10. Should biotechnology be used to create herbicide resistant crops? Explain.

Further Reading

Aldrich, R. J., and R. J. Kremer. 1997. *Principles in Weed Management,* 2nd ed. Ames: Iowa State University Press.

Charudattan, R. 2000. Current status of biological control of weeds. In G. G. Kennedy and T. B. Sutton, eds., *Emerging Technologies for Integrated Pest Management: Concept, Research, and Implementation,* pp. 269–288. St. Paul, MN: American Phytopathological Society.

Duke, S. O. 1996. *Herbicide-Resistant Crops.* New York: CRC Press.

Gressel, J. 1998. Biotechnology of weed control. In A. Altman, ed., *Agricultural Biotechnology,* pp. 295–325. New York: Dekker.

Holm, L., J. Doll, E. Holm, J. Pancho, and J. Herberger. 1997. *World Weeds: Natural Histories and Distribution.* New York: Wiley.

Radosevich, S., J. Holt, and C. Ghersa. 1997. *Weed Ecology: Implications for Management,* 2nd ed. New York: Wiley.

Ross, M. A., and C. A. Lembi. 1999. *Applied Weed Science,* 2nd ed. Upper Saddle River, NJ: Prentice Hall.

Weller, S. C., R. A. Bressan, P. B. Goldsbrough, T. B. Fredenburg, and P. M. Hasegawa. 2001. The effect of genomics on weed management in the 21st century. *Weed Science* 49:282–289.

Zimdahl, R. L. 1999. *Fundamentals of Weed Science,* 2nd ed. New York: Academic Press.

Toward a Greener Agriculture

Jonathan M. Shaver
Oklahoma State University

For five long days at the end of October 1998, "Mitch," a rare category 5 hurricane, pounded Central America. Torrential rains and high winds caused flooding and mudslides, leaving thousands dead or missing. In Honduras, nearly 20% of the nation's 5.3 million people were left homeless or unable to return to their homes because of flooding. As much as 60% of the nation's infrastructure and 70% of its crops were destroyed.

The Department of Lempira, located in the hilly terrain close to the border with El Salvador and with a population of 100,000, is one of the poorest and most isolated regions of Honduras. In 1990, its farmers barely produced enough maize, beans, and sorghum to feed their families. When Hurricane Mitch hit in 1998, these same farmers provided tons of emergency food aid to help their fellow citizens in other parts of the country.

In 1990, when the Lempira Sur project sponsored by the Food and Agriculture Organization (FOA) began, 72% of the people lived below the poverty level, and malnutrition was chronic. Most farmers used the slash-and-burn method of farming. Crops were grown on a cleared plot for 1 to 3 years until declining fertility caused the yields to drop and forced the farmers to move to a new area. In Lempira, most of the fields are on hillsides with slopes of up to 60 degrees. Much of the devastation of hurricane Mitch was the result of large-scale deforestation and cultivation of these sloping lands without any soil conservation methods.

This FAO-sponsored project began with the people living in the villages of Lempira. Through meetings and interviews, extension workers learned of the local farmers' experiments with planting trees to help prevent runoff and improve soil fertility. Based on these initial successes, the farmers and extension workers developed an agroforestry cropping system now called Quesungual, named for a village in Lempira (**Figure 18.1**). Such partnerships are a model for small, farm-based, rural development.

In this system, the farmers clear the land with a machete rather than burning the vegetation, which destroys valuable organic matter that otherwise protects the soil from drying out. The large trees, rather than being removed, are severely trimmed but left standing, allowing the sun to filter down to the crops below. Smaller trees are kept for firewood. The large trees remain good sources of fruit, and continue to provide organic matter (leaves) to the soil as well as shade for the crops. Most importantly, the extensive roots of the trees stabilize the soil. Prior to the 1998 hurricane, the farming system had already proven its worth by doubling the yield of maize and beans and providing

Figure 18.1 Soil conservation practices in Honduras. Crops such as maize are grown in cleared areas among selected native trees on the steep hillsides of Honduras. The Quesungual cropping system used by farmers in the department of Lempira, held the soil during the torrential rains caused by Hurricane Mitch, allowing these farmers to continue production and send food to their fellow citizens. *Source:* Food and Agriculture Organization.

high yields during the drought in 1997. As a result, the people of Lempira not only eat better, but they are also able to sell their excess products at the market, or to provide aid to those in need.

18.1 Many agricultural practices have adverse environmental and social effects both on the farm and away from the farm.

Genetically improved crop varieties and new management practices have kept food production ahead of population increases in the 20th century, but the agricultural practices required to sustain this level of production can have negative environmental consequences (see Chapter 3). Making agriculture sustainable in the future will require us to look for new and appropriate solutions. It is worth emphasizing that there are unsustainable aspects to both high-input modern agriculture and to the low-input agriculture of low-resource farmers as shown by the experiences of the Lempira farmers before 1990.

Soil Erosion. Clearing forests, growing crops on steep slopes, planting crops parallel to rather than perpendicular to the predominant slope, or planting on large open fields without protection from the wind can lead to soil erosion. Moderate to severe soil degradation affects almost 2 billion ha of arable cropland and grazing land. Every year, 25,000 million tons of topsoil are lost worldwide from arable lands. Soil erosion directly affects crop production because the topsoil, the most productive layer of soil, is lost, reducing nutrient levels, organic matter content, and water-holding capacity. The off-farm impacts of soil erosion are equally devastating, resulting in ruined aquatic habitats and clogged waterways; reservoirs silt up and riverbeds rise, increasing the risk of floods. In the United States, experts estimate the off-the-farm economic cost of soil erosion to be two to eight times greater than the on-farm costs of lost productivity. Just as in other industries, these costs are not counted as (food) production costs.

Decreased Water Availability. Irrigated agriculture has expanded dramatically in the past 40 years. The FAO estimates that 17% of the world's cropland is under irrigation and that 42% of all food is produced on this land. In India, 29% of all food is produced on irrigated land, in China 52%, and in Egypt 100%. The rate of world water use has increased fourfold in the last 40 years, with nearly 75% of that for crop irrigation. At this time 70% of all freshwater withdrawals are for agricultural purposes, but in the next 25 years industrial and home use of water will increase at the expense of agriculture. The primary source of irrigation water is not from surface waters such as lakes and rivers, but from underground water reserves called *aquifers*. Some aquifers are recharged from precipitation, but in many parts of the world, the rate of removal has exceeded the rate of natural recharge. As the water table falls, the cost of pumping becomes prohibitive, and land must be taken out of production. All irrigation methods are not equally efficient; whether a farmer uses furrow irrigation (**Figure 18.2**) or other methods such as overhead sprinklers depends on the cost of water to the farmer and the available technology.

Salinization. Years of irrigation will cause the level of salts to build up in the soil. All irrigation water contains some salts, and after irrigation the water is either used by the crop or evaporates, leaving the salt behind. If not enough water is applied or if the water table below the field is too close to the soil surface, the salts will not be washed past the root zone of the plant and accumulate in the soil (see also Chapter 3, Figure 3.12, and Chapter

Figure 18.2 Flood irrigation of cotton in the southern United States. Nearly 75% of global water use is for agricultural purposes. Many areas of the world that have a long growing season have inadequate precipitation to support crop growth. Irrigation from surface or groundwater sources makes food production possible. However, because of falling water levels, the cost of pumping and deepening of wells has become prohibitive. Overwatering is not only uneconomical and a waste of natural resources but also encourages diseases, leaches nutrients from the soil, and reduces the effectiveness of pesticide treatments. Efficient water use requires knowledge of soil water-holding capacity and needs of the crop according to the current growth stage of the plant.

11, Figure 11.7). Yields decrease under these conditions. If not corrected, soil salinization ultimately leaves the land unfit for agricultural purposes. Because of the rising cost of pumping irrigation water and because of salinization, 20% of irrigated land has been taken out of production since 1978, and the prediction is 20% more by 2020.

Fertilizer and Pesticide Contamination. The increased use of chemical fertilizers and pesticides, particularly herbicides, has increased the potential for groundwater contamination. The crop does not always use fertilizers efficiently, and analysts estimate that some 30–80% of applied nitrogen is lost and ends up in the environment. Similarly, pesticide applied to a crop may end up in ground or surface waters. In developing countries of Asia and South America, use of pesticides and fertilizers is still increasing (**Figure 18.3**), which may lead to similar problems in these countries. However, in the poorest countries, and especially in Africa, fertilizer use is very low—too low to replace nutrients lost through agriculture. Improper disposal of animal wastes from animal feedlots can also cause considerable surface water and groundwater pollution.

The major problem associated with pesticide use is not pesticide residues on fruits and vegetables, as is generally believed by the public at large, but runoff into lakes and streams and acute poisoning of farm workers. The latter problem is greatest in developing countries where farm workers may lack protective clothing and the necessary training to apply toxic chemicals safely.

Decreased Genetic Diversity. With expansion of crop production and urban populations into previously undisturbed lands, loss of unique species of plants and animals is substantial. Extinction of these plant and animal species represents a loss of the world gene pool, including potential sources of genes for food production or new medicines.

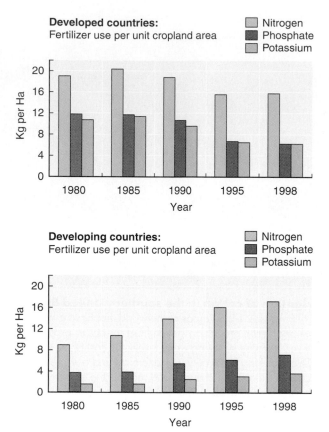

Figure 18.3 The rate of fertilizer use is decreasing in developed countries but is increasing in developing countries. In the past 20 years the global use of nitrogen fertilizers has increased by 33%, whereas total phosphorous and potassium fertilizer consumption has not changed, nor has the amount of cropland increased significantly. The slowing increase in fertilizer use has been caused by a decrease in the amount of fertilizer used per unit cropland area in developed countries. However, this decrease has been offset by an increase in amount of fertilizer used per unit cropland area in the developing countries. *Source:* Data from Food and Agriculture Organization.

Simplification of the cropping systems accompanied agricultural intensification in order to increase efficiency and reduce costs. Farms with landraces of different crops and the practice of crop rotation gave way to farms that grow a single genetically uniform crop on very large areas and that grow the same crop year after year (monoculture). This practice has economic benefits for the farmer, because it requires investment in fewer pieces of equipment and simplifies management. However, the downside is that the farmer is more prone to fluctuations in the market. In addition, if a particular crop becomes susceptible to a disease or insect pest, severe crop losses could result if conditions favor spread of the disease. As more and more producers adopt closely related modern varieties, genetic diversity of the crop begins to decrease. Farmers need incentives to participate in *in situ* conservation of plant genetic diversity.

At one time farmers in India grew more than 30,000 varieties of rice (*Oryza sativa*), but today 10 varieties cover 75% of the rice-growing area, and in the early 1990s just six varieties of maize covered more than 70% of the maize-growing area of the United States. Landraces may yield less in a given year, but their disease resistance and climatic adaptations give producers greater economic stability over time. Although one can decry this loss of genetic diversity, it is unrealistic to expect farmers to worry about it. The public policy response to this problem should be to reward farmers financially for germ plasm conservation.

Social Changes. A characteristic of modern agricultural systems is efficiency. Where land is plentiful or labor is scarce, farmers achieve efficiency by using machines rather than human labor (see Chapter 2). For example, in the United States the on-farm population dropped from 40% of the total population in 1900, to 2% today. In the older industrialized countries and in the newly industrialized countries of Asia (Singapore, South Korea, and Taiwan), there was adequate nonagricultural employment to support this population migration from the countryside to the city, but in most developing countries there are fewer alternatives to farm labor. This growth of the urban population necessitated changes in the food production/distribution system and other profound social changes. Similar changes are occurring now in developing countries and will continue into the 21st century.

18.2   Public policies and government assistance often promote nonsustainable agriculture practices.

Although resource-poor farmers in developing countries often find themselves without any assistance from the government, farmers in developed countries benefit greatly from government programs. Such programs aim to provide an inexpensive and safe food supply for the people and to maintain economic dominance in the global marketplace. The policies are influenced by lobbying groups who have specific objectives such as preserving the family farm, protecting specific markets (such as maize for ethanol production or sugar in the United States) or the interests of specific groups of farmers or corporations, or are attempting to protect taxpayer interests or the environment. Protests by farmers have occurred frequently in the past 20 years (**Figure 18.4**). In addition, environmental awareness by the public at large has generated new political pressures and has come to the forefront as a major influence on agricultural policies.

Environmental Protection Programs. The government has a major influence on agricultural practices, through economic policy and the setting of prices and mandates regarding how land can be used by farmers wishing to participate in government programs. One example of a government-sponsored environmental protection program is the Conservation Reserve Program (CRP), which encompassed 12 million hectares in 1999. Under the CRP, farmers voluntarily retire environmentally sensitive land for 10 to 15 years. In return, the government makes annual rent payments to producers and shares the cost of establishing approved conservation practices. A total of US$1.3 billion was paid out in 1999 at a cost of approximately US$110 per ha. Although it provides substantial environmental benefits, to have the largest impact taking environmentally sensitive land out of production must be combined with environmentally sound practices on the remaining land.

Production-Oriented Subsidies. In the United States, the policy that historically had the greatest impact on farming was the commodity program, under which the federal government guaranteed the farmer a minimum price for crops. Payments were determined by three factors: average area planted to the crop over several years, average yield of the crop over several years, and a target price. If the actual market price fell below the target, the government guaranteed to make up the difference to the farmer. To maximize payments from the government, therefore, farmers wanted to plant a large area that yielded a lot of

Figure 18.4 Farmer protests. The emergence of trading blocks such as the European Union and the North American Free Trade Agreement and the founding of the World Trade Organization in 1995 has resulted in many protests by different interest groups. Here, farmers watch as a truck dumps more than 5 tons of tomatoes in front of the headquarters of the autonomous government of Andalucia in Seville, Spain, in January 2000 during a demonstration by the Union of Small Farmers (UPA) to protest a European Union agreement to buy Moroccan tomatoes. *Source:* AP/Wide World Photo by Julio Munoz.

those crops that where subsidized. Because payments of grain were made on a yield basis, such subsidy programs encourage the excess use of fertilizers and pesticides to keep yields high. If yields were high, and even if it cost more to produce the crop than the crop was worth, the farmer needed to keep the multiyear average high, thus safeguarding future financial returns. Because returns to inputs diminish as yields go up, it takes more fertilizer and other inputs to raise the production from 6 to 8 tons per hectare than from 4 to 6 tons per hectare. Pushing production up to such a high level may not be cost effective, but taxpayers foot the bill. In effect, a higher target price subsidizes the less efficient and potentially damaging use of excess inputs. One result of the commodity program was that farmers did not readily switch to other crops or other types of farming practices. Because the federal programs were profitable, the farmers accepted the restrictions rather than try something different.

To address these concerns, the 1996 Farm Act introduced major changes in the commodity programs. The new rules replaced deficiency payments and supply controls with almost total planting flexibility, and decoupled income support payments for major crops so that payments were no longer linked to farmers' current decisions on production of specific crops. Farmers' planting and business decisions could now be guided more by supply and demand conditions than by terms and expectations of commodity policies. Passage of the 1996 Farm Act, which is in effect until 2002, occurred at a time when commodity prices were relatively high and the effort to reduce government involvement seemed advisable, but with the steady decrease in farm prices and a series of natural disasters, the amount of direct government payment to farmers has actually tripled since 1997, with the largest percentage in the form of emergency aid. The 2002 Farm Bill follows the 1996 Farm Act in giving flexibility to the farmer. This Farm Bill, however, institutionalizes emergency aid, adding "counter-cyclical" payments that will provide support to farmers when prices fall below a certain level. This Bill also provides additional funds for conservation programs, but the receipt of these funds is not necessarily tied to commodity payments for farmers. The U.S. agriculture policy must continue to be examined to ensure a safe and abundant food supply, providing adequate profit to producers and protection to the environment.

In the European Community, the Common Agricultural Policy regulates prices and price supports with similar objectives and effects. Subsidies are paid well above world prices, and environmental costs are not fully considered. In both Europe and the United States, an additional strong consideration is the family farm. In Europe, the city dwellers have close ancestors who were farmers, and still retain an affection for the romance of rural life. The urban taxpayers seem more than willing to keep price supports high to keep production local. This results in subsidies as high as in the more productive United States. In Japan, the desire for local production of rice rests not only on tradition but also on the historic tendency for self-sufficiency. Because of strong local political pressures to support rice farmers, the government of Japan has the highest price supports of the industrialized world.

18.3 Environmental accounting is a new way to formulate government policies.

As mentioned in the previous section, most government payment policies favor external inputs and technologies. In the short term this means that farmers moving from high-input to more resource-conserving practices will incur transition costs. However, if the situation were reversed and farmers (or any industry) were charged for the environmental costs (and able to pass these costs on to the consumer), rather than encouraged to support nonsustainable practices because of price incentives, then resource-conserving technologies would be more readily and easily adopted. Another approach would be to provide farmers subsidies, grants, credit, or low-interest loans to encourage the adoption of resource-conserving practices.

Economists typically estimate the cost of a commodity such as maize by its trading value on the market (such as price per ton). This price is determined both by its cost of production and by the laws of supply and demand (if demand is high and supply is low, the price rises). The questions now being asked are, Does the production cost really reflect the value of finite resources such as land and water? This cost cannot be internalized by the producer, but through which mechanism should consumers be made to pick up the cost? How do we account for the environmental costs?

Environmental accounting has become a part of the economic spreadsheet for many companies and has become an influential part of policymaking procedures. Adoption of environmental accounting techniques increases the visibility of environmental costs to company managers and gives companies the opportunity to

- Significantly reduce or eliminate environmental costs
- Improve environmental performance

- Gain a competitive advantage
- Earn public goodwill

The easiest costs for a company to assess are the environmental liability costs; in the United States those would be costs associated with complying with federal and state laws such as the Clean Air Act and Clean Water Act, cost of cleanups at Superfund sites, and disposal of hazardous waste or removal of underground storage tanks. Soil contamination, groundwater and surface water contamination, and air emissions are also relatively easy for policymakers to assess and use in making feasible rules and regulations about environmental activities. Other approaches to environmental accounting seek to place value on land or water by comparing possible economic uses of the land, such as forest, recreation, maize production, wheat production, or fallow. Then the economists model some measure of ecological stress (for example, nutrient depletion leading to transition from maize production to fallow), and place an economic value on this change. The calculation includes not only the loss of income from the yield of maize but also the cost of cleanup (if polluted) and the cost of regulatory adherence.

Farmers who adopt a more integrated and sustainable system of farming are internalizing many of the agricultural externalities associated with farming and so could be compensated for effectively providing environmental goods and services. Providing such compensation or incentives would likely increase adoption of resource-conserving technologies. Soil conservation, lower use of fertilizers, and implementation of IPM are all examples of practices for which farmers could be compensated.

Carbon sequestration provides another recent example. In December 1997, 160 nations signed the Kyoto Accord agreeing to implement policies that limit atmospheric emissions including carbon dioxide. In 2001, U.S. president George W. Bush indicated that implementing these agreements was not in the best interest of the United States. Carbon dioxide concentrations in the atmosphere could also be reduced by increasing carbon sequestration, which could be achieved by

1. Increasing organic carbon production (increasing plant biomass, trapping carbon within plants)

2. Decreasing organic carbon mineralization (managing crops and soils to reduce conditions that break down plant residues)

3. Reducing soil erosion (keeping the carbon trapped in the soil)

These goals can be accomplished using such practices as converting less productive farmland to native grasses or forest, or using no-till practices—valuable options in their own right. However, ecologists have as yet reached no agreement about the magnitude and duration of carbon sequestration that could be achieved with each practice. It should be noted that increased biomass production increases nitrogen need by the plant. A consortium of Canadian power companies is already paying farmers in Iowa for carbon credits called *carbon-dioxide emission reduction credits (CERCs)*, allowing the Canadian energy companies to offset their emissions.

Some countries, notably the Netherlands and New Zealand, have implemented "Green Plans." The Netherlands' National Environmental Policy Plan (NEPP), initiated in 1989, is an ongoing plan to achieve more sustainable development. The Netherlands is home to the world's busiest port (Rotterdam), has a US$45 billion a year chemical industry and a large agricultural sector. The problems and needs facing this country are similar to those facing the world community. The NEPP is a comprehensive ecosystem-based plan that integrates all areas of environmental concern, as well as public health and the economy. The plan includes more than 220 policy changes and transforms the way the Dutch government, businesses, and

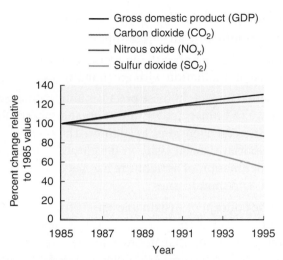

Figure 18.5 Sustainable policies do not counter economic growth. Since the inception of the Dutch National Environmental Policy Plan, relative level of greenhouse gases has decreased and gross domestic product (GDP) has increased. Contrary to many policymakers' beliefs, the Netherlands NEPP has shown it is possible to reduce harmful gas emissions by industries and still maintain economic vitality. The increase in CO_2 emissions is the result of the increased population and the increase in the number of privately owned vehicles. *Source:* National Environmental Policy Plan 3, VROM, February 1998. http://www.rri.org. Accessed November 1, 2001.

society approach the problems of environmental degradation. The first two phases of NEPP focused on industry's role in environmental degradation, but the most recent update, written in 1997, addresses such problems as overconsumption and other issues that involve the community at large. More than pure economics and policy, there is a general acceptance of such green plans because people place a value on the resources even if they are next to impossible to measure in economic terms. Most importantly, the projections show that the gross domestic product can continue to rise and that the economy will continue to expand, while at the same time society reduces the emission of gases (see **Figure 18.5**).

Assessing the value of land and water resources as they relate to ecosystem function is even more difficult. As a result, these "goods and services" of the environment are often given too little weight in policy decisions. It may be possible to put an economic value on ecosystem functions such as extractable raw materials, recreation, or food production, but it would be nearly impossible to apply a value to the functions of a healthy ecosystem such as the regulation of atmospheric chemical composition, soil formation, storage and retention of water, and buffering of the ecosystem during climatic fluctuations. These benefits are invaluable and are irreplaceable.

18.4 Grassroots movements have brought about an increased public awareness of agriculture's environmental impacts.

Plant growth and crop production are complex processes that depend on the interaction of many organisms. In general, the trend in modern agriculture has been to simplify the production practices by

- Substituting monoculture for crop rotation and intercropping
- Using herbicides and pesticides to combat pests rather than more complex biological control mechanisms

- Using genetically uniform plants that have a narrow genetic base
- Making fields bigger by eliminating all vegetation between them
- Using excess inorganic fertilizers rather than precise application or the more difficult to use organic manures in combination with green manures

The negative environmental impacts of these practices and the movement of people into areas that are more sensitive to human activity are of great concern. Recent global issues in food safety have further increased tensions. In addition, the economic plight of the small farmer has become an important social issue in developed and developing countries. Today the support or condemnation of agriculture has become a complex of biological, economic, social, political, and moral issues.

Confusion surrounds the concept of sustainable agriculture because sustainability and the means to achieve it often depend on a person's point of view. The U.S. Department of Agriculture has generated a definition that attempts to capture the farming as well as the societal aspects of sustainable agriculture. As defined in the 1990 Farm Act, sustainable agriculture means

> an integrated system of plant and animal production practices having a site-specific application that will (a) satisfy human food and fiber needs; (b) enhance environmental quality and the natural resource base upon which the agricultural economy depends; (c) make the most efficient use of nonrenewable resources and on-farm resources and integrate, where appropriate, natural biological cycles and control; (d) sustain the economic viability of farm operations and (e) enhance the quality of life for farmers and society as a whole.

Sustainable agriculture and the practices that it includes are known by many names—*alternative agriculture, regenerative agriculture,* and *biological farming.* No matter the name, sustainable agriculture should not be merely seen as a rejection of conventional agricultural practices, but rather acceptance of agricultural practices such as

- Integrated biological and cultural practices for pest control
- Tillage and cropping systems that minimize soil erosion and reduce water loss
- Reduction in use of inorganic fertilizers through more accurate and precise application and substitution with organic fertilizers and green manures
- Crop rotations to minimize buildup of weed, disease, and insect populations

These practices take into account the environmental and social consequences and try to reduce the negative impacts of intense agricultural production.

Developed countries are experiencing a transition in agricultural practices. Some of these changes, aimed at protecting food safety, water quality, and maintaining economic security, are legislatively mandated. Some producers adhere to the philosophy of sustainable agriculture for moral reasons, whereas others are looking to improve their profits by reducing the cost of inputs or by improving the market value of the products by meeting consumer demand for organic produce (**Figure 18.6**). Despite the absence of any scientific evidence, many consumers believe that organic farm produce is healthier and that such farming practices are more environment-friendly. The products of organic farmers are generally more expensive and the average consumer has not been willing to buy them. This increased concern of the public at large has caused farmers to be more aware of the adverse impacts of current practices. New best-practice recommendations are coming from researchers at publicly funded university and government research institutions.

Figure 18.6 Organic food sales have increased dramatically in the United States and Europe. Changing social attitudes have driven the demand for organic food and the development of government quality standards, even though no direct evidence exists that show such produce is healthier. *Source:* Courtesy of R. Morillon.

In making this transition, people need to maintain the benefits of modern agriculture while addressing its drawbacks. Sustainable agriculture does not mean a return to farming methods of the late 1800s. Sustainable agricultural practices are complex, incorporating traditional techniques that stress conservation with modern technologies such as certified seed, modern equipment for low-tillage practices, IPM that relies on biological control principles, and weed control that depends on a variety of techniques other than simply applying herbicides (see Chapter 17). An important goal is reduction of purchased inputs. Sustainable agriculture does not exclude use of genetically engineered crop strains, and crops could be engineered to address some of the practices that make agriculture unsustainable. The emphasis is on maintaining the environment, not on rules about what can or cannot be done. The most modern agricultural practices substitute management and knowledge for physical inputs: They use more information, highly trained labor, and management skills per unit production.

The poor in many parts of the world already practice a "low-input" agriculture that does not produce sufficient food and in addition is not sustainable because of the ecological damage it causes (see Chapter 5). Increased inputs are not necessarily the solution for these producers. These farmers need site-specific solutions that take into account the low capital, high labor availability, and diverse nutritional needs of the community. The marginal land on which these farmers are attempting to feed their families often has a low soil quality, steep slopes, and low water availability (**Figure 18.7**). Solutions in these regions must deliberately integrate and take advantage of native species and of beneficial biological interactions such as nitrogen fixation, and must have the support of the local community. In poor areas where sustainable solutions have been implemented, there has been a doubling or tripling of food production. In areas that benefited from the Green Revolution and in industrialized countries, it is predicted that more sustainable practices can be integrated while maintaining current yields.

Figure 18.7 Hillside farming is particularly prone to erosion. In this agroforestry demonstration plot in Burundi, the banana trees are intercropped with sweet potato. The bananas are planted on a small ridge to prevent water from running straight down the hillside. In this type of agroforestry, the native trees have been mostly cleared away, unlike the situation in Figure 18.1. *Source:* Food and Agriculture Organization.

18.5 Farming techniques to increase production and reduce environmental impacts.

A key feature of sustainable agriculture is the use of a variety of farming techniques that are appropriate for the specific situation and conditions of the farming operation. Generally natural processes are favored over external inputs. The farm is seen as a system where, for example, wastes from one component of the farm become inputs to another component. The goal of sustainable agricultural practices is to conserve existing on-farm resources such as nutrients, pest predators, water, and soil, or to use natural means that add more of these resources to the farm system.

Cropping Systems. Crop rotation is the recurring and planned alternating succession of crops on the same land in different years, and is the opposite of monoculture. Monoculture is planting the same crop on the same fields year after year. An example of a crop rotation appropriate for the U.S. Midwest is maize followed by soybeans; followed by a small grain such as barley, oats, rye, or wheat; followed by a perennial forage legume such as red clover or alfalfa. The economic and environmental benefits of crop rotations are well documented and are derived in part from the use of nitrogen-fixing legumes. Biological nitrogen fixation can supply all or much of the nitrogen required in some cropping systems. In our example, soybean, red clover, and alfalfa are legumes and can fix 45–225 kg of nitrogen per hectare per year, supplying their own needs and some or all of those of the next cereal crop (see **Table 18.1**).

Table 18.1	Nitrogen contributions of legumes to the subsequent crop.
Legume Crop	**Nitrogen Credit**
Alfalfa	
>80% stand	110–160 kg/ha
60–80% stand	70–110 kg/ha
<60% stand	0–70 kg/ha
Second year after alfalfa	
equals half of first-year credit	
Soybeans	0.5 kg for each 25 kg of harvest
Sweetclover (unharvested)	110–135 kg/ha
Red clover	35–70 kg/ha

Because legumes have nitrogen-fixing symbionts in their root nodules (see Chapter 12), they contribute nitrogen to the soil ecosystem even if the crop is harvested because a portion of the plant remains on the field. This unharvested nitrogen can be used as "credit" and reduces the need for nitrogen fertilizer on the subsequent crop.

Source: Adapted from R. E. Lamond, D. A. Whitney, L. C. Bonczkowski, and J. S. Hickman (1988), *Using Legumes in Crop Rotations*, Cooperative Extension Service, Kansas State University, Manhattan.

Even if the crop rotation did not include a legume prior to the maize crop, given the appropriate amount of nitrogen from other sources, the yield of maize in a maize-small grain rotation would be greater than if the maize had been grown continuously. The yield increase is an indirect result of other benefits conferred by a crop rotation in that the life cycle of pests adapted to a particular crop or cropping system is broken up by alternating the crops grown. For example, in wheat, which is a cool-season grass, cool-season grass weeds such as cheat (*Bromus secalinus*), jointed goatgrass (*Aegilops cylindrica*), and annual ryegrass (*Lolium multiflorum*) also benefit from the tillage, fertilization, and herbicide application that are needed to promote crop growth. As a result, these cool-season grass weeds thrive in continuously cropped wheat fields. These weeds could be easily controlled by rotating wheat with a crop that does not grow during the cool season, allowing the weeds to be mechanically controlled while there is no crop in the field, or by rotating in a cool-season nongrass crop, thus allowing use of an herbicide to control the prevalent weed population. The same advantage is true for diseases and insects that specifically target a crop. Another benefit comes from the observation that deep-rooted crops bring to the surface mineral nutrients that are then available to more shallow-rooted crops. Some crops have different nutrient needs that allow soil nutrients to be used more evenly.

The most widely used crop rotation in the United States is alternating maize and soybean. Although this rotation does provide the benefits of increased yield to both maize and soybean, the maximum benefit is not achieved, because both of these crops are row crops; that is, they are planted in straight, parallel rows anywhere from 50 to 100 cm apart. Crops that are closely sown (drilled), such as wheat, barley, oats, and rye, provide much better soil erosion control than row crops. Perennial forage legumes or grasses such as alfalfa, red clover, timothy (*Phleum pratense*), or bermudagrass (*Cynodon dactylon*) provide the maximum soil erosion prevention because once established, the soil is never exposed to wind or rain during the entire year. The maize-soybean rotation is the predominant crop rotation system in the United States, because these are the most profitable crops. Certain crops may be desirable in a rotation from an agronomic standpoint, but if there is no market and there are no government incentives, farmers won't use them.

In regions with an adequate growing season, it is possible to grow two or more crops on the same land within the same year, providing more yield per unit land area per year. This intense level of production depletes the soil of water and nutrients, and planning an optimal cropping system is crucial to sustainability of natural resources. Perennial crops such as forages or fruit trees are the most obvious examples of crops that can provide multiple harvests throughout the year. Sorghum (*Sorghum bicolor*), which is grown in the United States as an annual crop, is actually a perennial, and in the tropics and subtropics will regrow after being harvested. Sorghum is a multifunctional crop that can be used for food, feed, sugar, and biofuel. Double cropping is an example of growing two annual crops in the same field in the same year (**Figure 18.8**). A typical example is planting a short-season row crop such as soybean or sorghum into a field from which a small grain or forage has just been harvested. The soybean or sorghum is planted directly into the stubble of the previous crop. In this way, the soil is never exposed to erosion-causing elements.

In another system of multiple cropping, known as *intercropping,* farmers grow two, three, or even more crops on the same land at the same time. For mechanized systems, strip intercropping may be the most beneficial system to allow for mechanized planting, cultivation, and harvesting. In row intercropping, farmers plant two or more crops in the same row. In an intercropping system, the plants generally benefit one another. For example, the Native Americans planted the three sisters—maize, beans, and squash—where maize provided support for the beans, beans provided nitrogen for the corn and squash, and the squash covered the ground to conserve water and soil and suppress weeds. Other systems such as agroforestry do not necessarily use plants that provide mutual benefit but do maximize the use of land. In agroforestry (see Chapter 5), farmers plant a short grow-

Figure 18.8 Well-planned cropping systems can increase production while conserving natural resources. Double cropping allows farmers to grow two crops on the same land in the same year. Here sorghum is no-till planted into the stubble of recently harvested wheat. The wheat residue suppresses weeds, reduces evaporative water loss, and reduces soil erosion. No-till operations, however, also tend to rely more on chemical weed control than on mechanical control.

ing crop or a forage crop among a long-term crop such as fruit or nut trees or trees that can be harvested for firewood or lumber. This system often incorporates nitrogen-fixing trees and the leaves of the tree can be used as mulch to fertilize the soil, prevent weed growth, and conserve water and soil.

Integrated Nutrient Management. When crops are harvested, nutrients are invariably removed from the field. It is impossible to maintain crop production without adding nutrients.

Green manuring is a form of fertility maintenance that often depends on the nitrogen-fixing capacities of legumes (such as peas, beans, or perennial forages) that are tilled into the soil rather than harvested. In this way, all the mineral nutrients as well as the fixed nitrogen stay in the system, rather than being exported. A plant grown for green manure need not be a legume, but the use of a legume provides the added benefit of continuous cover, increased organic matter, *and* added nitrogen. The crop is often used as a mulch rather than being plowed under, so that it keeps the soil from being exposed to the forces of erosion (**Figure 18.9**).

Green manuring takes on a different form in Asian wetland rice cultivation. Here the rice fields are "inoculated" with the floating fernlike plant *Azolla pinnata* and with nitrogen-fixing bacteria. The symbiotic relationship between these two species is similar to that found in the nodules of legume plants and can provide more than 400 kilograms of nitrogen per hectare. When the Azolla plants die, they sink to the bottom and are slowly decomposed. This system provides an average rice yield increase of 700 kg/ha.

Animal manuring can make a significant contribution to soil fertility and to soil structure (see **Figure 18.10**). Manure contains nutrients essential to plant growth, and the organic matter improves the water holding capacity of the soil and reduces soil compaction problems. However, animal manures are generally low in the primary nutrients, nitrogen, phosphorus, and potassium. If, for example, 1 ton of cow manure contained 5.6 kg of nitrogen, 1.5 kg of phosphorous, and 3 kg of potassium, a farmer would need 19 tons of manure per hectare to provide enough nitrogen to produce a yield of 5 tons of maize per ha. Although a lot of animal manure is available in the United States, according to a 1995 study by the Council on Agriculture Science and Technology, only about 15% of the nitrogen needs of crops grown in the United States could be met using animal wastes. The actual amount of nutrient needs met using animal wastes is probably less, because intensive animal operations often have problems with storage and transport to locations where the waste can be used. In developing countries, manure is far more important. China, in particular, has been a leader in stimulating use of this on-farm resource.

Figure 18.9 Tomato crop grown with (top) and without (bottom) a mulching crop of vetch. Vetch is a legume, which adds nitrogen to the soil, giving rise to larger tomato plants that produce a more abundant crop. *Source:* U.S. Department of Agriculture.

Under intensive production systems, farmers can substitute increased knowledge and management for some nutrient requirements. The application of fertilizer should match the needs of the plant at the specific growth stages, but because of cost and time limitations, fertilizer is most often applied in fewer and larger doses, increasing the losses to the outside environment. The gathering and use of soil fertility levels and recommendations is a crucial means of reducing the amount of fertilizer needed (see **Box 18.1**).

Figure 18.10 Animal wastes provide a complete package of essential nutrients and increase soil organic matter. On a diversified farm, recycled animal wastes provide an inexpensive and readily available source of essential plant nutrients and increase the organic matter content of the soil, improving soil structure and water-holding capacity. Most animal wastes have a low concentration of nutrients, requiring heavy applications to meet plant growth needs. In most developed countries, adequate animal wastes are not available to meet plant growth needs. Like all fertilizing materials, the nutrient content of animal wastes should be measured before application. Mishandling organic fertilizers can also have negative environmental impacts. *Source:* Purdue University.

Conservation Tillage. How the soil is tilled, in which direction with respect to the slope of the land, how often, at which time of year, and whether crop residues are left on the field can all profoundly impact land productivity. Some tilling techniques combat water and wind erosion; others minimize weed problems. Tilling techniques must be adapted to the land, the crop(s), and the climate, and their effect on the land is independent of mechanization, unless extremely heavy machinery is used (**Figure 18.11**). The U.S. Department of Agriculture defined conservation tillage as any set of tillage and planting methods that leaves at least 30% residue cover (fodder from the previous crop) on the soil surface during the critical erosion period. The residue cover conserves soil and moisture, increases water percolation, and prevents weeds from growing. In high-input agriculture, conservation tillage also saves energy because most conservation tillage practices require fewer trips across the field. Conventional tillage practices, such as using a moldboard plow, allow mechanical weed control rather than using herbicides, aerate poorly drained soil, and bury crop residues to help control insects and diseases that would normally overwinter in the residue. However, when the residue is buried the soil surface is exposed to rain, wind, and sun, contributing to soil erosion and water loss.

Although conservation tillage practices offer the advantages of less soil erosion, greater water retention, and increased organic matter, some low-tillage systems require additional applications of herbicides to replace mechanical weed control. The residue cover can also harbor overwintering insects and disease-causing organisms, necessitating additional pest control later on. In addition, during cool wet springs farmers must delay planting because the cover keeps the soils cooler, inhibiting germination of the crop seed.

The area in the United States under conservation tillage practices, including no-till, mulch till, and ridge till, increased from 26% in 1990 to 36% (29.6 to 43.0 million ha) in 2000. The use of no-tillage or strip-till has increased most—from 6.8 to 20.6 million ha. Under a no-till or strip-till regime, seeds are planted into the soil together with the fertilizer, and weeds are controlled primarily with herbicides. The increase in no-till hectares

Box 18.1

Can Grain Production Be Increased by Using Less Fertilizer?

Precision farming, also known as *site-specific farming,* uses technologies such as GPS (global positioning systems), variable rate applicators, and computer databases to accurately place farm inputs such as fertilizer, herbicides, and manure in the amounts needed and at the specific places where they are needed.

Farmers and scientists have always known that soil conditions vary throughout the field. Some areas of the field are wetter than others, some areas have deeper topsoil, and some areas yield higher than others, and the farmer may or may not know the reason why. However, most fields have been treated as one homogeneous block—meaning that some parts get too little input to maximize production and some areas get more inputs than the plants can use.

To use the precision agriculture tools, the producer begins by determining the exact longitude and latitude of the field—that is, where on Earth this field is located. This is done using GPS equipment originally developed by the military to communicate with a series of satellites orbiting Earth. The field is then typically subdivided into blocks of 2–4 ha and soil samples are taken to collect data on nutrient level, pH, and organic matter levels. Yield monitors are installed on the harvesting equipment to record the exact location of the harvester and the yield at that location. When plotted, yield data show that yield variations across the field are continuous rather than in square blocks. Yield data are combined with soil test data and analyzed by agronomists to determine exact needs for each block of the field. Some blocks will be found to have lower yields because of inadequate nitrogen or phosphorus, or because of an improper soil pH. Variable-rate applicators, also equipped with GPS units, are used to apply the necessary amendments to improve the soil conditions within each block. The applicator adjusts the amount of input applied to the soil as it moves from one block to the next (see figure).

Knowing that soil conditions are highly variable and continuous *within* each block, scientists are working to develop an application system that allows for fertilizer ap-

A prototype of the sensor-based, variable-rate fertilizer applicator developed by engineers and agronomists to reduce the amount of inputs and increase yields.

plication, not on a hectare basis, but on a square meter basis. To take a soil sample every square meter in a field would be prohibitive, so scientists at Oklahoma State University are developing an on-the-go, sensor-based system that determines nutrient needs of the crop during the growing season. In this system the near infrared reflectance (NIR) of each square meter is determined just ahead of the fertilizer applicator. If there are just a few plants in the square meter, the sensor determines there is not a need to apply much fertilizer to this area. If there are a lot of plants with a low reflectance value, the sensor determines there is a need for more nitrogen. The scientists have shown that by measuring plant density and nitrogen needs of these plants on a square meter basis, rather than treating the field in large blocks, they can actually increase yields of the entire field while reducing the total amount of fertilizer.

is the result of improved planting equipment, adaptation of seed to cooler soil conditions and the advent of herbicide-resistant crops, which permit flexible in-season weed control. The majority of the remaining conservation tillage acres are under mulch till—a practice in which all the soil surface is disturbed for weed control and soil aeration, but with tillage tools that do not completely bury the residue such as sweep plows, chisels, or disks. This system permits more flexible nutrient and weed control methods.

Figure 18.11 Conventional tillage practices and cropping systems contribute to increased soil erosion. Regardless of the level of technology, poor agronomic practices contribute to increased environmental impacts. Fields on sloping lands should be terraced or vegetation left intact so as to reduce soil loss. Conservation tillage practices that leave residue on the soil to slow water flow and increase water infiltration should be used. *Source:* Food and Agriculture Organization.

Conservation tillage has not been universally adopted, because of the cost of modifying equipment or buying new equipment; in addition, some farmers will see reduced yields during cool wet springs or on poorly drained soils, and there are potential problems with perennial weeds such as johnsongrass (*Sorghum halepense*) or Canada thistle (*Cirsium arvense*), as well as with insects and disease. However, after at least three years most farmers actually begin to see a yield increase over that produced by conventional tillage. Often this is accounted for by increased soil biological activity, increased water retention, and the farmer's own adaptation. Even without an increase in yield, it is also important for the farmer to look at the net profit increase caused by fewer trips across the field, decreased labor, and time saved for other activities.

The benefits of contour cultivation to minimize erosion and leaving crop residues on the soil are well known to low-resource farmers (**Figure 18.12**). As mentioned at the start of this chapter, the farmers of Lempira do not burn the vegetation any longer but cut it down with machetes and plant their crops in this "rubble" of dead vegetation in the partial shade of some very large trees that were left standing. Contour cultivation and planting are also widely used in developed countries (see Figure 11.6). Planting crop rows perpendicular to the slope of the land reduces soil and water run-off. This practice is further enhanced with the use of terraces.

Increasing Genetic Diversity. The limited genetic diversity in crop species being grown across a particular region increases the risk of disease epidemics as well as the possibility of disease-causing agents to overcome the plant's resistance mechanism. If, for example, all the maize plants in a region are genetically identical, a pathogen able to attack one plant has an unlimited potential of spreading across the entire field, the entire county, state, or country. This is exactly what happened in 1970 to the U.S. maize industry, destroying 100% of the crop in some regions and reducing the total U.S. crop yield by 15%, demonstrating the potential threat of limited genetic diversity to food security.

Recent experiments in China show that growing a mixture of landraces of rice with different mechanisms of disease resistance to rice blast (*Manaporthe grisea*) can stop the spread of this devastating disease. Unimproved crop varieties, which are genetically heterogeneous, are called *landraces*. They are mixtures of genetically diverse plants with different forms of resistance. Why are these mixtures not preferred? Often the absolute yields are lower, but not in years when disease pressure is heavy. In the long term, the

Figure 18.12 Crop residues. Leaving the unharvested portion of the crop plant is a valuable means to recycle nutrients and increase soil organic matter. Crop residues left on the soil surface decrease the impact of raindrops and slow the flow of water on sloping land, thus reducing soil erosion. Surface residues also prevent weed growth, an important consideration for low resource farmers as shown in this picture. *Source:* Courtesy of CIAT.

yields of these landraces are higher than with homogeneous cultures. Harvesting may be a difficulty because of differing rates of maturity; furthermore, uniform quality cannot be guaranteed because some varieties may dominate the grain mixture. But in developing countries, where labor availability is high, hand harvesting is not an issue, and for noncommercial use, consistency of product is not as important. For areas that use harvesting machinery, experience has shown that the same benefits can be gained by planting mixtures in narrows strips and that a within-row mixture is not required.

Integrated Pest Management. Developed in the 1960s, and now widely accepted and adapted, integrated pest management, or IPM, is an example of a sustainable agriculture system that encompasses a variety of techniques aimed at controlling insect pests, as discussed in Chapter 16.

18.6 Is there room for GM technology in sustainable agriculture?

The goals of sustainable agriculture are to eliminate or reduce agronomic practices that degrade the environment, to minimize off-farm inputs while maintaining or increasing yields, and to increase the economic well-being of the community. There is no specific reason why GM technology must be incompatible with these goals.

The possible contradiction between sustainable agriculture and GM technology appears when one examines the most widely grown GM crops. In 2001, 68% of all cotton, 68% of all soybeans, and 26% of all maize grown in the United States were genetically engineered. These GM crops all represent examples that confer agronomic benefits to the producer, with the intended consequence of reducing the cost of inputs (insect control programs and herbicides). There are clear environmental benefits, such as a lower use of pesticides, a switch to biodegradable herbicides with lower levels of herbicides in the groundwater (**Figure 18.13**), and a switch to no-till or low-till agriculture. However, the genetic uniformity opens the door to rapid spread of diseases or pests should strains evolve that can overcome the plant's defenses. In the absence of a governmental system that taxes the environmental damage done by farming, for example, there may not be enough incentive for biotechnology companies to produce varieties that benefit the environment in new ways. Farmers would buy the more expensive GM seeds if they could then reduce their envi-

Figure 18.13
Percentage of Illinois water-monitoring samples containing major maize herbicides, 1999. The herbicides monitored were alachlor, metolachlor, atrazine, EPTC, butylate, acetochlor, and 2,4,D. These herbicides are not detected in most water samples. Under conventional maize, 12% of the samples had 2–4 parts per billion (PPB) of these herbicides and 4% of the samples had more than 4 PPB. Under glyphosate-tolerant maize, where glyphosate was the herbicide used, contamination of groundwater dropped dramatically. *Source:* Data from the Acetochlor Registration Partnership of Monsanto and Syngenta.

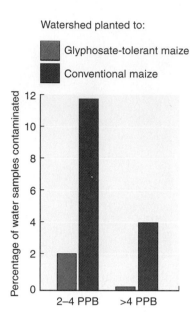

ronmental impact taxes because the GM seeds mitigated the impact of certain farming practices. Alternatively, the government could provide subsidies as it now does in the United States for erosion control. It is possible to create plants that extract certain toxic metals from the soil. However, this will only be done if the government mandates cleanup of toxic waste sites and provides funds to do so.

To be most beneficial to humanity, biotechnology researchers—in private and public sector alike—must broaden the application of biotechnology to work on plants that fit specific ecological niches and meet the nutritional needs of the people most in need of food. Such examples might be the development of an aluminum-tolerant wheat that will thrive in the low pH soils of the tropic regions or the recent development of a rice that contains the precursor of vitamin A (see Chapter 7). Currently, most biotechnology research is focused on a relatively small number of crops (maize, soybeans, rice, wheat, canola). To benefit people in developing countries, resources need to be devoted to the less widely grown but important crops such as cassava (*Cassava manihot*), millets, sorghum, yams, and pulses such as common bean and lentils.

The potential benefits of plant biotechnology are high, but are unlikely to be realized unless seeds are affordable. This will require heavy public investment by national governments and donors, as well as collaboration with the private sector in research and in distributing seed and technical advice.

18.7 An ecological approach could bring new solutions to sustainability.

Sustainable agricultural practices are actually more advanced than conventional high-input agricultural practices, because sustainable agriculture takes into account complex ecological interactions and attempts to assess benefits in terms of economic, environmental, and social impacts. Researchers and farmers have begun to look at a crop field as an ecosystem, often referred to as an *agroecosystem* (see also Chapter 5). The study of agroecology focuses on the processes that maintain natural ecosystems—nutrient cycling, predator–prey relationships, competition, and succession—and how these mechanisms can be enhanced for crop production. By understanding these processes, people can modify agroecosystems to increase production with fewer inputs and with fewer negative environmental consequences. This is seen as a means to achieve sustainable agricultural production.

According to well-known ecologist E. P. Odum, agroecosystems differ from natural ecosystems in several ways. In an agroecosystem,

- Productivity is enhanced by outside sources of energy such as human or animal power or fossil fuels.

- Diversity of plants and animals is greatly reduced.

- The dominant genetic variants are the result of artificial rather than natural selection.
- Level of input and balance of species are determined by external forces rather than internal forces.

The magnitude of the differences between an agroecosystem and a natural ecosystem depends strongly on the level and frequency of natural or human influence on the ecosystem. The final structure of the ecosystem is the result of endogenous biological and environmental features of the field and exogenous social and economic factors. Simplification has been a driving force of modern agricultural practices.

Agricultural systems can be described, studied, and designed using principles of ecology (see **Box 18.2**). Species diversity and genetic diversity within species are the basic components that promote a productive and stable ecosystem. Severe stresses such as a change in moisture, temperature, or light are less likely to harm the entire system, because in a diverse system there are alternative routes for the transfer of energy and nutrients. Therefore the system can adjust and continue to function after stress, with little or any detectable disruption. Similarly, internal biotic controls (predator–prey relationships) prevent the destructive oscillation of pest populations. Modern agricultural systems carry with them the disadvantages of an immature ecosystem similar to those found at the beginning of an ecological succession. In particular, these systems lack the ability to cycle nutrients, conserve soil, or regulate pest populations.

Two ecological hypotheses can explain the advantage of a polyculture system in reducing the level of insect pest populations. Compared to monocultures, polycultures provide more diverse pollen and nectar sources that can attract natural enemies of the crop pest and increase their reproductive potential. In addition, diversity of plant species increases the diversity of all herbaceous insects, which can serve as a food source for natural enemies, making them less likely to leave during times when the primary pest species are rare. A second hypothesis suggests that insects have a more difficult time locating and reproducing on their preferred hosts when these preferred plants are spatially dispersed and masked by the confusing visual and chemical stimuli presented by associated nonhost plants.

In his book *New Roots for Agriculture*, ecologist Wes Jackson proposed that for the Great Plains region of the United States, a permanent and stable agroecosystem of perennial plants should be established that mimics the prairies that once existed there. Prairies contained C_3 and C_4 grasses (see Chapter 10) as well as legumes and composites. These plants thrive at different times of the year and have different root systems and different nitrogen requirements. Some rely on symbionts for nitrogen and phosphate, others do not. A similar mix of perennial species, suitable for harvesting, could produce a stable ecosystem.

Although the prospect sounds attractive, many problems would need to be solved to make it a reality. Perenial crop plants would have to be bred specifically for this system, but currently most breeding efforts are directed to annuals. Because of their short life cycle, annual plants can be improved faster through traditional breeding. Because of the relatively lower yields, additional acres would have to be used for food production. In addition, it is not clear how fertilizers would be handled in such a permanent system. If the harvest removes the seeds with their abundant nutrients, these nutrients would have to be replaced in some way. A legume would help with nitrogen, but most other nutrients, including phosphorous and potassium, would still have to be replenished.

Box 18.2

Managing Biodiversity to Maximize Cattle Production and Wildlife Habitat

Rangelands are the dominant land classification throughout the United States (61%) and the world (70%), consisting primarily of native plants. The primary use of rangelands in the United States is for cattle production. Most techniques of rangeland management attempt to increase livestock production by decreasing the inherent diversity of plant species and variability of growth stages found on the rangeland, by controlled grazing, and by favoring the most productive, palatable forage species for domesticated cattle. However, this decreased diversity of plant species cannot provide the appropriate variety of habitats to support native animal and bird species. According to the North American Breeding Bird Survey, 70% of the 29 bird species characteristic of North American prairies have declined between 1966 and 1993. These grassland species are declining faster than any other group of terrestrial birds in North America.

Before European settlement of North America, bison were the primary herbivores on the Great Plains. Bison promoted plant diversity across the landscape by their intense, patchy grazing patterns, resulting in some areas that were in an early stage of natural plant succession and some regions that were late in plant succession. This diversity of plant species across the landscape was enhanced by naturally occurring fires. By studying bison grazing patterns in protected rangelands, researchers have shown that bison actually follow the fire pattern—heavily grazing the fresh new growth that appears on burned areas following fire. The differential grazing of the burned patches results in the accumulation of fuel and an increased probability of fire in the previously unburned patches—promoting use of all the land area over time. This pattern provides the bison with a continuous supply of fresh new growth, and because the burned and unburned areas are at different stages of plant succession, there is ample diversity of habitat to support the wildlife community.

This pattern of grazing provided the bison with adequate nutrition to maintain high reproductive and growth rates throughout the year. In contrast, under conventional rangeland management practices cattle require protein supplementation to maintain high reproductive and growth rates, especially during the winter dormant season.

To increase plant heterogeneity that supports increased wildlife and to maximize the use of rangelands for cattle production, scientists are now attempting to mimic this pattern of fire and grazing that occurred before European settlement of North America. Using patch burning to simulate prairie fires (see figure) and using moderate grazing intensity, these scientists have shown that like bison, domestic cattle preferentially graze the burned areas, and the level of cattle production can be maintained or even enhanced compared to conventional grazing patterns. In addition, habitat diversity noticeably increases. That is, some areas of the range are green and lush; other areas are tall and filled with a lot of dead grass. Depending on the time since burning, there is variation in the plant species. This increase in plant diversity and plant growth stages has led to a measurable increase in the number and diversity of insects and of birds. By mimicking the interactions of the native rangeland ecosystem, scientists have shown that it is possible to profitably raise domestic cattle while maintaining a thriving wildlife population.

Patch burning is used to simulate natural burning on the open prairie. The new grass that grows in these burned areas is selectively grazed by cattle, allowing other areas to grow up, providing diverse habitats for wildlife.

18.8 On-farm research allows all parties to find the best solutions.

M. S. Swaminathan, the former director-general of IRRI, has set up a new research institute in Madras, in southern India. According to Swaminathan,

> The greening of agriculture requires the greening of both technology and public policy. Producing more food and agricultural commodities from less land, water and energy is a task that will call for the integration of the best in modern technology, with the ecological strengths of traditional farming practices.

Scientists such as Swaminathan realize that a new approach is needed in which agricultural development is guided by new principles.

High-input agriculture, which supports 1.2 billion people, is epitomized by practices common in the developed countries. These countries are rich in natural resources and have other industries that employ a majority of their citizens. Specific policies support the current agricultural practices and provide subsidies to support the agricultural system. The Green Revolution was based primarily on the development of agricultural packages that included modern varieties of seeds, fertilizers, and pesticides, typical of high-input agriculture (see Chapters 4 and 14). Today, some 2.3–2.6 billion people are fed through agricultural systems characterized by the modern technologies brought about by the Green Revolution. However, an estimated 1.9–2.2 billion people have not benefited from these advances (see Chapters 2 and 4). These farmers lack the capital needed to purchase the inputs required to fully exploit the land for optimal crop production. As a result many people in the world have no choice but to move into more and more marginal lands to increase overall production. Many subsistence farmers in developing countries face a dilemma—either they must eke out a living with dignity on marginal land, thereby contributing to deforestation and land degradation, or they must go to the rapidly growing cities where there are few jobs and where life for them is often degrading.

Farmers are innovators. Most advances in agriculture have occurred as a result of experimentation (deliberate or not) by farmers—trial and error, observations, and the accumulation of knowledge over time. With the formalization of agricultural research, people have made rapid changes in agriculture that were necessary to keep pace with global population increases, but in bringing about changes, they have moved from a farmer-derived, bottom-up system to a top-down system in which agricultural research institutions or governments prescribe solutions that may or may not be acceptable or available to the farmer.

Practicing farmers have the most intimate and detailed knowledge of the farming system, ecological requirements, past failures, and community needs, and they will help to adapt new technology (**Figure 18.14**). The needs and solutions are often more complex than can be predicted by someone outside the system. For example, scientists from the International Crops Research Institute for Semi-Arid Tropics (ICRISAT) working at the edge of the Thar Desert in northwestern India organized a workshop for women farmers to help in selecting pearl millet (*Pennisetum americanum*). Women were invited because they play a prominent role in selecting and maintaining seed, and for their comprehensive knowledge about millet storage and use. In this study, the women's own varieties were grown in a blind experiment alongside commercial varieties and the selections of ICRISAT plant breeders. Although the ICRISAT varieties had higher seed yield, the women chose the local varieties that they felt were more adapted to the erratic environmental conditions,

Figure 18.14 Appropriate solutions require farmer input. The combined expertise of an extension worker and the long established experience of the indigenous farmer provide a fertile interaction for increasing local food production using feasible, adaptable, and acceptable crops and traditions. *Source:* Courtesy of CIAT.

had better forage qualities, and tasted better, demonstrating that consistent harvest and utility were more important than increased seed yields.

There are several problems with the top-down approach of agricultural development. First, the large-scale adoption of "technology-based" agriculture can lead to loss of knowledge about traditional farming methods, which people developed over many centuries and adapted to their physical, biological, and social environments. Second, another danger in ignoring traditional agriculture is that these methods may be applicable for designing sustainable systems in developed countries. Third, with the widespread adoption of new modern crop varieties, highly adapted and local varieties are often lost forever. This diverse germ plasm can be (could have been) the solution to many problems in both the developing and developed parts of the world.

A primary reason why agricultural relief projects fail is a lack of support from the local community. It is therefore crucial to get the farmers involved from the start (**Figure 18.15**). Otherwise, even when farmers are convinced to support a program or a practice (often through payment or price breaks), support will disappear if the program ends when the money runs out.

Figure 18.15 Discussion between a research director from an international research institute and local farmers in the Andean hillsides of Colombia. Such discussions help identify the needs of the farmers and provide valuable input for designing on-farm research projects. *Source:* Courtesy of CIAT.

However, if the people feel ownership of the program, they are more likely to keep it going after the transition support has left. Many international research organizations have discovered that their research efforts and recommendations for increasing yields and reducing environmental degradation were not adopted because (1) the farmers felt that other problems were of greater concern than those identified by the research organization; (2) the solutions meant additional cost and provided no short-term benefit such as fodder, fuel, food, or income; and (3) the farmers had little sense of technology ownership, because they had no part in developing the solutions. As a result, the spontaneous "farmer-to-farmer" dissemination of the knowledge was lost, requiring continuous education and promotion by extension educators. In southwestern Colombia, more "participatory" methods were used in a project where farmers were slow to adopt soil conservation recommendations such as terraces, strip cropping, or grass strips. Looking at demonstrations and surveys, researchers found that farmers were looking for a barrier plant that could stop erosion but that also had value in itself. Unlike the grasses chosen by the researchers, these farmers preferred sugarcane, because it has multiple uses, including as sugar and as forage for livestock. The same farmer-derived solutions drove success in Lempira, Honduras. It is hoped that including both the farmer and the community in identifying the applicable and realistic solutions will help to raise the productivity of these regions of the world that have until now been bypassed by the research achievements of the 20th century.

CHAPTER
SUMMARY

Improvements in agricultural practices have made it possible to feed the rapidly growing world population. However, these rapid changes have had unintended negative consequences—environmental and social—both on and off the farm. Society is increasingly concerned about these negative effects of agriculture. The agricultural practices and policies used in developing countries to increase production were modeled after those of developed nations. These methods were most useful in regions with adequate natural resources where one or two genetically improved crops could be grown in large areas. A large segment of the world population is still in need of solutions to increase food production. Researchers and policymakers need to examine current practices and the policies that support them, in both the developed and the developing nations, with the goal of making agriculture more sustainable. Many practices of sustainable agriculture are already in use, and more are being developed for the different regions of the world. A farm or a farming region can be considered as an ecosystem, and the laws that govern ecosystems are applicable. It may be possible to rely less on purchased inputs and construct more complex systems with more species that produce high yields. Genetic engineering has a place in sustainable agriculture if one keeps in mind the ultimate goal: producing enough food without eroding the resource base for that production. Everywhere, but especially in the developing countries, innovations must start with the farmers and with on-farm research.

Discussion Questions

1. Diversity of species and diversity within species is critical to consistent and durable crop yields. In what ways is diversity being lost, both off the farm and on the farm?

2. What are the social and environmental ramifications of only 2% of the U.S. population being involved in production agriculture?

3. In your community, what is the major source of water? What are the major uses of water in your area? What is the most important use of water—residential, industrial, agricultural? How should the price of water be determined? Explain.

4. As a consumer, how do your shopping habits promote nonsustainable or sustainable agricultural practices?

5. How can genetic engineering be used to promote sustainable production practices?

6. Do you believe the federal government should supplement farmer incomes, or should the business of farming be based on pure competition? On what ideas do you base your answer?

7. What would be the positive and negative effects of incorporating the costs of transition to sustainable practices into the cost of your food? Should farmers be given tax breaks and subsidies to convert to more sustainable production practices, or should legislation be passed to fine farmers for not converting to more sustainable production practices? Explain your reasoning.

8. If you needed to account for the environmental effects of a specific crop production practice, how would you go about measuring the impact and positive or negative value of that practice on ecosystem functions such as regulation of atmospheric chemical composition, water filtration, or soil formation?

9. The disadvantages of growing a single crop over very large acreages year after year are widely understood. Why, then, does this type of cropping system continue to dominate in developed countries?

10. From a producer point of view, what are the advantages and disadvantages of conservation tillage? Of intercropping? Of crop rotation?

11. What endogenous biological and environmental factors, and exogenous social and economic factors influence the structure of an agroecosystem?

12. How is the development of top-down solutions for farmers in developing countries similar to participation in government-sponsored conservation programs by farmers in developed countries?

13. Describe why you agree or disagree with the following statement: "Farmers in developing countries need greater access to fertilizers and better seed to produce more food."

14. Just as people are saving the seeds of locally adapted landraces handed down from generation to generation, how can they also save the knowledge about locally adapted farming practices?

Further Reading

Altieri, Miguel A. 1995. *Agroecology: The Science of Sustainable Agriculture*. Boulder, CO: Westview Press.

Bunch, Roland. 1985. *Two Ears of Corn: A Guide to People-Centered Agricultural Improvement*. Oklahoma City, OK: World Neighbors.

Centro Internacional de Agricultura Tropical. 2000. *CIAT in Perspective, Anatomy of Impact*. Cali, Colombia: CIAT.

Conway, Gordon. 1998. *The Doubly Green Revolution: Food for All in the 21st Century*. Ithaca, NY: Comstock.

Costanza, Robert, Ralph d'Arge, Rudolf de Groot, Stephen Farber, Monica Grasso, Bruce Hannon, Karin Limburg, Shahid Naeem, Robert V. O'Neill, Jose Paruelo, Robert G. Raskin, Paul Sutton, and Marjan van den Belt. 1997. The value of the world's ecosystem services and natural capital. *Nature* 387:253–260.

Food and Agriculture Organization Web site. 2000. http://www.fao.org/. Accessed October 28, 2001.

Holmes, Bob. 1993. Can sustainable farming win the battle of the bottom line? *Agriculture* 260:1893–1895.

International Center of Agricultural Research in the Dry Areas. 1995. *Farm Resource Management Program, Annual Report*. Aleppo, Syria: ICARDA.

International Crops Research Institute for the Semi-Arid Tropics. 2000. *Science with a Human Face, Annual Report*. Andhra Pradesh, India: ICRISAT.

International Institute of Tropical Agriculture. 1999. *IITA Annual Report*. Ibadan, Nigeria: IITA.

National Research Council. 1989. *Alternative Agriculture*. Washington, DC: National Academy Press.

Pretty, Jules N. 1995. *Regenerating Agriculture*. Washington, DC: Joseph Henry Press.

Tillman, David. 1999. Global environmental impacts of agriculture expansion: The need for sustainable and efficient practices. *Proceedings National Academy of Sciences* 96(11):5995–6000.

U.S. Department of Agriculture-Economic Research Service Web site. 2000. http://www.ers.usda.gov/. Accessed October 14, 2001.

Wood, Stanley, Kate Sebastian, and Sara Scherr. 2001. *Pilot Analysis of Global Agroecosystems*. Washington, DC: World Resources Institute and IFPRI.

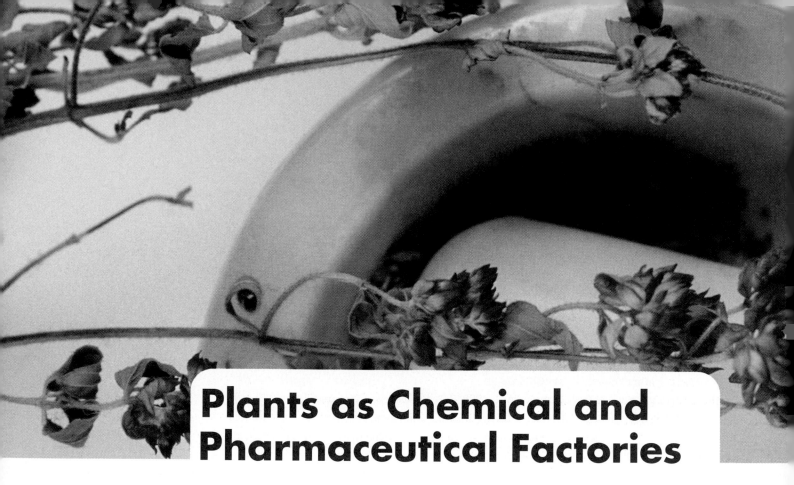

Plants as Chemical and Pharmaceutical Factories

John Ohlrogge
Michigan State University

Maarten J. Chrispeels
University of California, San Diego

Plants are amazing chemical factories! Provided only with the simplest and most inexpensive inputs of carbon dioxide, sunlight, water and minerals, plants produce thousands of sophisticated chemical molecules with different structures. A team of organic chemists starting with the same raw materials could never accomplish the same results in their lifetimes. Some of the chemicals that plants produce for their basic metabolic processes are the same ones found in all living organisms; amino acids, sugars, nucleic acids, and lipids are very similar or identical throughout the plant, animal, and microbial worlds. However, plants differ from other organisms in the vast diversity of additional products they produce. At least 50,000 different chemical structures have been characterized so far within the plant kingdom, and even the majority of plant species have not yet been closely analyzed! A single plant species produces only a fraction of this number of chemicals, but there are hundreds of thousands of different plant species, many of which produce chemicals that are unique to each one.

These so-called secondary metabolites (already discussed in Chapter 16) apparently are not essential to the life of the cell, and are often produced only in certain cell types or at certain times. They can be classified into broad molecular types based on their distinguishing structural features (**Table 19.1**). Why do plants produce these thousands of different chemical structures? For the vast majority of secondary products their specific function is unknown. In many cases, scientists have good evidence that they act as defense molecules or as deterrents against feeding by animals or microorganisms (see Chapter 16). Another possibility is that some compounds have no real function and represent the results of randomness in the evolutionary divergence of traits that are neutral for survival. There may be little or no negative impact on evolutionary fitness if a plant species uses some of its carbon and energy to produce additional structures.

In many cases, plants produce only small amounts of a secondary metabolite, but these metabolites can have extremely potent biological activities. Products that represent less than 1% of the weight of a leaf or root can make the organ toxic to a potential grazer. In other cases, plants produce secondary metabolites that accumulate to very high levels often in spe-

Table 19.1 Diversity of chemical structures produced by plants

Molecular Class	Number of Known Structures	Examples
Isoprenoids	>25,000	Menthol, turpentine, rubber
Alkaloids	>12,000	Caffeine, nicotine
Phenolics	>8,000	Vanillin, anthocyanins
Fatty acids and derivatives	>500	Castor oil, urushiol (poison ivy toxin)
Miscellaneous	>5,000	Sorbitol, guar gum

cific cell types. The plant secretory glands found in specialized epidermal hairs of species such as mint represent a cell type able to produce large amounts of relatively pure organic compounds (such as menthol). Interestingly, epidermal hairs of other plant species also represent some amazing chemical factories. For example, hairs of some tomato species secrete sticky glucose fatty-acid esters that can represent 25% of the leaf dry weight, and cotton fibers represent specialized cellulose fiber factories that derive from epidermal hairs.

19.1 Harvesting biochemical diversity: Can (green) plants replace (chemical) plants?

One of the major goals of 21st-century society will be to develop renewable and sustainable resources to replace limited petrochemical reserves. This goal is schematically illustrated in **Figure 19.1**, and the following discussion outlines some factors that will affect the development of a more bio-based chemical industry.

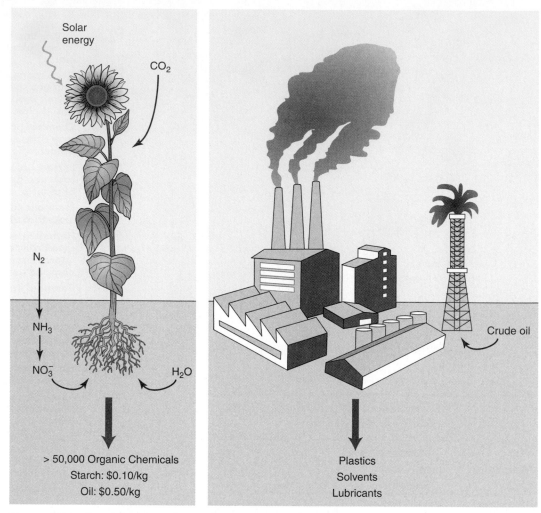

Figure 19.1 Can plants replace plants? In green plants the inputs are carbon dioxide and solar energy, in chemical plants the input is petroleum.

Today people think of agriculture largely as a source of food for human consumption and feed for animals. There are of course many other uses for crops, such as the production of cotton, linen, or jojoba oil, but these represent only a small fraction of overall agricultural production. In the United States, nonfood and nonfeed uses of field crops account for only 10 to 20% of the total value of field crops. Thus, despite the amazing potential of plants, the agricultural products that humans now use are much more limited in chemical variety and in human applications. Seeds, harvested for their food value, constitute the economic value of most crops. Therefore, agricultural crop production largely takes the form of the starch, proteins, and oils that are the major constituents of field crop seeds.

For most of human civilization, plants provided a much wider range of products. Many ancient civilizations had highly evolved uses of plants to provide medicines, dyes, and other chemicals. One hundred years ago, a major proportion of human clothing, fuel, dyes, medicines, construction materials, and industrial chemicals was still derived from plants, and these products represented very substantial aspects of the agricultural economy. This situation changed dramatically over the past 50 to 100 years. Synthetic rubber and syn-

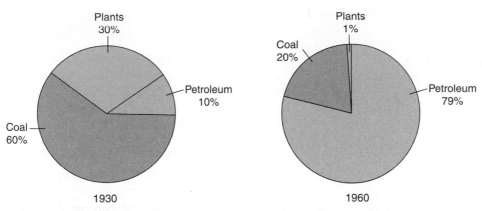

Figure 19.2 Change in the primary sources of industrial chemicals in the United States between 1930 and 1960. Note the rise of oil and the disappearance of plants and decreased importance of coal over this 30-year period. As of 2000, petroleum provides over 95% of organic chemicals used in the United States.

thetic fibers for clothing (nylon, polyesters, and so on) are just two examples of petroleum-derived products that were developed to replace products once obtained exclusively from agriculture. Over a 30-year period—between 1930 and 1960—the sources of industrial chemicals changed from coal and plants as the dominant raw materials to petroleum as the new feedstock (**Figure 19.2**). As of 2000, over 95% of organic chemicals that society uses are derived from petroleum, not from plants.

Why has the chemical industry switched to petroleum as the almost exclusive source of starting raw materials? Two major factors brought this about. First, the development of the automobile, which led to enormous demand for inexpensive liquid fuel, coincided with the discovery of vast petroleum reserves. Second, the chemical industry invented new processes that permitted the inexpensive conversion of oil into more valuable chemicals that replaced many products that society had long derived from agriculture. Furthermore, during the time the chemical industry was rapidly expanding, petroleum was a much less expensive raw material than agricultural products. For example, in the 1940s and 1950s, a kilogram of maize cost four to five times more than a kilo of crude oil. The development of the chemical industry, and in particular the plastics industry, is an amazing scientific and industrial success story that has led to the creation of an industry that produces US$110 billion in products per year in the United States, compared to US$90–100 billion for major U.S. field crops.

Technological advances and economic factors during the 20th century led people to replace many nonfood agricultural products with petroleum-derived products. Many believe this trend will be partially reversed in the 21st century by genetic engineering of plants and the continuing decline in the price of agricultural products (see **Box 19.1**). Over the past 50 years agricultural products have become less expensive because of the increased productivity of crops and decreased labor input, whereas the price of crude oil, even if one ignores the sharp peak in the 1980s, has been steadily rising (**Figure 19.3**). Relatively speaking, plants are a bargain. Additional factors that could bring about a switch from oil to plants may be the environmental, political, and social benefits that could accompany such a changeover. Are the trends shown in Figure 19.3 likely to continue? Petroleum is a limited commodity produced in a relatively small number of countries, and production at present levels cannot be sustained. Therefore, its cost is expected to continue to rise in the future and known supplies will likely be exhausted within the next 75–100 years, unless automobiles use a different source of energy. In contrast, over the past 50 years, agricultural

Box 19.1

Benefits of Using Plants to Produce Chemicals

Several additional factors may help encourage the broader development of plants as chemical factories.

"Green Chemistry" Potential for Reducing Greenhouse Gases and Environmental Damage

Using plants rather than petroleum to produce chemicals can help the environment in different ways. First, global warming is believed to be caused largely by huge amounts of carbon dioxide released from burning fossil fuels. When crops are grown to produce chemicals they fix carbon dioxide, helping remove the excess from the atmosphere. Second, production of chemicals from plants will in many (but not all) cases use less energy and release less carbon dioxide than producing the same chemicals from petroleum. And third, many plant-produced alternatives to petrochemicals are biodegradable, which, in the case of plastics, can reduce their impact on landfills and/or esthetics. Balanced against these positive benefits are concerns that expanding agriculture to produce chemicals could eliminate more undeveloped or recreational land.

Reduced Dependence on Imported Petroleum

The United States and most other nations must pay huge sums of money to import petroleum. These costs lead to negative trade balances that leave less money within the economy for economic investments, growth, and social needs. Using local agricultural systems to replace some of the imported petroleum will leave more money in the local economies and overall more evenly distribute capital throughout the world.

Building a Renewable/Sustainable Economy

Sustained economic growth depends on having a secure supply of raw materials. The growing world population and its increasing demands on a decreasing supply of petroleum can lead to inflationary or other negative economic trends. Using agriculture as an alternative to petroleum will help build material resources that are renewable, more diversified, and sustainable.

Diversification and Enhancement of Rural Economies

The tremendous success of agriculture in producing surplus commodities at low cost would seem at first to benefit rural economies. Paradoxically, however, this success and the resulting low prices for farm products have made it harder and harder for small farmers all over the world to make a living from their farms. In 1998, the number of U.S. farms was only 35% of the number in 1930. Similar trends have occurred all over the world, and one result has been migration of major segments of population from rural areas into overcrowded urban centers. If agriculture can produce new products with higher value, this will increase the profitability of small farms, bring more money into rural economies, and help to slow the concentrations of populations in cities.

production has continued to increase slightly faster than demand and crop prices have generally declined. Whether this trend continues, it seems almost certain that for the next 50 years or more agricultural raw materials will be less expensive than petroleum, a very different situation from when the chemical industry began its major expansion. A detailed comparison of some recent prices of chemicals from plants and from petroleum is in **Table 19.2**.

Table 19.2	Approximate cost of phyto- versus petrochemicals		
Phytochemical	**US$/kg**	**Petrochemical**	**US$/kg**
Corn	0.10	Crude oil	0.12
Starch	0.11	Gasoline	0.2
Vegetable oil	0.55	Benzene	0.5
Glucose	0.22	Plastic	1.0
Ethanol	0.30		

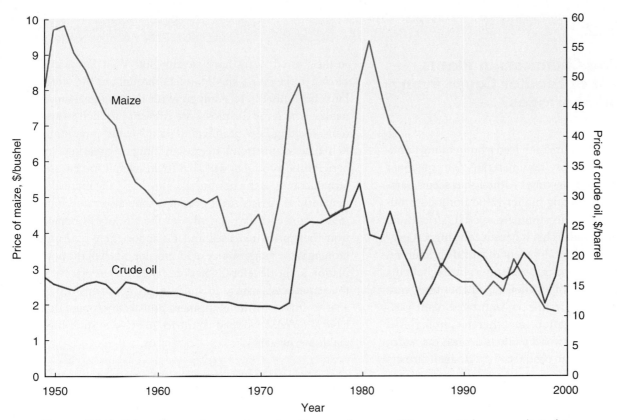

Figure 19.3 Prices for maize and crude oil over the past 50 years. Values are adjusted to constant 1996 dollars.

If plant material is now less expensive than petroleum, why has the chemical industry not switched back to plants? A major reason is that the industry has over 50 years of experience using petroleum as its raw material and has invested huge sums in chemical plants that convert fossil hydrocarbons into plastics and other consumer goods. Additional comparisons of the advantages and disadvantages of fossil fuels versus plants as sources of raw material for the chemical industry are shown in **Table 19.3**.

Table 19.3 **Industrial chemicals from fossil versus plant sources**

Petroleum-Based Raw Material	Plant-Based Raw Material
50–100 years of chemical experience in industry	Less chemical experience
Raw material is imported, and costs have increased	Raw material is domestic, and costs have declined
Petrochemical industry is major source of pollution	Less pollution
Chemical diversity is limited: Primarily highly reduced carbon	Chemical diversity is almost unlimited

Box 19.2

Will Producing Chemicals in Plants Raise the Cost of Food or Cause Even Greater Food Shortages?

How can agriculture both feed a burgeoning population and supply raw materials for consumer goods at the same time? Although it seems paradoxical at first, producing higher-value nonfood products in plants is likely to stimulate overall agricultural output, including foods. This is because the biggest negative factor hindering development of world agricultural systems has been low food prices. Farmers throughout the world, but particularly in less developed countries, cannot afford to invest in fertilizers, improved plant varieties, equipment, irrigation, or other technologies if surplus commodities on world markets depress the return they receive for their farm products. Just as small farmers

in the United States have steadily sold their farms and moved to the cities, small farmers throughout the world have been unable to compete with the huge agribusinesses, which use the most cost-efficient production systems. Making new products in plants could potentially help reverse this trend, because finding new markets for agricultural products will lead to an overall increase in farm incomes. If the new markets are large, the increased demand for agricultural products may also raise food prices. However, farmers all over the world will benefit from these price increases, and it is important to note that farming is the major source of income for much of the population in the developing nations. With higher prices for their products, farmers in undeveloped countries will be able to invest in the equipment and technologies that have allowed developed countries to achieve such high yields per hectare.

19.2 Gene manipulation will let crop plants produce specialty chemicals and permit production of biologically based plastics.

The surplus capacity for agricultural production in industrialized countries and the innovation of plant genetic engineering are leading to new ideas about how people might use agriculture to benefit society. A recent coalition of government, industry, and academic groups suggested that the United States should attempt to use biological sources such as crops to supply 10% of industrial chemical needs by the year 2020 and to increase this to 50% by 2050. Can these ambitious goals be achieved? Ultimately, agriculture will produce almost all biological resources, and plant genetic engineering will be an essential tool for increasing use of renewable resources. An underlying concept of plant genetic engineering is that limitations to large-scale production and use of valuable plant chemicals can be overcome by transferring the genes for their biosynthesis from wild plants into crop species. In this way, much previously untapped chemical biodiversity can be harnessed for society. What are the prospects that biotechnology can create new uses for crops, and could this affect the cost of food (see **Box 19.2**)?

Currently, the major process for producing chemicals biologically is microbial fermentation. Worldwide, approximately 15 million tons of products such as ethanol, citric acid, lactic acid, and others are now produced by fungi or bacteria grown in huge 10,000-liter (or larger) stainless steel fermentors. The energy (carbon) source for growing the microorganisms in these fermentors is almost always derived from plant material, usually from maize starch. The microbes are always very highly selected or have been genetically engineered so that they efficiently convert feedstock (glucose) into end product (ethanol or other). Maize is the number one U.S. field crop, and the ability to produce maize starch at a price of less than 15 cents per kg has led to many uses of starch other than as food. Roughly 8% of the U.S. maize crop, or 14 million tons, is used in fermentation processes (**Figure 19.4**). Approximately 1 billion U.S. gallons of ethanol are produced in this way for automobile fuel (compared to 130 billion gallons of gasoline used per year in the United States). By increasing demand for maize, production of fuel ethanol raises the price of maize (about 1 cent per kg of grain) and helps provide additional markets for excess U.S. agricultural capacity. Improvements over the past 10 years in ethanol yields and recovery have converted the overall ethanol production process from a net energy consumer to a positive energy balance. However, ethanol's ability to compete with low-cost petroleum still requires tax advantages to be practical.

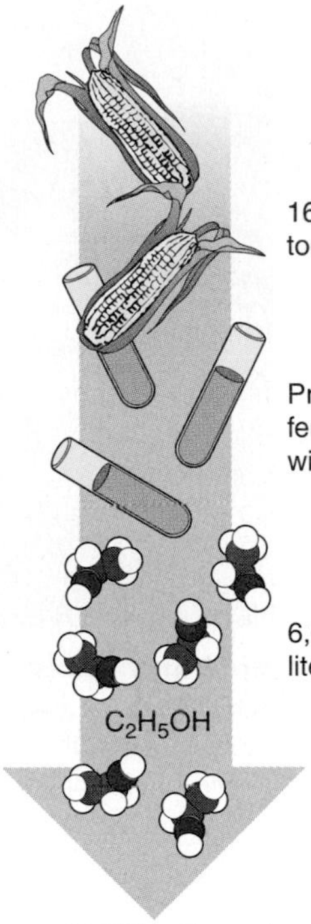

The production of millions of tons of ethanol from maize has demonstrated the feasibility of large-scale conversion of crops into new products. In addition to ethanol, some new fermentation products are now becoming available; this will lead to higher-value products from agriculture that can fully compete with petroleum. Large-scale production facilities have recently been established for two precursor molecules

16.9 million tons of maize

Processed via fermentation with novel enzymes

6,000 million liters of ethanol

C_2H_5OH

60 billion liters of blended gasoline

Figure 19.4 Conversion of maize to ethanol by large-scale fermentation. Approximately 7% of the U.S. maize crop is used for production of 3.5 billion liters of ethanol for use in blended fuels for cars and trucks.

that can be polymerized to make plastics. A plastic with the trade name NatureWorks is produced by CargillDow Polymers. The NatureWorks process is based on the fact that bacteria that are fed glucose derived from starch can produce very high yields of lactic acid. After recovery from the fermentor, technicians polymerize the lactic acid to polylactic acid (PLA), a biodegradable plastic with many attractive properties for products such as carpet fibers and thin films. By using a biologically produced feedstock (glucose), PLA uses 30 to 50% less fossil fuel than is required to produce conventional plastic resins. In 2001, a large-scale facility began production of 150 thousand tons per year. Analysts expect that up to 1 million tons of this plastic may be produced in 2004 at a cost comparable to most petroleum-derived plastics.

For another example, fermentation of glucose can produce 1-3 propane diol, which after combining with terephthalic acid can be polymerized to produce a plastic useful for fiber production. DuPont will market this bio-based plastic under the name Sorona. One key to economic production of both of these bio-based plastics has been genetic engineering of bacterial strains to achieve very high yield of the precursors for polymerization. In the case of PLA, almost 70% of the carbon from the plant glucose feedstock ends up in the final plastic.

An excellent application of PLA plastic films would be as a replacement for the polyethylene films now used to solarize soil for vegetable production. Covering the soil with plastic film (**Figure 19.5**) reduces water evaporation, prevents weeds from growing, and, most importantly, kills harmful nematodes that accumulate when the same land is used year after year. This new method to prevent damage from nematodes could replace chemical fumigation of soils with methyl bromide, a potent pesticide that kills harmful and beneficial insects alike. If these films were made from a biodegradable plastic, the farmer could simply plow them under at the end of the growing season while preparing the new planting bed.

Figure 19.5 Transparent polyethylene film is applied to solarize the field of an organic vegetable farm in the San Joaquin Valley of California. If polyethylene films could be replaced by biodegradable PLA plastics, they could simply be plowed under at the end of the growing season, instead of being hauled off to landfills. *Source:* Courtesy of J. Stapleton.

19.3 Chemical production within a plant may offer economic advantages over fermentation.

Although fermentation can produce some chemicals at a relatively low cost, the expense of building and maintaining large fermentors increases the cost of the products. For this reason, most products sell for several-fold higher prices than starch or glucose. An ideal process would produce the desired chemical in the plant, where the major energy input is sunlight, and would require only minimal further processing after extraction from the plant. The production of the biodegradable plastic polyhydroxyalkanoate (PHA) in plants represents one effort to reach this goal. Many species of bacteria synthesize and accumulate biodegradable plastics called *polyhydroxyalkanoates* at levels up to 80% of their dry weight. One such PHA, called *polyhydroxybutyrate* (PHB), is produced commercially by fermentation with the bacterium *Alcaligenes eutrophus*. PHB combines biodegradability with water resistance, making it a suitable polymer for a wide variety of uses. The major drawback of bacterial PHB is its high production cost, making it substantially more expensive than synthetic plastics and thereby restricting its large-scale use. Could large-scale production of such plastics in plants rather than in bacteria lower their cost? In 1992, scientists at Michigan State University demonstrated that expression in the plant *Arabidopsis* of genes from the bacterium *Alcaligenes eutrophus* could lead to production of the plastic polyhydroxybutyrate (PHB) in plants (**Figure 19.6**). This landmark advance demonstrated that plants can be engineered to produce entirely new chemical products using genes from distant organisms. However, growth of the plants was stunted. In a 1994 study, three genes from this bacterium were engineered so that the proteins they encode would end up in the chloroplasts. The genetically engineered *Arabidopsis* plants produced PHB inclusions exclusively in their chloroplasts (**Figure 19.7**). The amount of PHB detected in fully expanded leaves of the plants was up to 10 mg/g fresh weight (about 14% of dry weight). No significant deleterious effect on growth or seed yield was detected in these plants. In response to these discoveries, industry has begun to explore production in plants of polyhydroxyalkanoate polymers that have physical properties superior to those of PHB.

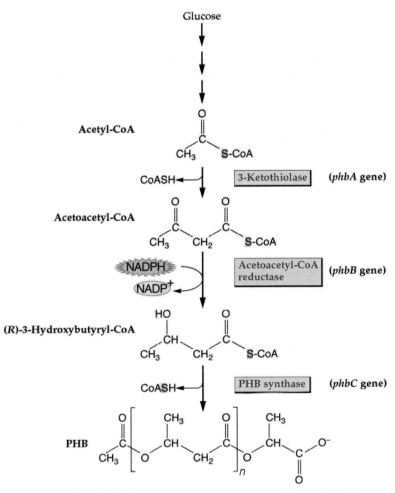

Figure 19.6 Pathway of polyhydroxybutyrate (PHB) synthesis in chloroplasts of genetically modified plants. The pathway begins with acetyl-CoA, which is the normal substrate for fatty acid synthesis in chloroplasts. Arabidopsis was transformed with three genes from the bacterium *Alcaligenes eutrophus.* Introducing the *phb A, phb B,* and *phb C* genes from the microorganism, modified so that the enzymes encoded by these genes are transported into the chloroplasts, then allows the chloroplasts to make PHB. *Source:* B. B. Buchanan, W. Gruissem, and R. L. Jones, eds. (2000). *Plant Biochemistry and Molecular Biology* (Rockville, MD: American Society of Plant Physiologists), Figure 10.74.

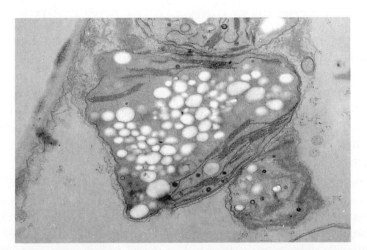

Figure 19.7 Electron micrograph of a chloroplast in a transgenic Arabidopsis plant, showing the presence of PHB globules. The plants accumulated PHB up to 14% of the leaf dry weight. Despite this large investment of energy into PHB, they grew normally. *Source:* B. B. Buchanan, W. Gruissem, and R. L. Jones, eds. (2000). *Plant Biochemistry and Molecular Biology* (Rockville, MD: American Society of Plant Physiologists), Figure 10.75.

19.4   Extraction, purification, and energy costs can greatly influence the success of plant-produced chemicals.

Although at first it seemed that producing plastics directly in plants could offer the best approach toward a renewable, "green," chemical industry, many factors determine the balance of costs and benefits. To recover plastic from plants will require new extraction facilities that use organic solvents to extract the plastic. Calculations of the energy inputs needed to grow and harvest the crops, to operate the extraction facilities, and to recover the solvents suggest that the amount of fossil fuel consumed may actually be higher than if the same plastic were produced from petroleum. Such calculations require a number of assumptions that depend on future energy costs. If the energy used to run the extractions were derived from plant material, such as the crop residues after the plastic is extracted, the balance could again swing in favor of the plant-based production system. These considerations emphasize the complexities in judging the balance of costs and benefits for each chemical production system.

Plastic production via fermentation offers one useful comparison of how extraction and purification costs impact the cost of two biodegradable plastics produced biologically. PHA plastic (biopol) produced in bacteria at a level of 80% of their dry weight costs US$4–5/kg. In contrast, polylactic acid (PLA) plastic is produced by chemical polymerization from lactic acid that is first produced by bacteria. PLA plastic can be produced for US$1–2/kg. This substantial price difference is in large part caused by the extraction and purification costs associated with recovering PHA polymer from bacterial cells, whereas the precursor of PLA is soluble and can be recovered from the fermentor at low cost. Thus, although all steps of PHA synthesis occur biologically, the extraction/recovery costs raise the final cost of the product above that of PLA, which requires two biological production steps (in plants and in bacteria), followed by a chemical polymerization. In many cases, producing a monomer or precursor in plants that is easily recovered, rather than producing a complex polymer, will lower overall costs. If the desired product can be recovered using existing plant-processing technology (such as oil extraction or wet milling), the economics of chemical production in plants may be even more favorable.

19.5   Plant oils can be engineered for new industrial uses.

Different plant species use different polymers, usually oil or starch, to store energy needed for seedling growth (see Chapter 9). In oilseeds, up to 50% of the seed dry weight is in the form of oil. Some plants also produce large amounts of oil in other organs. For example, olive, avocado, and oil palm fruits contain high levels of oil. Here oil probably serves as an animal attractant, to aid in seed dispersal. Plant oils are triacylglycerols, also called *triglycerides* (see Chapter 7) and consist of three fatty acids attached to glycerol (**Figure 19.8**) that accumulate in discrete subcellular organelles called *oil bodies*. Oil bodies are droplets of vegetable oil surrounded by a monolayer of phospholipids.

Vegetable oils are a major commodity, with a well-developed commercial infrastructure for production and use, supplying both food and industrial needs. World vegetable oil production, from soybean, palm, rapeseed (canola), and other crops is over 100 million metric tons per year and is valued at approximately US$50 billion in annual oil sales. The primary use for vegetable oils today is in the food industry and includes salad oils, margarine, and oils used in frying and baking. In most plants, the same five or six fatty acid structures that are also found in the phospholipids in the cell membranes occur in the triacylglycerols found in seeds. These 16- and 18-carbon fatty acids (primarily palmitic, oleic, and linoleic acids) are the major constituents of the vegetable oils that are consumed as foods.

In addition to food uses, about 30% of vegetable oils that are produced today are used by the oleochemical industry for hundreds of products such as soaps, detergents, paints, lubricants, and polymers. In many cases, the fatty acid composition of the oils used for these applications differs from that found in edible oils, and these different structures lead to special applications. For example, the tropical oils from coconut and palm kernel are rich in lauric acid, which is a 12-carbon saturated fatty acid. The properties of lauric acid lead to a balanced solubility in water and oil, which makes it ideal for production of soaps and detergents. As a result, the United States imports up to US$400 million of these tropical oils for use largely in soap and detergents. Thus, although these oils are edible, their major use is not for food.

Although most plant species store oils with the same five or six fatty acid structures in their seeds, analysis of seeds from more than 10,000 different plant species has revealed several hundred fatty acid structures that are considered "unusual." The chain lengths vary from 8 to 24 carbons and the number of unsaturated double bonds varies from zero to 4, or more. In addition, hydroxy, epoxy, acetylenic, cyclic, and conjugated unsaturated groups occur. As with lauric acid, the unique properties of these fatty acids have led to a number of industrial applications (**Table 19.4**).

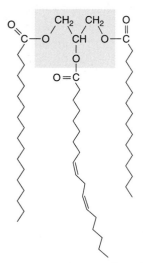

Figure 19.8 Space-filling and conformational models of the structure of tri-acylglycerol, the major form of carbon and energy storage in oilseeds. In each case, the glycerol portion of the molecule is at the top. The fatty acid on the left has 18 C atoms and no unsaturated bonds (stearic acid); the fatty acid in the center also has 18 C atoms and two unsaturated bonds (linoleic acid); the fatty acid on the right has 16 C atoms and no unsaturated bonds (palmitic acid).

For over 100 years, chemists have explored plant fatty acid structures as interesting alternatives to petroleum. Thus, the chemical industry has a rich knowledge about the properties and chemical potential of plant oils. However, in most cases these unusual fatty acids are not available in large quantities and at a low enough cost to compete with petroleum alternatives in large-scale uses. The desirable fatty acid structures are generally

Table 19.4 Some specialty uses of plant fatty acids and oils

Lipid Type	Example	Major and Alternative Sources	Major Uses	Approx. U.S. Market Size (10³ t)	10⁶ US Dollars
Medium chain (C8–C14)	Lauric acid	Palm kernel, coconut, Cuphea	Detergents	640	320
Long chain (C22)	Erucic acid	Rapeseed, Crambe	Lubricants, nylon, plasticizers	30	80
Epoxy	Vernolic acid	Epoxidized soybean oil, Vernonia	Plasticizers	64	64
Hydroxy	Ricinoleic acid	Castor bean, Lesquerella	Lubricants, coatings	45	40
Trienoic	Linolenic acid	Flax	Coatings, drying agents	30	45
Low melting solid	Cocoa butter	Cocoa bean, illipe (*Shovea stenoptera*)	Chocolate, cosmetics	100	500
Wax ester	Jojoba oil	Jojoba	Lubricants, cosmetics	0.35	

produced only in wild plant species that produce low yields if planted in fields. Although efforts to domesticate some of these species are under way, this may take decades and may meet insurmountable barriers. Plant gene technology offers the potential to move genes that control production of these high-value fatty acids into high-yielding and well-developed oilseed crops such as canola.

High-Lauric Canola Oil: A Success Story in Genetic Engineering of Oils

The first commercial product to result from changing the composition of a plant seed via genetic engineering is high-lauric-acid canola oil. Lauric acid is found at a very high level in tropical oils, but could temperate crops be genetically engineered to produce the same high level of this short fatty acid? Scientists at Calgene, a California biotechnology company, discovered the biochemical pathway responsible for lauric acid synthesis. They used extracts of seeds from the California bay tree; these seeds accumulate high levels of lauric acid–containing oils. The scientists cloned the gene for the critical enzyme in the pathway, introduced it into canola, and dramatically changed the spectrum of fatty acids produced in the canola seeds (**Figure 19.9**). In 1995, industry achieved the first commercial production of a genetically engineered oil by extracting 500 tons of oil from canola seeds engineered to produce an oil with 40 to 50% lauric acid.

Figure 19.9 Genetic engineering of canola oil that is high in lauric acid, a fatty acid with 12 carbon atoms. By introducing a single gene from the California bay tree, the canola oil was changed from containing 60% oleic acid to 60% lauric acid. This new canola oil resembles the oil found in coconut and oil palm. *Source:* Courtesy of T. Voelker, Calgene/Monsanto.

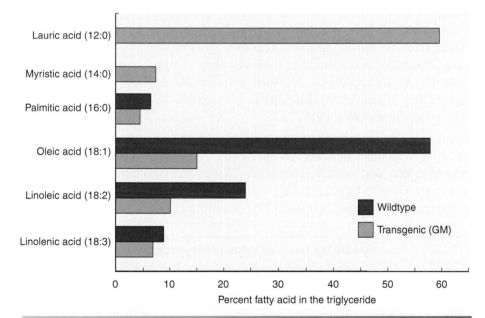

19.6   High-oleic soybean oil has been genetically engineered with improvements for both food and nonfood uses.

Soybeans are the largest source of vegetable oils in the world, and in the United States soybean oil accounts for about 70% of vegetable oil consumed. Most soybean varieties produce an oil rich in polyunsaturated fatty acids (about 50% linoleic acid or 18:2 and 10% linolenic acid or 18:3), and these fatty acids make the oil unstable and easily oxidized so that it becomes rancid. When heated, the oil develops objectionable flavors and odors. Thus unprocessed soybean oil is unsuitable for many applications, so it is chemically hydrogenated. This process adds to the cost of the oil and also introduces side reactions such as conversion of double bonds from the *cis* to *trans* configuration creating *trans*-fatty acids.

The biosynthesis of polyunsaturated fatty acids in plants is catalyzed by a series of enzymes; in the first step, an enzyme converts oleic acid (18:1) to linoleic acid (18:2). After the gene for this enzyme was isolated, molecular biologists succeeded in suppressing the expression of the gene in soybean. This decrease of the 18:1 fatty acid to 18:2 conversion step almost completely eliminated polyunsaturated fatty acids in the soybean oil.

The new transgenic soybean oil has 85% oleic acid, one of the highest oleic acid contents found in nature. The absence of polyunsaturated fatty acids eliminates the need for hydrogenation to stabilize the oil. Furthermore, an unanticipated benefit of the oleic increase was that the saturated fatty acid content of the oil fell from approximately 15% to less than 8%. The new soybean oil has a composition similar to olive and other high-oleic oils, which are considered to provide greater health benefits, compared to other plant and animal oils. The fatty acid trait was stable in field trials, and the oil yield of the crop was identical to the control lines. Thus, neither the transformation process nor the major change in fatty acid composition was detrimental to the high yield of the soybean line. This example is also instructive because it demonstrates how quickly some discoveries can be translated into new crops. With the resources of a major corporation, genetic engineers needed only five years from gene isolation to a field-tested transgenic soybean crop ready for commercialization of an industrial product. Marketing a food product would require additional safety tests mandated by U.S. regulatory agencies.

Vegetable oils have long been known to have useful properties as lubricants, and because they are biodegradable are ideal for applications where harm to the environment must be avoided. However, the tendency of the oils to break down or polymerize as a result of oxidation limits their use. With its very low polyunsaturated fatty acids, high-oleic soybean oil has an oxidative stability more than 10 times greater than most vegetable oils. As a result, it can substitute for mineral oil in many applications such as marine engines, chain saw lubricants, and other applications where oil spills are particularly damaging. A number of other industrial applications may also become possible, because chemical additions to the double bond can lead to polymers and other products that have desirable properties for certain plastics.

The Challenge of High-Level Production

In the two cases of high-lauric acid canola oil and high-oleic acid soybean oil, manipulation of only one gene produced a new commercial product in transgenic plants. However, these two cases may not be representative of the types of metabolic engineering that will be needed for most new products. A number of other attempts to engineer oilseed fatty acid composition have been less successful, because the amount of the desired product

Table 19.5	The challenge of unusual fatty acid production in transgenic oilseeds. Specialty fatty acids are found at high levels in nature, but are produced at low levels in transgenic plants	

Fatty Acid	Level in Native Plant	Level in Transgenic Plant
$18:1^{D6}$ (petroselinic)	85%	<10%
$16:1^{D6}$	80%	<10%
Cyclopropane	50%	<5%
Ricinoleic	90%	17%
Acetylenic	70%	25%
Epoxy	60%	15%
Conjugated	65%	17%
Lauric (+10:0)	65% (+25%)	50–60%

was too low (**Table 19.5**). Although the genes were isolated from plants or other organisms that accumulate the unusual fatty acids at levels of 50 to 90% in their oils, when engineers transformed the genes into a crop plant, accumulation was much less. The activity of the introduced enzyme has generally not been limiting, so other factors probably limit product accumulation. In at least one case, the new fatty acid induces its own breakdown. Thus, it is clear that plant metabolic engineering will be technically challenging.

Furthermore, even if high-level production of a chemical is achieved, the costs of producing a new product in plants may be higher than production by existing methods. Such costs fall into several categories: First, the costs of crop breeding approximately doubles for each independently segregating gene that must be maintained in the breeding population. Second, any yield penalty associated with the transgene will add to the cost of the end product. Third, identity preservation of a GM crop adds to storage, transport, and processing costs. And fourth, perhaps most importantly, the cost of special extraction or processing of a new product can substantially increase the price of an end product. Plant metabolic engineering will be most successful if industry can recover the products at low cost and high yield.

19.7 Potential impact of large-scale chemical production in temperate and tropical crops.

If biotechnology can overcome the problems just discussed and many new chemicals are produced in plants, what will be the impact on agriculture? It could be huge, because of the area of land needed (**Table 19.6**). Producing a chemical such as ricinoleic acid in a crop, rather than importing it, could be easily accommodated in the United States with little impact on land use or commodity prices. In contrast, producing the bulk of adipic acid for nylon manufacture could require 10 million hectares of canola. In comparison, only 2 million hectares of maize are needed to produce 1 billion gallons of ethanol. Large-scale production in plants of plastics or their monomers could greatly exceed the use of crops for ethanol production.

At present, genetic engineering of oilseeds is largely confined to temperate crops such as canola and soybean. The immediate goal is to expand the range of fatty acids available from crop species so that the commercial uses of plant fatty acids can be expanded. In the future, genetic engineering of perennial tropical species such as oil palm will increase in importance. Oil palm, which can grow year round, is capable of producing up to 100 bar-

Table 19.6	Production of industrial chemicals with contrasting market size from fossil fuel versus GM crops	
Raw Material	Ricinoleic Acid: A "smaller market"	Adipic Acid: A "large market"
U.S. market size	30 million kg; US$60 million	1 billion kg; US$2 billion
Approx. cost of chemical	US$1–2/kg	US$2/kg
Production area needed for GM oilseed	0.3 million ha soybean; 0.1 million ha rapeseed	28 million ha soybean; 10 million ha rapeseed

rels of oil per hectare per year. This production capacity is 3- to 10-fold higher than yields obtained from most annual temperate crops. Palm oil is currently the second largest type of vegetable oil worldwide (**Figure 19.10**), mostly coming from Malaysia, Indonesia, and the Philippines. However, in a number of tropical areas in Africa and South America oil palm production could be expanded. As genetic engineering of oil palm advances, this tree may produce a wide range of chemicals at a cost competitive with petroleum. Because oil palm can be grown in several less developed countries, such expansion could have im-

Figure 19.10 Oil palm produces 3- to 10-fold more vegetable oil per hectare than oilseed crops grown in temperate climates. However, it requires hand labor for harvesting and a new plantation requires six or more years before becoming productive. (1) oil palm; (2) fruiting spadix; (3) detail and cross-section of fruit with seeds.

portant social benefits by bringing new agricultural income and exports to these areas. However, the governments will have to weigh the benefits of such cash crops against the need for food production (see Chapter 5).

19.8 Will "plants as factories" biotechnology hurt the economies of developing countries?

Experts such as the Kenyan economist Calestous Juma warn that transferring production of high-value plant metabolites from farms in developing countries to biotech crops in developed countries will have a negative impact on the economies of the developing countries. They will lose an important source of foreign exchange, and many farmers will lose their livelihoods. In his book *The Gene Hunters: Biotechnology and the Scramble for Seeds* (1989, Princeton, NJ: Princeton University Press, p. 143), Juma writes,

> The direction of this research is aimed at reducing dependence on imported raw materials. The impact on the countries exporting this product will be profound and irreversible. Over the years, large sections of the population have organized their lifestyles around the production of these crops. This is going to be changed by current developments in biotechnology. The impacts will not be only economic, but will have long-run political implications and the attempt by communities to reorganize themselves in response to the changed conditions. It is, therefore, in the interest of raw material exporters to closely monitor current trends in biotechnology and the use of genetic resources and modify their internal policies in anticipation of potential long-term effects.

One possible impact of biotechnology is illustrated by the present production of vanilla. Although the plant *Vanilla planifolia* is indigenous to Central and South America, commercial production of vanilla occurs largely in Madagascar, the Comoro Islands, and Reunion. Madagascar alone accounts for 75% of the world market (about US$80 million) and earns more than US$50 million in foreign exchange. Vanilla beans account for 10% of Madagascar's export earnings. The production of vanilla involves over 70,000 small landholders (owners of small farms) whose incomes depend on vanilla exports. The current price of vanilla is about US$70–75 per kg. By the two criteria mentioned in the previous section—price per kilogram and size of the market—vanilla is presently not threatened by genetically engineered crops. Yet research to try to change this is tempting, because the structure of vanilla is simple and may require modification of only one or two genes to produce vanilla in seeds.

Another example of the potential relocation of production from developing countries to developed countries is provided by the newly discovered sweet proteins such as thaumatin. Certain tropical fruits and leaves of tropical shrubs contain small proteins that are 1,500–3,000 times sweeter (based on weight) than sucrose. Some companies have established plantations in the tropics to produce these proteins. They can be used as sugar substitutes and advertised as "natural" sweeteners that are possibly safer than synthetic products such as saccharin and aspartame. Because they are proteins, each is encoded by a single gene, and such genes can easily be isolated and introduced into bacteria, yeast, or plants, which will then produce the sweet protein. Work on the genes of sweet proteins has attracted capital from some of the world's biggest companies, indicating the importance they attach to this emerging technology. Tate and Lyle, the British company that established plantations in Liberia, Ghana, and Malaysia to produce thaumatin, has also introduced the thaumatin gene into yeast by genetic engineering, allowing the company to manufacture the sweetener in large fermentors. The cheaper process will eventually displace the more expensive one.

It is important to note that the relocation of industries and production processes from one country to another has been going on for hundreds of years. The development of new chemical technology for producing synthetic rubber had an immense impact on the economics of rubber tree plantations in countries such as Malaysia. In some cases, the economy of a country can accommodate such changes, particularly if soil and climate conditions are favorable. In Malaysia, oil palm plantations replaced many of the rubber tree plantations and the agricultural economy remained strong. Many countries that were industrial leaders in the 20th century have seen their manufacturing industries move production facilities to locations with the lowest cost of labor and raw materials. Many products (such as electronics) once primarily produced in North America and Europe are now predominantly produced in (other) Pacific Rim countries. Thus the biotechnology industry is not unique or particularly to blame for long-standing trends where economic and political forces, rather than social considerations, drive relocation of production systems. Although in the short term the most developed countries may be the first to benefit from agricultural biotechnology, in the long term all countries with good climate, soil, and water will derive increased income and economic growth from the production of chemicals in plants.

19.9   Starch and other plant carbohydrates have a wide range of industrial uses.

Maize, wheat, rice, and other grains consist of 50 to 70% starch by dry weight. Potato and some other root crops are even richer in starch. Starch can be recovered at very low cost and in high purity by wet milling of grains. In this process maize is soaked 30 to 40 hours in water and then starch, gluten (protein), fiber, and the germ are physically separated. Very large-scale processing plants can recover starch at a price of less than 20 cents per kg. The other parts of the grain can be used as food or animal feed. This price for a complex organic polymer is substantially lower than for plastic polymers derived from petroleum (such as nylon or polystyrene), and the price comparison emphasizes the inherent efficiency of photosynthesis and large-scale agriculture. This low price has stimulated decades of research to find new uses for starch. For example, in the food industry starch is used as a thickening or gelling agent. The conversion of starch to high-fructose syrup is another important application. In the United States, more of this sweetener is now used as compared to sucrose (sugar beet or cane sugar), primarily because it is cheaper to produce. Starches are also used in hundreds of nonfood applications such as paper manufacture and adhesives and have been incorporated into plastic films to increase their breakdown in landfills. Chemical modifications of starch have led to dozens of other uses, such as absorbents. For example, Superslurper is a starch/acrylonitrile copolymer capable of absorbing 1,500 times its own weight in water and is used in diapers, bandages, and seed coatings (see Chapter 9). However, the major industrial use of starch has been for fermentation to produce ethanol and other chemicals described earlier.

Starch is a mixture of two glucose polymers: amylose, which is linear, and amylopectin, which is highly branched (see Chapter 7). The different properties of these two polymers and their relative proportions in starch determine the physical properties and uses of starch. A relatively small number of enzymes are involved in producing starch from glucose, and the relative activities of these enzymes determine the ratio between amylose and amylopectin. Genetic engineering of these enzymes might allow starches to be designed for specific end uses. However, almost all mutants in starch biosynthetic enzymes, although often providing useful starch properties, also lead to reductions in overall starch content and crop yields. Therefore, to date, genetic engineering of starch composition has not reached commercial success.

A number of bacterial and fungal carbohydrate polymers have important applications, and if plants could produce these in high yield the costs of production might be less than from fermentation. For example, xanthan gum—a glucose polymer produced by the bacterium *Xanthomonas campestris*—is used as a thickener and for other food and nonfood applications. Each year 20 million kg of xanthan gum are produced, with a price of US$6 to US$8/kg. This market size is sufficient to stimulate efforts to engineer plants to produce xanthan gum rather than starch.

19.10 Specialty chemicals and pharmaceuticals can be produced in plants.

Most chemicals and products discussed so far can be considered "commodities." Such products are characterized by large market volume and low price. Both the agriculture and the petrochemical industries have excelled at producing low-cost commodity products and in some cases these two sectors compete with each other to capture markets. In such raw material markets, profit margins are small, and manufacturers often choose one supply source over another based on a price difference of 1% or less.

Specialty chemicals represent a different situation. These chemicals are more expensive to produce and almost always have a smaller volume of sales. Many thousands of chemicals fall into this category. For example, the Aldrich Chemical Company lists over 25,000 chemicals in its catalog and collectively these specialty chemicals have annual sales of many billions of dollars. Many flavors and fragrances, both synthetic and biologically produced, fall into this class.

What are the prospects that biotechnology will lead to alternative or improved plant sources for such compounds? Although scientists can envision schemes where plants are engineered to produce thousands of specialty chemicals, you have seen that metabolic engineering is not always as straightforward as expected. Despite much research in the past 10 years focused on new products, only a few processes have been commercialized. This is both because of complexities in engineering biochemical pathways and because of economic considerations. Economists estimate that research and development for a new product requires at least a 10-year investment and that this investment can only be recovered if the product has an annual market of at least US$100 million. Only a small subset of new plant products falls into this category. For example, both jasmine and ajmalicine, components of perfume, are quite expensive (jasmine costs US$5,000 per kg and ajmalicine costs US$1,500 per kg), but the total market for jasmine is estimated to be only US$500,000 and for ajmalicine, US$5 million. Spearmint oil, in contrast, has a substantial market (US$100 million), but the relatively low price (US$30 per kg) means that new plant production systems must produce high yields. Currently, these considerations limit industry investments in research to products with very large market values. However, in the future, as more experience is gained in biotechnology, the process of rational plant metabolic engineering will become more cost effective and will be extended to a wider range of products.

Pharmaceuticals represent a type of specialty chemicals. As much as 25% of the pharmaceutical drugs that people use are either produced in plants or originated in plants. For example, morphine and codeine (and opium) are alkaloids derived from the opium poppy. Other legal and illegal plant-derived drugs consumed in large quantities include caffeine, nicotine, heroin, and tetrahydrocannabinol (from marijuana). The structures of several important plant-derived drugs are shown in **Figure 19.11**. Some medically important structures once isolated from plants are now produced synthetically, because synthesis provides

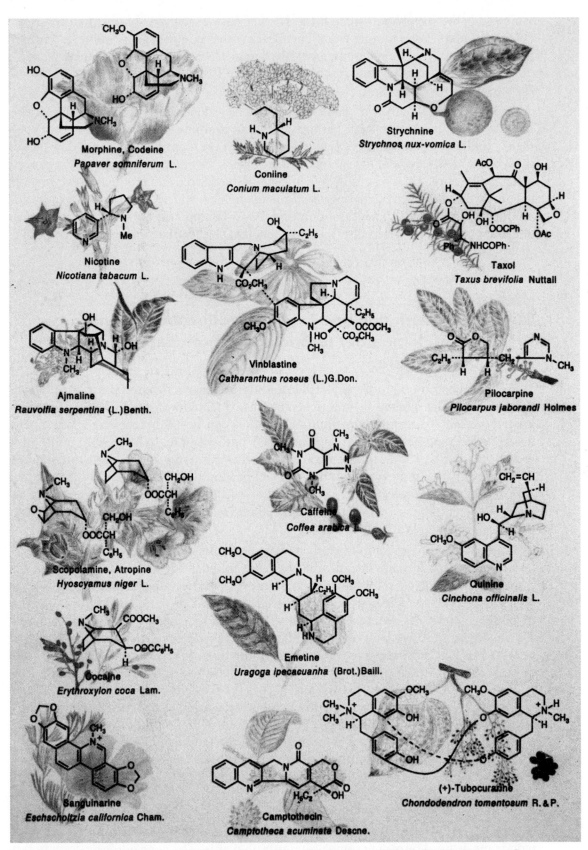

Figure 19.11 Structures of biologically active alkaloids and the plants that produce them. *Source:* Kutchan, T. M. 1995. Alkaloid biosynthesis—The basis for metabolic engineering of medicinal plants. *Plant Cell* 7:1059–1070.

a more reliable supply and sometimes at lower cost. However, some structures are so complex—taxol, for example—that chemical synthesis, although possible, is prohibitively expensive. However, in some cases, the complex structure makes chemical synthesis difficult. The anticancer drug, taxol, can be made from a precursor isolated from the European yew tree.

Can genetic engineering of plants result in increased yields of products such as taxol? In many cases, the biosynthetic pathways of these complex structures are still not completely understood. In addition, geneticists have not yet identified the genes that encode the enzymes that carry out the various synthetic steps. Overexpression of the first enzyme in a pathway sometimes results in higher levels of the desired end product. However, even in some well-characterized pathways efforts to engineer higher flux (in both plants and microorganisms) have been disappointing. A promising strategy may be to discover global regulators, such as transcription factors that can regulate many steps in a biosynthetic pathway.

19.11 Fermentation and in-plant production systems are complementary technologies; both depend on agriculture.

Microorganisms are often the preferred production system for producing small water-soluble molecules such as organic acids, because during fermentation the compounds are released into the growth medium, where they can be easily recovered. Citric acid, used in both food and nonfood applications is one example. In 1998, 200 million kg, with a value of US$500 million, were produced by fermentation in the United States. Although this is a large and attractive market, producing high concentrations of citric acid or other small molecules in a plant might be difficult, because if the compound accumulates to high levels in the cells the metabolic and osmotic consequences could be disastrous for the plant.

Similarly, production of industrial enzymes now takes place largely by fungal or bacterial fermentation, with the enzymes being secreted into the medium. Tons of enzymes such as proteases, lipases, glucose isomerase, and pectinases can be produced at costs as low as US$3 to US$10 per kg. Subtilisin, a protease produced by bacteria, is added to laundry detergents, where it constitutes the largest single use of industrial enzymes. The detergent industry uses over 80 million kg of subtilisin worth US$500 million per year. At first glance it seems that large-scale production of these proteins in crops would be even cheaper than by fermentation. For example, soybean meal, which is 48% protein, sells for US$0.21 per kg and one hectare of soybeans can produce >1,000 kg of protein. If an industrial enzyme such as subtilisin could be produced at a level of 20% of the soy meal protein, then a single 100-hectare farm could produce 20,000 kg of enzyme. However, the crucial cost consideration in this example is purification of the product. Separating the desired enzyme from other proteins in the seed can be expensive, and therefore the fermentation process, although it requires more expensive inputs, often produces at a lower cost, a sufficiently pure enzyme, because the proteins are easily extracted from the medium.

A group of Canadian scientists has developed one ingenious solution that may help reduce the costs of purifying some proteins from plants. The gene of the desired protein is engineered by fusing it to the gene that encodes oleosin. Oleosin is an oil body surface protein with a long hydrophobic anchor and a cytoplasmic terminus found in the seeds of canola and other oil crops. The protein to which oleosin is fused will be firmly attached to the oil bodies. When oil is extracted from the seeds, the protein is recovered with the oil, where it is separated from the bulk of other proteins. The desired protein product can then be cleaved from its target sequence and recovered in relatively high purity (**Figure 19.12**).

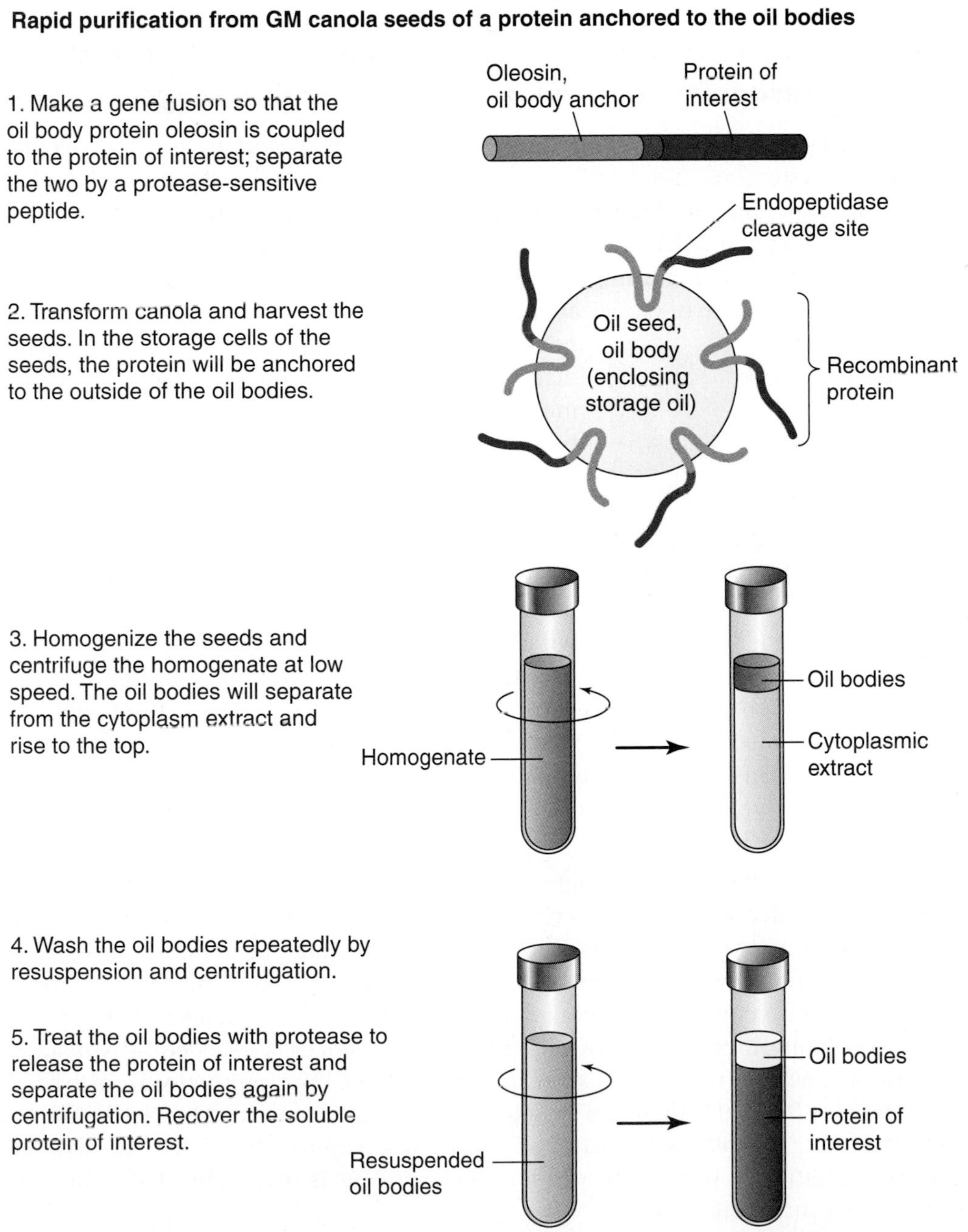

Rapid purification from GM canola seeds of a protein anchored to the oil bodies

1. Make a gene fusion so that the oil body protein oleosin is coupled to the protein of interest; separate the two by a protease-sensitive peptide.

Oleosin, oil body anchor

Protein of interest

Endopeptidase cleavage site

2. Transform canola and harvest the seeds. In the storage cells of the seeds, the protein will be anchored to the outside of the oil bodies.

Oil seed, oil body (enclosing storage oil)

Recombinant protein

3. Homogenize the seeds and centrifuge the homogenate at low speed. The oil bodies will separate from the cytoplasm extract and rise to the top.

Homogenate

Oil bodies

Cytoplasmic extract

4. Wash the oil bodies repeatedly by resuspension and centrifugation.

5. Treat the oil bodies with protease to release the protein of interest and separate the oil bodies again by centrifugation. Recover the soluble protein of interest.

Resuspended oil bodies

Oil bodies

Protein of interest

Figure 19.12 Factories of the future. Seeds of canola can be genetically engineered to produce specialty oils as well as proteins that are easily purified if they are anchored to the oil bodies.

There are other situations where producing industrial proteins in plants is likely to be cost effective. In some cases, the desired use of the enzyme does not require purification. For example, cell wall polymers cannot be digested by monogastric animals and therefore have no nutritive value (see Chapter 7). Adding enzymes that help break down cellulose or hemicellulose can free up more calories and increase feed efficiency. For example, a thermostable endoglucanase has been expressed in transgenic potato at levels up to 25% of leaf-soluble protein. A small amount of crop containing this enzyme could be mixed with other animal feed to partially digest cellulose, thereby enhancing feed use.

19.12 Plants are ideal production systems for diagnostic or therapeutic human proteins needed in large amounts.

Mammalian genes can readily be expressed in plants, and plants are likely to become the production systems of choice for some mammalian proteins that are needed as diagnostics for detecting human diseases or as therapeutics for treating human and animal diseases. Research in "biopharming" is aimed at producing monoclonal antibodies, blood plasma proteins, peptide hormones, and cytokines in plants. The reasons that plants are emerging as the system of choice are purity and cost. Small amounts of human recombinant proteins can now be produced in cultured mammalian or insect cells, but many medical applications of monoclonal antibodies or plasma proteins require the production of as much as 1,000 kg of pure protein per year. Scaling up the production of antibodies by mammalian cells would require a capital investment of US$100 million, and the antibodies would cost thousands of dollars per gram. Extraction from crop plants, in contrast, can be done in a factory that costs US$10 million, and the antibodies will cost US$200 per gram. About 20 companies worldwide are now doing research to eventually produce recombinant human proteins in plants. Likely production systems are the leaves of tobacco plants, the endosperm of maize seeds, potato tubers, or the stems of sugar cane. The targets could include not only therapeutic proteins but also proteins such as collagen that are used in the cosmetics industry. Factories where human proteins made in plants will be purified have already been built (**Figure 19.13**).

Figure 19.13 Factory to process tobacco genetically engineered to express human proteins and therapeutics. This state-of-the-art facility belonging to the Large Scale Biology Corporation is located in Owensboro, Kentucky. The construction of the 35,000-square-foot facility was finished in 1998; it is able to process 3 tons of fresh tobacco leaves per hour. The proteins can be produced with GMP certification, and can therefore be used as therapeutic drugs for human use under FDA guidelines.

Proteins could be produced after stable transformation of a plant or by transient expression after infecting the leaves with a virus that carries the gene of interest. One important aspect that scientists must consider is that if human proteins are made in seeds, the crops must be grown in relative isolation so that the human genes do not show up in maize used for human or animal consumption. Although the proteins might be harmless, their consumption could have unintended consequences. Genes could spread if, for example, pollen from maize plants that are producing the human proteins were to be blown to other fields. The fields therefore need to be isolated, so that there is no chance that pollen from the transgenic plants may cause the genes to spread to other fields. An advantage of using a vegetatively propagated crop such as sugar cane is that it eliminates the possibility of pollen spreading genes.

The problem of gene spread can also be circumvented with transient expression systems based on plant viruses that infect leaves. Transient expression means that the proteins are made for only a few days, but the goal is to have very high levels of expression during that short period. Viruses have small genomes consisting of only a few genes. It is sometimes possible to replace some of the viral genome with a segment of nucleic acid that carries the information to make the active fragment of an antibody molecule. When this is done, the virus may still be able to replicate and reproduce itself when it infects a plant. Whether or not this is possible depends on the type of virus. Thus, when a leaf of a plant is infected with a genetically engineered RNA virus that carries the active segment of a human gene, the virus will spread throughout the plant, making billions of copies of itself. If viral infection is done before the plant flowers, then the virus does not spread to the pollen. The plant cells will translate the messenger RNA and produce the antibody fragment. Engineering the virus and allowing the plants to make the protein can all be accomplished in two months. Such a rapid production system can be used to create antibodies that are customized for a particular patient—a requirement for treating non-Hodgkin's lymphoma, a type of cancer that attacks certain white blood cells.

The body uses antibodies to fight diseases (see later discussion), but antibodies also have many other uses in medicine. They can be used therapeutically, to deliver a killer drug specifically to cancer cells, for example, and also to diagnose diseases. Antibodies are directed at antigenic determinants (or epitopes) at the surfaces of proteins or other molecules. For example, an antigenic determinant could be a few amino acid side chains displayed on the surface of a protein. Antibodies directed at antigenic determinants displayed on the surface of cancer cells specifically bind to those cancer cells. Different antibodies will home in on different types of cancer cells. Such specific binding can be used to image the cancer, or to target the cancer cells for destruction. Antibodies produced in plants could also provide protection against intestinal pathogens such as hepatitis viruses, *Helicobacter pylori*, or toxic *E. coli*. They could target respiratory pathogens (rhinoviruses and influenza), sexually transmitted diseases, and dental caries, and could be used as contraceptives. Clinical trials with antibodies produced in plants are already under way, and these novel plant-derived pharmaceuticals may be in the marketplace by 2005.

19.13 Plants can be used to deliver edible vaccines for serious diseases of humans and domestic animals.

In humans and other warm-blooded animals, the primary task of the immune system is to defend against invading organisms—particularly against pathogenic bacteria and viruses. Most infections are initiated in the mucosal surfaces that line the digestive tract, the respiratory tract, and the urino-reproductive tract; these surfaces also contain the first line of defense—the mucosal immune system. Killing the invaders by engulfment is the function of phagocytic cells such as the macrophages in the bloodstream. Engulfment can occur after the pathogens have been covered with antibodies—proteins produced by the immune system—that are directed at specific antigenic determinants on the surface of the invaders. Producing antibodies is the function of specialized cells of the immune system. The unique features of antibodies are that they recognize a single molecular structure (often a protein) and bind to it with very high affinity. A common type of antibody is the immunoglobulin G (IgG) protein, consisting of two heavy (large) polypeptide chains and two light (small) polypeptide chains.

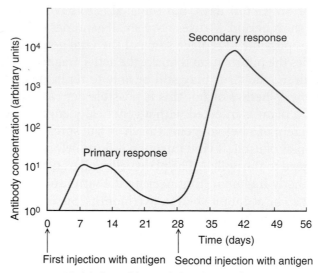

Figure 19.14 Responses to initial and later injections of an antigen. At day 0, antigen was injected into a mouse and the response was followed by measuring levels of serum antibody to the antigen. At day 28, when the initial response had subsided, a second immunization was done with antigen. The secondary response to antigen was faster and greater than the initial response to antigen. Notice that the vertical scale is logarithmic and that the secondary response is 1,000 times greater than the primary response.

When a foreign invader comes calling the first time, the body's response in terms of antibody production has a considerable lag time and may not be strong enough to overpower the invader, so the animal may get very sick or even die. If the immune system has been exposed to the antigenic determinants on the invader's surface in the past, and now faces a second invasion, then the response is faster and more powerful (**Figure 19.14**). That is the basis of vaccination. Because detection of an invader can occur in the mucosal membranes or in the bloodstream, vaccines can be effective whether taken orally or injected. For example, there are effective oral and injected vaccines against the polio virus. Vaccination makes people immune because it triggers the immune system to produce antibodies by simulating that first invasion, not with virulent pathogens, but with a greatly attenuated strain of bacteria or viruses. Alternatively, immunity through vaccination can be achieved by injecting or ingesting only the antigenic determinants (often parts of proteins that are the antigens) that are displayed on the surface of the invading organisms. Such vaccines are called subunit vaccines. The use of subunit vaccines increases vaccine safety by circumventing the need to use live bacteria and viruses. However, subunit vaccines are not as stable and need to be refrigerated, increasing their cost. Safe vaccines for many important diseases such as cholera and hepatitis are well beyond the reach of the health care system in many developing countries. Scientists realized that plant genetic engineering could solve the problem of providing low-cost vaccinations to people in developing countries, a realization that prompted the idea of expressing the antigens in plants and creating edible plant vaccines. Various pharmaceutical companies are targeting quite a few diseases for vaccine production in plants (**Table 19.7**).

Some of the antigenic determinants on the surface of viruses and bacteria are proteins made by the pathogen. The genes encoding these antigens can be isolated from the pathogens and expressed in plants using standard recombinant DNA technology. In addition, it is possible to incorporate an antigenic determinant (usually a short stretch of

Table 19.7 Production of vaccines in plants

Potential Application	Protein Expressed in Plants
Dental caries	*Streptococcus mutans* surface protein
Cholera and *E. coli* diarrhea	*E. coli* heat-labile enterotoxin
Hepatitis B	Hepatitis B antigen
Hoof-and-mouth disease	Hoof-and-mouth viral antigen
Malaria	Malarial B-cell epitope
Influenza	Hemagglutinin
Rabies	Rabies virus glycoprotein
HIV	HIV epitopes gp41 and gp120

Note: Experiments on the preceding proteins have been reported in the scientific literature.

amino acids) into an existing plant protein or into a plant virus. If the plant makes the antigenic determinant, then eating the plant (under the right conditions) will amount to vaccination. If the antigenic determinant is incorporated into a plant virus and expressed at the surface of the viral particle, the plant virus can be used as a vaccine (plant viruses are quite harmless to humans and animals; when eating vegetables, people consume large numbers of various plant viruses). Researchers conducted the first human clinical trials with transgenic plant-derived vaccines in the late 1990s. They showed that such vaccines are protected from digestion in the human intestinal tract and stimulate both the mucosal and systemic immune responses. Vaccinating people against cholera by feeding them bananas, or vaccinating animals against foot- or hoof-and-mouth disease by feeding them sugar beets could well be a reality by 2010.

CHAPTER
SUMMARY

GM technology can transform traditional crops from their current role of providing low-cost food and fiber toward a much more diverse and profitable role of producing an array of higher-value chemicals. This vision arises from three simple observations. First, plants are efficient and diverse chemical factories that use sunlight and carbon dioxide to produce an extremely wide range of organic chemicals. Second, modern agriculture excels at producing oils, starch, protein, and fiber in surplus quantities for commodity markets at extremely low cost. And third, basic research advances in biochemistry and molecular genetics have created unparalleled new opportunities to engineer plant metabolism to produce entirely novel, value-added products, such as biodegradable plastic.

Production of both commodity and specialty chemicals in plants is in its infancy but can be expected to expand as scientists gain more experience with the complexities of metabolic engineering. As more agricultural production is devoted to nonfood uses, food prices may increase slightly, but this trend is likely to be reversed as more efficient agricultural practices emerge and plants are engineered to thrive on marginal lands.

Using transgenic plants as edible vaccines will greatly reduce the cost of vaccines and enable developing countries to combat some major diseases prevalent there. In developed countries, transgenic plants will soon be used to produce a variety of mammalian proteins (such as antibodies) for therapeutic and diagnostic purposes.

Discussion Questions

1. What are some ethical considerations in using agriculture to produce chemicals for plastics rather than food?

2. What are some major uses of plant material to produce chemicals today?

3. Many chemicals such as lactic acid and citric acid can be produced by fermenting sugar derived from maize starch. Why is it unlikely that new crops will be developed for large-scale production of these chemicals?

4. What are some factors other than scientific "know-how" that influence future uses of plant products?

5. Discuss some scientific and commercial limitations on making commercial products in transgenic plants.

6. What problems could arise with producing antibodies in maize (a cross-pollinator) as opposed to soybean (a self-pollinator) and sugar cane (no pollination required)?

7. How should the government regulate crops that produce edible vaccines?

Further Reading

Croteau, R., T. M. Kutchan, and N. G. Lewis. 2000. "Natural products (secondary metabolites)." In B. B. Buchanan, W. Gruissem, and R. Jones, eds., *Biochemistry and Molecular Biology of Plants*. Rockville, MD: American Society of Plant Physiologists.

Frost, J. W., and K. M. Draths. 1995. Biocatalytic syntheses of aromatics from d-glucose: Renewable microbial sources of aromatic compounds. *Annual Review of Microbiology* 49:557–559.

Gerngross, T. U., and S. C. Slater. 2000. How green are green plastics? *Scientific American* (August):37–41.

Hitz, B. 1999. Economic aspects of transgenic crops which produce novel products. *Current Opinion in Plant Biology* 2:135–138.

Kinney, A. J. 1998. Plants as industrial chemical factories—new oils from genetically engineered soybeans. *FETT-LIPID* 100:173–176.

Ohlrogge, J. B. 1994. Design of new plant products—engineering of fatty acid metabolism. *Plant Physiology* 104(3):821–826.

Poirier, Y., C. Nawrath, and C. R. Somerville. 1995. Production of polyhydroxyalkanoates, a family of biodegradable plastics and elastomers in bacteria and plants. *Biotechnology* 13:142–150.

Richter, L. J., Y. Thanavala, C. J. Arntzen, and H. S. Mason. 2000. Production of hepatitis B surface antigen in transgenic plants for oral immunization. *Nature Biotechnology* 18:1167–1171.

Slattery, C. J., I. H. Kavakli, and T. W. Okita. 2000. Engineering starch for increased quantity and quality. *Trends in Plant Science* 5(7):291–298.

Somerville, C., and D. Bonetta. 2001. Plants as factories for technical materials. *Plant Physiology* 125:168–171.

Szmant, H. H. 1986. *Industrial Utilization of Renewable Resources*. Lancaster, PA: Technomic Publishing.

Web sites:

National Biobased Products and Bioenergy Coordination Office. *Developing and Promoting Biobased Products and Bioenergy*. Available at http://www.bioproducts-bioenergy.gov/pdfs/presidentsreport.pdf. Accessed October 25, 2001.

Office of Industrial Technology, Department of Energy. *Technology Roadmap for Plant/Crop Based Renewable Resources 2020*. Available at http://www.oit.doe.gov/forest/pdfs/ag_technology_roadmap.pdf. Accessed October 25, 2001.

Urban Myths and Scientific Facts about the Biosafety of Genetically Modified (GM) Crops

C. Neal Stewart, Jr.
University of Tennessee

Sarah K. Wheaton
University of Rhode Island

With the exceptions of western Europe, Japan, and parts of Latin America, genetically modified (GM) crops are fast joining agriculture throughout the world today and will play an increasingly important role in global food production. Both India and China have dramatically increased investment in molecular technologies, to increase their agricultural productivity. Molecular techniques, including gene transfer into crops, are the basis of the next logical development in agronomy and plant-breeding research (see Chapter 14). Although this technology can be viewed as an extension of traditional breeding (which also entails moving genes around), some people emphasize that transgenes are novel; that is, they do not originate from sexually compatible or closely related plants and instead can be derived from a range of organisms. For example, the *Bt* toxin genes from the bacterium *Bacillus thuringiensis* can and do function in plants after considerable modification of the nucleotide sequence of the gene. In fact, about half of the cotton and nearly one third of the maize grown in the United States in 2000 was genetically modified with *Bt* for insect resistance.

Because many transgenes are new to agriculture and might result in novel phenotypes, prudence dictates that people examine the risks before wide-scale deployment of transgenic crops. Some maintain that this breaking of the species barrier is so novel that the products (GM crops) pose uncertain risks to health and the environment. They want GM crops banned and maintain that people do not know enough about the consequences of introducing these foods in the human food chain and the plants into global environment (**Figure 20.1**).

The development of agriculture as a science and its continual use and implementation of technology have clashed with a more idealized view in which a purity of purpose is considered on a par with scientific fact. We will examine some urban myths that have arisen during debates over GM crops and that have been propagated by activists,

**Figure 20.1
Public protest in the streets of Oakland, California.** Groups opposed to and in favor of GM crops hand out information at a peaceful rally. Activism and protests are part of every democracy. *Source:* Courtesy of P. Lemaux.

the popular media, and even some scientists. We then discuss some of the real biosafety facts and concerns and how they are being handled.

Part I. Urban Myths about GM Crops and GM Foods

Urban myths arise because many people distrust new technologies that they do not understand and over which they have no control. The less control they can exert, the greater is their perception that it may be risky (see later, discussion of risk). Furthermore, distrust of large companies stems from the fact that people see such companies as less subject to national regulations (local control) and have not always been averse of putting financial gain ahead of public welfare. So there are sound reasons why urban myths about GM crops have developed. Furthermore, organizations that depend on voluntary donations to meet their large payrolls and entities that will secure a larger market share if GM technology fails (such as the organic food industry) are not averse to helping propagate such myths by ignoring some facts and emphasizing uncertainties.

The discussion of biotechnology risks in the media and by environmental organizations has not included the broader context of current agricultural practices. Problems ascribed to GM crops are often not unique to GM crops. Therefore, the consumer has increasing misperceptions about the dangers of GM crops and the role of biotechnology in agriculture. The discussions of GM crops have, in fact, added to urban people's mystification about how food is produced; no wonder the public feels a growing sense of alienation regarding food production. In this increasingly urban society, farming is romanticized, on the one hand, by the picture of the family farmer communing with the earth, and simultaneously demonized, on the other hand, by an image of giant agribusiness corporations treating livestock inhumanely and polluting pristine environments.

Today, only the smallest fraction of our population has any firsthand understanding of and appreciation for the strict constraints (biological, economical, and so on) of agriculture, information that was once considered familiar territory—literally in people's backyards. Many consumers want to be informed, but find unbiased information hard to come by. When distorted examples of research and development in biotechnology are added to the mix, it is not surprising that people feel threatened (see also the discussion of hazard and outrage in Chapter 7). In the course of the popularization of misconceptions about agricultural biotechnology, various myths have arisen.

Myth 1: The monarch butterfly is endangered by *Bt* corn.

The beautiful monarch butterfly (**Figure 20.2**) has become a powerful rallying symbol for the forces opposed to GM technology. They maintain that *Bt* maize threatens the butterfly population. This assertion is based on a study by J. Losey and colleagues of Cornell University, published in the prestigious journal *Nature,* that raised this possibility without providing evidence for it. The news that *Bt* maize pollen would kill larvae of the monarch butterfly (*Danaus plexippus*) circled the globe in days. Because milkweed (*Asclepias syriaca*), the only food for these larvae, grows in and around maize fields, and maize pollen could conceivably drift to, and land on, milkweed leaves, the Cornell researchers dusted milkweed plants with *Bt*-containing maize pollen, to look for toxicity to the larvae. They observed decreased feeding, growth, and survival rates in exposed larvae compared to larvae that ate leaves dusted with nontransgenic corn pollen. The authors concluded that *Bt* maize could endanger monarch populations feeding on milkweed near *Bt* maize. Several other scientists immediately questioned the validity of this study, arguing that its methods were not reproducible, that the "no choice" feeding strategy for the larvae did not represent true conditions, and that the pollen levels used were artificially high. Extensive follow-up studies by Mark Sears and colleagues have now shown that survival of monarch butterfly populations is not endangered by the planting of *Bt* maize in the United States, and that the impact of *Bt* maize is likely to be small (Sears et al., 2001).

Milkweed, the food of the monarch butterflies, is undesirable in both maize and soybean fields, and farmers try to eliminate it by the usual methods, including herbicides, which will always minimize monarch larva food supply in U.S. maize and soybean belts. Because milkweed patches are more frequent and larger on roadsides than in crop fields, perhaps scientists need to assess the impact of nonagricultural technologies, such as automobiles, on monarch populations. Although people are not going to stop using cars, having data about relative mortality rates to monarchs could be useful in the overall risk assessment to the species. One source (Monarch Watch, at the Web site www.monarchwatch.org) suggests that perhaps as much as 10% of the monarch butterfly

Figure 20.2 Monarch butterfly and larva. Activists opposed to GM technology and the media claimed that the beautiful monarch butterfly is threatened by extinction by *Bt* maize. The claims resulted from a misinterpretation of results published in the journal *Nature.* Extensive follow-up studies showed that *Bt* maize does not imperil the monarch population. *Sources:* Left photo, Iowa State University Entomology Image Gallery; right photo, Cornell University press release, photo by Kent Loeffler.

habitat in the U.S. maize belt is actually in maize fields. Would this habitat proportion be relevant to the monarch butterfly population levels?

Biotechnology currently represents a potential risk to monarchs, but how can people assess its relative importance to monarch butterfly survival? Although automobiles, tropical habitat loss, increasing exotic bird populations, and global climate change might have equivalent or greater actual effects, biotechnology risks are currently being assessed in a vacuum. The questions are simple, but the answers are complex.

Myth 2: GM plants will create superweeds.

Many people who feel generally unfavorable toward biotechnology have evoked the superweed idea, arguing that biotechnology will create weeds that are more invasive and damaging than our current weeds.

Weeds are the scourge of agriculture (see Chapter 17), and people have created plenty of superb weeds by moving weeds from one continent to another (such as the Russian thistle or tumbleweed) (**Figure 20.3**), by agricultural practices (monocultures; herbicide usage), and by hybridizing crops with native plants. The myth of the superweeds is best exemplified by the notion of genes for herbicide tolerance flowing from GM crops to related weedy plants. With only a few GM crops—namely, canola (usually *Brassica napus*)—engineered for herbicide tolerance, it is possible that transgenes could move from crop to weed in Canada and the United States. To minimize this occurrence, GM canola is not cultivated where its close relatives (such as field mustard, *Brassica rapa*) are dominant weeds. On a worldwide scale, there are other scenarios: Genes can flow from cultivated rice to wild rice, from cotton to wild cotton, from maize to teosinte.

Herbicide-tolerant weeds already exist as a result of natural selection after years of herbicide usage. Are these superweeds, or must a weed be transgenic to be a superweed? If herbicide-tolerant field mustard did arise, would it be worse for the farmer than it was before tolerant GM crops facilitated better weed control? These questions are in many ways rhetorical. Today there are no GM weeds, but in five or ten years, there will be. What novel traits will the transgenes confer to the host? How will they interact with the host plants' genetics, physiology, and ecology? This is a much more complex situation than can be addressed by the unilateral approach of equating the development and evolution of GM crops to the creation of superweeds. It is highly unlikely that GM crops will make weed control more difficult in the future or that transgenic weeds will invade pristine environments, any more than other crops have in the past.

Figure 20.3 Tumbleweed is a "superb" weed, if not a "superweed." Agricultural practices, and especially the movement of seeds from one continent to another, have contributed to the emergence of weeds that create serious difficulties for farmers. The tumbleweed or Russian thistle (*Salsola kali*) originally came from southern Russia and arrived in the United States as a contaminant of flax seed in 1877. Gene flow from crops to weeds will occur, but is very unlikely to create superweeds.

Figure 20.4 Are there genes in this food? Maize and soybeans were the main GM crops grown in the world in 2001. Because products derived from these two crops (for example, starch, protein, oil) are used in most processed foods, up to 70% of all products in U.S. supermarkets contain some GM ingredients. When the crops are processed, GM crops are not kept separate from traditionally bred varieties. Keeping separate production streams would add 10% to the cost of the ingredients. There are no good reasons for separating ingredients that come from GM crops; they are substantially equivalent.

Myth 3: GM foods have genes, whereas normal foods do not.

Opinion polls of people's attitudes toward GM crops in the late 1990s showed that many respondents believe that only GM crops have genes, whereas other crops do not. Plants vary quite dramatically (more than 50-fold) in the amount of DNA each cell contains; there is less variation in the number of genes and many plant species probably contain 25,000 to 40,000 genes (humans have 30,000 genes). Each cell has two copies of these genes. GM crops contain 2 to 3 additional genes (**Figure 20.4**).

Myth 4: There are fish genes in tomatoes and rat genes in lettuce; transgenes will change the fundamental vegetable nature of plants.

As part of exploratory research, scientists may perform experiments with particular crops, using many different genes to assess their performance. The media focused an inordinate amount of attention on one basic research project, in which an arctic fish gene was isolated and inserted into plants with the hope that it would confer freeze and frost protection. Such antifreeze properties would have great benefit to farmers who routinely lose crops to cold weather. In addition, antifreeze properties could extend the growing range of certain crops, such as tomatoes and citrus. However, when this particular fish gene was expressed in plants, it was ineffectual in providing frost/freeze protection. So, although there was, for a time, a fish gene in a tomato, it never made it into a jar of tomato sauce. Nevertheless, the distaste that people have for the smell of fish oil could be transposed to how they would feel about fruits with fish genes and has been cleverly exploited by people opposed to GM technology by creating the fish-shaped strawberry (**Figure 20.5**) as a symbol of their movement.

Another myth concerns plants with pig genes. No porcine genes are being evaluated by researchers for transfer into plants. The reason for the furor surrounding the subject and the reason why it will not happen are the same: Religious groups for whom pork products are forbidden make it economically unsound to transfer pig genes into crop plants. Even if scientists were to discover a pig gene that conferred salt tolerance or some other useful crop trait, it would not be commercialized because of economic considerations and the public's perception of what a gene from an animal does to the nature of a plant. This is an area where biotech companies are likely to follow public perceptions even though the religious authorities that have addressed the question agree that introducing a single animal gene does not alter the vegetable nature of a plant. A pig gene in a petunia does not a petunia pig make!

Figure 20.5 The hybrid fish-strawberry logo has become a powerful symbol of the anti-GM crop groups. This type of gene transfer is highly unlikely to ever be carried out commercially, but may be done by researchers to understand how genes function. Opinion surveys show that the public does not support using animal genes to improve crops, even though there is no scientific reason not to do so.

Myth 5: GM foods are not natural.

Human food plants do not occur in "nature" and generally cannot survive in natural environments, because their fitness has been changed by mutations, especially those that affect seed dispersal (see Chapter 13). Their continued existence depends on human intervention. Opponents of GM technology stress the idea that GM crops are unnatural because genes from any organism can be transferred through GM technology. This is indeed correct. Unfortunately, one cannot equate natural with good. HIV is just as natural as vitamin C. Mother's milk is as natural as cyanide. People need to examine, analyze, and regulate food products rather than determine if they are "natural." Gene transfer between organisms that are very distantly related does occur in nature as exemplified by the crown gall disease, in which a segment of bacterial DNA incorporates into plant DNA. Plant-breeding methods requiring embryo rescue or radiation are equally "unnatural," but have yielded more than 2,000 crop varieties that are generally accepted by traditional and organic farmers worldwide.

Myth 6: When you transform plants, you don't know what you are doing to the DNA.

Traditional breeding may introduce into a crop plant thousands of genes from wild relatives or related species. Even after many backcrosses, a couple of hundred genes will remain, and the breeders don't know which ones. With plant transformation, the genetic engineers know precisely which two or three genes they are introducing. With wide crosses (between species) and embryo rescue (see Chapter 14), one also does not quite know how the genes eventually rearrange and line up. With radiation breeding (see Chapter 14), the scientists expose the DNA to powerful radiation, causing many random chromosome breaks, point mutations, and deletions of DNA segments. They do not have a clue what has happened to the DNA. However, subsequent breeding and selection eliminate deleterious genetic accidents so that the resulting cultivars have the desirable properties the breeder seeks. Similarly, creation of GM crops involves years of breeding and selection after the initial transformation. In any case, the plant genome harbors a considerable number of transposons (mobile DNA) that can cause gene duplications and generally reshuffle a small number of genes in each generation. Thus DNA is not as stable as scientists thought 20 years ago, and this gene shuffling is a major driving force of evolution. GM crops, like other elite crops, need to be tested over and over again to make sure they retain their important characteristics.

Myth 7: This debate is not about economics but about food purity: Food suppliers will demand "GM free" foods.

Some prominent food handlers and processors have made public announcements that they will no longer use GM varieties in their food products, such as Gerber baby food, although most grain handlers accept without question those GM varieties and hybrids that are labeled for sale in both North America and Europe. Iceland Group, a large food distributor and processor in the United Kingdom, had a policy of carrying only organic vegetables. However, because of low sales, their policy has changed to sell both organic and traditionally grown foods, according to a BBC news story in December 2000. Although there have always been factions inclined toward one extreme or another, in today's market, consumers for the most part do not seem to respond strongly to food politics. Perceived value is what carries the most weight for consumers when they are considering the varieties, whether GM, organic, or something in between.

However, different sets of consumers have different priorities. Those with more disposable income may buy organic produce known to be GM free. Larger corporations—whether farms, food processors, or food retailers—that deal exclusively or largely in organic foods, realize they can increase their market share by supporting the notion that GM foods are probably unsafe or at the very least are not being adequately tested by government agencies. Thus, there is an important economic aspect to the many demonstrations and advertising campaigns. Similarly, the agricultural biotechnology companies that have invested billions of dollars in GM technology are defending their economic stake. Information to help consumers make informed decisions is presented by all parties to this debate (**Figure 20.6**).

Figure 20.6 Informational brochures explaining the risks and benefits of GM crops are available to the public and produced by various organizations.

Myth 8: Low-resource farmers in developing countries will not benefit from biotechnology.

The first and second generations of GM crops were developed for industrialized societies, where the principals would be able to recoup their investments through high-value products. So far the major benefits have gone to the companies that developed the technologies and produced the products and to the farmers in the form of reduced production costs. Not to be ignored, however, are the benefits to farm workers, especially those who apply pesticides (**Figure 20.7**). *Bt* crops require less frequent pesticide applications (especially *Bt* cotton), and field workers who apply synthetic pesticides run considerable health risk in doing so. These risks are greatest in developing countries, where safety rules are frequently ignored.

Bringing biotechnology to developing countries has come about primarily through humanitarian efforts. Many multinational agricultural biotechnology companies have "noncommercial" projects aimed at the crops of the developing countries. Three examples of projects that will help poor farmers and consumers are the "Golden Rice" already discussed in Chapter 7, the virus-resistant potato lines being created in Mexico, and the virus resistant sweet potatoes now undergoing field trials in Kenya. The latter two projects are going forward with "donated" technology. Another effort that is underway is the development of GM plants to detect buried explosives (Neal Stewart, unpublished), in order to locate landmines in developing countries, in some of which (Afghanistan) they pose a substantial threat. Such humanitarian efforts targeted to the developing countries would not be possible without technology from the developed world and its subsequent global dissemination. Paradoxically, the same is true for the misinformation and melodrama that accompany the advent of any new technology; they are often exported in similar ways from the developed countries to the developing countries.

Whether people like it or not, the choices made in the developed world are crucial to the well-being and future of the developing world. Developing nations have bene-

Figure 20.7 Major health benefits could accrue from wider use of *Bt* crops to farm workers who spray insecticides in developing countries. The major problem with pesticide use is not pesticide residues on produce, but environmental damage and health problems for field workers. *Source:* Courtesy of Eugene Hettel, IRRI.

fited significantly from the students of their nations who trained in biotechnology fields in the United States, Europe, Japan, and Australia. These scientists returned home with the tools and information to convince leaders in their own countries that technology can play an important role in providing solutions to ever-increasing food production and environmental challenges. Many in the developing world see the efforts of organizations from the developed world (such as Greenpeace) to deny them the benefits of agricultural biotechnology as arrogant and misguided and as yet another expression of colonialism.

Myth 9: Antibiotic resistance genes used to produce transgenic crops will horizontally transfer into microbes and thus exacerbate problems of antibiotic resistance in human and animal pathogens. Transgenes will move from plants to gut microflora to humans.

This hot topic has grabbed public attention because it has married a real problem (antibiotic resistance in medicine) with the current controversy over GM crops. Although it is true that most GM crops have been produced using antibiotic resistance genes, that fact does not imply significant risk. Antibiotic resistance genes help protect transgenic plants in the presence of a drug that technologists administer to kill off untransformed cells. For example, the *nptII* (neomycin phosphotransferase) gene has been used for selection against the drug kanamycin. The FDA has approved *nptII* for this very practice, and no data suggest that *nptII* or any other gene can move intact from a plant into microbes such as those found in the human gut. In fact, scientists have performed numerous experiments to try to instigate that exact event, but have never succeeded. It is not surprising that this event does not occur, or over time gut microbes would become plants, or scientists would at least find in microbes genes that look like plant genes, which they do not. Genes can move between bacterial species (mainly via plasmids), and even from bacteria to plants (such as the case of *Agrobacterium tumefaciens*) (**Figure 20.8**), but movement in the other direction seems extremely unlikely. A compelling argument against the remote possibility of the movement of plant genes (or transgenes) into bacteria is the dissimilarity between bacterial genes and plant genes. Plant genes contain introns, whereas bacterial genes do not (although most transgenes are made from intronless cDNAs). The preferred genetic codon usage in plants is different from that of bacteria, and they use different kinds of regulatory sequences (promoters and terminators) as well. It is almost inconceivable that a large piece of DNA could withstand digestion in the human gut, but if it did, the intact DNA, including the promoter and terminator, would need to transfer to a microbial cell. That cell would then have to integrate the gene into its genome or plasmid, and the researchers would have to use kanamycin to select for the gene. Statistically, all this is so unlikely that one need not lose sleep over it! Although most scientists consider using antibiotic genes for plant transformation safe, all major biotechnology companies are now adopting technologies that remove those genes after the plants have been transformed and before they are released to the farmers as new crop lines.

Myth 10: GM crops are not adequately tested or regulated.

In all countries, government agencies regulate the products of technology that are sold to consumers. The reason is that technology is not inherently safe, and regulations are put

Figure 20.8 Crown gall disease caused by *Agrobacterium tumefaciens* is an example of unusual gene transfer between organisms. In this natural process, the bacterial pathogen transfers some of its genes to plant cells, which then grow out into a gall. *Source:* With the permission of the American Phytopathology Society.

in place to assess safety. These agencies all require that certain tests be done, and agency scientists then review the evidence and make decisions based on their understanding of the results. This procedure is followed for drugs, for example. Industry generally funds drug tests, which are often carried out by university scientists. For such research projects, strict conflict-of-interest guidelines are needed so that the scientists have no financial interests in the companies whose productivity they are testing. Government scientists then scrutinize the data and make recommendations. The situation for GM crops and GM foods is very similar. In the United States three agencies are involved: the U.S. Department of Agriculture, the Environmental Protection Agency, and the Food and Drug Administration. The amount of information submitted for approval of a single product is truly staggering. Large companies generally perform tens of thousands of analyses to show that GM products are compositionally and nutritionally equivalent to conventional plant varieties. Globally, tens of thousands of field tests have been conducted during the past 15 years to establish the safety of GM crops, and hundreds of food safety tests and animal feeding studies have been done during the past 10 years. For example, as of 2001 the data on Monsanto's herbicide-resistant soybeans have been examined by and approved by 31 regulatory agencies in 17 countries. Furthermore, leading national and international scientific authorities have concluded that biotech products are as safe for people and for the environment as conventional plant varieties. The real problem is actually the reverse: These crops are so highly regulated that small companies and nonindustrial entities (such as universities) that would like to develop GM crops lack the financial resources to do the tests required. In the past, noncommercial agencies released many improved plant varieties, but that is unlikely to be the way of the future for GM crops, because of required testing and the high degree of regulation.

Part II. Health and Environmental Risks of GM Crops and GM Foods

There are risks in growing and eating GM crops, as there are risks inherent in any human activity. These risks are not new or specific to GM crops, but derive from risks already existing in agriculture. Every year, new genes are expressed in novel crop varieties without being questioned; nor does the public hear much about the varieties that arise from random mutagenesis caused by chemicals or irradiation. These are accepted technologies with which people have achieved a certain level of comfort, and that were introduced at a time when technology was not scrutinized as it is today (**Table 20.1**) (see Chapter 14). Unknown factors exist in any new technology, and until a technology is thoroughly understood the risks cannot be completely characterized; so they are often misunderstood and overemphasized. The perception that special risks are associated with GM crops results from a combination of misinformation and fear (**Table 20.2**); true risk and perceived risk can be quite different from each other. **Figure 20.9** compares the actuarial risks of common activities that have been observed over time. There have been no documented injuries, illnesses, or deaths caused by the use of GM varieties in agriculture, from which one can infer that they are relatively low risk. However, GM foods are not specifically labeled in the United States, so it is not possible to conduct large epidemiological studies. All safety

Table 20.1	**Examples of cultivars and/or species originating from spontaneous mutation, induced mutation, somaclonal variation, and interspecific hybridization**

Species	Source	Trait
Capsicum annum (pepper)	Gametoclonal variation	Reduced seed number
Lycopersicum esculentum (tomato)	Somaclonal variation	*Fusarium* race 2
Nicotiana tabacum (tobacco)	Gametoclonal variation	Potato virus y resist
Zea mays (corn)	In vitro selection	Imidazilinone resistance
N. tabacum (tobacco)	Somatic interspecific hybridization	Nicotine content, blue mold, black root rot
Hordeum vulgare (barley)	Mutation	Proanthocyanidin free, beer stabilizing factor
Brassica napus, B. rapa (canola)	Spontaneous and induced mutation	Low erucic acid and glucosinolates, edible oil source
Triticale (×Triticosecale)	Interspecific hybridization	New cereal species (human made)
Triticum aestivum (wheat)	Interspecific hybridization	20+ disease-resistant cultivars

Table 20.2	Why some activities—in this case, eating GM food—have greater perceived risk perception than actual risk
Factor	**Example**
Coerced rather than voluntary	Everyone must eat GM food, if it is unlabeled
Industrial rather than natural	Big multinational hybrids versus landraces
Dreaded rather than not dreaded	Unknown risks (cancer?) stigma of dread
Unknowable rather than knowable	Only experts know risk, and they debate
Controlled others/controlled those at risk	Big multinational compared to individual
Untrustworthy rather than trustworthy	Multinational compared to small farmer
Unresponsive versus responsive management	Open versus arrogant and remote

Source: Adapted from Peter M. Sandman (1994), in Ruth A. Eblen and William R. Eblen, eds., *Encyclopedia of the Environment* (Boston: Houghton Mifflin), pp. 620–623.

studies are conducted in the laboratory with animals and with foods that are spiked with high levels of the proteins and genes that are novel to the GM food.

Since the first commercialized GM crop, the FlavrSavr® tomato (**Figure 20.10**), was introduced in 1994, people have learned a great deal about the real risks and the substantial benefits of GM crops. The benefits of growing GM crops far outweigh the few measurable risks discovered to date. The potential dangers that do exist are specific to particular crops and transgenes and are not associated with the process of plant transformation as a whole.

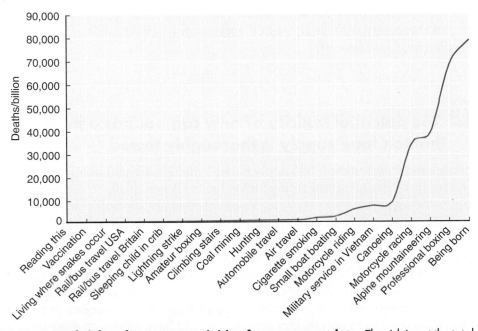

Figure 20.9 Real risks of common activities from actuary data. The risk is not the total number of deaths, but rather the total number divided by the number of people engaged in each activity. *Sources:* W. Stannard, *Insurance,* October 25, 1969; E. E. Pochin (1974), Occupational and other fatality rates, *Community Health* 6:2–13.

Figure 20.10 The first GM crop in the marketplace: FlavRSavR® tomatoes marketed under the MacGregor label. *Source:* Courtesy of William Hiatt, Calgene, Davis, California.

The risks of GM crops can be classified into two general categories: food safety risks and ecological risks. The food safety issues of GM food revolve around allergenicity and toxicity of the proteins to humans. There are several categories of ecological risks:

- Resistance management of insects to *Bt* and other plant-produced toxins
- Gene transfer and persistence
- Nontarget effects

Naturally, none of these risks exist in isolation; they must be viewed in the context of existing agricultural systems. Nature did not invent agriculture, people did; people should therefore analyze how one artificial construct (GM crops) affects and is affected by another existing and necessary artificial construct (agriculture), as well as how it might affect unmanaged ecosystems (nature).

20.1 The potential toxicity of new compounds entering the human food supply is thoroughly tested.

Any compound entering the food supply in the United States and many other countries is subject to specific regulatory scrutiny for food safety. A potentially toxic transgenic product such as *Bt* toxin must pass the same standards that are applied to any chemical pesticide product. Exceptions to this type of testing occur when the gene product (protein) expressed in a transgenic plant is found to be substantially equivalent to an ingredient or compound already existing in the food supply. Regulatory agencies such as the Federal Drug Administration use the doctrine of substantial equivalence to determine if a GM product is compositionally or nutritionally different from the original product. Synthesis of normal dietary components such as vitamins A and E would be exceptions to this rule; however, even these common dietary products would have to be tested for bioavailability and for any unexpected effects that could have occurred during crop transformation. When a plant overproduces innate compounds or when the transgene product has a known level of toxicity, it is necessary to conduct toxicity testing. An example of the former would be the overproduction of proteinase or amylase inhibitors for insect resistance. Plants already produce such inhibitors as part of their defense arsenal. Because they are endogenous to plants and could offer sublethal insect control, this class of proteins might be desirable to use in GM crops (see Chapter 16). However, because these compounds are natural antibiological agents, testing would be necessary to determine levels safe for human consumption. Toxicity testing also must be performed for all proteins not found in the human diet. For instance, green fluorescent protein (GFP) has numerous potential applications because of its visible fluorescence (**Figure 20.11**). It could be used to monitor the movement of transgenes to unintended weedy hosts, or track disease and stress responses with GFP that is produced using disease- or stress-inducible promoters. These would be valuable tools for agriculture, but would entail that GFP enter the human food chain, requiring that the potential toxicity of GFP be determined.

Some scientists have argued that protein products and the downstream metabolites are not the only potential source of toxicity in transgenic plants. They hypothesize that secondary pleiotropic or secondary mutagenic effects, resulting from gene expression or integration, could cause unforeseen hazards, including toxicity of secondary metabolites or lowering of important nutrients. In two documented cases, traditional crop improvement strategies have led to the appearance of unacceptable levels of toxic products: glycoalkaloids in potatoes and carcinogens in celery. These were detected only after breeders had released the new varieties to the public. Thus, the problem of unusual changes in some of the thousands of chemicals that crops contain is not specific to GM crops but is the consequence of genetic change, by whatever means. One method for assessing these potential problems would be quantitative measurement of thousands of metabolites—called *metabolic profiling*—to assure that any GM variety is materially equivalent to its nontransgenic counterpart.

The study that caused the European backlash against GM foods was initially introduced to the public in a British television interview with Arpad Pusztai, a well-known

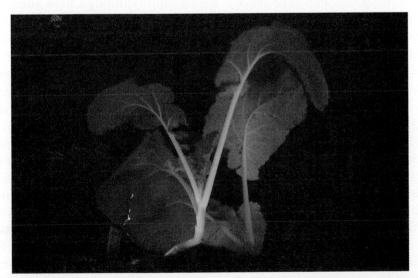

Figure 20.11 Green fluorescent protein (GFP) expression in canola. The plant on the left expresses the jellyfish gene that encodes GFP, a protein that can be imaged with the proper equipment. This and other genes could be used to follow the spread of genes in plant populations. *Source:* Courtesy of Matt Halfhill.

biochemist/animal nutritionist from Aberdeen, Scotland. This study was subsequently published in the British medical journal, *The Lancet* (Ewen and Pusztai, 1999). Pusztai and his colleagues examined the effect of feeding rats transgenic potatoes that produced a lectin found in the bulbs of snowdrop (a nonfood plant). This lectin protein has insecticidal properties. The control feed included either wild type potatoes or wild type potatoes spiked with lectin. The researchers reported that only rats fed transgenic potatoes showed signs of intestinal damage and lowered immune response, and they concluded that the genetic transformation process itself caused the observed complications. This study has been heavily criticized for its lack of a control group fed transgenic potatoes that did not express the lectin gene, as well as for the imbalanced diet used overall. Because potatoes are protein deficient, they are a poor choice of food as the sole nutritional source; this kind of imbalanced diet could itself damage experimental subjects. Other reports have contradicted Pusztai's conclusion that transformation itself is a suspect technology. On the positive side, Pusztai emphasized the need for long-term nutritional studies with mammals in evaluating certain transgenes. In addition, the study showed the difficulty of evaluating GM crops. Spiking the food with a novel protein is relatively easy and requires only that the researchers know how to purify the protein, which is usually not too difficult. However, will such a study get the same results as one in which the protein is made by the crop itself and present in the cells? If the crop is a poor source of nutrition when fed by itself, and the level of the new protein is low—as it usually is—the experiment becomes even more difficult if not impossible to do.

20.2 Special testing ensures that novel proteins are not allergenic.

A major concern of people with food allergies is the possibility that genetically modified crops could introduce allergens into the food supply. Although food allergies are not completely understood, there is enough information about them to generate a limited list of common food allergens and standard characteristics that are used for defining food allergies (**Box 20.1**). Certain proteins have short stretches of amino acids on their surfaces that cause mammals to produce a special class of immunoglobulins called IgE, which are responsible for the allergic reactions. These short peptides can be identified, and researchers have determined the amino acid sequences of more than 200 food proteins with allergenic sites. No common amino acid motif or consensus sequence has been discovered. If a compound is known to be allergenic, then the process of evaluation is simplified; proteins that are not normally allergenic will not suddenly become allergenic when expressed in a transgenic plant. For instance, no known allergy to the iron-carrier-protein plant ferritin exists; therefore, transgenic iron-enriched rice that expresses ferritin poses no allergenicity risk. If a gene product is already an allergen, then it will remain an allergen when expressed in a transgenic plant. When researchers introduced the gene encoding brazil nut albumin into soybean to increase its methionine content, they found that serum from brazil nut–allergic people reacted with extracts of the transgenic soybean (**Figure 20.12**). This became apparent when Pioneer Hi-Bred International, the company that produced the transgenic soybean, tested it for food safety, because the same subjects were not allergic to soybean. The FDA and Pioneer came to the same conclusion: The transgenic soybean variety would carry a significant allergy risk and should not be commercialized.

Box 20.1

Some Facts About Food Allergies

Allergy (food allergy): Any adverse reaction to an otherwise harmless food or food component (a protein) that involves the body's immune system. To avoid confusion with other types of adverse reactions to foods, it is important to use the terms "food allergy" or "food hypersensitivity" only when the immune system is involved in causing the reaction.

Frequency: According to the U.S. National Institutes of Health, approximately 5 million people in the United States (5 to 8% of children and 1 to 2% of adults) have a true food allergy.

Most common food allergens: Eight foods account for 90% of human allergic reactions. They include peanuts, tree nuts (walnuts, pecans, Brazil nuts, cashews, and so forth), fish, shellfish, eggs, milk, soy, and wheat. Peanuts are the leading cause of severe allergic reactions, followed by shellfish, fish, tree nuts, and eggs.

Severe reaction (anaphylaxis): Medical researchers estimate that as many as 100 to 200 people die each year from food allergy–related reactions; approximately 50 people die from insect sting reactions. In highly allergic people even minuscule amounts of a food allergen (for example, 1/44,000 of a peanut kernel) can prompt an allergic reaction.

Naturally, risk assessment is considerably more complicated when the allergenicity of a transgenic protein is unknown. Once again, GFP is a good example. Although there are no known allergies to GFP, might it induce allergies if people routinely ingest GM foods expressing GFP over a long period of time?

One typical characteristic of food allergens is that they are not easily broken down in the gut. Testing a protein's stability during the digestive process is one way of identifying potential allergens; if a protein is degraded in the gut then it may not reach immune cells and cause a hypersensitivity response. For this reason, proteins that are stable in the human gut require extensive examination. The Aventis Starlink® *Bt* Cry9 maize variety found in Taco Bell taco shells in the summer of 2000 was a good example of a product that people feared could contain new allergens. *Bt* Cry9 is more stable in digestion than the other *Bt* toxin proteins in commercial crops, so the EPA took the precautionary measure of approving the Starlink® maize only as animal feed (pigs and cattle do not generally have food allergies). After discovering the maize in human foodstuffs because the farmers, grain elevator operators, and others down the food production line were either unable or unwilling to segregate the GM Starlink® maize, the EPA made the decision not to approve GM crops only for animal consumption in the future—once again, as a precaution against the recurrence of such problems.

One procedure that can be performed to assess whether a recombinant protein might be allergenic is to compare peptides of the recombinant protein to those of known allergens. Novel proteins with significant sequence similarities can be tested for reactivity with serum from subjects who are allergic to the homologous allergen. Although these tests may not be completely comprehensive in identifying potential allergens, the limited variety of allergenic foods (Box 20.1) suggests that the vast majority of transgene proteins will be safe for consumption. All those now in the marketplace have been thoroughly tested. It is interesting to note in this respect that traditional foods that contain known potent allergens, such as peanuts, and that are responsible for a number of deaths in the United States every year, are not labeled as being potentially life threatening.

Figure 20.12 Large Brazil nuts amid (much smaller) soybeans. To make soybeans more nutritionally complete, a gene from Brazil nut was transferred to soybean. Many people are allergic to Brazil nut, and it turned out that the chosen gene encoded a major allergen from Brazil nut. When it was discovered that people who are allergic to Brazil nuts, were now allergic to the GM soybeans, the project was stopped. This episode is always cited by opponents of GM technology as evidence of the dangers. It also clearly shows that the regulatory process works and that potential problems can be identified.

The other side of the coin is that GM technology can and will be used to eliminate allergenic proteins from some major crops by suppressing expression of genes that encode those proteins. Projects are under way to make hypoallergenic soybean and wheat.

20.3 Transgenic volunteer crop plants could become a nuisance in agriculture.

One important purpose of GM crop technology is to improve a crop's agricultural performance. Toward this end, it would be useful for some crops to acquire broader abiotic and biotic tolerances, allowing them to be grown in new geographic areas or under new conditions. Some people have argued that with crops such as alfalfa (*Medicago sativa*), canola (*Brassica napus* and *Brassica rapa*), sunflower (*Helianthus annuus*), and rice (*Oryza sativa*), all of which possess one or more weedlike characteristics, transgenic and novel traits could allow the crop itself to become more weedy and invasive. Generally, cultivated crop species contain few of the characteristics of weedy species (**Box 20.2**). In fact, most or all of the modifications associated with GM varieties are meant to enhance their productivity under intensive agricultural management. Such changes are not only less likely to make a crop species weedy but would tend to reduce its competitive capability in nonagricultural circumstances.

Weed volunteerism is an agricultural problem in which uncollected seeds from last year's crop germinate and grow within the crop currently being grown in the same location (see Chapter 17). Canola, to date, has been modified with at least three distinct herbicide resistance genes (two via genetic engineering and one through mutagenesis), and

Box 20.2

Characteristics of Weeds and/or Weedy Relatives of Economic Species

A weed is defined as an unwanted plant, especially a wild plant, growing where it is not desired by humans. In addition, weeds may be characterized by

- Seed production early in their life cycle
- High fecundity by seeds or vegetative structures
- Long-lived seeds, seed dormancy, and/or asynchronous germination
- Adaptation to coexist and be spread with crop seeds
- Production of allelochemicals that suppress the growth of other plants
- Adaptations such as prickles, spines, or thorns that aid dispersal or repel predators

- Parasitism of other plants
- Storage organs or seed reproduction that promote survival in harsh environments
- High photosynthetic growth rates and/or extensive root systems

Source: H. Baker (1965). Characteristics and modes of origin in weeds, in H. G. Baker and G. L. Stebbins, eds., *Genetics and Colonizing Species* (New York: Academic Press), pp. 147–168.

volunteers of these varieties could become a particular nuisance to agriculture. Individual plants combining all three herbicide resistance genes and expressing resistance to several herbicides did arise as a result of crossing in the field of one farmer in Canada who decided to grow all these varieties in close proximity. Such individual plants are at a selective advantage and will make weed control more difficult. More stringent regulatory requirements by the USDA have been applied to certain transgenic crops that have the potential for increased invasiveness and damaging volunteerism.

Once again, this problem is not unique to GM technology, because troublesome weeds have arisen in the past as a result of hybridization between crops and weeds. For example, the sugar beet industry in Europe was severely depressed at the end of the 20th century by the emergence of the weed beet, a cross between the sugar beet (*Beta vulgaris* subsp. *vulgaris*) and the sea beet (*Beta vulgaris* subsp. *maritima*) (**Figure 20.13**).

Figure 20.13 Weed beet. The weed beet is a hybrid between the sugar beet and a related wild species of the sugar beet. In this view, the weed beet has completely overtaken the crop. *Source:* Courtesy of Detlef Bartsch, Aachen University of Technology.

20.4   It is virtually certain that transgenes will flow from GM crops to other related plants.

Intraspecific hybridization occurs readily when wind- or insect-pollinated transgenic crops are grown in close proximity to nontransgenic varieties, and the agricultural practice of annually saving harvested seeds can unintentionally allow transgenic material to persist from one year to the next. Crops such as maize have the potential to pass genes to adjacent conspecifics (members of the same species) whether the crop is GM or a conventional variety, and many organic farmers worry about this possibility. Although in the United States a growing organic farming industry seems to coexist peacefully with conventional farmers using GM crops, organic farmers are opposed to the adoption of GM crops generally, either for their own use or use by conventional farmers. The term "genetic pollution" has been coined for the spread of transgenes from their home crops to the surrounding plants, whether crops or wild plants. Such spread worries producers of non-GM crops such as organic farmers. In the past the organic farming industry has had rigorous standards for pesticide overspray and trace "contaminants" in its products and seeds, and threshold limits for trace transgenes will also need to be established.

There is also concern that GM crops might rapidly accumulate several fitness-enhancing traits (transgene stacking) and that this could lead to new and unforeseen problems. This issue of unintended consequences will persist until scientists gain more first-hand knowledge about transgene stacks themselves. What would be the interactions, for example, among gene products that confer drought and aluminum tolerance, insect resistance, and increased nitrogen use efficiency? Scientists will have to assess ecology and physiology of such "superplants" individually, just as they do now. A related development in plant biotechnology is metabolic engineering—the ability to transform plants with several genes that make new metabolic pathways.

A more immediate problem is that of hybridization between closely related species. The most difficult problem would be if a weed species could receive transgenes directly from a related crop being grown nearby; these transgenes, if expressed, could then increase the fitness of the weed in nature. In a worst-case scenario, the weed could become more invasive and competitive, and in a relatively short time could damage natural ecosystems. People have to go back to the list of weedy traits and ask if it is likely that a transgene will exacerbate or promote weedy ability. Such evaluation is part of the government's regulatory system.

Interspecific hybridization depends on several concurrent circumstances to allow gene flow between related species. The crop must have some naturally occurring wild relatives growing near land under cultivation. Crops such as maize and soybean have no relatives in the United States and Canada; therefore, they represent no risk of interspecific gene flow. It is important to note that there may be unintentional movement of transgenic plants from the United States to other countries. Scientists can only speculate about the ramifications of transgenes introgressing from maize into teosinte, a wild relative of maize and a global treasure that originates and grows in Mexico (see Chapter 13). Sunflower (**Figure 20.14**), alfalfa, Brassica crops, and rice are all crop species that do have wild relatives near cultivation areas; these species complexes have all been the focus of gene flow studies in the United States. The wild and domesticated species involved must share a degree of sexual compatibility, and distantly related species do sometimes share enough genome similarity to produce viable progeny. They must occur in close enough proximity to allow transfer of viable pollen, and they must flower at the same time as well.

The variable homology of the genomes between related species leads to a wide range of possibilities for the introgression rate of a transgene, or any other gene, after the F_1 hybrid generation. Meiotic abnormalities caused by the distant relationship between parental genomes can decrease rates of introgression into new genotypes, so the production of initial hybrids does not at all guarantee that the transgenes will move into weeds. Unequal pairing during meiosis can cause chromosomes to be lost or disrupted, which results in higher rates of infertility and decreased rates of seed production. Recombination, an important process in the incorporation of foreign DNA, is diminished by the unstable chromosome configurations of hybrids produced by distant relatives. In contrast, hybrids produced by closely related species have been shown to combine fitness indices (seed production, pollen fertility, biomass, and so on) that parallel the parental species. In this situation, the hybridization barrier between species can be very low, and introgression of a transgene is likely.

Figure 20.14 Wild sunflowers in Nebraska. There is little doubt that gene flow from GM sunflowers to wild sunflowers is likely to occur in the future. The questions are, Is there likely to be ecosystem damage from such gene flow? And is this damage worse than present ecosystem damage from agriculture? *Source: Courtesy of A. A. Snow, Ohio State University.*

In particular crops, it is virtually certain that transgenes will flow from crop to weed. For example, canola, *Brassica napus*, hybridizes easily with birdseed rape or field mustard, *Brassica rapa*. Transgenic interspecific hybrids have been produced between transgenic canola modified with herbicide resistance and insect resistance genes, and wild *B. rapa*. After only one backcross, many of the progeny are morphologically and cytologically similar to the *B. rapa* parent. After another generation, the progeny are essentially *B. rapa* with a transgene; transgenic weeds in three generations! The transgenes have also been found expressed in the weedy genetic background. A *Bt* transgene in canola could have the same expression level when placed in the *B. rapa* genome through introgression. However, note that two types of herbicide-resistant canola are available: GM and bred by traditional means. The problem of herbicide-resistant weeds is not limited to GM canola but exists with the traditionally bred variety as well.

Just because transgenic weeds will arise does not mean that such weeds will be weedier or more invasive; the possibility for increased fitness of transgenic hybrids and backcrosses depends on the nature both of the transgene and of the environment. For example, weeds that contain a transgene conferring resistance to a herbicide would be a nuisance to agriculture, but would have little effect in a nonagricultural environment where the herbicide is absent. In contrast, an insecticidal *Bt* transgene in a weed host could alter natural ecology by giving transgenic weeds a selective advantage if a key insect had been historically critical to limiting the weed's survival. Transgenes that provide fitness-enhancing characteristics under natural conditions have the greatest potential to disrupt the balance of established ecosystems. However, most weeds already seem to have better insect resistance than their elite crop counterparts. Does insect herbivory presently limit weed populations? In gardens, insects seem to prefer tomato plants to weeds; will adding a *Bt* gene make a difference? How much weed fitness increase from transgenes should be tolerated? Norman Ellstrand and colleagues (Ellstrand, Hand, and Hancock, 1999) have suggested a 5% fitness increase, at which point they believe significant economic impacts might occur that would outweigh the benefits gained from the transgenic crop.

20.5 Effects on nontarget organisms are difficult to investigate.

Transgenic crops that express insecticidal transgenes to control agricultural pests may also affect nontarget organisms, and there are several different ways in which this could occur. An insect might eat a transgenic toxin in a food source it does not typically encounter. To use a much-touted example, the monarch butterfly could be impacted directly by feeding on parts of a crop plant that it does not usually eat, in this case, *Bt* maize pollen that has landed on milkweed plants (milkweed is the sole food of monarch larvae) adjacent to or in maize fields, as discussed earlier. A different nontarget effect involves interactions that occur through three trophic levels. A. Hilbeck and colleagues in Switzerland found that the lacewing (*Chrystoperla carnea*), an insect predator (**Figure 20.15**), suffered higher mortality rates from feeding on European corn borers (*Ostrinia nubilalis*) reared on *Bt* maize compared with those fed the non-*Bt* variety. However, no field studies on plant and insect population systems have been performed to determine if a GM plant (*Bt* maize, in this case) would have a significant impact on biodiversity in a farm setting. In another type of trophic interaction study—this one performed in Great Britain by T. Schuler, G. Poppy, and colleagues—insect behavior experiments using "choice" feeding (the insects can choose their food from a number of selections) showed that a parasitic wasp (*Cotesia plutellae*) preferentially selected *Bt* canola leaves as a food source habitat when the leaves had been damaged by *Bt*-resistant diamondback moths (*Plutella xylostella*). That is, the plant damage drew the wasp to the location of the moth larvae. If one were simply looking at whether the plant were transgenic (*Bt* canola versus non-*Bt* canola), the conclusion would have been that *Bt* decreased parasitism by the wasp. But the inclusion of *Bt*-tolerant larvae in the experiment uncovered the fact that the key factor was plant damage. The parasitic wasp experienced no reduction of reproductive success from exposure to *Bt* when it consumed *Bt*-resistant moth larvae, and could, in fact, help constrain the spread of *Bt*-resistant pests through natural predation.

Figure 20.15 Lacewings have a voracious appetite for insect larvae. Commercial insectaries breed lacewings and sell them to farmers for biocontrol of insect pests.
Source: Courtesy of Matthias Meier.

Another possible nontarget effect of *Bt* crops has to do with the *Bt* toxin contained in root exudates from *Bt* maize. Numerous studies have shown that soil organisms rapidly degrade *Bt* toxin. Nevertheless, G. Stotzky and colleagues of New York University have demonstrated that soils in which *Bt* transgenic maize was grown contain *Bt* protein that is not degraded. When tobacco hornworm (*Manduca sexta*) larvae were fed on this *Bt*-containing soil, the larvae suffered higher mortality rates than larvae that fed on control soil. Earlier studies demonstrated that the *Bt* protein, like many other proteins from root exudates, binds tightly to clay soil particles. The high sensitivity of tobacco hornworm to *Bt* permitted these low levels of *Bt* to be detected. The studies that show the rapid degradation do not eliminate the possibility that a fraction of *Bt* protein survives degradation when bound to clay particles. No one has examined the effect of this bound *Bt* on the soil ecosystem. Tobacco hornworm in the laboratory can be forced to eat soil particles, but whether insect larvae living in the soil ingest soil particles at similar levels is not known. The many questions raised by the various studies made to date demonstrate a clear need to analyze possible nontarget effects caused by genetically modified crops. However, for such research to be more relevant, researchers need to extend and examine the findings in the context of current agricultural practices.

Possible negative side effects must also be weighed against the positive effects of an insect control regime that uses insecticidal transgenic plants instead of chemical insecticides. For example, *Bt* cotton requires three or fewer insecticide treatments per year, a substantial reduction from the five to twelve annual sprays needed to control pests in nontransgenic cotton fields. Plantings of *Bt* cotton alone reduced pesticide use in the United States by over 900,000 kg during 1997. The overall reduction of pesticides results in lower costs (**Table 20.3**) and a safer working environment for the farmer, and a dramatic drop in amount of chemicals added to the environment. The decrease in broad-spectrum insecticides brought about by using specialized insecticidal transgenic plants also benefits

Table 20.3	Comparison of farmers' costs of growing traditional and *Bt* cotton in the United States in 1999	
	Standard	**Bt**
Total insecticide cost (US$/acre)	178	109
Yield (lbs of lint/acre)	933	975
Return (US$/acre)	1081	1187

Note: Analysis of 17 different studies showed wide variation but an average of $106/acre net return. The United States has just over 6 million hectares of cotton, so the potential savings are substantial if this trend continues. In 2000 *Bt* cotton was grown on 2.2 million hectares.

nontarget insect populations. Insect biodiversity is encouraged, as is natural pest control through enhanced predator–prey interaction. Using fewer insecticides because of using GM crops can have many advantages for the environment, for the farmer, and especially for the farm workers, who currently deal with constant or repeated exposure to subtoxic levels of chemicals.

20.6 Careful management of GM crops is needed to avoid the emergence of resistant insect strains.

Obviously, evolved resistance to transgenic proteins by insect pests limits the usefulness and longevity of any insecticidal transgenic crop variety. The diamondback moth, an important pest of Brassica crops worldwide, was the first documented insect to develop resistance to *Bt* sprays in open-field populations. David Heckel (1994) has shown that *Bt* resistance in another insect, *Heliothis virescens* (tobacco budworm), is linked to several different genes on different chromosomes; resistance to *Bt* is not likely to result from a single recessive gene. Currently, no dominantly inherited *Bt* resistance genes have been documented, but they would severely limit the effectiveness of future *Bt* crops. Various resistance management strategies have been proposed to delay the onset of resistance, and the method commonly used at present is the deployment of a high *Bt* expression in transgenic plants coupled with a nontransgenic refuge planting; this is called the high-dose/refuge strategy. The high dose kills all *Bt*-susceptible insects, and the refuge allows *Bt*-susceptible pests to survive on the nontransgenic material and mate with *Bt*-resistant individuals that might arise out of the high-dose fields. The goal of this strategy is to keep the recessive *Bt* resistance genes at low levels, and thus limit the rate that the entire population will become *Bt* resistant. The effectiveness of this strategy depends on refuge size, refuge design (refuge plants mixed with transgenics or separate from them), the quantity of pesticides used for spraying the refuge, and the rate of migration of insect pests. Several scientists widely recognized for their sustained contributions to insect control strategies such as Fred Gould at North Carolina State University, Bruce Tabashnik of the University of Arizona, Tony Shelton at Cornell University, and David Andow at the University of Minnesota have contributed their expertise to formulate control strategies (Tabashnik, 1994; Shelton, Tang, Roush, Metz, and Earle, 2000) and to detect resistant insects when they arise (Andow and Alstad, 1998). Everyone agrees that *Bt* crops must be deployed with care to assure that the resource of unique *Bt* toxin proteins is not squandered. People have learned that the chemical insecticide treadmill, where insects become resistant to each insecticide in turn, so that every insecticide must be replaced by another, ad infinitum, is not the paradigm they want to follow with transgenic crops.

CONCLUSION

GM crops are fast becoming a part of agriculture throughout the world, but as with any new technology, opposition has surfaced that questions the safety and the appropriateness of this technology. Moreover, acceptance does not mean that there are no unresolved issues or that the technology is risk free. No technology is risk free. In the last few years, a number of "urban myths" about genetic engineering of plants have sprung up, and opposition groups have made effective use of powerful environmental symbols. The companies that developed this technology were caught by surprise that the technology, which has real potential benefits for farmers and consumers alike, was not more readily accepted. In Europe certainly, governments seemed to be more willing to listen to those perpetuating the urban myths than to the scientists who understood the technology. Much of the opposition stems from a general uneasiness that ordinary people lose out when there is general agreement between multinational companies, international organizations, and national governments on how society should develop and how new technologies should be applied. That some companies have knowingly harmed the public interest, and that governments have sometimes failed in their evaluation of what constitutes a public health or environmental danger, sustain this opposition.

Some technological scares have later proved to be nonissues for the public. For many years people thought that air flight would never be valid transportation, and when microwave ovens were first produced, people were afraid to use them for fear of radiation damage. Indeed, there are risks involved in using these and almost all tools. As people come to understand the real risks and benefits of a technology, and as they become educated and familiar with it, then they are in a position to judge it and accept it according to its merits. GM plants have an important role to play in developing an agricultural system that can serve ever-growing global food needs. When people move beyond urban myths, they will experience the fruits of this technology.

This does not mean that there are no unresolved issues. Every technology can be improved, and GM crop technology will be gradually improved as people learn more about the problems the technology creates. Technologies are generally not without problems, nor are they absolutely safe.

Further Reading

Andow, D. A., and D. N. Alstad. 1998. F-2 screen for rare resistance alleles. *Journal of Economic Entomology* 91:572–578.

Beringer, J. E. 1999. Cautionary tale on safety of GE crops. *Nature* 399:405.

Ellstrand, N. C., S. C. Hand, and J. F. Hancock. 1999. Gene flow and introgression from domesticated plants into their wild relatives. *Annual Review of Ecology and Systematics* 30:539–563.

Ewen, S. W. B., and A. Pusztai. 1999. Effects of diets containing genetically modified potatoes expressing *Galanthus nivalis* lectin on rat small intestines. *Lancet* 354:1353–1355.

Gianessi, L. P., and J. E. Carpenter. 1999. *Agricultural Biotechnology: Insect Control Benefits.* Washington, DC: National Center for Food and Agricultural Policy.

Hansen, L. C., and J. J. Obrycki. 2000. Field deposition of *Bt* transgenic corn pollen: Lethal effects on the monarch butterfly. *Oecologia* 125:241–248.

Hartzler, R. G., and D. D. Buhler. 2000. Occurrence of common milkweed (*Asclepias syriaca*) in cropland and adjacent areas. *Crop Protection* 19:363–366.

Hashimoto W., K. Momma, H. J. Yoon, S. Ozawa, Y. Ohkawa, T. Ishige, M. Kito, S. Utsumi, and K. Murata. 1999. Safety assessment of transgenic potatoes with soybean glycinin by feeding studies in rats. *Bioscience. Biotechnology, and Biochemistry* 63:1942–1946.

Heckel, D. G. 1994. The complex genetic basis of resistance to *Bacillus thuringiensis* toxin in insects. *Biocontrol Science and Technology* 4:405–417.

Hilbeck A., M. Baumgartner, P. M. Fried, and F. Bigler. 1998. Effects of transgenic *Bacillus thuringiensis* corn-fed prey on mortality and development time of immature *Chrysoperla carnea* (Neuroptera: Chrysopidae). *Environmental Entomology* 27:480–487.

Hodgson, J. 1999. Monarch *Bt*-corn paper questioned. *Nature Biotechnology* 17:627.

Kuiper, H. A., P. J. M. Noteborn, and A. C. M. Peinenburg. 1999. Adequacy of methods for testing safety of genetically modified foods. *Lancet* 354:553–564.

Losey, J. E., L. S. Rayor, and M. E. Carter. 1999. Transgenic pollen harms monarch larvae. *Nature* 399:214.

Nordlee, J. A., S. L. Taylor, J. A. Townsend, L. A. Thomas, and R. K. Bush. 1996. Identification of a brazil-nut allergen in transgenic soybeans. *New England Journal of Medicine* 334:668–692.

Palm, C. J., D. L. Schaller, K. Donegan, and R. J. Seidler. 1996. Persistence in soil of transgenic plant produced *Bacillus thuringiensis* var. *kurstaki* delta-endotoxin. *Canadian Journal of Microbiology* 42:1258–1262.

Saxena, D., S. Flores, and G. Stotzky. 1999. Insecticidal toxin in root exudates from *Bt* corn. *Nature* 402:480.

Sears, M. K., R. L. Hellmich, D. E. Stanley-Horn, K. S. Obenhauser, J. M. Pleasants, H. R. Mattila, B. D. Siegfried, and G. D. Dively. 2001. Impact of *Bt* corn pollen on monarch butterfly populations: A risk assessment. *Proceedings of the National Academy of Sciences USA* 98:11937–11942.

Sims, S. R., and L. R. Holden. 1996. Insect bioassay for determining soil degradation of *Bacillus thuringiensis* subsp. *kurstaki* CryIA(b) protein in corn tissue. *Environmental Entomology* 25:659–664.

Schuler, T. H., R. P. J. Potting, I. Denholm, and G. M. Poppy. 1999. Parasitoid behavior and *Bt* plants. *Nature* 400:825–826.

Shelton, A. M., J. D. Tang, R. T. Roush, T. D. Metz, and E. D. Earle. 2000. Field tests on managing resistance to *Bt*-engineered plants. *Nature Biotechnology* 18:339–342.

Tabashnik, B. E. 1994. Evolution of resistance to *Bacillus thuringiensis*. *Annual Review of Entomology* 39:47–79.

Ye, X., A. Al-Babili, A. Kloti, J. Zhang, P. Lucca, P. Beyer, and I. Potrykus. 2000. A (β-carotene) biosynthetic pathway into (carotene-free) rice endosperm. *Science* 287:303–305.

INDEX

O

obesity, 29, 158
oil bodies, 521
oleosin, 521
omega-3' fatty acid, 158
organic food, 165, 483
organic matter, 273, 305–306, 308–310, 325
organic residues, 308
osmoprotectants, 294
osteoporosis, 166
outcrossing, 225
overgrazing, 110

P

palm oil, 514–515
papaya ringspot virus, 382
parasitic wasps, 441
parasitism, plant-on-plant, 312–313
parenchyma, 186
particle gun, 148
pasteurization, 177
pastoralists, 55
patent protection, 45
pathogen resistance, 403
pathogenesis-related genes, 405
pathogenic fungi, 400–401
pathogens, 390
pearl millet, 495
PEP carboxylase, 246
pericycle, 188, 189
pest control, 414–442
pest outbreaks, 420, 422
pest resistance, 67, 344, 386
pest-resistant crop, 431–432
    varieties of, 428
pesticide resistance, 421
pesticide spraying, 414
pesticides, 59, 421, 422
    natural, 423
pests,
    evolution of, 432
    resistance of, 437
petal, 184–185, 194
petrochemicals, 504
petroleum, 503–504
pharmaceuticals, produced in plants, 518, 520
phenolic compounds, 312
phloem, 187, 249
phloem loading, 250
phloem transport, 251
phospholipids, 156, 157

photoassimilates, 249
photoperiod, 201, 203, 343–344
photoperiod response, 387
photoperiodism, 192
photoprotection, 254, 256
photorespiration, 264
photosynthesis, 240–268, 320
    C3 cycle, 245–246
    C4 cycle, 246–247
photosynthetic apparatus, 243
photosynthetic efficiency, 254, 255, 258, 260
photosystem, 243–245
phylloxera, 431–432
phytate, 174
phytic acid, 216
phytochemicals, 167
phytochrome, 206
phytoestrogens, 168–169, 173
Phytophtora sojae, 306
Pierce's disease, 399
pig production, 29
pigeon pea, 114
pigment, 243
pineapple, 254
pistil, 194
plant breeding, 59, 360–388
Plant Patent Act, 44
Plant Variety Protection (PVP), 357
Plant Variety Protection Act (PVPA), 233
plants,
    as factories, 500–525
    cultivated, 19
    life cycle of, 192–193
plantain, 120
plasma membrane, 190, 286
plasmid, 141, 142
    replication of, 147
plastic films, 507
plastoquinone, 245
pollen culture, 379
pollen grains, 194
pollination, 368–369
pollution,
    genetic, 546
    global, 71
polyethylene film, 508
polyhydroxybutyrate, 508–509
polymerase chain reaction (PCR), 322
polymorphism, 382
polypeptide, 130, 131
polyploids, 352
polysaccharides, 154, 155
polyunsaturated fatty acids, 513